Dam Disaster Monitoring from Space
Grand Ethiopian Renaissance Dam

Maged Marghany
President
Innovative Information Sdn. Bhd.
Safanad Information Technology
Kuala Lumpur, Malaysia
Faculty of Agriculture
Department of Oceanography
Malikussaleh University
North Aceh Regency, Aceh, Indonesia

CRC Press
Taylor & Francis Group
Boca Raton London New York

CRC Press is an imprint of the
Taylor & Francis Group, an **informa** business
A SCIENCE PUBLISHERS BOOK

***Cover Credit*:** Image provided by the author.

First edition published 2026
by CRC Press
2385 NW Executive Center Drive, Suite 320, Boca Raton FL 33431

and by CRC Press
4 Park Square, Milton Park, Abingdon, Oxon, OX14 4RN

CRC Press is an imprint of Taylor & Francis Group, LLC

Library of Congress Cataloging-in-Publication Data (applied for)

ISBN: 978-1-032-35534-4 (hbk)
ISBN: 978-1-032-35537-5 (pbk)
ISBN: 978-1-003-32732-5 (ebk)

DOI: 10.1201/9781003327325

Typeset in Times New Roman
by Prime Publishing Services

Dedication

To my mother, Faridah
whose unwavering support has been the foundation of my journey.

To Captain Prof. Dr. Mohamed Ibrahim Mohamed
my esteemed professor, whose mentorship has been a guiding light throughout my journey.

And to
Richard Feynman and Michio Kaku
who taught me that the measure of an academic institution lies not in its fleeting pursuit of rankings, but in its dedication to nurturing and valuing the contributions of its scholars. Their legacy underscores the importance of integrity and genuine advancement in the academic world.

Preface

In this work, we embark on an exploration of the Grand Ethiopian Renaissance Dam (GERD), a landmark construction project with implications far beyond the typical water resource disputes. This endeavor reflects my ranking among the world's top 2% of scientists, as recorded by Stanford University—a recognition bestowed over five consecutive years.

In surveying existing literature on GERD, it became evident that current analyses often concentrate primarily on water shortages impacting downstream nations, especially Egypt, with limited attention to the broader consequences on agriculture and associated industries. However, GERD's significance extends beyond water scarcity: the dam's potential structural vulnerabilities, compounded by its placement on tectonically active fault zones, suggest a risk of catastrophic failure. This possibility necessitates a comprehensive framework to model and predict the impacts of a potential breach on both Upper and Lower Egypt.

This book aims to bridge several gaps in the current understanding of GERD. Notably, previous studies have often neglected advanced image-processing techniques and remote sensing, which could enhance our predictive capacity. Research discourse on GERD, particularly regarding the groundbreaking studies by Heggy, has sometimes been marred by unproductive academic rivalry. Furthermore, literature has not fully leveraged remote sensing technologies, such as vegetation normalization indices or microwave radar sensors, which could greatly enhance environmental monitoring. Consequently, this book introduces an innovative suite of image technology tools based on computer-aided processing to simulate realistic scenarios surrounding GERD's stability.

In doing so, this text fills a unique niche by integrating advanced quantum computing algorithms, microwave remote sensing, and an analysis of GERD's fluctuating dynamics. New satellite technologies, such as Sentinel-1, ICEYE, and TerraSAR-X, are harnessed to deepen our understanding of radar imaging and model the instability risks associated with GERD. Additionally, this work introduces the Marghany Hologram-TomoSAR algorithm, an original four-dimensional predictive tool designed to assess breach likelihood. The book further addresses the ongoing discourse about the biblical prophecies concerning GERD, reflecting a view prominent in international media narratives. To complete the analysis, both civil engineering and military solutions to mitigate the risks posed by GERD are also briefly examined.

The book's progression begins in Chapter One, which lays the foundation by discussing the fundamentals of dam structures. This includes definitions, structural types, and the construction processes involved in dam building. It outlines critical aspects such as river diversion, dam foundation preparation, and post-construction stabilization, emphasizing the complex planning and execution required. The chapter also addresses potential failure causes—overtopping and foundation issues—that may lead to disastrous outcomes if poorly managed. Furthermore, it reviews the environmental repercussions of dam construction, covering changes to hydraulic systems, temperature shifts, ecosystem disturbances, and even geological phenomena such as earthquakes, emphasizing the delicate equilibrium between engineering and environmental stewardship.

Chapter Two provides an in-depth examination of GERD itself, focusing on its geographic placement within the tectonically active East African Rift System, an area known for volcanic activity. This chapter contextualizes GERD within regional hydropolitics, discussing how it may reshape Nile Basin water allocation, and explores historical treaties like the 1959 Water Sharing Agreement. Moreover, the chapter evaluates GERD's potential benefits, such as power generation, against the risks of unilateral water management, underscoring the need for collaborative water-sharing strategies.

In Chapter Three, prior to the application of advanced remote sensing technologies to GERD, the concept of 'quantized remote sensing' is introduced. This approach, based on the quantized blackbody model, utilizes thermal infrared bands in ASTER data to analyze material compositions in GERD's construction. By applying Kirchhoff's law to quantify emissivity variations in materials like granite, concrete, and limestone, this chapter highlights the critical role that spectral distinctions play in monitoring GERD's structural integrity.

Building on these foundations, Chapter Four delves into quantum spectral modeling, offering a new methodology for acquiring spectral reflectance signatures around GERD. Through quantum spectral computing, concepts like quantum entanglement and quantum teleportation are applied to enhance imaging precision and geological analysis. This chapter lays the groundwork for subsequent explorations of sedimentation issues within the GERD reservoir.

In Chapter Five, we introduce the 'quantized Marghany clay saturation algorithm' to determine sediment concentrations, revealing alarming levels of saturation within the GERD reservoir. These findings emphasize the necessity for sediment management strategies to ensure the dam's sustainability and effective functioning.

Chapter Six presents methods for automatic detection of lineaments and faults using multispectral data. Here, a combination of quantum fuzzy B-spline and quantum edge detection methods provides detailed geological mapping, essential for evaluating the dam's safety amidst ongoing tectonic shifts.

Moving into Chapter Seven, the text transitions to microwave remote sensing, elaborating on radar altimeter systems and their utility in assessing water height variations. This section underscores the importance of radar technology in monitoring GERD's structural health.

Chapter Eight extends this analysis with interferometric altimeter radar applications, demonstrating their accuracy in measuring water levels in nearby lakes. Chapter Nine then introduces the 'Marghany Quantized Multiobjective Algorithm' (MQMA), which evaluates GERD's operational efficiency through turbulence analysis, an innovative indicator of the dam's energy output and functional status.

With Chapter Ten, the novel Along-Track Interferometric Synthetic Aperture Radar (SAR) processing technique is examined. Through quantum mechanics-based phase measurements, this chapter uncovers GERD's operational limitations, indicating its current underutilization.

The application of advanced radar interferometry techniques, particularly Persistent Scatterer Interferometry (PSI) combined with Quantum Convolutional Neural Networks (QCNN) for phase unwrapping, is addressed in Chapter Eleven. This fusion of methodologies accurately assesses GERD's structural deformation, critical for long-term stability assessments.

Chapter Twelve introduces the 4D Marghany Hologram-TomoSAR algorithm and the Marghany Quantum Automaton Algorithm for 4D Phase Unwrapping (MQAA). Through high-resolution TerraSAR X-band data, these algorithms forecast potential GERD breaches and predict downstream impacts on infrastructure, urban centers and heritage sites. Further, the Marghany Quantum Tsunami Theory is presented, positing that stress from GERD could initiate geological shifts, possibly affecting the Red Sea and Indian Ocean regions. The chapter concludes with strategic recommendations, including military considerations, to address GERD's potential threats.

Additionally, it reflects on Isaiah's prophecies, suggesting that these insights may offer a historical perspective on GERD's regional implications.

In sum, this book is a pioneering integration of scientific analysis, technological advancement, and theoretical speculation, providing a multifaceted perspective on GERD. It not only advances monitoring techniques but also invites reflection on the broader historical, environmental, and geopolitical forces shaping this monumental infrastructure project.

Distinguished Prof. Dr. Maged Marghany
President

Innovative Information Sdn. Bhd.
Safanad Information Technology
Kuala Lumpur, Malaysia
Malikussaleh University
North Aceh Regency, Aceh, Indonesia

Contents

Preface ... v

1. Fundamental Principles of Dam Construction and Their Environmental and Societal Impacts ... 1

1.1 What are Dams? ... 1
1.2 The Elements of a Dam ... 2
1.2.1 Water Retaining Structure ... 2
1.2.2 Water Releasing Structure ... 3
1.2.3 Water Conveying Structure ... 4
1.3 What are the Different Types of Dams? ... 4
1.3.1 Arch Dam ... 4
1.3.2 Buttress Dam ... 4
1.3.2.1 Deck Slab Buttress Dam ... 5
1.3.2.2 Multiple Arch Buttress Dam ... 6
1.3.2.3 Massive Head Buttress Dam ... 7
1.3.2.4 Multiple Dome Buttress Dam ... 8
1.3.2.5 Columnar Buttress Dam ... 8
1.3.3 Embankment Dams ... 8
1.3.3.1 Rolled Fill or Placed Earthfill Dams ... 9
1.3.3.2 Hydraulic Fill Dam ... 9
1.3.3.3 Classification of Embankment Dams ... 9
1.3.3.4 Elements of Embankment Dams ... 10
1.3.4 Gravity Dams ... 11
1.3.4.1 Types of Gravity Dams ... 12
1.4 Dam Constructions ... 15
1.4.1 Diversion of the River ... 15
1.4.2 Basement ... 16
1.4.3 Constructing Dams ... 16
1.4.4 Post Dam Construction ... 16
1.5 The Physical Factors Regulating Dam Type Selection ... 16
1.5.1 Topography ... 17
1.5.2 Geological and Foundational Conditions ... 17
1.5.3 Spillway ... 20
1.5.3.1 Types of Spillways ... 21

1.5.4 Hydrology ... 26
1.5.5 Material Accessibilities ... 26
1.5.6 Earthquake and Landslide ... 26
1.6 The Primary Causes of Dam Failure ... 27
1.6.1 Reservoir Filling, High Water and Overtopping ... 27
1.6.2 Dam and Levee Overtopping ... 31
1.6.3 Backward Erosion Piping ... 31
1.6.4 Foundation Failure ... 33
1.6.5 Location and Geology ... 33
1.6.6 Maintenance and Monitoring ... 34
1.7 The Negative Impact of the Dam on the Environment ... 37
1.7.1 The Impact of the Dam on the Hydraulic System ... 37
1.7.2 Impacts on Temperature and Chemical Characteristics of the Reservoir ... 38
1.7.3 The Impact of the Dam on the Atmospheric System ... 38
1.7.4 The Dam Can Cause Earthquakes ... 39
1.7.5 Dam Impact on Health ... 40
1.7.6 Dam Can Causes Coastal Erosion ... 40
1.8 The Effect of the Dam on the Earth's Rotation ... 41
1.9 Conclusion ... 42
References ... 43

2. Grand Ethiopian Renaissance Dam Construction Plans and Surrounding Geological Environment ... 46
2.0 Introduction ... 46
2.1 A Point of View of the Nile River Basin ... 46
2.2 The Significance of GERD ... 51
2.3 The Grand Ethiopian Dam (GERD) ... 51
2.4 GERD Construction Characteristics ... 53
2.5 The Reservoir of GERD ... 55
2.6 GERD Spillways ... 57
2.7 Drainage System and Grouting Operations ... 57
2.7.1 Operation Designing of Curtain Grouting and Bulb Grouting ... 58
2.7.2 Contact/Consolidation Grouting ... 58
2.7.3 Drainage System ... 59
2.8 Filling Scenarios of GERD ... 60
2.9 Intense Criticism from Scientists Over Heggy's Study: A Call for Scientific Integrity and Respectful Discourse ... 63
2.10 General GERD Impacts on the Environment ... 64
2.11 Geological Features of Ethiopia ... 65
2.12 Is a New Ocean Forming in Africa? ... 73
2.13 Conclusion ... 75
References ... 76

3. Utilizing Quantized Maxwell's Equations for Remote Sensing and Imaging of the Grand Ethiopian Renaissance Dam (GERD) ... 79
3.0 Introduction ... 79
3.1 Advanced Remote Sensing Technology ... 79
3.2 A Novel Definition of Remote Sensing ... 80
3.3 The Difference Between Marghany's Remote Sensing Definition and the Classical One ... 81
3.4 Mechanism of Quantum Remote Sensing Imaging ... 82

3.5 Quantized Maxwell Equations 84
3.6 Hilbert Space Description of Quantized Maxwell's Equations 88
3.7 Derivative Spectral Signature from Quantized Maxwell's Equations 89
3.8 Quantized Blackbody Radiation 89
3.9 Derivative Blackbody Radiation Energy from Quantized Spectral Signature 91
3.10 Quantized Kirchhoff's Laws 93
3.11 The Role of Spatial Dimensions in Satellite Remote Sensing for GERD Monitoring and Analysis 96
3.11.1 Spatial Resolution 97
3.11.2 Temporal Resolution 98
3.11.3 Spectral Resolution 98
3.12 Conclusion 99
References 99

4. Quantum Spectral Signature of the Grand Ethiopian Renaissance Dam Construction and its Surrounding Geological Features 101
4.0 Introduction 101
4.1 The Importance of Geological Investigations for Dam Site Selection 102
4.2 How Geological Investigations Impact Dam Safety? 103
4.3 Emergency Response Planning 103
4.4 Some Famous Dams Built on Challenging Geological Sites 104
4.4.1 Hoover Dam (USA) 104
4.4.2 Three Gorges Dam (China) 104
4.4.3 Jinping-I Dam (China) 105
4.4.4 Itaipu Dam (Brazil/Paraguay) 106
4.4.5 Kariba Dam (Zambia/Zimbabwe) 106
4.5 The Role of Instrumentation in Monitoring Dam Behavior 107
4.6 Remote Sensing Multispectral and Hyperspectral Monitoring of Dam Behavior 109
4.6.1 Key Differences Between Multispectral and Hyperspectral Sensors 110
4.6.2 Comparing MSI and HSI Systems 110
4.6.3 Practical Implications 111
4.7 Quantization Techniques for Multispectral and Hyperspectral Data 111
4.8 Spectral Reflectance Quantum Image Formation (SRQIF) 116
4.9 Entanglement Shapes the Formation of Spectral Libraries for GERD 117
4.10 Multispectral and Hyperspectral Data Storage in Quantum Computing 119
4.11 Modeling Quantum Spectra Signature of GERD Construction and Its Surrounding Environment 120
4.12 Quantum Spectra Signature of the Surrounding Environment of GERD 121
4.13 Comparison Between Quantum Spectra Signature of GERD Construction in Landsat-8 and ASTER Data 123
4.14 Quantum Spectral Signatures Accurately Identifies GERD Construction and its Surrounding Environment 124
4.15 Conclusion 127
References 127

5. Quantum Support Vector Machine for Simulating Sedimentation Concentration in the GERD Reservoir 129
5.0 Introduction 129
5.1 Understanding the Importance of Reservoir Sedimentation 130
5.2 Reservoir Sedimentation Mechanism 130
5.3 Sediment Types in GERD Reservoir 132

5.4 The Sources of Sedimentation in the GERD Reservoir 133
5.4.1 Watershed Characteristics 133
5.4.2 Soil Composition of the Blue Nile Basin, Ethiopia 135
5.4.3 Sediment Yield 136
5.5 Optical Remote Sensing of Soil Investigations 136
5.6 Methods for Calculating Cation Exchange Capacity in the Laboratory 138
5.7 Sentinel-2 Satellite Data Acquisition 140
5.8 How Clay Potential Percentage in Remote Sensing Data Can be Retrieved 141
5.9 Quantized Marghany Clay Saturation Algorithm for Retrieving Sedimentation Occurrence in GERD Reservoir 143
5.10 Support Vector Machines 150
5.11 Quantum Support Vector Machines 153
5.12 QSVM entangles Quantized Marghany's Clay Saturation Algorithm 154
5.13 Quantum Inspired Algorithm for Retrieving Suspended Sediment Concentration from Sentinel-2 Data 155
5.14 Conclusion 158
References 158

6. Advanced Quantum Image Processing for Automatic Lineament and Fault Detection along GERD Using Multispectral Remote Sensing Data 161
6.0 Introduction 161
6.1 Lineament 161
6.2 The Different Types of Lineaments 166
6.3 The Fundamental Differences Between Positive and Negative Lineaments in Terms of their Topographic Expression and Formation Processes 168
6.4 Satellite Remote Sensing and Image Processing to Detect Lineament Features 168
6.5 Challenges in Extracting Geological Features from Remote Sensing Data 169
6.6 How Multispectral Remote Sensing Data Identify Lineaments? 171
6.7 Can Digital Elevation Models be Used for Lineament Delineation? 176
6.8 The Foremost Interrogation 178
6.9 Fuzzy B-Splines for Digital Elevation Model Reconstruction 179
6.10 Fuzzy Quantum Entanglement for DEM Reconstruction 182
6.11 Quantum-Based Edge Detection for Lineament Automatic Mapping 182
6.12 Conclusion 187
References 187

7. Fundamentals of Microwave Altimetry: Analyzing Satellite Data 190
7.0 Introduction 190
7.1 Radio Detecting and Ranging 190
7.2 The Altimeter Microwave 191
7.3 How the Electromagnetic Spectrum Functions in Radar Altimeter Operation? 192
7.4 Principles of Altimeter 193
7.5 Radar Altimeter Frequencies 195
7.6 Radar Altimetry Pulses 199
7.7 Deriving Mathematical Formulas of Radar Pulses 200
7.8 How Can Radar Altimetry Equation be Derived from Pulse Print F_p? 202
7.9 Altimetry Echo Shape Determination 203
7.10 Deriving Waveform above River and Dam Reservoir 205
7.11 Water Height Estimation in River and Dam Reservoirs Using Delay Time 206
7.12 Range Ambiguity 207

7.12 Conclusion 210
References 210

8. Modeling Water Height Level Variations in Lake Millennium Using Advanced Interferometry Altimeter Satellite Data 212
8.0 Introduction 212
8.1 Microwave Altimeter Measurements of Inland Water Bodies 212
8.2 Time Delay versus Water Height and SNR 215
8.3 Precision in Height Retrieval 217
8.4 Estimating Surface Height with Radio Altimeter 220
8.5 Numerical Example of Estimating Water Level Anomalies 221
8.6 Delay-Doppler Radar Impact on Altimeter Return Signals 222
8.7 Altimeter Sensors 224
8.8 Principles of Synthetic Aperture Radar Altimeter Interferometry 225
8.9 Altimeter Interferometry Technique 227
8.10 InSAR Precision Procedures Altimeter Scheme 228
8.11 Delay-Doppler Altimeter 230
8.12 CRYOSAT-2 SIRAL Data Acquisitions 231
8.13 Cycle of Water Heights Across Lake Millennium 233
8.14 SARIn – An Accurate Tool 237
8.15 Conclusion 238
References 239

9. Unveiling Dam Turbulence and Efficiency Using Marghany's Quantum-Optimized Multi-Objective Algorithm in SAR Imaging 241
9.0 Introduction 241
9.1 Turbulence Enhances Dam Operation Monitoring 241
9.2 Turbulence Generation Mechanisms within and Around the Dam 243
9.2.1 Flow Over Spillways 243
9.2.2 Obstructions and Structural Features 245
9.2.3 High Flow Velocities 245
9.2.4 Shear Force 246
9.2.5 Flow Direction Changes and Turbulence Generation 246
9.2.6 Interaction with the Riverbed 247
9.3 Turbines Connected to a Dam Produce Turbulent Flow 247
9.4 Turbines Enhance the Operation and Efficiency of Dams 249
9.5 Advanced Remote Sensing Technology Helps Monitor and Track Turbulence Due to Turbine Functionality 249
9.6 Microwave Data Offers Advantages Over Optical Data in Assessing Dam Turbulence Flows as an Index of Dam Operation and Efficiency 250
9.7 The Synthetic Aperture Radar (SAR) 251
9.8 Quantized Marghany Speculation for High-Resolution SAR Data 253
9.9 Radar Imaging Mechanisms for Dam-Generating Turbulence 255
9.10 Marghany Quantized Multi-Objective Algorithm for Detecting Dam Turbulence in SAR Images 259
9.11 Evaluating Dam Efficiency Through Marghany Quantized Multi-Objective Algorithm 262
9.12 Tested High-Resolution SAR Image of GERD 263
9.13 Conclusion 270
References 270

10. Quantum Along-Track Interferometry: Modeling Downstream Flows and Water Discharge Using TanDEM-X Data 273
10.1 Dam Discharge 273
10.2 The Fascinating Process Behind Releasing Water from a Dam 275
10.3 The Difference Between Streamflow and Discharge in the Context of Hydrological Studies 276
10.4 Types of Dam Discharges 278
10.4.1 Spillway Discharge 278
10.4.2 Sluice Gate Discharge 279
10.4.3 Outlet Works Discharge 279
10.4.4 Powerhouse Discharge (Hydropower Release) 279
10.4.5 Bottom Outlet or Drainage Discharge 280
10.4.6 Flood Discharge 281
10.4.7 Environmental Flow Release 281
10.4.8 Emergency Discharge 281
10.5 Do Dams Influence the Discharge of Water: Do They Increase or Decrease It? ... 282
10.6 Mathematical Explanation of Water Discharge from a Dam 282
10.6.1 Water Level Rise and Pressure Increase 283
10.6.2 Structural Overloading of the Dam 284
10.6.3 Potential Dam Break 284
10.7 Challenges in modelling GERD Discharge 285
10.8 Principles of Interferometric Synthetic Aperture Radar (InSAR) 286
10.9 Deriving the Mathematical Equations of Along-Track Interferometry for Water Discharge from a Dam 289
10.10 Quantum of Along-Track Interferometry 291
10.11 Quantum Hopfield Algorithm for ATI Phase Unwrapping 292
10.12 Quantum ATI Hopfield Algorithm Application to TanDEM-X Satellite 295
10.13 Approvals for Precision of qHop Along-Track Interferometry Technique 300
10.14 Conclusion 301
References 302

11. Quantum Convolutional Neural Network for Monitoring Deformation of the Grand Ethiopian Renaissance Dam Using Persistent Scatterer Interferometry Technique 305
11.0 Introduction 305
11.1 Persistent Scatterer Interferometry (PSI) 305
11.2 How Does PSI Operate? 306
11.3 Interferometric Processing 307
11.4 Simulation of InSAR Coherence 308
11.5 Quantized Coherence SAR Simulation 311
11.6 Relation to Multiscale Entanglement Renormalization Ansatz (MERA) 312
11.7 Persistent Scatterer Selection 316
11.8 Simulation of PSI Using Functional Model 321
11.9 Quantum Convolutional Neural Network (QCNN) for Phase Unwrapping 323
11.10 Multi-Qubit Gates for Phase Manipulation 325
11.10.1 Controlled-NOT Gate (CNOT) 325
11.10.2 Controlled Phase Gate (CPHASE) 325
11.11 InSAR Generated from PALSAR-1 SAR and Sentinel-1A Data 327
11.12 PSI Generated Using QCNN from MultiSAR Data 328
11.13 Spatial Variation of Deformation Velocity Across the Nile Basin and GERD 334

11.14 Why Tunnels Cause Land Deformation? 340
11.15 Why QCNN Deliver Accurate PSI Deformation Patterns? 341
11.16 Conclusion 344
References 344

12. Four-Dimensional Marghany Hologram-TomoSAR Algorithm and Quantum Machine Learning for Monitoring and Forecasting Dam Failures 347
12.0 Introduction 347
12.1 The Difference Between TomoSAR and Traditional SAR Techniques 347
12.2 Mathematical Models Behind TomoSAR 350
12.3 Super-Resolution Enhancement 355
12.3.1 Regularized Super-Resolution Problem 355
12.3.2 Types of Regularization: Sparse and Edge-Preserving Norms 355
12.3.3 Total Variation (TV) Regularization (Edge-Preserving) 357
12.3.4 Extension to Nonlinear and Adaptive Regularization 357
12.3.5 Lagrange Multiplier Interpretation and Dual Problem 357
12.4 Procedures of Achieving TomoSAR 357
12.4.1 Preprocessing and Registration 358
12.4.2 False Alarm and Detection Probabilities 358
12.4.3 Minimization of Terms for Estimation 358
12.4.4 Application of the Generalized Likelihood Ratio Test (GLRT) 358
12.5 Mathematical Construction of 4D from TomoSAR 363
12.5.1 Problem Setup: TomoSAR for 4D Reconstruction 364
12.5.2 Incorporating Quantum Algorithms 364
12.6 Derivation of the Marghany 4D Quantum Phase Unwrapping Algorithm 365
12.6.1 Quantum State Representation 366
12.6.2 Quantum Entanglement for Correlated Measurements 366
12.6.3 Quantum Fourier Transform (QFT) 366
12.6.4 Quantum Automaton for Phase Unwrapping 367
12.7 Four-Dimensional Marghany Hologram-TomoSAR Algorithm for GERD Monitoring 368
12.8 Simulation of Sedimentation Volume Using 4D Marghany Holographic TomoSAR Algorithm 371
12.9 Quantum Machine Learning for Forecasting Dam Breaching 375
12.9.1 Quantum Data Encoding 375
12.9.2 Defining the Quantum Machine Learning Model 376
12.9.3 Cost Function and Training Process 377
12.9.4 Quantum Optimizer 377
12.9.5 Forecasting Future States 378
12.10 HAARP Triggers GERD Breach 379
12.10.1 Influence of Potential Pathways of HAARP on GERD 379
12.10.2 Seismic Induction by Electromagnetic Waves 380
12.10.3 Climate Amplification Through Indian Ocean Dipole (IOD) Dynamics 381
12.11 4D Marghany Holographic-TomoSAR Algorithm for Forecasting GERD Breaching 382
12.12 Marghany Quantum Tsunami Speculation: Dynamics of the Red Sea and Indian Ocean Tsunamis Triggered by the GERD 395
12.13 How 4D Marghany Hologram TomoSAR Algorithm Detects, Monitors, and Forecasts the Impact of GERD Collapse on Downstream Nations? 403

12.14 Coherent Response Strategies Utilizing Conjunctions to Optimize Breach Flow Management ... 405
12.15 The Most Effective Military Strategies to Address the GERD Challenge ... 405
12.16 Understanding the Construction of the Grand Ethiopian Renaissance Dam Through the Lens of Isaiah's Prophecy ... 406
12.16.1 Prophecies of Hardship and Judgment (Isaiah 19) ... 407
12.16.2 The Coming of a Savior and Global Recognition of God (Isaiah 19:19–22) ... 407
12.16.3 A Vision of Peace and Unity Among Nations (Isaiah 19:23-25) ... 407
12.17 Conclusion ... 408
References ... 409

Index ... ***413***
About the Author ... ***419***

Chapter 1

Fundamental Principles of Dam Construction and Their Environmental and Societal Impacts

1.1 WHAT ARE DAMS?

Dams are structures that are typically built across rivers to keep back and retain water in a lake or reservoir. Dams are often constructed with concrete or natural materials such as Earth and rock, and in scenarios such as the well-known dams of the Hoover Dam (Figure 1.1) and the Three Gorges (Figure 1.2), are complex engineering ventures with lengthy construction timings.

Figure 1.1 Hoover Dam construction.

DOI: 10.1201/9781003327325-1

Figure 1.2 Three Gorges Dam construction.

Dams have been an important part of the water infrastructure for millennia and have served a variety of purposes. Reservoirs retain rainwater before it is filtered and prepared for human consumption. Hydroelectric power plants create electricity that can be generated the moment the reservoirs grab the heavy rainfall. In addition, reservoirs regulate the flow of water into rivers after heavy rain to prevent flood disasters. Dams can be used to boost upstream water levels and improve navigation conditions. Dams control irrigation and water management. During dry periods, water can be stored in reservoirs and used to irrigate land [1].

1.2 THE ELEMENTS OF A DAM

Dams, as stated previously, are constructions designed to resist or block the influx of water in a stream or river. A wide-span barrier/wall is constructed in front of the flow of water, allowing a controlled volume of water to flow through while the excess water is preserved or kept for a future crisis [2]. Prior to a piece of detailed information being given about dam types, the structural elements of dams need to be covered in depth. In terms of the structural parts of dams, they are practically identical over all kinds of dams; different dam designs may modify the designs of the dam elements. As a consequence, a dam structure can indeed be divided into three parts: (i) the water-retaining structure, (ii) the water-releasing structure, and (iii) the water-conveying structure [1, 3].

1.2.1 Water Retaining Structure

The water retention structure is the dam's walled construction that resists water flow and enables only a restrained amount to flow downstream. The face of the dam is the side of the dam that faces downstream. The crest of the dam is the plane top surface and the crest is a roadway that spans the dam. The dam's crest is built on top of the spillway. This includes the dam's overall length, which is the dam's overflow portion. The succeeding elements of abutments; dam's base; crest; cut-off; cut-off wall; face, upstream; parapet walls; the heel of the dam; and the toe of the dam (Figure 1.3) originate in the dam's water-retaining section [1–3].

Abutments are the valley sides of dams that are designed with masonry wall work. The purpose of the abutment is to provide stability to the wall of the dam. Therefore, the base of the dam is the entire width of its basement, which diminishes with its vertical ascent. Consequently, the uppermost planar surface of the dam is known as the crest. It is the dam's overflow section. The dam's parapet walls, therefore, are built on the dam's crest. The purpose of the parapet wall is to ensure the safety of tourists. [1, 4].

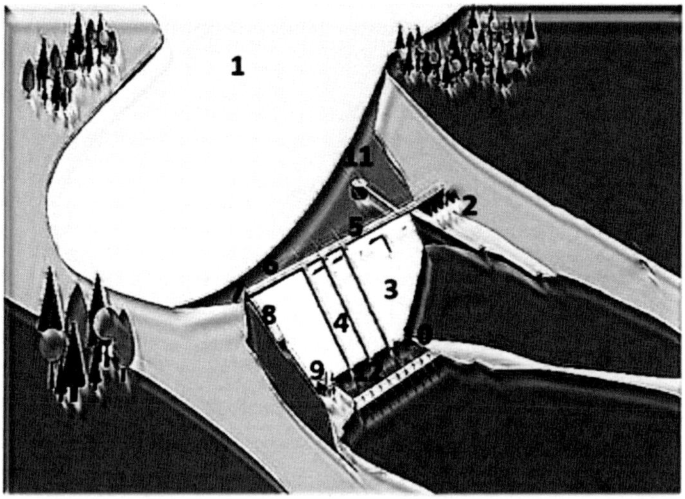

Figure 1.3 Dam construction elements.

Subsequently, the dame's cut-off is an impermeable substance that blocks water from leaking through its foundation. In this sense, the cut-off wall is a wall of impervious material, such as wood panels, concrete, or steel which is built into the core of the dam. This wall aims to prevent seepage as seepage undermines the structural integrity of the dam. Furthermore, the downstream side of the dam is known as its face. Upstream of a dam, on the other side, is the level of a reservoir remanded back by the dam. The upstream water flows through the channel forward towards the downstream in this perspective. In this sense, the heel of the dam refers to the point at which the upstream face confronts its foundation. Consequently, the toe of the dam is the point where the downstream of the dam reaches the ground [2–4].

1.2.2 Water Releasing Structure

The water-releasing structure is a section of the dam that permits water to flow downstream, which is termed as the spillway of the dam. The mechanism of the spillway allows for controlled water quantity. A spillway mainly consists of five gates (i) crest gate/ spillway gate (ii) flap gates (iii) outlet gate (iv) radial gate and (v) slide gate. The crest gates/ spillway gates are situated on the crest of the spillway. These function of the gates is to enable a partial amount of reservoir water to pass through. Therefore, the flap gates are flow control gates that are hinged at the top. The outlet gate, accordingly, regulates the flow of water from the reservoir. A radial gate, subsequently, has radial arms and curved upstream plates. On the other hand, the slide gate releases and closes by sliding to control the water flow [3, 5].

Spillways are classified into two types (i) auxiliary spillways and (ii) ogee spillways. Auxiliary spillways are built along or above the main crest of the spillway. They are intended for emergency water release to avert dam destruction. A dam rupture or damage can result in a disastrous flood. An ogee spillway, therefore, is an overflow spillway in which the crest, downstream slope, and bucket are built in an 'S' shape or ogee form of a curve. The form is intended to keep high-velocity water from damaging the downstream area at the bottom of the dam. [1, 3, 5].

1.2.3 Water Conveying Structure

Conduits are essential components of a dam's exterior design, specifically constructed to channel water from reservoirs either through, around or beneath an embankment dam. These conduits facilitate controlled water flow, playing a critical role in maintaining reservoir levels and managing overflow. In many dams, conduits also serve as spillways, allowing excess water to be safely discharged, especially during times of high inflow or flooding. By directing water away from vulnerable areas, spillway conduits aid in protecting the dam structure itself as well as downstream areas from potential flood damage [1, 5].

1.3 WHAT ARE THE DIFFERENT TYPES OF DAMS?

1.3.1 Arch Dam

An arch dam curves in the shape of an arch (Figure 1.4), with the crest of the arch facing back into the water, to form an extremely strong structural structure that is resistant to water pressure behind the dam. Arch dams are typically utilized in narrow, steep-sided valleys because they are composed of concrete and require adequate rock support for foundations and sides [3–5]. In other words, arch dams are stabilized by the canyon walls in which they are constructed. On both sides of the canyon, notches are excavated into the canyon walls, and the dam is constructed to fit into these notches. The shape of the arch is curled upstream, into the reservoir (Figure 1.4) and transfers the force exerted by the water in the reservoir to the canyon walls [1, 6].

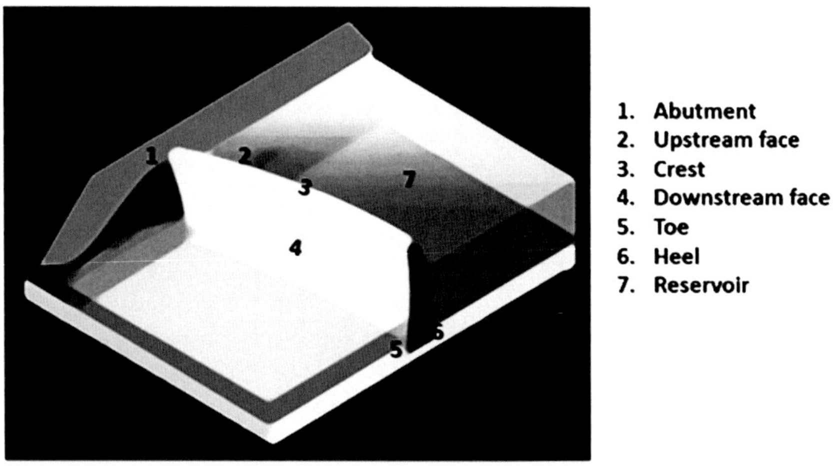

Figure 1.4 Arch dam construction.

1.3.2 Buttress Dam

Buttress dams contain triangular-shaped walls, or buttresses (Figure 1.5), positioned at regular intervals on the downstream side to resist water pressure from the upstream side. They must be built on stable rock and are typically constructed of concrete or masonry. Due to the overwhelming spacing between the buttresses, they can be extremely competent in terms of materials construction [1, 3, 5, 7].

Needless to say, the buttresses are placed at regular intervals to deflect the force that the water puts on the dam down to the basement of the dam. Buttresses, while commonly triangular, can take on a variety of shapes based on the architecture and location of the dam. Therefore, there are five varieties of buttress dams, the most prevalent of which are deck slabs and multiple arches. Buttress dams can involve a deck slab buttress dam, multiple arch buttress dams, massive head buttress dams, multiple dome buttress dams and columnar buttress dams [3, 7].

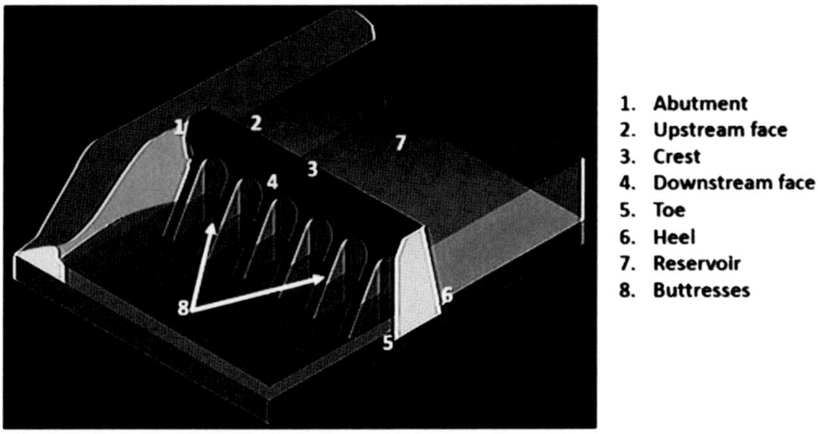

Figure 1.5 Buttresses dam construction.

1.3.2.1 Deck Slab Buttress Dam

A deck slab is supplied in this sort of dam, which is supported by buttress corbels. This type of dam is built to modest heights, often ranging from 20 to 50 m. The slab offered is 40° to 55° inclined to the horizontal. This inclination is required to sustain the dam and support a dead load of stored reservoir water, as well as to prevent the dam from sliding owing to self-weight. In this situation, each deck slab unit on two neighboring buttresses functions as a single, independent unit (Figure 1.6). As a result, if one unit is affected or damaged, there is no need to be concerned about the other units [1, 3].

Figure 1.6 Example of deck slab buttress dam.

Furthermore, there are three varieties in the case of deck slab dam: (i) the fixed deck slab (ii) the free deck slab (iii) and the cantilever form. In the instance of a fixed deck slab type dam, deck slab, and buttresses are casted monolithically. Therefore, free deck slab buttress dams are also known as simple deck slab buttress dams. They are is built when the foundation soil is exceedingly weak. In this circumstance, reinforcement is placed at the downstream face of the deck slab. Cantilever-type buttress dams (Figure 1.7), on the other hand, are cantilevered at both ends and have reinforcement at the upstream face [5, 7].

Figure 1.7 Cantilever buttress dams.

1.3.2.2 Multiple Arch Buttress Dam

A multiple-arch dam is made up of a sequence of arches, and in this case, arch slabs are given on the upstream face of the dam. Buttresses hold these slabs in place. The buttress wall can be built as a single stiffening wall or as a double hollow wall. This style of dam is preferred for heights greater than 50 m. Multiple arch buttress dams are more flexible and stable than deck slab-type buttress dams (Figure 1.8).

Figure 1.8 Multiple arch buttress dam construction.

The biggest disadvantage in this instance is that the adjacent buttress units are interdependent. As a result, if one unit fails, the entire series of arches suffer severe consequences. To address this problem, a solid foundation should be laid for each buttress wall. In this understanding, the thickness of the arch varies with greater spans and remains constant with shorter spans. The center angle of the arch should be between 180° and 150°. The distance between buttresses should be 15 to 21 m. Larger spacing can, nevertheless, be accomplished by taking some considerations into account [4, 7].

1.3.2.3 Massive Head Buttress Dam

No slab or arch is built at the upstream face of a massive head buttress dam; instead, that buttress head is enlarged and merged with an adjacent buttress head. As a possible consequence, all buttress heads form a powerful water-supporting surface (Figure 1.9). Buttress heads can be enlarged in a variety of shapes, including circular, diamond, and square (Figure 1.10). As a result, these shapes can withstand a lot of water pressure. Copper strips are used to reinforce the connecting of heads. The construction of a large head buttress dam is simpler than other varieties since no reinforcement is required; instead, mass concrete is put entirely over the dam body. It is heavy and resists slippage due to its massive concrete body [1, 4, 8].

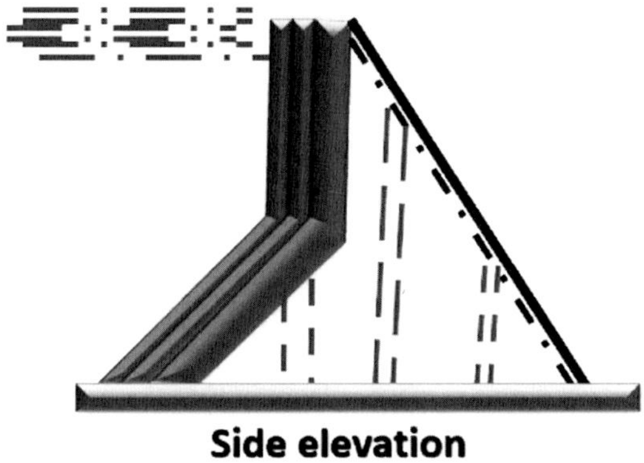

Figure 1.9 Massive head buttress dam side elevation.

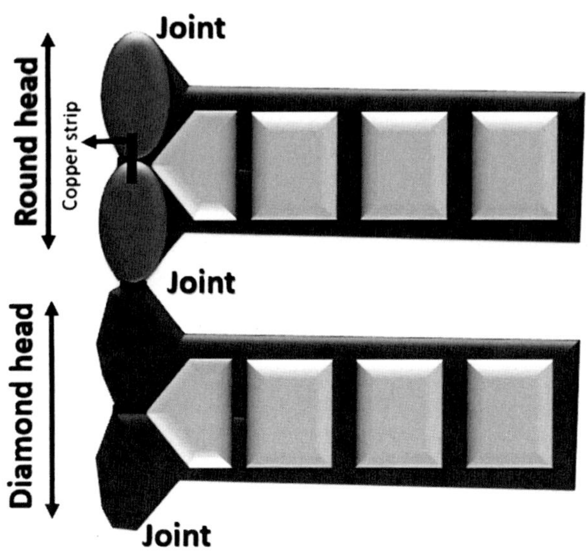

Figure 1.10 A variety of shapes of massive head buttress dam constructions.

1.3.2.4 Multiple Dome Buttress Dam

The multiple dome buttress dam (Figure 1.11) is, therefore, similar to the multiple arch type buttress dam, except that domes are used instead of arches. Aside from that, nearly all the features are the same. The presence of such domes has an impact on the distance between the buttresses. As a result, longer spans can be provided while the number of buttresses can be reduced.

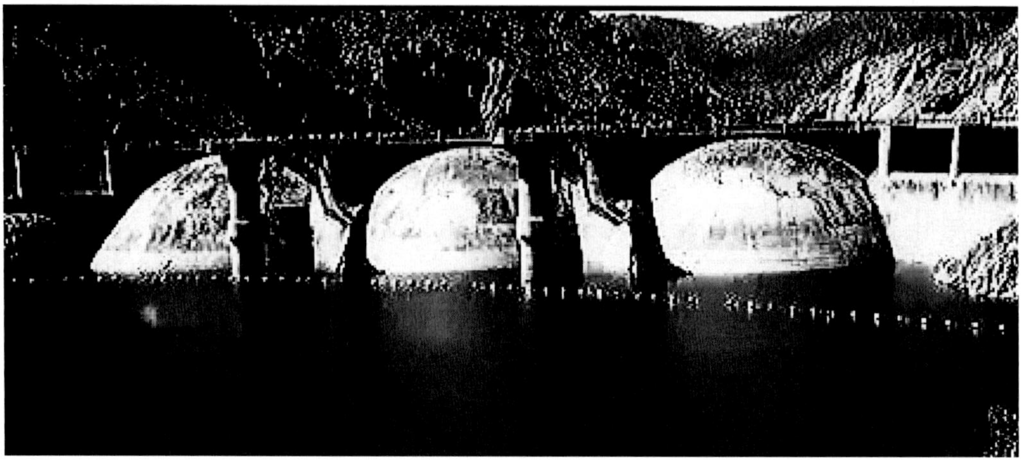

Figure 1.11 Multiple domes buttress dam construction.

1.3.2.5 Columnar Buttress Dam

However, inclined columns support the deck slab of the columnar buttress dam (Figure 1.12). It is typically a modification of a deck slab sort buttress dam. In this regard, it involves identical robust and steady bases. It also entails that only professional dam experts build the buttresses. Hence, it is not a broadly depleted dam as the foundation pressures are higher; therefore, either a solid rock foundation, spread footings, or a continuous mat may be required, which raises the building cost.

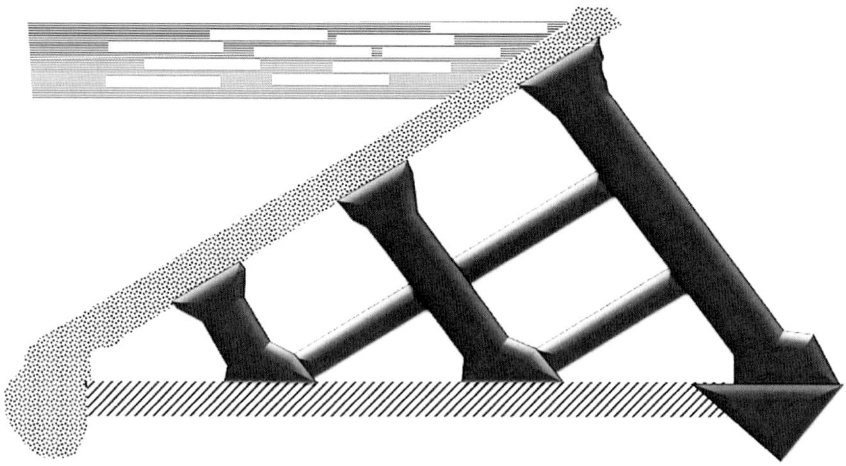

Figure 1.12 The construction shape of the columnar buttress dam.

1.3.3 Embankment Dams

Embankment dams (Figure 1.13) seem to have been conventionally highly prevalent in the United Kingdom and typically situated to places that have heavy valleys. They are usually designed using natural materials such as compacted soil or rocks that are often quarried or dug locally. An embankment dam has a hill or bank shape in the cross-section, with a central core built of an impermeable material, such as clay soils or concrete, to impede water flow [1, 4, 9]. Embankment dams, therefore, are classed as rolled fill or placed earth-fill dams, or hydraulic fill dams, regardless of the type of construction.

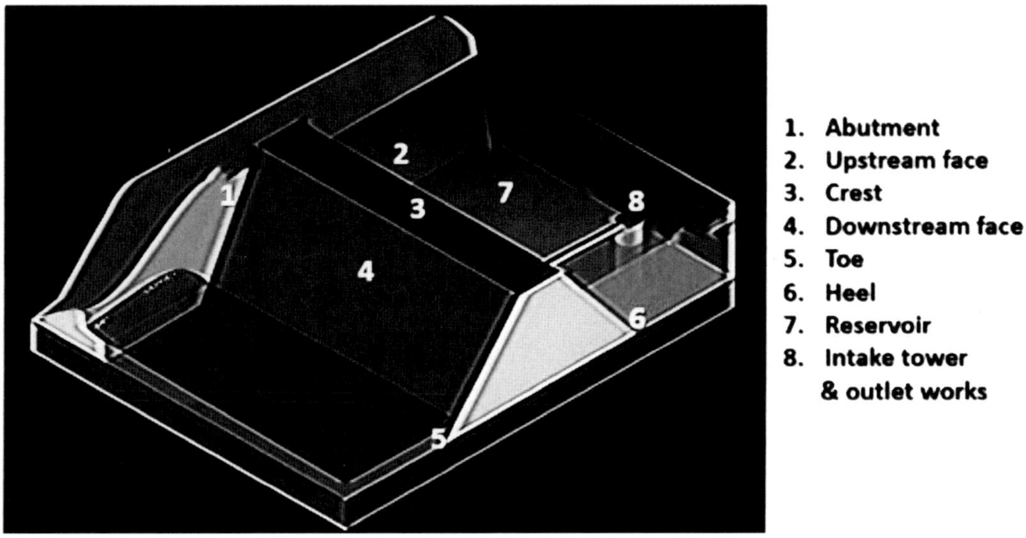

Figure 1.13 The construction shape of an embankment dam.

1.3.3.1 Rolled Fill or Placed Earthfill Dams

The largest component of the embankment is created in successive layers that are mechanically compacted in the rolling fill method. Construction materials are almost always obtained from building sites, often known as borrow areas. In this sense, materials from the borrowed regions are transported to the embankment site and dispersed to the desired thickness; water is added, and the material is compacted with power-driven rollers. In this regard, natural materials that have not been treated are utilized [7, 9].

1.3.3.2 Hydraulic Fill Dam

In hydraulic fill dams, the embankment materials are typically mixed with water. To achieve this, a soil-water suspension consisting of about 85% water is delivered to the site and allowed to settle. By using this method, construction can be carried out quite effectively with proper settling techniques and suspension management. However, challenges arise due to the segregation of coarse particles, which can make this approach less suitable. As a result, the earth-fill type is often the preferred choice in these situations [1, 3, 7, 9].

1.3.3.3 Classification of Embankment Dams

Therefore, embankment dams are classified into three types: (i) diaphragm (ii) homogenous and (iii) zonal [1, 9].

1.3.3.3.1 Diaphragm

A large amount of the embankment in the diaphragm dam is made of precious materials. In other words, the diaphragm, therefore, might be made of soil, cement concrete, bitumen, or other materials. A thin diaphragm of impermeable material is used to create a water barrier. In this view, the diaphragm can be situated in the center as a vertical core or on the upstream face like a blanket. If the core of the Earth is used as a diaphragm, its thickness should be less than 3 m or less than the height of the embankment.

1.3.3.3.2 Homogenous dam

A perfectly homogenous dam is made of a single sort of material, except for the material used for slope protection. Plenty of the low- to moderate-height dams are practically identical. The material chosen for such dams should be sufficiently impermeable and the slope should be relatively level to meet stability criteria. Homogeneous dams of 6 to 8 m in height must have some form of downstream drain to help lower pore water pressure in the downstream area of the dam and regulate any seepage. Toe drains, horizontal blanket drains and chimney drains are examples of drains found in homogenous dams.

1.3.3.3.3 Zonal

This is a more common structure of an embankment dam, with pervious and impermeable substances included to construct it. The impermeable material, known as the core, is in the middle, flanked by pervious material zones known as shells or casings. The shells provide support and protection to the center core. The upstream shell provides stability against rapid draw-down, while the downstream shell serves as a drain to keep the seepage line under control.

Sands, gravel, cobbles, rocks, or mixtures of these substances are used in zonal dam construction. In this regard, the thickness of the core is determined by substance accessibility and design considerations including stability and seepage. If a diversity of soils is accessible in a defined location, there would be a huge array of materials available to be chosen for the various portions of the embankment. In such scenarios, the dam's height would be unconstrained, and the zoned kind would invariably result in a more cost-effective section. A rockfill dam is one in which the majority of the structure is made of rock [1, 3, 5, 7, 9].

1.3.3.4 Elements of Embankment Dams

The foundation, shell, and core of an embankment dam are the three essential components (Figure 1.14). Additional appurtenances are added to enable the fundamental components to function efficiently, depending on the sort of dam.

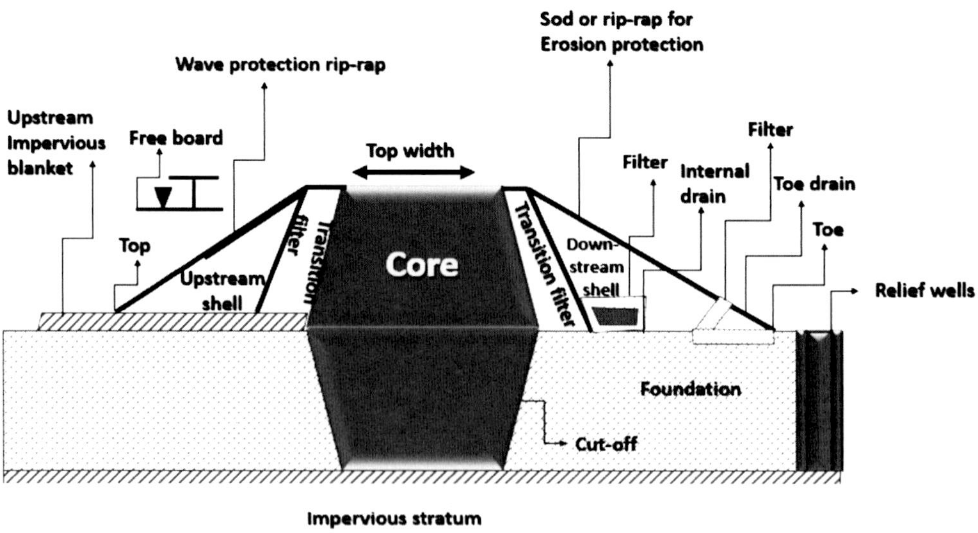

Figure 1.14 Components of an embankment dam.

1.3.3.4.1 Foundation

The foundation of a dam is often the only supporting material that can sustain vertical and horizontal forces. The foundation can allow or block water flow depending on the foundation material, i.e., whether it is soil or rock. If the rocks are free of faults, joints, seams of soft shale or clay, etc., they make ideal foundation material. Dams can also be supported by sands and gravels, but suitable measures to limit seepage must be employed. To avoid liquefaction, fine sands with relative densities less than roughly 65% should be compacted by vibration. If no steps are taken to speed up the consolidation of clay foundations, major stability and settling issues will arise. Foundations, therefore, require a flat slope because of their low shear stress. In this sense, dam design slopes should be up to 15 m high to provide low shear strength or stress [2, 7, 9].

1.3.3.4.2 Casing

The shell, therefore, provides stability and protects the core. The casing can be made out of any relatively pervious material that isn't prone to cracking under normal atmospheric conditions. The upstream and downstream slopes of the casing must be determined depending on material availability, foundation condition, height, and dam type. For stability, the upstream ranges from 2:1 to 4:1, and a flat one is selected for materials with low permeability. The typical downstream slope is between 2:1 and 2.5:1 [1, 3, 5, 8].

1.3.3.4.3 Core

The core operates as an impermeable barrier, preventing water from freely passing through the dam's body. Consequently, soils with significant compressibility, swelling, shrinking and organic content are not suited for cores. The core may be positioned centrally or inclined upstream depending on the availability of material, the terrain of the site and diversion issues. The top level of the core should be set one meter above the maximum water level. This requirement is imposed to prevent seepage by capillary siphoning. The minimum top width of the core should be 3 m, and the final thickness must be determined depending on practical concerns; which include the availability of adequate impermeable material, pipe resistance, permissible seepage across the dam, and the presence of alternative materials for the casing, filters, etc. [1, 4, 7].

1.3.4 Gravity Dams

A gravity dam is a concrete structure with a triangle cross-section (Figure 1.15) that spans a large valley. The most prevalent type of concrete dam resists the force put on it by the reservoir by using its weight. The biggest and heaviest concrete dams are gravity dams. They have a massive foundation and can only be built on rock. The question then is what are the gravity sections of some embankment dams?

Gravity is what keeps gravity dams in place. The dams are usually composed of concrete, masonry, or both and have a triangle cross-section. They must be constructed on solid rock and are suitable for either wide or narrow valleys. Therefore, the mechanical function of a gravity dam demonstrates that the water barrier is structured in such a way that the water is held back by its weight. The water pushes on the wall in a horizontal direction (Figure 1.16). The water tries to push the wall away, but the dam's foundation, which is hefty and vast, holds the water back. The gravity of the wide-span walls, or its weight, are the only reason for its success [3, 7, 9].

Spillways are components of a dam and are exercised to safely and controllably transport overflow water over, around or through a dam. There are various kinds of spillways. The overflow spillway permits water to flow over and down its frontage wall as it is lower than other dam sections and is bowed to regulate the flow of water. Gravity or buttress dams, therefore, are the most prevalent configurations [9–12].

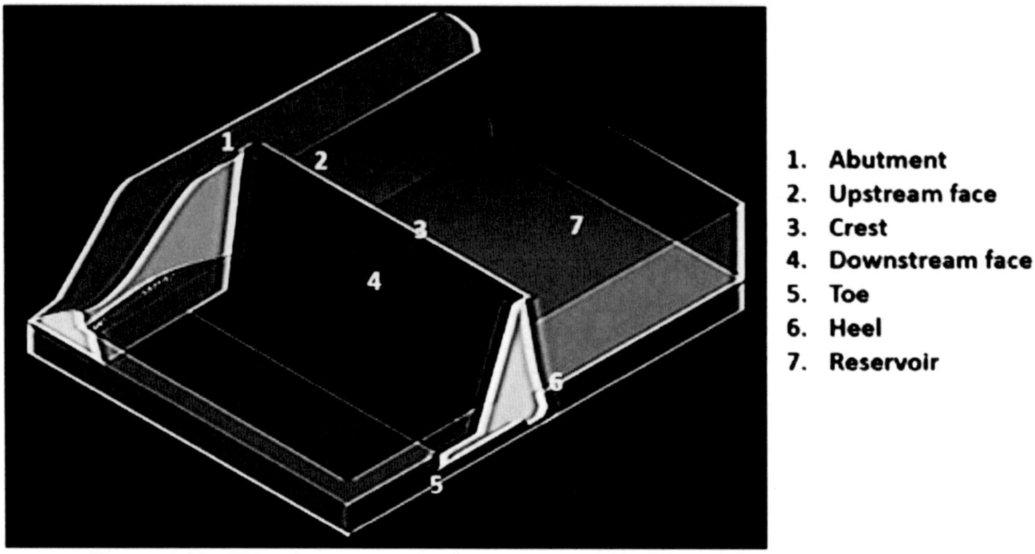

Figure 1.15 Triangle cross-section between toe and hell in gravity dam construction.

Consequently, the side-channel spillway is positioned a short distance upstream of a dam and diverts water into a side-channel before flowing into the river downstream of the dam. They are generally used in conjunction with embankment dams. Subsequently, the shaft spillway has a round hollow tower located near the dam in the reservoir. When the water level rises, it runs through a funnel at the top of the shaft, where it is diverted downstream [1, 8, 10, 12].

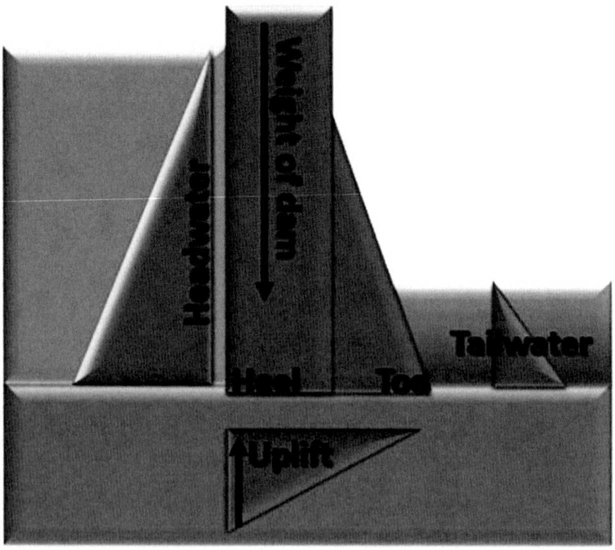

Figure 1.16 Mechanical operator concept of the gravity dam.

1.3.4.1 Types of Gravity Dams

1.3.4.1.1 Concrete dams

Concrete dams are gravity dams that are often constructed in huge blocks separated by joints to make construction easier and reduce thermal stress. After the dam has cooled, the joints are grouted. The axis of the dam is perpendicular to the transverse joints. A concrete dam is a significant construction project that necessitates a large amount of concrete, which necessitates the usage of a lot of cement. Cement manufacturing emits CO_2, which is harmful to the environment and it is also a high-cost technique.

The arch-gravity dam, which combines the qualities of both the arch and gravity dams is one of the most common types of concrete dams. The arch-gravity dams are curved dams that rely on the arch principle for strength, but are significantly thicker and feature a filled core rather than traditional arch dams. Thanks to their tremendous weight and the improved strength of an arch over a straight-line structure they are able to keep the dam in place and keep the water out [1, 3, 13].

Some concrete dams contain multiple minor arches or buttresses, a pair of massive arches, or other modifications, but the majority, if not all, are variations on one of three fundamental designs. Many concrete dams are built to leverage water flow to create electricity through hydroelectric power plants, and many have spillways for discharging water when needed. Flowing water turns into gigantic turbines, which provide up to 20% of the electricity that the world requires (Figure 1.17) [3, 14].

Figure 1.17 Utilization of concrete dams for power generation, water management and flood control.

1.3.4.1.2 Roller compacted concrete (RCC)

Roller-compacted concrete is a type of concrete used to build concrete dams that use large rollers to squeeze the concrete during construction (Figure 1.18). This approach is being used to build a lot of newer dams but the designs are still the same [1, 8, 13].

In their unhardened states, RCC dams are built of composite construction material with no slump consistency. RCC is composed of cement/fly ash, water, sand, coarse aggregate, and typical additives, but it contains significantly less water than ordinary concrete. In its unhardened stage, this type of drier concrete can withstand a vibratory roller during compaction. When compared to ordinary concrete, the use of reduced cement content and fly ash results in less heat generation during curing. RCC dams are often constructed in thin, horizontal lifts, allowing for quick construction [5, 8, 10, 13].

Because of the heat generated by fly ash, hydration is substantially less than that generated by Portland Cement hydration. Therefore, partial substitution of fly ash for Portland Cement is an important feature of RCC dam construction. This, in turn, minimizes heat loads on the dam and the possibility of thermal cracking. RCC is a mixture of cement/fly ash, water, sand, aggregate,

Figure 1.18 Dam construction exploiting RCC technique.

and standard additives, however, it contains significantly less water In other words, richer RCC mixtures are commonly used to enhance water tightness and bonding at lift joints, along with smaller maximum aggregate sizes, stricter construction restrictions, special lift joint treatments, upstream membranes, and unique facing mixtures. For instance, the upstream face of Georgia's Hickory Log Creek Dam is made of 6-foot-high by 16-foot-long precast concrete panels with a geomembrane entirely bonded to the downstream face. Six galvanized steel rods secure each panel to the dam (Figure 1.19) [8, 13].

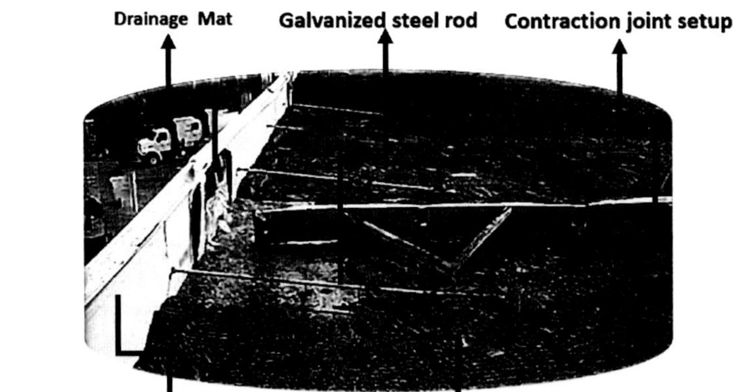

Figure 1.19 The mixtures used in RCC construction.

This is crucial for dam stability because a richer RCC mixture with more cementitious content and a wetter consistency improves the dam's strength and durability. It enhances the bonding between aggregates and ensures better hydration, contributing to a more stable and long-lasting structure. The middle spillway of the upstream section of the vertical face and downstream stepped face were slip-formed employing common concrete. The richer RCC mix delivered a stronger tensile strength, lowering the dam›s cross-section. Furthermore, the richer mixture and the upstream conventional concrete experienced improved seals and eliminated seepage at lift joints [3, 8, 14].

The resulting mix was drier and had little to no droop. RCC is installed in the same way that paving is: the material is delivered by dump trucks or conveyors, spread by small bulldozers or specially modified asphalt pavers and compacted by vibrating rollers.

RCC was initially and primarily used for backfill, sub-base, and concrete pavement construction, but it is increasingly being employed to create concrete gravity dams due to the low cement content and usage of fly ash, which generates less heat during curing than conventional mass concrete installations. Roller-compacted concrete dams have significant time and cost advantages over conventional mass concrete dams, including faster concrete placement, reduced material costs, and lower post-cooling and formwork expenses [1, 13, 15].

1.4 DAM CONSTRUCTIONS

Dam construction is extremely challenging and necessitates the participation of a diverse team of specialists from various disciplines. Civil engineers are keystone players in deciding the proper dam design for the site as well as developing technical drawings detailing the construction process. Civil engineers engage with engineering geologists and hydrologists on the technical aspects of the site as well as the required parameters depending on the level or volume of water involved. On the other hand, the requisite piping, valves, and floodgates are designed by mechanical engineers. Geotechnical engineers further evaluate if the rock or soil beneath the planned dam is strong enough to support the weight, as well as the potential permeability. A measurement of how easily a fluid may flow through a porous media assists in monitoring dam stability. Permeability is determined by the physical features of the medium, such as grain size, porosity, and pore shape [1, 10, 11]. This is the standard sequence of events for dam construction.

1.4.1 Diversion of the River

The flow of water in a river or stream is redirected to generate a dry region in which the dam is built (Figure 1.20). Lower flows, therefore, would be diverted through tunnels or channels created

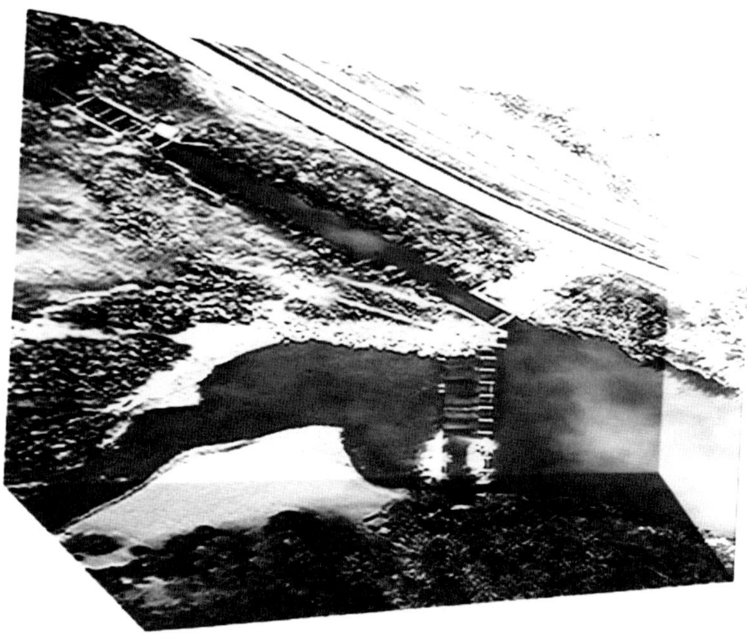

Figure 1.20 Example of river diversion.

around the dam sides, which will be excavated with explosives when necessary. On the other hand, higher flows could be too challenging to redirect using separate channels, thus a dry pit is built on one side of the river, ensuring that other areas remain open and accessible for water flow. The dam is developed in sections, with dry regions completed in succession. The dam has opening gates to enable the river to flow through smoothly [1, 3, 6, 10].

1.4.2 Basement

The basement is constructed under the natural ground level, with thinner soils or rock excavated and, if required, reinforced with higher strength. To prevent water leakage, cracks and fractures in the rock foundations must be plugged with grout. Holes are drilled into the rock and grout is filled into them, spreading out and bridging fractures [1–5].

The reasons why a dam is constructed with such a thick basement are many. A dam is often used to prevent (restrict) the flow of water. Water mass weighs 62.4 pounds per cubic foot. Every increase in the level of water contained by the dam corresponds to an increase in the mass that must be resisted by the dam. The more the mass that must be held back, the deeper the water behind the dam. A dam is essentially a wall. Every wall has its mass, and to withstand the mass of the water, the wall must be of a particular thickness, or it will overbalance and fail. One strategy to overcome this mass overbalancing is to build a larger wall. This is most easily accomplished by raising the thickness of the dam/wall as depth increases [4, 6, 8, 10, 13].

1.4.3 Constructing Dams

Since concrete dams require a considerable amount of ready concrete, a concrete batching plant is frequently erected on-site. The concrete is subsequently transported to the dam using either a conveyor belt system or trucks and cranes.

Pouring the concrete into a formwork mold created in the desired shape of the dam is the traditional way of using it. The dam is constructed up to 1 or 2 m at a time, with the concrete cured before the next part is formed on the top. Another option is to spread a concrete mix and compact it with rollers. The dam is raised in 600 mm stages at a time. Low concrete walls are built initially on the upstream and downstream faces, with concrete being spread in thin layers in between the walls before being compacted with rollers.

Embankment dams, for instance, are built in a succession of thin levels from the bottom up. Bulldozers distribute fill material in a thin layer, usually 300 mm thick if employing ground, or up to 1 m deep if using rock fill. The core of the dam is similarly built in layers to retain the same height as the remainder of the dam. When the entire height is reached, a protective coating is laid on the upstream face. This guards against wave damage and often provides waterproofing [2, 6, 10, 14].

1.4.4 Post Dam Construction

If the reservoir was not filled during construction, it can be filled once the dam is completed (in the case of high-flow rivers). Valves and floodgates must be thoroughly tested and dam behavior must be monitored.

1.5 THE PHYSICAL FACTORS REGULATING DAM TYPE SELECTION

The most important question is what physical factors need to be considered before selecting the type of dam construction. In this regard, to ensure economical and appropriate designs for physical factors such as topography, geology and foundation conditions, available materials, hydrology, and seismicity, experts from several disciplines, including planners, hydrologists, geotechnical, hydraulic, and structural engineers, and engineering geologists, must collaborate and be in agreement. Moreover, the site and kind of dam should be carefully studied during the early stages of planning and design. It is only under extraordinary circumstances that only one form of dam or appurtenant structure is suitable for a specific dam site. In general, preliminary drawings and estimates for many types of dams and appurtenant structures are required before determining which is the most suitable and cost-effective. As a result, it is critical to recognize that the project is likely to be overly expensive unless judgments about site selection and dam type are based on proper research [1, 6, 10, 15].

Essentially, the key factors that influence dam style are grounded on a cost comparison of the various dam designs. The next sections address key physical considerations in dam design.

1.5.1 Topography

The shape of the surface surrounding the dam site and the reservoir area, along with the availability of nearby resources are important topographic factors that need to be considered when designing a dam (see Figure 1.21). Topography largely influences the initial design choice for the dam. For instance, a narrow stream nestled between steep and rocky cliffs requires either a rockfill or concrete spillway dam. In contrast, an earthfill dam would be more suited for flat, rolling plains. In cases where the conditions are mixed, a composite structure might be the best option. Essentially, topography plays a crucial role in determining the construction of a dam.

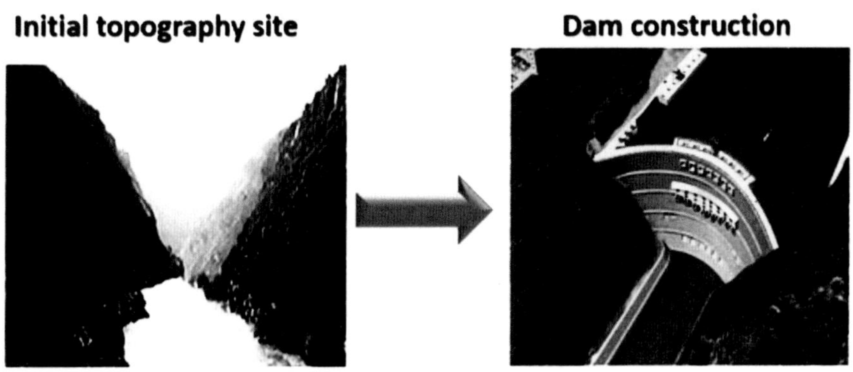

Figure 1.21 Topographic selection for dam construction.

The selection of appurtenant constructions, therefore, could also be dictated by topography. For instance, once there are natural saddles, a spillway through a saddle could be practicable. If the reservoir rim is greater than the height of the dam and is steady, a chute or tunnel spillway may be necessary. This means the kind of dam can be influenced by spillway issues. In a deep, steep-walled canyon, a concrete dam with an overflow spillway may be more cost-effective than providing a spillway for a rockfill dam [2, 6, 8, 10, 14].

1.5.2 Geological and Foundational Conditions

Therefore, geologic concerns must be studied to determine the effectiveness of different kinds of rocks and soil as foundation and construction materials. The geology of a dam site commonly dictates the sort of dam that is appropriate for that site. The strength, thickness, and inclination of strata, permeability, fracturing and faulting are all key factors in dam design. The relevant sections deal with some of the most prevalent foundations.

Competent rock foundations (Figure 1.22), which are free of notable geologic defects, have relatively high shear strengths, and are resistant to erosion and percolation, limiting the sort of dams that may be erected on them. The overall cost or material economy should be the deciding factor. The removal of disintegrating rock, as well as the grouting of seams and cracks is frequently required. Weaker rocks (Figure 1.23), such as clay shales, certain sandstones, weathered basalt, and so on, can pose considerable challenges to dam design and construction and can strongly impact the type of dam chosen [1, 6, 9, 11, 15].

Figure 1.22 Rock dam foundations.

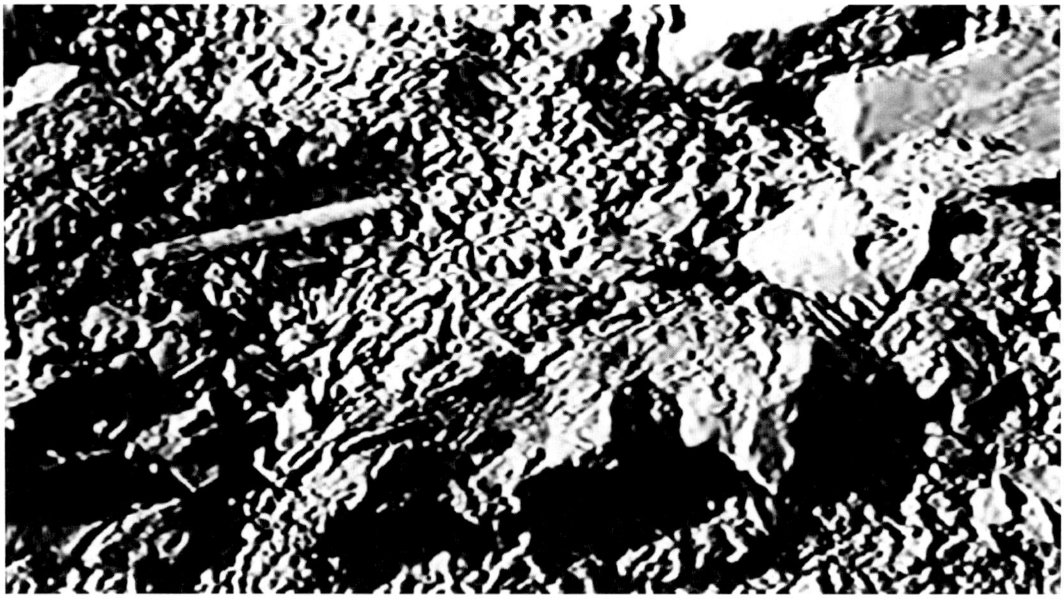

Figure 1.23 Weaker rocks.

One of the most significant geological and foundation conditions is the gravel foundation. Gravel foundations are ideal for earthfill or rockfill dams (Figure 1.24) if adequately compacted. Since gravel foundations usually experience significant rates of water percolation, particular efforts must be taken to provide effective seepage control or efficient water cutoffs or seals. Consequently, silt or fine sand foundations, if properly designed, can be used for low concrete gravity dams and earthfill dams, although they are generally not practical for rockfill dams. The non-uniform settlement, probable soil collapse upon saturation, uplift forces, piping prevention, excessive percolation losses and erosion protection of the foundation at the downstream embankment toe are all design considerations. However, uniform foundations of the types described above cannot be found, necessitating the adoption of a nonuniform foundation of rock and soft material if a dam is to be built. Nonetheless, such situations are frequently mitigated by unique design characteristics. Even common dam sites have unique challenges that necessitate the selection of proper treatment by expert engineers.

Figure 1.24 Rockfill dam.

Clay foundations (Figure 1.25), therefore, can be used to sustain earthfill dams. However, due to their lower shear strengths, they require very flat embankment slopes. Clay foundations beneath dams can also solidify. On account of the demand for smoother slopes and the tendency of clay foundations to sink significantly, building a rockfill dam on a clay foundation is usually not cost-effective. Clay foundations are not usually suited for concrete gravity dams. To establish the consolidation characteristics of the foundation strata and their ability to support the superimposed load, tests of the foundation material in its natural form are frequently required [8, 10, 15].

Figure 1.25 Clay soil for dam reconstruction.

1.5.3 Spillway

A spillway is a hydraulic structure constructed at a dam site to convey excess water from a reservoir once it has been overfilled (Figure 1.26). Spillways are categorized into various classifications depending on the configuration of the control structure, a conveyance channel, and a terminal structure.

Figure 1.26 Example of spillways.

The expenditure of constructing a massive spillway is generally a considerable factor in the budget of the project. In such instances, merging the spillway and dam into a single structure may be preferable, implying the use of a concrete spillover dam. For some cases, where excavated soil from various spillway channels can be used in the dam embankment (Figure 1.27) an earthfill dam may be useful. Small spillway requirements frequently encourage the selection of earthfill or rockfill dams even in narrow dam sites [1, 6, 16].

Figure 1.27 Spillway channel in the dam embankment.

Constructing overload concrete spillways on Earth or rock embankments has conventionally been avoided because of the more conservative design assumptions and extra effort required to avoid breakdowns. Uneven settlements in structures often arise from the differential consolidation of the embankment and foundation once reservoir loads are applied. This phenomenon can compromise structural integrity, highlighting the critical need for precise engineering considerations during construction.

To address this challenge, it is essential to implement special provisions to mitigate potential damage. For instance, cracking in the concrete or the opening of joints must be prevented, as these defects could lead to leakage from the channel into the fill. Such leakage poses a significant risk, potentially causing piping or washing away of the surrounding material, which could escalate into severe structural failure.

Therefore, to ensure long-term stability and functionality, completing the embankment in its entirety is a non-negotiable requirement. A fully constructed embankment acts as the primary line of defense against these issues, reinforcing the importance of comprehensive planning and meticulous execution in large-scale infrastructure projects.

One of the most prevalent and preferable spillway options is that of using a channel excavated from one or both abutments outside the dam's boundaries or at some point far outside the dam. When such a position is selected, the dam can be of the non-overflow kind, which expands the options to include earthfill and rockfill constructions. In contrast, failing to locate a spillway site away from the dam necessitates the selection of a dam type that can contain an overflow spillway. The overflow spillway can then be located since it only occupies a fraction of the major river channel, in which scenario the remaining dam can be constructed of earth, rock, or concrete such as the Olympus Dam. In the case of Olympus Dam, the pre-Cambrian age rocks used in the concrete gravity section foundation include granite, schist, and gneiss. This foundation is frequently joined due to compressional folding and reverse faulting. Sand, gravel, and cobbles up to 15 feet deep lie above decayed, fractured, and shattered granite in the embankment base. Lake Estes, located below the Estes Power Plant, is formed by the Olympus Dam, which was built over the Big Thompson River. The Estes Power Plant can satisfy daily changes in electricity demand thanks to afterbay storage in Lake Estes and forebay storage in Marys Lake [1, 3, 6, 6, 10, 17].

1.5.3.1 Types of Spillways

Section 1.2.2 briefly mentioned the different types of spillways. The various types of spillways are: straight drop spillway, ogee spillway, shaft spillway, chute spillway, side-channel spillway, siphon spillway and labyrinth spillway

1.5.3.1.1 Straight drop spillway

A straight drop spillway is composed of a low height weir wall with a downstream face that is generally or precisely vertical (Figure 1.28). When the water level of the reservoir rises over the regular pool level, the excess water descends continuously from the crest of the weir, which gives rise to the terms "Straight Drop Spillway" and "Free Overfall Spillway". On the downstream side, an artificial pool (Figure 1.28) with a concrete apron and a low secondary dam is made to

Figure 1.28 Straight drop spillway.

prevent erosion of the downstream bed by falling water jets. To prevent pulsating and fluctuating consequences, the underneath of a sliding jet should be properly maintained.

On the crest of the weir, an overhanging projection is sometimes constructed to prevent minor discharges from entering the face of the weir wall. Straight-drop spillways work well with thin arch dams, earthen dams and bunds.

1.5.3.1.2 Ogee spillway

The appellation ogee spillway (Figure 1.29) refers to the shape of the downstream face of the weir. It is the best version of a straight-drop spillway. In this scenario, the downstream face of the weir is built to match the shape of the lower nappe of a freely falling water jet, which is ogee-shaped. The downstream face's ogee form is based on the projectile design principle. Because the shape of the lower nappe of the water jet is not constant for all water heads, the shape derived for the maximum head is considered when building an ogee spillway.

Consequently, if there is excess water, it is spontaneously disposed of through the ogee spillway along its ogee-shaped crest. It is, therefore, also known as an overflow spillway. In this view, ogee spillways are most highly demanded in gravity dams, arch dams, and buttress dams, which are often positioned within the dam body in gravity dams.

1.5.3.1.3 Shaft spillways

Shaft spillways (Figure 1.30) are spillways that are mainly composed of a vertical shaft accompanied by a horizontal conduit. The superfluous water flows to the vertical shaft, then to the horizontal conduit and lastly to the channel downstream. The shaft might be either artificial or natural. Drilling for the natural shaft is always practical whenever the hard rocky layer on the upstream side is detectable. Consequently, the horizontal conduit either runs through the structure or foundation of the dam.

Figure 1.29 Ogee spillway form.

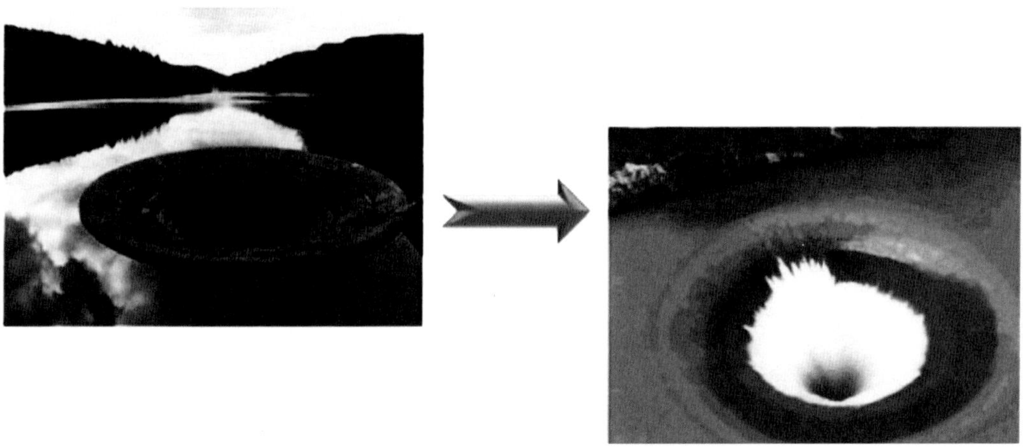

Figure 1.30 Shaft spillway forms.

In major projects, the entrance hole of the vertical shaft is specifically formed and referred to as the morning glory or glory hole of the spillway. As a result, the shaft spillway is also known as the morning glory spillway or the Bell Mouth spillway. Shaft spillways are recommended when there is an insufficient area to accommodate other forms of spillways such as ogee spillways and straight drop spillways.

1.5.3.1.4 Chute spillway

A chute spillway (Figure 1.31) is a form of a spillway that disposes excess water from upstream through a steeply sloped open channel. It is mainly constructed at one end of the dam or independently away from the dam on a spontaneous saddle on a riverbank.

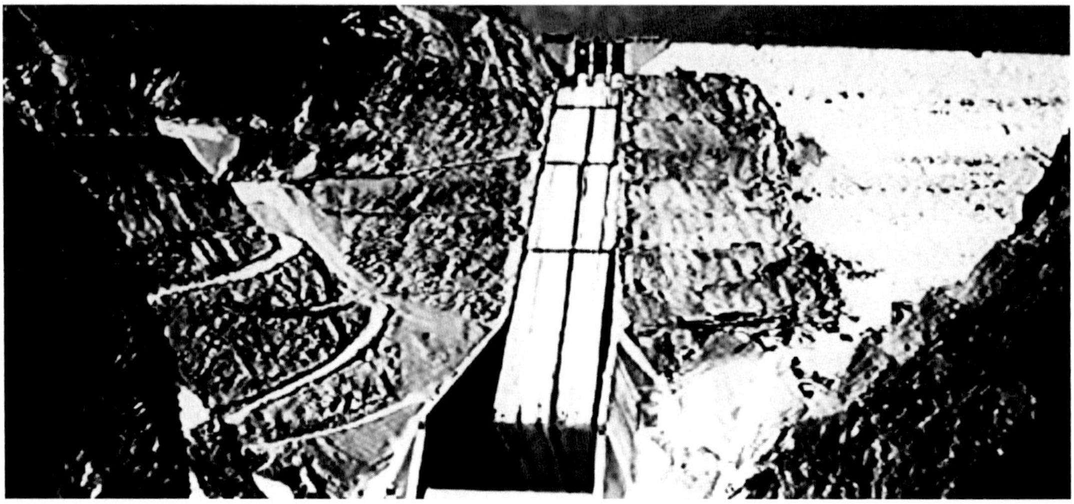

Figure 1.31 Chute spillway assembly.

Chute spillways are appropriate for gravity dams, earthen dams, rockfill dams, and other similar structures. However, they are recommended only when the river valley is quite narrow. The water flows downstream of the river in an abruptly sloped chute, trough, or open channel. The term 'Chute spillway' can also refer to a trough spillway or an open channel spillway. Therefore, the inclination of the chute spillway is constructed so that the flow is always in catastrophic condition. Energy dissipaters can be installed on the bed of the chute spillway to release energy from the falling water.

1.5.3.1.5 Side-channel spillway

The only difference between a side-channel spillway (Figure 1.32) and a chute spillway is that the crest of a side-channel spillway is positioned on one of its sides, whereas the crest of a chute spillway is located between the side walls [16–18]. In other words, unlike the chute spillway, the water overflowing from the crest is turned 90° and runs parallel to the crest of the side-channel spillway. When flanks of adequate width are not available, side-channel spillways are favored over chute spillways, usually to prevent extensive cutting. Therefore, the angle of turn of the water flow after passing over the weir crest can also be maintained between 0 and 90°.

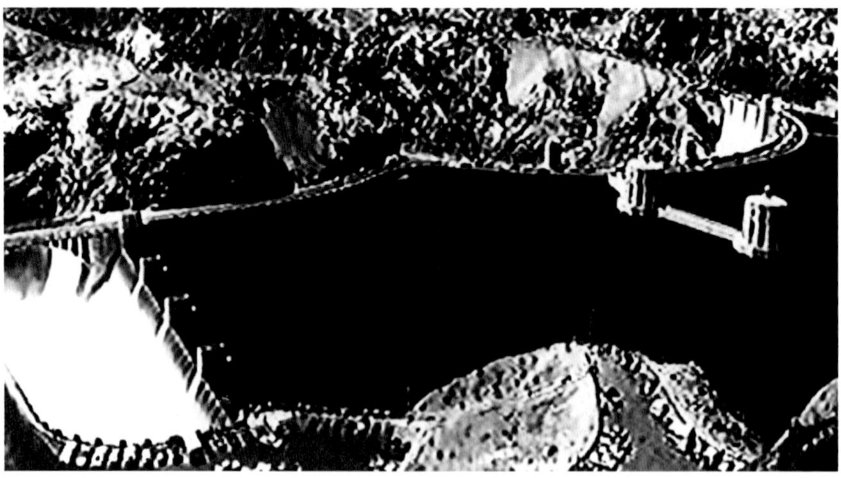

Figure 1.32 Side-channel spillway structure.

1.5.3.1.6 Syphon spillway

A siphon spillway (Figure 1.33) is a form of spillway that directs excess water downstream via an inverted u-shaped tunnel. It is usually arranged inside the dam or over the crest of the dam. Air vents are placed at the bent portion of the top passageway in both types of siphon spillways to prevent water from entering when the water level is below the typical pool-level. When the level rises over the regular pool level, water enters the conduit and is expelled downstream by siphonic action.

Figure 1.33 Construction of siphon spillway.

1.5.3.1.7 Labyrinth spillway

Finally, a labyrinth spillway (Figure 1.34) is a design spillway in which the weir wall is designed in a zigzag form to improve the effective length of the weir crest following the channel width. This improvement boosts the discharge capacity of the weir, enabling larger water flow at low heads to be easily conveyed downstream [1, 3, 10, 17].

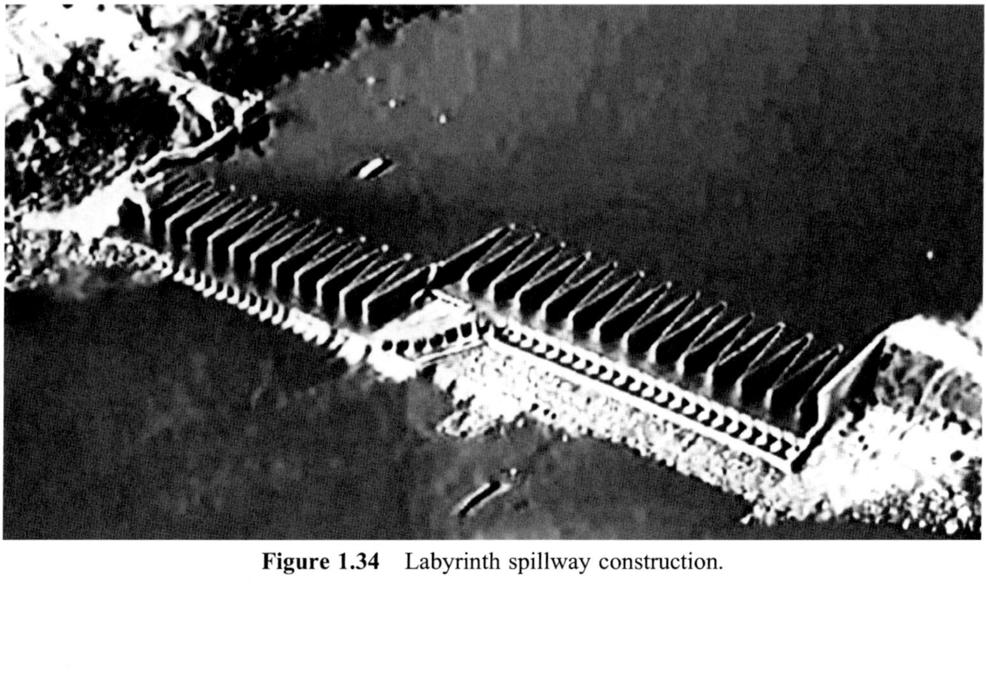

Figure 1.34 Labyrinth spillway construction.

1.5.4 Hydrology

The hydrologic and economic considerations that control the type of dam and appurtenant structures are inextricably linked. Streamflow characteristics and precipitation may have a significant impact on building costs by impacting water treatment and diversion and lengthening construction time. When extensive tunnels are necessary for diversion, converting them to tunnel spillways may be the most cost-effective spillway option [1, 8, 18].

1.5.5 Material Accessibilities

Dam materials that may be accessible or be near the site are: (i) soils for embankments (ii) bock for embankments and riprap and (iii) concrete aggregate (sand, gravel, crushed stone).

The elimination or reduction of transportation charges for construction materials, especially those that are used in massive amounts significantly reduces the entire cost of the project. The most cost-effective dam is commonly one for which a prominent quantity of materials can be acquired within a commutable distance of the site.

The provision of appropriate sand and gravel for concrete at a reasonable cost nearby and possibly even on the property to be bought for the project is a consideration that favors the choosing of concrete construction. The availability and quality of rock for rockfill is a factor that influences the decision to build a rockfill dam. A natural resource that minimizes project costs while maintaining efficiency and quality of the final structure should be exploited [16–18].

1.5.6 Earthquake and Landslide

If a dam is in an earthquake zone, its design must account for seismic forces. Earthen dams and concrete gravity dams are the structures best adapted to withstand earthquake shocks without causing harm. Therefore, large waves can be generated from the ground slopes of the dam collapsing and the lake [19–21].

Many massive dams have been constructed in seismically active areas around the world, perhaps in Japan, the western United States, New Zealand, the Himalayas, and the Middle East. For instance, the Tokachi earthquake in 1968 damaged 93 dams on Honshu, the largest island of Japan; all of these were embankment dams of a remarkably low height [22].

Despite extensive research on the distribution of seismic activity, the measurement of significant ground motions, and dam responses to such motions, dam earthquake design remains imprecise. The features of powerful ground vibrations at a certain location cannot be predicted and all dams have some degree of freedom, imperfect elasticity, and imprecise damping. Nonetheless, computers and model challenges hold up the prospect of future advancement. The response of a concrete dam to any specified ground motion can now be modeled; it has been done for the Tang-e Soleyman Dam in Iran and the Gariep Dam in South Africa [20, 22].

Since concrete dam foundations are often keyed into bedrock, concrete dams typically do not experience significant accelerations when shaken by earthquakes; as a result, concrete dams have an outstanding safety record in terms of withstanding seismic pressures. With the significant exception of earthfill dam construction adopting hydraulic fill techniques, the safety record for embankment dams is also excellent. When hit by a seismic event, such dams retain a huge amount of water within their soil structure, making them prone to liquefaction of the saturated soil. When a substantial amount of hydraulic fill 'slipped' during an earthquake in 1971, the Van Norman Dam (or lower San Fernando Dam) in Los Angeles partially collapsed. Engineers have recently realized that big artificial reservoirs can cause earthquakes that would not occur in the absence of the reservoirs. Reservoir-induced earthquakes can be triggered by increased water weight or, more commonly, by increases in groundwater pore pressure. In this understanding, the possibility of tsunami occurrence is remarkably high especially if the dam is constructed in a seismic active zone either by the massive weight of the dam or its reservoir impact. This will be addressed in the next chapters.

The factors discussed, along with more conservative construction details, have influenced the selection of alternative spillway design solutions. These details include increased lining thickness, more reinforcement steel, cutoffs, joint treatment, drainage, and preloading. Such techniques involve constructing the structure above or under the natural material of the foundation or the dam as a tunnel.

1.6 THE PRIMARY CAUSES OF DAM FAILURE

Dams are incredibly beneficial and their failure, while unexpected, can be catastrophic. When dams collapse, floods, tsunamis, and mudflows frequently inundate downstream areas. Failures commonly occur in fatalities, loss of downstream structures, land denudation, severe pollution, and widespread disturbance of downstream ecosystems, economies and communities. Dam failure is often accompanied by the loss of dam-related functions which include irrigation, flood control, electricity generation, and mining operations. Large dams are frequently massive capital assets that are extremely costly to lose [1, 12, 24].

Water dams have failed almost as long as humans have constructed them, and several of the key provisions, including the danger of overtopping and the destructive impacts of spillway erosion are familiar to almost everyone from our childhood days of playing with shovels and streams. The majority of water dam failures occur in minor dams; however, large dams have also failed on occasion. The challenge lies not in finding evidence of dam failure, but in determining which pieces of evidence are reliable and applicable. Water dam failures happen around once per 10,000 years and primarily in smaller dams. Tailings dams fail far more frequently, at a rate of around one in 1,000 every year, or three to four per year globally [24–26].

1.6.1 Reservoir Filling, High Water and Overtopping

Post-reservoir filling, the water level in dams fails either during the filling process or immediately afterward. Consequently, in well-established water dams, dam failures are primarily associated with overfilled reservoirs, however, dramatic failures have also arisen under normal circumstances. Overtopping is the most prevalent cause of failure, and most failures occur in smaller dams, which are also significantly more numerous than huge dams. Over the last 75 years, water overtopping is responsible for approximately 30% of global dam failures [1, 12, 21, 23, 26].

Tailings dams can fail at any point during their lifespan, even though they are continuously strengthened over time, due to factors such as unforeseen structural weaknesses or environmental conditions. The dams are constructed higher as the volume of tailings impoundment increases. This implies that the dam is being erected and the reservoir is being filled for the entire duration of the mine or industrial plant's operation to which it is connected. When a mine or mill closes, the tailings dams are usually left in place indefinitely to contain the waste. Because the dam is no longer being operated it may be less vulnerable to failure at this point; nevertheless, if monitoring and maintenance decrease, it will become more vulnerable. Embankment dams, in particular, are extremely vulnerable if they are breached, as streaming water quickly cuts through them. Some concrete dams can withstand overtopping better than others, and temporarily transform into man-made waterfalls [1, 12, 26].

Consistent with the above perspective, the mechanism of overtopping can be explained as follows. Due to erosion, the overflow forms its breach channel, resulting in inflow concentration and an improved capacity to erode and transfer cover material. The gullying of the new channel allows the overtopping flow rate to gradually grow at first, then accelerate as the erosion front migrates upstream towards the crest. Analyses of dam failure by erosion in this scenario reveal two distinct phases. The first is the initiation phase, during which water runs over the embankment, damaging the downstream face and part of the crest but having no effect on the rate of discharge because it is still regulated by inflow at the upstream face of the embankment.

The main question is what are the overtopping phases. The first phase involves the initial running water seeping into the crest layers (Figure 1.35). The second phase has an overflow that generates the first signs of shearing on the surface. Resistance to shearing of these layers reduces as shearing action rises, depending on grain size distribution, starting state of compaction and the evolution of soil water content (Figure 1.36). Consequently, in the third phase, erosion triggered by particle detachment forms a new flow channel, concentrating elements that promote shear stress and turbulent flow (Figure 1.37) [2, 12, 24, 27].

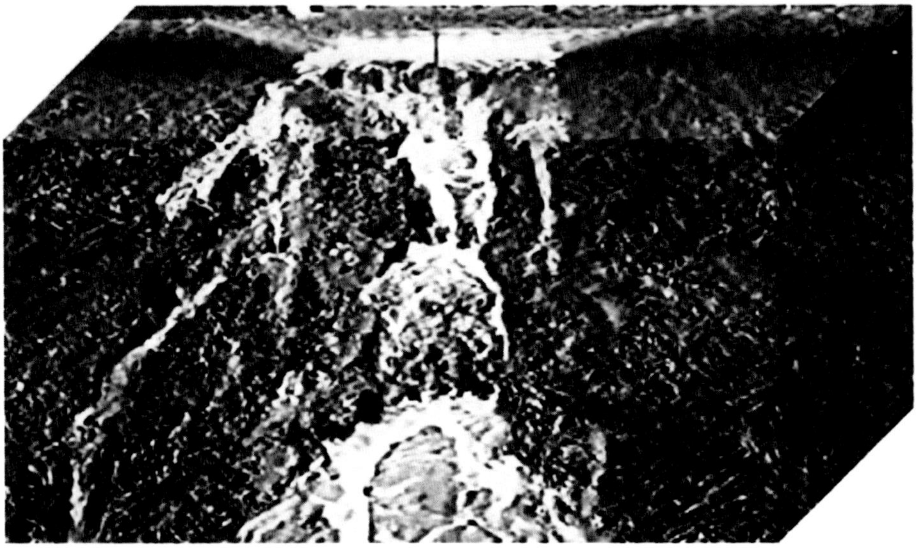

Figure 1.35 The initial phase of overtopping.

Figure 1.36 The second phase of overtopping contains soil.

Figure 1.37 Erosion due to turbulent flow in the third phase of overtopping.

The fourth phase is reached when erosion begins to distress the crest line. In this circumstance, breach formation occurs and the rate of breach outflow turns out to be superior (Figure 1.38). As a result, a wide new breach channel is created and a steep bank is formed as a function of the thalweg deepness in the embankment dam. The fifth phase is developed during gullying and weak slope stability gets hold of borderline levels. As a result, the breach becomes wider owing to dam failure and flow erosion which causes a series of bursts (Figure 1.39). Lastly, the sixth phase is reached once a flow erosion arises in the thalweg of the breach channel and creates a large scale of turbulent outflow downstream (Figure 1.40). However, not all overtopping failures involve all six mechanisms, but rather a combination of them are triggered by dynamical characteristics specific to each case [27–30].

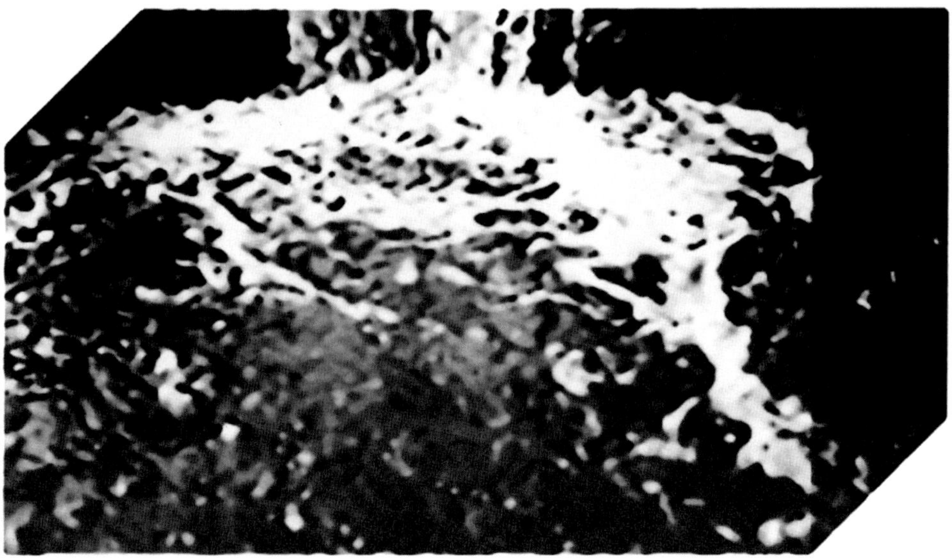

Figure 1.38 Massive water breach outflow.

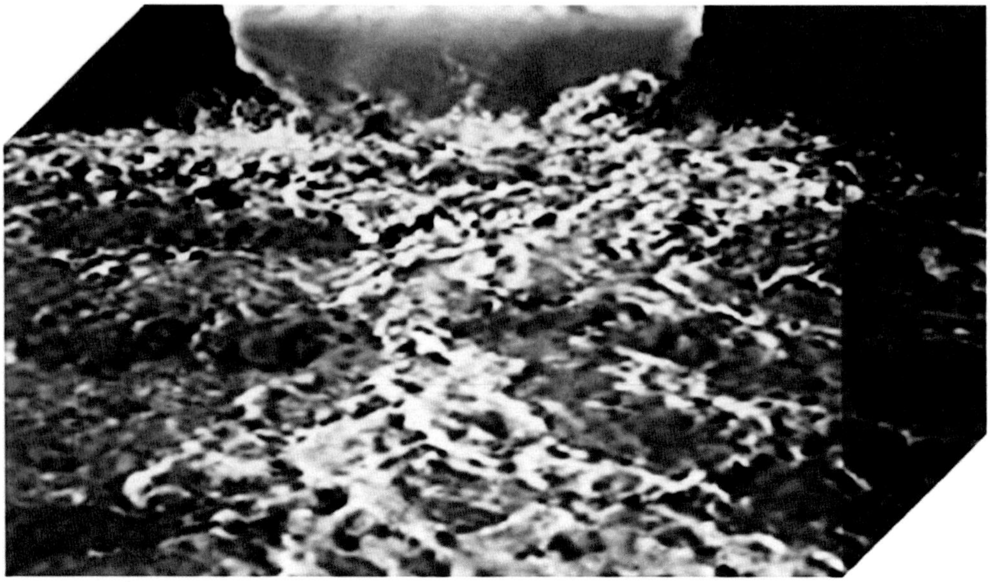

Figure 1.39 Wider breach due to flow erosion and dam failure.

Figure 1.40 The large scale of turbulence flows downstream due to continued erosion flow.

Consistent with the above perspective, overtopping flow or waves could also erode and ruin the top of a less erodible embankment, and then regressive (head cut) erosion may begin in the more erosive embankment body from the area where the blanket is destroyed. Generally, overtopping can occur for a variety of reasons, including excessive rainfall or the failure of an upstream dam, large inflows into the reservoir caused by excessive rainfall or the failure of an upstream dam, extreme waves and surges, inadequate design, construction, and maintenance of the structure, debris blockage in the spillway and flood channel, and settlement of the embankment crest [31–34].

Needless to say, the majority of worldwide dams have the potential for overtopping. In other words, it is the major cause of embankment dam failure worldwide. Furthermore, overtopping failures are projected to become more common as a result of considerable increase in global warming [35–39], which, as predicted by climatologists, will result in high variability and extremes in precipitation patterns. This highlights that spillways and flood control structures designed using projected flood return periods based on historical hydrological data may fail disastrously if there is a significant change in the frequency and intensity of precipitation events.

1.6.2 Dam and Levee Overtopping

What is levee overtopping? When water overtops the crest of the levee, it is considered that the levee has failed. Levee overtopping occurs when floodwaters simply reach the lowest crest of the levee system, or when severe winds generate huge swells (a storm surge) in the ocean or river water, causing waves to break over the levee [35–40].

Overtopping arises when the still water level (SWL) and wind stress configuration surpass the crest of the dam. Overwash occurs when wind blow and wave run-up combine to generate a water level higher than the crest of the dam (Figure 1.41). Predominantly, a greater surface area upstream (fetch) is essential to enable wind to cause waves that would be driven towards and overtop the embankment [1, 34, 39, 41].

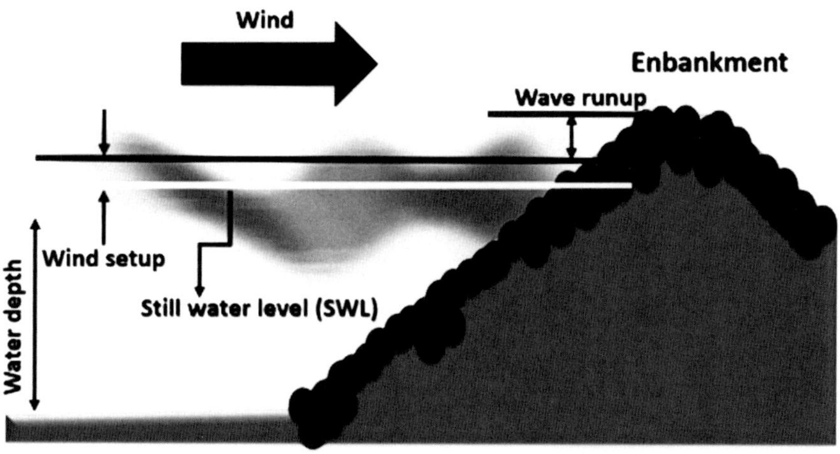

Figure 1.41 Wind stress creates levee overtopping.

Most embankment dams and levees cannot endure sustained overtopping of a foot or two feet or more without failing. Due to the sturdy nature of the construction and the rock base, most concrete dams can more often than not tolerate a certain level of overtopping; nevertheless, some may be vulnerable due to jointing and fracturing in the rock foundation. Due to variations in design criteria, levee overtopping is expected to occur at a significantly higher intensity than dam overtopping.

During or after high water, trained observers evaluate levees for the presence of craters on the downstream side of the water-retaining structures. However, the process of flooding can be very fast, especially if a pipe from a previous high flood was already present. A levee failure typically occurs without warning. Furthermore, due to the high tide, the eroded sand may be carried away [37].

1.6.3 Backward Erosion Piping

For cohesive water-retaining structures built on sandy aquifers, backward erosion piping is a common failure mechanism. Backward erosion pipework can develop beneath levees, dams, locks, and other structures that contain the water of rivers, seas, reservoirs, canals, and other bodies of water in various types of homogeneous, stratified and heterogeneous soils. Several mechanisms can demonstrate the backward erosion pinging as illustrated in Figure 1.42. The question now is: what is a predictable keystone phenomenon to understand the backward erosion of piping mechanisms? The answer is the horizontal groundwater flow across a sandy layer, which is nourished upstream throughout the water reservoir and has an ejection stage downstream where the groundwater can flow spontaneously. In such circumstances phase one is achieved and leakage is inevitable. Consequently, either local disruption or the absence of the downstream top layer can be the keystones of the discharge phase. Near the defect, the high local hydraulic gradients create vertical stress on the sand grains and heavy seepage concentrations are accumulated. In this circumstance, phase two is developed at the beginning of the erosion owing to the ejection of sand grains, which is termed piping initiation. Therefore, the second phase develops as the erosion progression resumes to enlarge in the upstream direction, forming a shallow sand layer around the pipes [40]. However, the pipes do not break because the cohesive and substantial nature of the overlying material provides strong support and prevents failure.

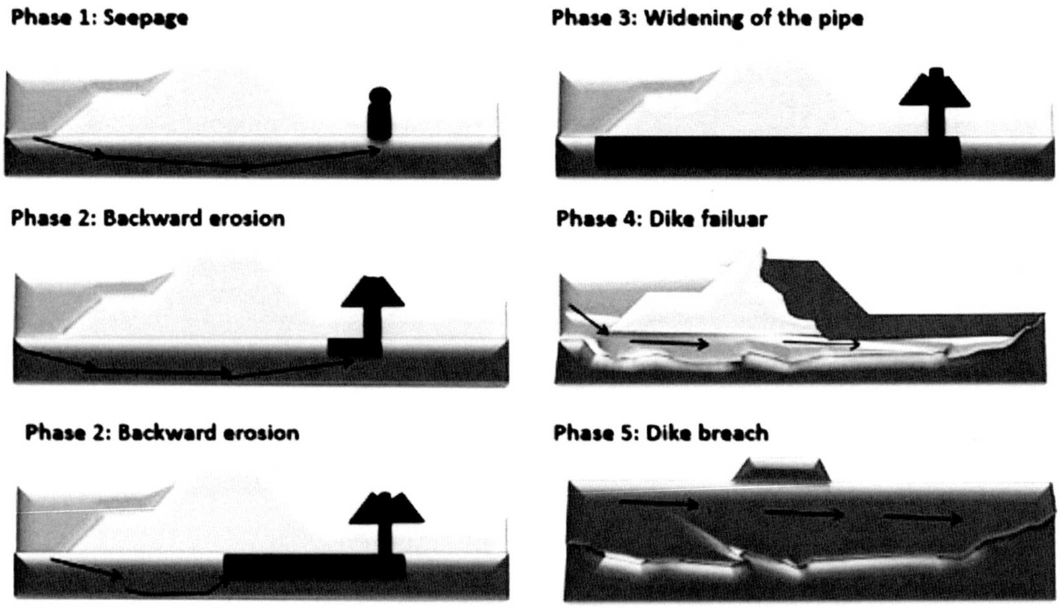

Figure 1.42 Mechanisms of backward erosion piping.

The transition between the second to third phases takes place as the pipes eventually grow to a stage where they create a precise alliance between the upstream and downstream sides of the construction. The fourth phase develops rapidly accompanied by extreme erosion due to huge stream water flow, forming thickening and broadening of the pipe. In this circumstance, the fifth phase develops since the pipe bursts causing a discriminatory collapse, which causes downcutting and lateral widening. In this scenario, the pipe initiation and erosion development stages are perhaps established in several days or weeks, whereas the subsequent embankment breaching possibly will withstand just a few hours or less [41]. This too can be termed as a failure of the levee.

Backward erosion piping is, without a doubt, one of the primary triggers of levee failure around the world. Because of its gradual onset, it is commonly too late to intervene when the first predictors present. As a consequence, the implications could be catastrophic.

1.6.4 Foundation Failure

Defects in the foundation and structure of a dam commonly occur worldwide. The differential settlement, sliding and slope instability, high uplift pressure and uncontrolled foundation seepage are all examples of foundation and structural faults. Differential settlement can lead to cracks and the weakening of layers in the embankment, which can lead to internal erosion and piping failure. Excessive foundation seepage through porous substrata is frequently accompanied by the creation of sand boils, which resemble miniature volcanoes ejecting water and silt and, if not managed, can lead to piping failure. Slope instability and sliding collapse can be caused by high uplift pressure and saturation of foundation materials. The piping failures produced by foundation faults are similar to those described in the previous subsection, although the initial breaches may involve the entire embankment height.

Concrete arches have broken spontaneously in several circumstances. Notably, the Malpasset arch dam in France suddenly collapsed from its foundations during a heavy rainstorm in 1959, killing 421 people. In 1920, the St. Francis Arch dam in California collapsed, although under normal weather and load circumstances, killing an estimated 470 people. Among other flaws, the St. Francis dam was built on a weak sedimentary rock and immediately across a recognized fault. The dam is thought to have collapsed due to foundation failure and no witnesses survived the event. Moreover, concrete gravity dams, which hold back water by sheer bulk (as do embankment dams), have moved as units off their foundations in some situations due to the force of the water held.

Another foundation hazard has been identified for the Mosul embankment dam in Iraq. It is made of karst gypsum, which is water-soluble. The foundation rock has cavities and has dissolving over time, making the foundation weaker and weaker [41–44].

1.6.5 Location and Geology

Incorrect dam location and engineering can be significant contributors to failures. Geological occurrences such as landslides and earthquakes are rarely associated with dam collapse as dams have successfully weathered earthquakes, landslides and reservoir tsunamis in the past. Dams were the first buildings to be systematically constructed to withstand earthquakes.

Figure 1.43 Reservoir tsunamis of the Vajont dam, Italy.

Reservoir tsunamis, are not well-acknowledged in scientific literature. A review of literature by researchers discovered no examples of dams structurally breaking as a result of reservoir tsunamis. However, the Vajont dam (Figure 1.43) remained intact but functionally 'failed and was rendered inoperable' when it was overtopped by a tsunami.

In the case of the reservoir tsunami of the Vajont dam, the entire bulk slid northward at up to 30 ms^{-1}. The mass closed the gorge to a depth of up to 400 m and traveled up the opposing bank for up to 140 m. After a maximum of 45 seconds, the landslide mass stopped moving. The reservoir had 115 million cubic meters of water at the time. A wave of water was forced up the opposing bank, destroying Casso, 260 m above lake level, before overtopping the dam by up to 245 m. In this sense, the water was believed to be 30 million cubic meters in volume, then descended more than 500 m over the communities of Longarone, Pirago, Villanova, Rivalta and Fae destroying them. A total of 2,500 people were killed., How the floodwater after the dam collapsed washed away the valley bottom, where the communities once stood is demonstrated in Figure 1.44. The flood wave swept down the Vajont valley, which can be seen in the upper right corner of Figure 1.44.

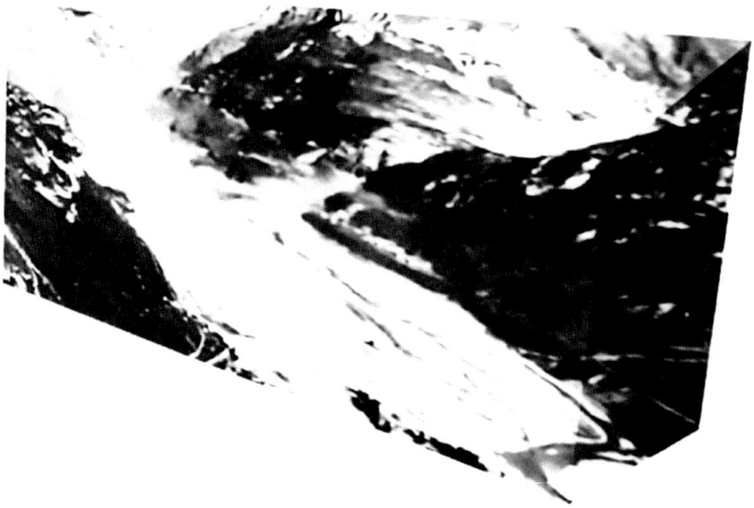

Figure 1.44 Damage to Vajont valley after the Vajont dam collapsed.

The critical question is what causes reservoir tsunami or landslide dam failure? A weak foundation layer of limestone approximately in alliance with a thin clay layer of 5 to 15 cm is the keystone for landslide and tsunami occurrences across the embankment dam such as Vajont. Excessive water level increases the stress in the clay layers, plummeting the operative regular strength and hence the shear resistance [44]. In this case, resistance to the movement was formed by the chair-like configuration of the shear surface. Consequently, the shear stresses increased as the water in the jointed limestone was discharged into the drain owing to the increment of hydraulic pressures as a function of dropping the level of the reservoir. Consequently, failure struck resulting in strength loss. The speed of flow was most likely instigated by frictional heating of the pore water in the clay layers [40–44]. Needless to say, earthquakes can make dams shake, instigate landslides, shift foundations and cause reservoir tsunamis.

1.6.6 Maintenance and Monitoring

Finally, dam sustainability depends on maintenance. Dams are massive but all they demand is regular maintenance. Concrete cracks (Figure 1.45) must be caulked with concrete pavers; earthen hillsides disintegrate periodically and must be managed; and spillways must be properly maintained and functioning. Cracks arise in concrete dams mostly as a result of concrete shrinkage induced by temperature changes. These cracks might form inside the dam's body or externally on its surface.

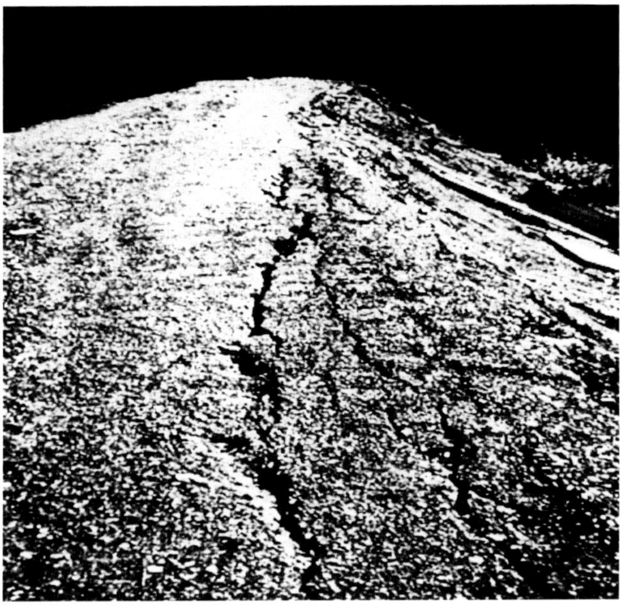

Figure 1.45 Longitudinal concrete crack.

A sort of cracking that arises parallel to the dam crest is known as longitudinal cracking (from abutment to abutment). Such cracks can indicate poor dam compaction and gradual crest settling or a combination of both. It could also be a sign of regular traffic by animals or people along the crest.

Transverse cracking (Figure 1.46) is similar to longitudinal cracking and can occur as a result of insufficient compaction and settlement. Transverse cracking, on the other hand, occurs perpendicularly to the crest (from upstream to downstream) rather than in the parallel. These types of fissures, if left uncontrolled, can lead to sinkholes, low regions, slides and more settling. They are more typically a sign of an impending sinkhole caused by seepage pipes [40–46].

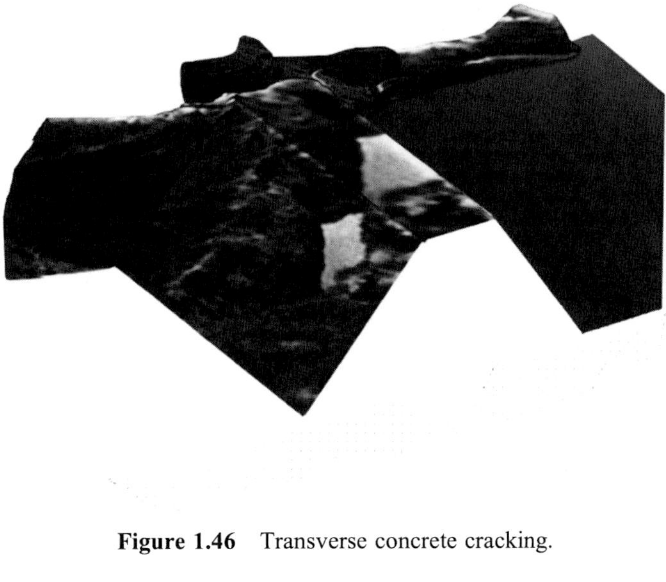

Figure 1.46 Transverse concrete cracking.

Consequently, sinkholes are sometimes represented as circular cracks that grow around a hole (Figure 1.47). They are generally triggered by piping or poor settling beneath the damaged portions, which creates voids for the upper surface material to fall through. The area affected by this issue is denoted by the exterior cracks surrounding the hole. Filling the pit with rock and gravel is a temporary emergency fix.

Figure 1.47 Sinkholes crack formation.

Lastly, slides are also referred to as displacements, sloughs, and slips (Figure 1.48). They can cause sinkholes, water flow restrictions, bigger displacements and surface erosion/maintenance concerns. In severe circumstances, they can cause a dam to fail. Seepage, plumbing, settling, compaction, and cracking problems that are not adequately maintained can all lead to slides.

Figure 1.48 Slide cracking.

If left unchecked, a crack can readily develop into vertical displacement, which can easily end in dam failure in the worst-case scenario. Seasonal changes can exacerbate cracking when water begins to enter the fracture via natural events or overtopping.

Several contemporary dams have the option to live eternally if maintained properly, yet all dams are susceptible to complete failure. If humans suddenly disappeared the largest embankments in the world and all concrete dams would presumably remain standing as compared to all other man-made structures, they would not do so in efficient working conditions.

The ability of the operators of the dams and their commitment to sustain the dam is thus an important consideration in dam safety. Most water dams, which provide long-term benefits (for instance, electricity generation, flood control, and agricultural water) must be maintained by societies and dam owners. Many of these dams are also located in places where a failure could give rise widespread property destruction and life casualties downstream.

Excessive and poorly maintained vegetation is a common embankment dam maintenance issue. It can obscure important dam components, concealing other flaws and making normal inspections more difficult. Vegetation may also make it difficult to access some portions of the dam. Trees and plants with deep roots can create seepage paths through the dam. This frequently results in pipework and internal erosion, endangering the dam's structural stability. Furthermore, vegetation facilitates animal invasion and burrowing. It should be recognized that well-maintained grass can be exploited as erosion protection on the dam's crest or on its downstream side. This, however, may result in grazing cattle, which can lead to distortion and cracking. Furthermore, improperly cut or unmanicured grass would merely add to surplus vegetation.

Tailings dams, on the other hand, are constructed to hold pollutants and serve as cost-cutting techniques and disagreeable restrictions rather than as value-generating assets. This produces a risk dynamic that the International Commission on Large Dams (ICOLD) particularly mentions in a study as problematic. Tailings dams are occasionally required to be maintained long after the associated profit-producing mines and industries have closed down, the relevant organizations have relocated or dissolved. In such cases dam maintenance becomes an entirely undesirable cost. As a result, economic and psychological incentives could work together to promote proper long-term maintenance of tailings dams [43–46].

1.7 THE NEGATIVE IMPACT OF THE DAM ON THE ENVIRONMENT

Aside from the benefits of (i) managing stream regimes and thus preventing floods, (ii) acquiring residential and irrigation water from stored water, and (iii) generating energy, dams have a wide range of harmful and beneficial environmental effects. Dams have the potential to cause significant harm to living creatures, in addition to the benefits of addressing basic societal needs and raising living standards. Therefore, the most crucial consideration here is to clearly distinguish between short-term, long-term and permanent effects. Organizations comprised of biologists, engineers, hydrologists, social scientists and other professional groups must plan and attend environmental impact assessment studies to determine the environmental consequences of setting up a dam.

1.7.1 The Impact of the Dam on the Hydraulic System

The main hydraulic consequence of this dam design is that the assembly basin discharges into a stationary reservoir instead of directly flowing into a stream bed. An immediate change occurs downstream when a stream dries partially or completely once the reservoir starts to store water. The hydrological balance can break down during these temporary or periodically recurring time intervals. This collapse can lead to irreversible effects, such as the death of organisms, the disappearance of species, and significant structural changes in water-dependent ecosystems.

The decay of dead flora and fauna in the new water body accelerates. Thus, upstream water is polluted with a lack of oxygen in deeper areas, which becomes dark for a long period of time, and generally smells rotten owing to sulphurous hydrogen discharge. The Aswan Dam is an excellent demonstration of this case. Aside from the Nile River ecology, this dam has altered the feeding patterns of the Eastern Mediterranean [45–47].

1.7.2 Impacts on Temperature and Chemical Characteristics of the Reservoir

Similar to temperature variations, huge changes occur in the chemical composition of river water. The fluctuations in water temperature occur within the reservoir as a result of the water depth that was able to seep through the downstream gates and the exchange of water at a constant temperature. From a chemical and biological standpoint, the river will act as a cold climate river since the water will always be cold, even in summer, if the gate depth is below the reservoir's thermocline. If the flowing water is at surface water temperature, it will behave like a hot climate river. Needless to say, thermal fluctuations that, are influenced by water depth in the dams, may result in the generation of annual thermal layers.

Water containing oxygen and even sulphurous hydrogen may appear in the deeper parts of the river depending on the depth of the reservoir. Once water flows downward, vital changes in the downstream section may take place depending on the depth of the dam gates. The chemical changes of organic matter deposited in discrete layers of the river where oxygen is scarce would resurface, preceded by abrupt gas discharges. Such an event induces an abrupt influx of various chemical nutrition components to the biosystem, as well as a variation in the water quality. Likewise, nitrogen in the air is dissolved at severe exposure levels in the cascading downstream portion of the water. Water that is approximately 150% nitrogen-saturated can be harmful to fish [15, 48].

1.7.3 The Impact of the Dam on the Atmospheric System

The critical issue is how does a dam affect the atmosphere. Dams can create greenhouse gases. Because of the lack of oxygen, methane (a particularly potent greenhouse gas) is released from the decomposition of plant materials at the reservoir's bottom. It is gradually emitted into the environment leading to global climate change. In other words, excess water is pumped onto the banks, which commonly has a huge amount of plant growth. These plants suffocate and perish. Bacteria in the water then degrades the plants, producing carbon dioxide and methane, which is eighty-six times more toxic than CO_2. These gases surge to the reservoir's surface and are discharged into the atmosphere. More of these gases are created if the upstream river water contains excess nitrogen or phosphorus from fertilizer runoff. These nutrients enable naturally produced algae in the river to flourish, but soon deplete the water's oxygen supply. When algae die, it decomposes like other organic stuff releasing even more of these gases.

Dams and reservoirs are generators of greenhouse gases that contribute to global warming. It is speculated that the surface of reservoirs has emitted roughly 20% of all man-made methane emissions.

When a forest is cut down to create accommodation for a dam and reservoir, the trees are no longer available to absorb the carbon dioxide deposited by fossil fuels. It is believed that rotting vegetation in the river enables the dams to release approximately a billion ton of greenhouse gases every year. This amounts to approximately 1.3% of annual anthropogenic which is caused by humans and added as harmful planetary emissions. In this sense, determining which rivers seem to be more likely to become high-methane reservoirs could aid us in halting greenhouse gas production in the future.

Naturally, variations in moisture percentage, temperature, and air body movements generated by large stationary water bodies change the microclimate of a region. In other words, evaporation has an impact on the surrounding microclimates of a dam, disturbing natural temperature changes, ecosystems and habitats. Greater evaporation, for example, in the vicinity of a major dam alters the moisture concentration in the air, resulting in greater heavy rainfall. Furthermore, regional climate shifts might be noticed. These changes may not appear to be hazardous to human health, but they are noticeable in the affect they have on many plants and animals. Their secondary consequences have an impact on human life [48].

1.7.4 The Dam Can Cause Earthquakes

Humans must be blamed for creating earthquakes on Earth. The Earth is massive, as are the tectonic plates. It doesn't seem possible that what humans do to the Earth would have any effect on those massive plates. Nevertheless, evidence, which blames humans for earthquake occurrences is accumulating.

This means reservoir-triggered quakes, the quick filling of dams, as well as their depth, may trigger earthquakes. The interval between reservoir filling and the huge quake would have enabled reservoir water to sink deeper into the crust, where it could weaken a fault. And the greatest risk of triggering a quake occurs when the water level is falling rather than rising.

Large reservoirs have the potential to cause earthquakes. This phenomenon is known as reservoir-induced seismicity, and it is caused by high water pressure in micro-cracks and fissures in the foundation units beneath and near the reservoir. Water within rock masses under enormous hydrostatic pressure lubricates faults that are already under tectonic strain but are prevented from slipping by rock plane friction. It is well understood that it is primarily determined by the type of structural geology and lithology of the surrounding rocks. However, predicting when and where a reservoir-induced earthquake would take place is extremely challenging. However, ICOLD suggests considering Reservoir Triggered Seismicity (RTS) for reservoirs with depths more than 100 m [45–47]. Consistent with the United States Committee on Large Dams (USCOLD), reservoir-induced seismicity (RIS) should be investigated for reservoirs deeper than 80–100 m. In this sense, the reservoir capacity, as well as reservoir depth, have a crucial role in triggering earthquakes.

The phenomenon of reservoir-generated seismicity (RIS) mostly relates specifically to reservoir filling times, which can be calculated for a reservoir only when a timeframe has expired. Small-magnitude earthquakes, in this context, can occur immediately after a reservoir is filled or when there are sudden fluctuations in the reservoir's water level. These events are primarily attributed to stress changes in the underlying bedrock, the collapse of karst caves or mining pits, and natural seismic activity. For such events to occur, the initial stress in the area must already be close to a critical failure point. Even a minor change in the strength characteristics of a fault plane—caused by the presence of reservoir water—can trigger seismic activity. The amplitude of these reservoir-triggered seismic (RTS) events may gradually increase over time, eventually leading to a major shock.

Strong earthquakes can have a wide-ranging impact. According to recent surveys, there are at least 100 incidents of earthquakes caused by reservoirs. The worst situation may be the 7.9-magnitude Sichuan earthquake on May 12, 2008, which killed an estimated 90,000 people. This earthquake has been related to the construction of the Zipingpu Dam, a 156-meter-high concrete-faced rockfill dam with a reservoir capacity of 1120 hm^3. Once the 7.9 magnitude Wenchuan earthquake struck, a reservoir was blamed as the main cause. The Zipingpu Dam had been constructed only 500 m away from the tectonic plate and 5.5 km from the quake's epicenter. The much-maligned Three Gorges Dam was ruled out as being too far away. The timing was perfect as the Zipingpu reservoir began filling in December 2004, and the water level had risen by

120 m in two years, which contributed to the earthquake. The blockage of many hundred million tons of water by the dam would not have been sufficient to create an earthquake of that magnitude but could have depressed the fault and increased the stresses on it.

1.7.5 Dam Impact on Health

Initially, decomposing organisms trigger an increase in nutritional compounds in water over a short period. Accordingly, the BOD (Biological Oxygen Demand) value of water increases. The stationary layers along the reservoir depth are used to execute anaerobic decomposition media. Thus, the lake turns dark and stinks. The subsequent, huge growth in phytoplankton is noticed, which is fed by the increased amount of nutrients. In addition to the plants that cover the water's surface as enormous green-dark masses, macroflora grows on the water's surface. These growths can be hazardous not only to the inhabitants of the lake, but also to those fishing, enjoying a boat ride and even to the dam gates and turbine propellers. The macroflora generated here can sometimes serve as a source of disease vectors. Simultaneously, the increase in water plants causes larger evaporation losses than conventional evapotranspiration.

The dam, therefore, is a significant impediment for the animals swimming from one end of the river to the other. The dam's existence implies death for fish species that spend portions of their lifecycle in the spring or in floodwater, and other portions in the intersections where the river converges with the sea [48]. Certain marine fish migrate to fresh water and swim up to springs to lay eggs, and subsequently return to the sea with new juvenile fish. A dam can disrupt the life cycle of these organisms and cause mass mortality. Unquestionably, building plenty of dams destroys the proficiency of waterways to sustain vital ecosystems, which leads to unsustainable health and integrity of systems of the Earth.

To recap, dams cause a variety of environmental changes with various extents of impact. It is hard to anticipate the interconnections between these impacts and identify which favorable and negative consequences would arise. This prediction should indeed be determined for every dam and reservoir independently. Conversely, it is inappropriate to understand the consequences entirely negatively.

1.7.6 Dam Can Causes Coastal Erosion

The serious question is: How does a dam cause coastal erosion? Rivers transport sediment down their streams, causing depositional features including river deltas, alluvial fans, braided rivers, oxbow lakes, levees, and coastal coastlines to form. The construction of a dam interrupts the flow of sediment downstream, triggering downstream erosion of certain sedimentary depositional settings and increasing sediment buildup in the reservoir. While the rate of sedimentation varies by dam and river, all reservoirs eventually lose water-storage capacity when "living storage" space is exchanged for sediment. Reduced storage capacity reduces the ability to generate hydroelectric power, reduces the availability of water for irrigation, and, if left unaddressed, may eventually result in the dam and river expiring. In this view, sediment entrapment in reservoirs reduces sediment delivery downstream, which has a detrimental influence on channel shape, aquatic ecosystems, and delta land elevation management. Alternative to dam removal, there are other ways to reduce reservoir sedimentation. In this regard, the flushing flow process involves partially or emptying a reservoir behind a dam to dissolve and transfer silt accumulated on the bottom. Flushing flows restore natural water and sediment fluxes in the river downstream of the dam, although they are less expensive than dismantling dams or building bypass tunnels, which prevent riverbed erosion while increasing morphological variety downstream of the dam [49–51].

Since all dams reduce sediment load downstream, a dammed river is extremely sediment-demanding because it will not have enough sediment. Because there is less silt to deposit, the rate of sediment deposition is considerably reduced, but the rate of erosion remains virtually constant. The water flow erodes the river banks and riverbeds, endangering shoreline habitats, deepening the riverbed and narrowing the river over time. This results in a degraded water table, lower water levels, uniformity of river flow, and consequently less ecological variety, support for animals, and less sediment reaching coastal plains and deltas. This induces coastal erosion since beaches are unable to restore what is eroded by waves without the sediment deposition of supporting river systems [50]. Downstream channel erosion of dammed rivers is related to riverbed morphology, which differs from directly examining sedimentation levels since it is susceptible to specific long-term conditions for each river system. For example, the eroded channel may result in a lower water table level in the affected area, affecting bottomland crops like alfalfa or corn and leading to a lower yield. The Three Gorges Dam of China has resulted in a new balance of erosion and sedimentation in the river's lower reaches over ten years [51]. The dam's upstream effects have also been related to the effects on the tidal zone [49, 52].

1.8 THE EFFECT OF THE DAM ON THE EARTH'S ROTATION

It is believed that one rotation of the Earth concerning a point on its central axis is intended to take 23 hours, 56 minutes, and 4.09 seconds. This means that each day, we plan for approximately four minutes more of performance than we are truly able to squeeze in! Perhaps we could delay the Earth's rotation to create a full 24 hours day—or even add an hour to make it a 25-hour day. Or we could perhaps simply be more efficient. Therefore, the great speculation to slow down or speed up rotation of the Earth must be considered from the point of view of the construction of mega-project dams. Dams can influence the Earth's rotation speed. In particular cases, dams would cause the Earth to spin slower, while in others, they would force the Earth to speed up.

Consistent with speculation, how can the same structure make the Earth spin slower in certain circumstances but faster in others? The simple answer is that dams modify the rotation speed of the Earth because of inertia; so, when mass is shifted closer to or further away from the Earth's center, its inertia is completely modified. In other words, if the mass is moved further from the center of rotation, the moment of inertia is changed [53, 55].

Dams store a massive amount of heavy, dense water at a reasonable height, further away from the Earth's coreSome dams, such as the Chinese Three Gorges Dam, accommodate 27 million cubic meters of water in an innovative man-made reservoir that is higher than sea level. Instead of all the waves crashing down to sea level, it is trapped further away from the Earth's core [54].

The Three Gorges Dam, which stands almost 600 feet tall and is nearly 1.5 miles long and the Gorges Reservoir has a surface area of 400 square miles and stretches 370 miles upstream from the dam. The water is more than 39 trillion kilograms in weight (42 billion tons). The rotation of the Earth is influenced by a phenomenon known as 'the moment of inertia', which is the inertia of a rigid rotating body concerning its rotation. The moment of inertia of an object about a given axis explains how difficult it is to change its angular motion about that axis. The greater the distance of the mass from its axis of rotation, the slower it spins. With this understanding, we realize that 39 trillion kilograms of hovering water 175 meters above sea level increases the moment of the inertia of the Earth, slowing its rotation. However, the influence will be negligible. NASA scientists determined that shifting such a mass would only add 0.06 microseconds to the length of the day and make the Earth slightly rounder in the middle and flat on top. It will also cause the pole location to change by around two cm (0.8 inches). Therefore, it should be noted that a change in the mass of any object on the Earth relative to its axis of rotation would affect its moment of inertia, albeit most adjustments are too slight to be detected. Consequently, the next chapters will explain in detail how the increment in the number of dams on the Earth can slow down its rotation [54–57].

This speculation will deliver a critical question: how does the Grand Ethiopian Renaissance Dam (GERD) impact the rotation of the Earth besides the Three Gorges Dam?

From the point of view of the author, The Grand Ethiopian Renaissance Dam (GERD) and the Three Gorges Dam (TGD) represent significant engineering feats that have profound implications for the Earth's rotational dynamics. Firstly, both dams contribute to a substantial redistribution of water mass across the Earth's surface. This redistribution is crucial because it alters the moment of inertia of the Earth, which directly influences its rotation speed. Consequently, as water is stored in these reservoirs, it can shift the angular velocity of the Earth, analogous to a figure skater pulling in their arms to spin faster.

Furthermore, the construction of large reservoirs like GERD and TGD modifies the local gravitational field. This alteration in gravity occurs due to the concentration of mass in specific areas, which can lead to slight changes in how gravity acts across different regions. Therefore, these changes can affect the rotational dynamics of the Earth, resulting in measurable impacts on its angular velocity.

Moreover, research has indicated that filling large reservoirs can lead to observable changes in Earth's rotation. For instance, studies have documented that the filling of the TGD has caused a slight acceleration in Earth's rotation due to its massive water storage capacity. Similarly, as GERD continues to fill over time, it is anticipated that it will produce comparable effects on the Earth's rotational speed.

While both GERD and TGD share mechanisms through which they influence Earth's rotation, ongoing research is essential to fully understand GERD's specific impacts. The interplay between human engineering projects and planetary systems underscores a fascinating area of study within geophysics and hydrology, revealing how our actions can resonate on a global scale.

1.9 CONCLUSION

This chapter offers an in-depth examination of dam construction principles, presenting a detailed overview of various dam types, their structural elements, and the specific engineering and environmental considerations that shape their design and functionality. It addresses the complex interplay of technical factors—topography, geology, hydrology, and material availability—that influence the selection of dam types, as well as the structural choices that affect a dam's performance and resilience over time. The environmental impacts of dams are also explored, from changes in hydraulic systems and local atmospheric conditions to seismic risks that emerge from large-scale water displacement. Additionally, the chapter investigates the causes of dam failures, underscoring risks such as overtopping, erosion, foundation instability, and the critical importance of ongoing maintenance to ensure structural safety and longevity.

In synthesizing these elements, the chapter underscores the sophisticated balance required in dam construction and management, where engineering precision, environmental stewardship, and safety protocols converge. Effective dam management depends on meticulous planning, strict adherence to construction standards, and a comprehensive understanding of the environmental consequences. This integrated approach to dam construction and operation highlights the critical importance of multidisciplinary expertise in fostering sustainable infrastructure that serves societal needs while protecting surrounding ecosystems.

Furthermore, the chapter delves into the fascinating notion that large dams, such as the Grand Ethiopian Renaissance Dam (GERD) and the Three Gorges Dam (TGD), possess the capacity to subtly influence the Earth's rotation. By redistributing massive amounts of water at substantial elevations, these structures alter the Earth's moment of inertia, producing minor adjustments in its angular velocity. This phenomenon, governed by principles of rotational dynamics, suggests that as mass is shifted farther from or closer to the Earth's axis, it can result in a minute deceleration or acceleration of planetary rotation. For instance, the filling of the TGD has been documented to increase Earth's moment of inertia, resulting in a slight lengthening of the day by microseconds and a minor shift in the poles.

Additionally, the chapter explores how large reservoirs impact local gravitational fields, as the dense concentration of water affects gravitational stability and regional stress points. These insights reveal an intriguing intersection between human engineering and planetary science, where infrastructure developments have a measurable impact on Earth's dynamics on a global scale. Future research into the effects of GERD and similar mega-dams promises to deepen our understanding of these influences, underscoring the potential cumulative impact of multiple large-scale dams on Earth's rotational behavior.

In conclusion, this chapter not only illuminates the complexities of dam construction but also emphasizes the broader implications of these structures, revealing how the engineering of water reservoirs can resonate at the planetary level. This knowledge calls for careful consideration of both the immediate and far-reaching effects of dam infrastructure, highlighting the imperative for sustainable development practices that account for impacts spanning from local ecosystems to global geophysics.

References

[1] Turpin, T. 2008. Dam. Reaktion Books.

[2] Hager, W.H., Schleiss, A.J., Boes, R.M. and Pfister, M. 2020. Hydraulic Engineering of Dams. CRC Press.

[3] Jackson, D.C. 2017. Dams. Routledge.

[4] Tanchev, L. 2014. Dams and Appurtenant Hydraulic Structures. Boca Raton: CRC Press.

[5] Hughes, A. 2008. Dam. Thomas Telford Ltd.

[6] Herzog, M.A. 1999. Practical Dam Analysis. Thomas Telford Ltd.

[7] Jansen, R.B. 2012. Advanced Dam Engineering for Design, Construction, and Rehabilitation. Springer Science & Business Media.

[8] Wahlstrom, E. 2012. Dams, Dam Foundations, and Reservoir Sites (Vol. 6). Elsevier.

[9] Bertram, G.E. 1973. Embankment Dam Engineering. John Wiley & Sons.

[10] Khagram, S. 2018. Dams and Development: Transnational Struggles for Water and Power. Cornell University Press.

[11] Jobin, W.R. 2017. Sustainable Management for Dams and Waters. CRC Press.

[12] Zhang, L., Peng, M., Chang, D. and Xu, Y. 2016. Dam Failure Mechanisms and Risk Assessment. John Wiley & Sons.

[13] Chen, S.H. and Chen, S.H. 2015. Gravity dams. pp. 283–395. *In*: S.H. Chen (ed.). Hydraulic Structures. Springer, Berlin, Heidelberg

[14] Izrailovich, B.V., Isaakovich, V.A., Grigorievich, G.E., Alexandrovich, L.Y. and Borisovich, M.Y. 2021. Concrete Gravity and Arch Dams on Rock Foundation. CRC Press.

[15] Wittke, W. 1990. Foundations of gravity and arch dams. pp. 534–566. *In*: W. Wittke (ed.). Rock Mechanics: Theory and Applications with Case Histories. Springer, Berlin, Heidelberg.

[16] Dubler, J.R. and Grigg, N.S. 1996. Dam safety policy for spillway design floods. Journal of Professional Issues in Engineering Education and Practice. 122(4): 163–169.

[17] Jalili Kashtiban, Y., Saeidi, A., Farinas, M.I. and Quirion, M. 2021. A review on existing methods to assess hydraulic erodibility downstream of dam spillways. Water. 13(22): 3205.

[18] Ferdowsi, A., Mousavi, S.F., Farzin, S. and Karami, H. 2020. Optimization of dam's spillway design under climate change conditions. Journal of Hydroinformatics. 22(4): 916–936.

[19] Tanyaş, H., van Westen, C.J., Allstadt, K.E., Anna Nowicki Jessee, M., Görüm, T., Jibson, R.W., et al. 2017. Presentation and analysis of a worldwide database of earthquake-induced landslide inventories. Journal of Geophysical Research: Earth Surface. 122(10): 1991–2015.

[20] Rodrıguez, C.E., Bommer, J.J. and Chandler, R.J. 1999. Earthquake-induced landslides: 1980–1997. Soil Dynamics and Earthquake Engineering. 18(5): 325–346.

[21] Sassa, K., Fukuoka, H., Scarascia-Mugnozza, G. and Evans, S. 1996. Earthquake-induced-landslides: distribution, motion and mechanisms. Soils and Foundations. 36: 53–64.

[22] Shao, X. and Xu, C. 2022. Earthquake-induced landslides susceptibility assessment: A review of the state-of-the-art. Natural Hazards Research. 2(3): 172–182.

[23] Budimir, M.E.A., Atkinson, P.M. and Lewis, H.G. 2014. Earthquake-and-landslide events are associated with more fatalities than earthquakes alone. Natural Hazards. 72: 895–914.

[24] Abdulamita, S. Demetriua, Aldeaa, A., Neagua, C. and Gaftoia, D. 2017. Ambient vibration tests at some buttress dams in Romania, X International conference on structural engineering. Procedia Engineering. 199: 2196–2201.

[25] Abdo, Fares 2008. Roller-compacted-concrete dams: design and construction trends. Hydro Review. Retrieved 2017-04-16.

[26] Berga, L. 2003. Roller Compacted Concrete Dams. Taylor & Francis. ISBN 978-90-5809-564-0.

[27] Nawy, E.G. 2008. Concrete Construction Engineering Handbook. CRC Press, Boca Raton, Fla.

[28] Schrader, E. and Rashed, A. 2002. Benefits of Non-Linear Stress-Strain Properties & Membranes For RCC Dam Stresses. Roller Compacted Concrete (RCC) Dam Construction in the Middle East. Jordan University of Science and Technology (JUST) and Technische Universitaet Muenchen (TUM).

[29] Scuero, A, and Vaschetti, G. 2003. Synthetic geomembranes in RCC dams: since 1984, a reliable cost effective way to stop leakage. pp. 519–529. *In*: Luis Berga (eds). RCC Dams - Roller Compacted Concrete Dams: Proceedings of the IV International Symposium on Roller Compacted Concrete Dams, Madrid, Spain, 17-19 November 2003- 2 Vol set. Swets & Zeitlinger, Lisse, Netherlands, Lisse.

[30] Hansen, K.D. 1996. Roller-compacted concrete: a civil engineering innovation. Concrete International. 15(3).

[31] Smoak, W.G. 1994. Crack repairs to upper stillwater dam, Concrete International. 13(2): 24–32.

[32] Cowan, G.L., Ferguson, K.A. and Fares Y. Abdo. 2006. New RCC Dam Replaces 70-Year Old Concrete Dam, Portland Cement Association Publication PL464, Skokie, Ill.

[33] Shannon, D.A., Allen, J.C. and Hass, R.D. 2007. Design/Build Approach Big Success for Pine Brook RCC Dam, Portland Cement Association Publication PL466, Skokie, Ill.

[34] Parsons, James M., Randall P. Bass, Charles M. Kahler and Rodolfo T. Ruiz-Gaekel. 2008. Hickory log creek: building a roller-compacted-concrete dam. Hydro Review. 27(1): 36–45.

[35] Holderbaum, R.E., Judy, R.L. and Schweiger, P.G. 2008. Elkwater Fork Dam, Portland Cement Association Publication PL468, Skokie, Ill.

[36] Paul, G.S. 2001. Facing Systems for Roller Compacted Concrete Dams & Spillways. Portland Cement Association, Skokie, Ill.

[37] Marche, C. 2005. Embankment dam overtopping and collapse: an innovative approach to predict the breach outflow hydrograph. WIT Transactions on the Built Environment. 84.

[38] ASCE/EWRI Task Committee on Dam/Levee Breaching. 2011. Earthen embankment breaching. Journal of hydraulic engineering. 137(12): 1549–1564.

[39] Milly, P.C.D., Wetherald, R.T., Dunne, K.A. and Delworth, T.L. 2002. Increasing risk of great floods in a changing climate. Nature. 415(6871): 514–517.

[40] Vandenboer, K. 2019. A study on the mechanisms of backward erosion piping (Doctoral dissertation, Ghent University).

[41] Ponce, V.M. 1982. Documented cases of earth dam breaches. Research Reports SDSU Civil Engineering. Series, No. 82149. San Diego State University. CA. 43.

[42] Hendron, A.J. and Patten, F.D. 1985. The Vaiont Slide. US Corps of Engineers Technical Report GL-85–88.

[43] Jaeger, C. 1980. Rock Mechanics and Engineering. Cambridge University Press.

[44] Kiersch, G.A. 1964. Vaiont reservoir disaster. Civil Engineering. 34: 32–39.

[45] Costa, J.E. 1985. Floods from dam failures. 85(560). US Geological Survey.

[46] Müller, L. 1964. The rock slide in the Vaiont valley. Felsmechanik und Ingenieur-Geologie. 2: 148–212.

[47] ICOLD. 2010. Selecting seismic parameters for large dams. Guidelines, Revision of Bulletin 72, Committee on Seismic Aspects of Dam Design, International Commission on large Dams, Paris.

[48] Tahmiscioğlu, M.S., Anul, N., Ekmekçi, F. and Durmuş, N. 2007. Positive and negative impacts of dams on the environment. *In*: International Congress on River Basin Management. 759–769.

[49] Ly, C.K. 1980. The role of the Akosombo Dam on the Volta river in causing coastal erosion in central and eastern Ghana (West Africa). Marine Geology. 37(3–4): 323–332.

[50] Malini, B.H. and Rao, K.N. 2004. Coastal erosion and habitat loss along the Godavari delta front—a fallout of dam construction(?). Current Science. 87(9): 1232–1236.

[51] Warrick, J.A., Stevens, A.W., Miller, I.M., Harrison, S.R., Ritchie, A.C. and Gelfenbaum, G. 2019. World's largest dam removal reverses coastal erosion. Scientific Reports. 9(1): 13968.

[52] Yokoo, Y. and Udo, K. 2016. Connectivity between sediment storage in dam reservoir and coastal erosion: implications through zonal mappings of monitoring data. Journal of Coastal Research. (75): 725–729.

[53] Richards, K.S. and Reddy, K.R. 2007. Critical appraisal of piping phenomena in earth dams. Bulletin of Engineering Geology and the Environment. 66: 381–402.

[54] Dam, T.G. 2019. Three Gorges Dam. The Sanxia. Encyclopaedia Britannica. Available at: https://www.britannica.com/topic/Three-Gorges-Dam [Accessed 20 March 2023].

[55] Wang, L., Chen, C., Zou, R. and Du, J. 2014. Surface gravity and deformation effects of water storage changes in China's Three Gorges Reservoir constrained by modeled results and in situ measurements. Journal of Applied Geophysics. 108: 25–34.

[56] Wang, L., Chen, C., Du, J. and Wang, T. 2017. Detecting seasonal and long-term vertical displacement in the North China Plain using GRACE and GPS. Hydrology and Earth System Sciences. 21(6): 2905–2922.

[57] Wang, L.J., Wu, L., Hou, X.Y., Zheng, B.H., Li, H. and Norra, S. 2016. Role of reservoir construction in regional land use change in Pengxi River basin upstream of the Three Gorges Reservoir in China. Environmental Earth Sciences. 75: 1–15.

Chapter 2

Grand Ethiopian Renaissance Dam Construction Plans and Surrounding Geological Environment

2.0 INTRODUCTION

The Grand Ethiopian Renaissance Dam (GERD) is the main bone of contention in the Nile Basin. The conflict not only involves political issues, but also impinges on human life along the Nile Basin. The GERD conflict is part of a long-running rivalry between Egypt and Sudan, the downstream governments, and Ethiopia, as well as the upstream riparian over-absorption of the Nile's waters, believed to be a lifesaver for masses of Egyptians and Sudanese. Egypt and Sudan would experience a water crisis because of the completed dam's continuous operation, according to the 1959 Water Sharing Agreement. Egypt would undoubtedly continue to lose out three times as much as Sudan does. Developed countries in the Nile Basin need to exploit natural resources sustainably and progressively to improve their water infrastructure systems.

We devote this key chapter to the GERD's innovative design and completed construction; the geological background of the unique environment surrounding GERD, and its potential impact on the entire Nile Basin environment. This can aid in monitoring the dam by employing remote sensing technology to establish applicable engineering solutions to mitigate its detrimental impact on Ethiopia, Sudan and Egypt.

2.1 A POINT OF VIEW OF THE NILE RIVER BASIN

An understanding of the Nile Basin would help in properly establishing a balance between GERD's considerable advantages and potential disadvantages. The following chapter's debates provide accurate knowledge to examine the general impacts of GERD on likely climate change; the environmental ecosystem; and the plate tectonic features of the Nile River Basin.

Both cooperation and conflicts are events that spring along the Nile River Basins as the river crosses independent political boundaries. However, livelihood dependence is common to the remarkable diversity of distinct species and nations since ancient times. The nations of the Nile River Basin continue to rely on it for their livelihood and conservancy.

DOI: 10.1201/9781003327325-2

Egypt and Sudan have been historically possibly dependent on the the Nile River for survival. The Nile provides sustainable livelihood and productive ecosystem to the countries downstream as it flows from the wet equatorial lakes and highlands of Ethiopia to the arid desert regions of Northeast Africa [1, 3].

Globally, the longest river is the Nile, which spans nearly 6,700 km, covering over 35° of latitude and drains a specific area of over 3-million-kilometer squares—one-tenth of Africa's total land mass. The Nile River Basin naturally contains an extraordinary variety of natural zones. It originates in equatorial zones which are richest in varieties of wetlands, woodlands, lakes, and tropical forests with high mountains. This natural unique basin has powerfully aided an enormous variety of sustainable livelihoods, an extraordinary abundance of ecological systems, and an incredible diversity of developed civilizations for millennia [1, 3, 5].

The Nile has many tributaries: the White Nile, the Blue Nile, Tekeze (Atbara), and Baro-Akobo (Sobat). The central equatorial African zone contains the Great Lakes, the primary source of the White Nile [4, 6, 7]. On the other hand, the fertile highlands of Ethiopia are particular sources of Baro-Akobo (Sobat), Tekeze (Atbara), and the Blue Nile (Abay) (Figure 2.1).

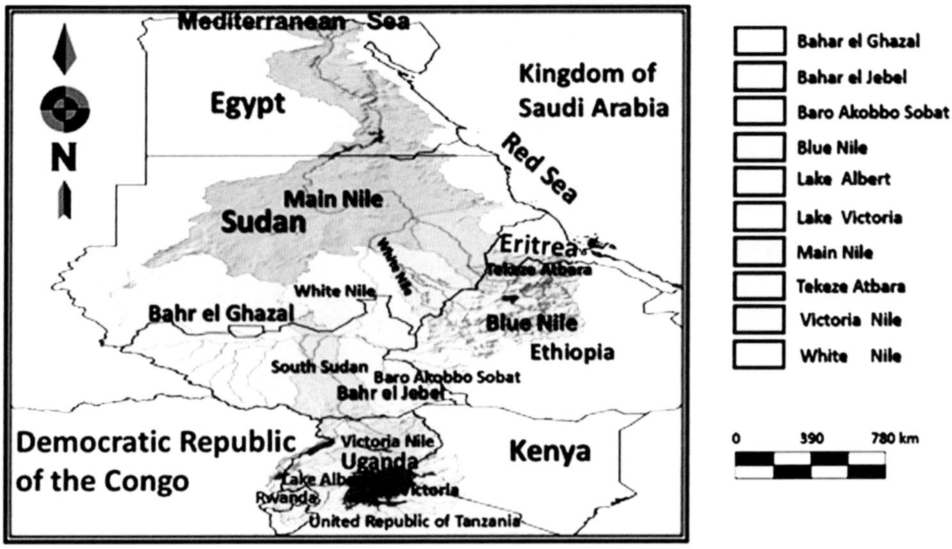

Figure 2.1 The Nile Basin tributaries.

The equatorial lakes and the Bahir el Ghazal basin (Figure 2.2) are the other sources of the Nile River. The Baro-Akobo (Sobat) crosses the White Nile as it flows from southern Ethiopia. The Equatorial Plateau delivers 15% of the flowing water that falls into the Nile Basin, whereas 85% of it typically originates in the Ethiopian Plateau. The Kagera River, which flows across Burundi, Rwanda, Tanzania and Uganda before pouring into Lake Victoria is a crucial route originator [2–5]. The Nile River traverses eleven countries: Tanzania, Ethiopia, Sudan, South Sudan, Eritrea, Burundi, Congo, Egypt, Kenya and Rwanda (Figure 2.1).

Two main sub-systems comprise the Nile Basin: (i) the Equatorial Nile, and (ii) the Eastern Nile. Ten sub-basins are formed within the basin: Lake Victoria, Victoria Nile, Lake Albert, Bahr El Ghazal, Bahr El Jebel, Baro-Akobo-Sobat, the White Nile, the Blue Nile, Atbara, and the Main Nile. These sub-basins contain five major physiographic zones, each with a unique topography, drainage system and geomorphology. At a height of almost 3,000 km above sea level Victoria Lake serves as the primary source of the Nile River. With an elevation of less than 300 m, the Nile delta is the lowest location in the Nile basin (Figure 2.3). The Nile also flows in arid zones, as shown in Egypt's territoriality. It has a massive delta on the Mediterranean Sea (Figure 2.4).

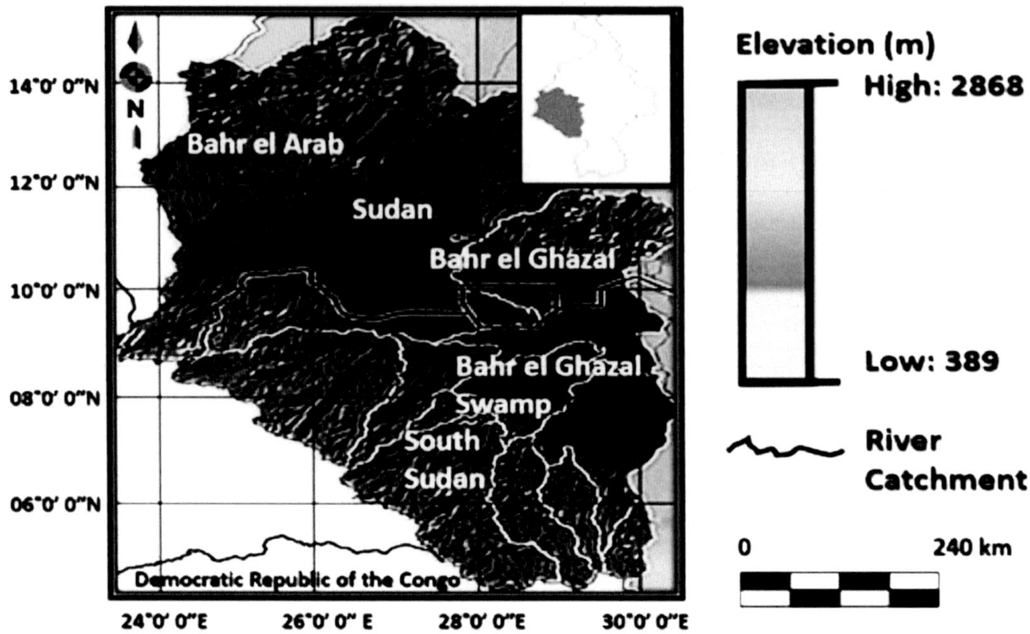

Figure 2.2 Bahir el Ghazal basin.

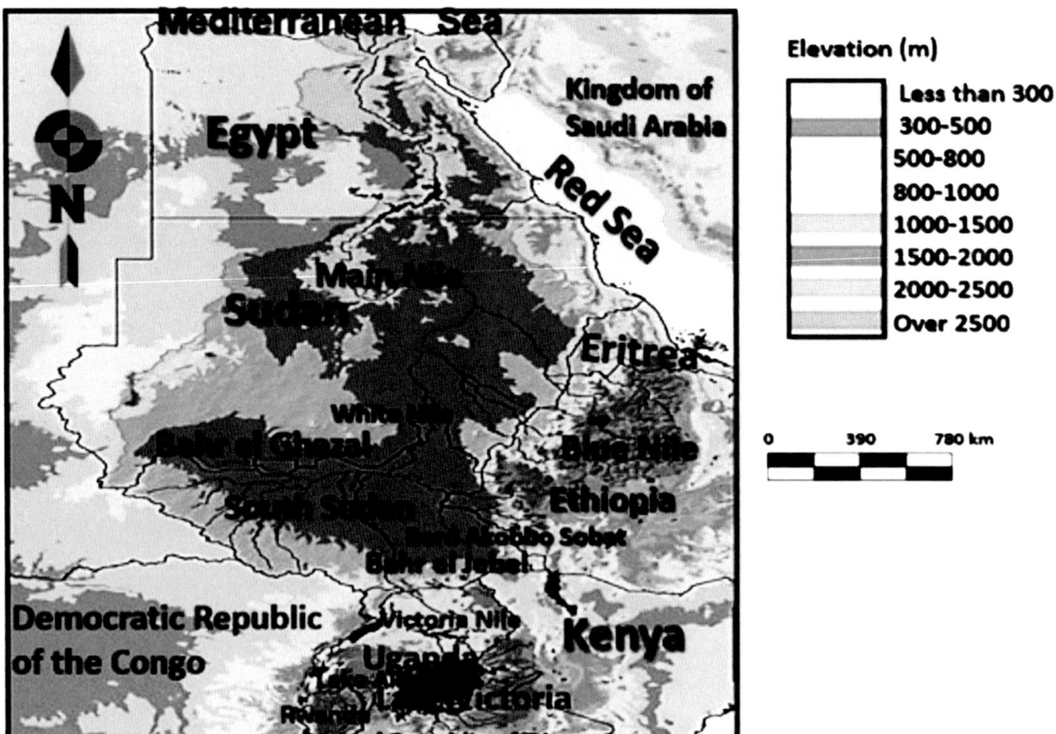

Figure 2.3 Digital elevation of the Nile basin.

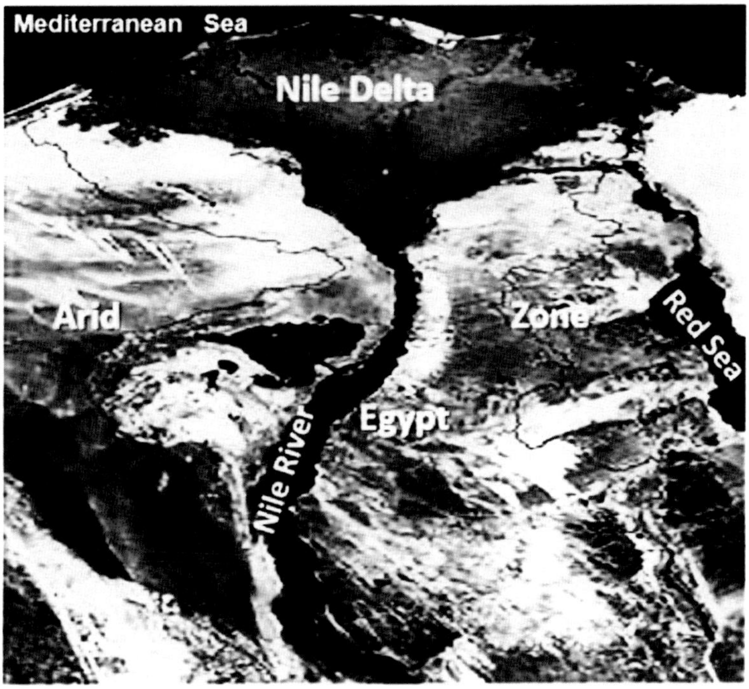

Figure 2.4 The Nile Delta flows across an arid zone.

The physiographic zones are then divided into five categories: uplands (plateaus and mountains), open water (lakes, both natural and man-made), marshes and swamps, lowlands, and deserts. Surface, slope, soils, topography, and vegetation are more or less uniquely combined in each physiographic region. While the latter three physiographic zones mostly span the middle and lower subbasins, the first two physiographic regions are mostly in the upper subbasin. In this scenario, 3,176,541 km^2 of land is drained by the catchment area, which occupies nearly one-tenth of the continent's surface. The Sudd Wetland (Figure 2.5), Lake Victoria (Figure 2.6), the seventh wetlands that have been listed with Ramsar, and a variety of flora and wildlife are just a few of the distinctive characteristics of the Nile River basin that set it apart from other major river basins in the globe.

As a consequence, numerous millions of human races in the Basin are reliant on agriculture as one of their main sources of income. It engages approximately 80% of the laborers and accounts for a majority of the population of the Nile Basin. Approximately 6,000,000 hectares of agricultural land are inundated across the river basin, with over 97% of these territories in Egypt and Sudan.

The construction of the GERD expresses the downstream governments' disrupted and occasionally demoralizing collaboration and the contemporary sociopolitical and economic circumstances, which call for more aggressive infrastructure development. Given the enormous population and extensive infrastructure reliant on Nile water, the Nile could become an increasingly scarce resource in the future. In that case over 1.6 billion people would experience drought circumstances by 2025 [15]. More than ever a frank and honest debate and collaboration are reasonably required to chart the proper path toward an egalitarian distribution of flowing water and potential benefits before the precarious regional hydro-politics degenerates into a water conflict. However, GERD is roughly half the reserve storage of the High Aswan Dam, far exceeding the Blue Nile River's annual yield. Synchronization of all dams in series would limit the consequences of GERD refilling for a common shared purpose of minimizing negative repercussions and boosting regional cooperation [4, 6, 8, 15].

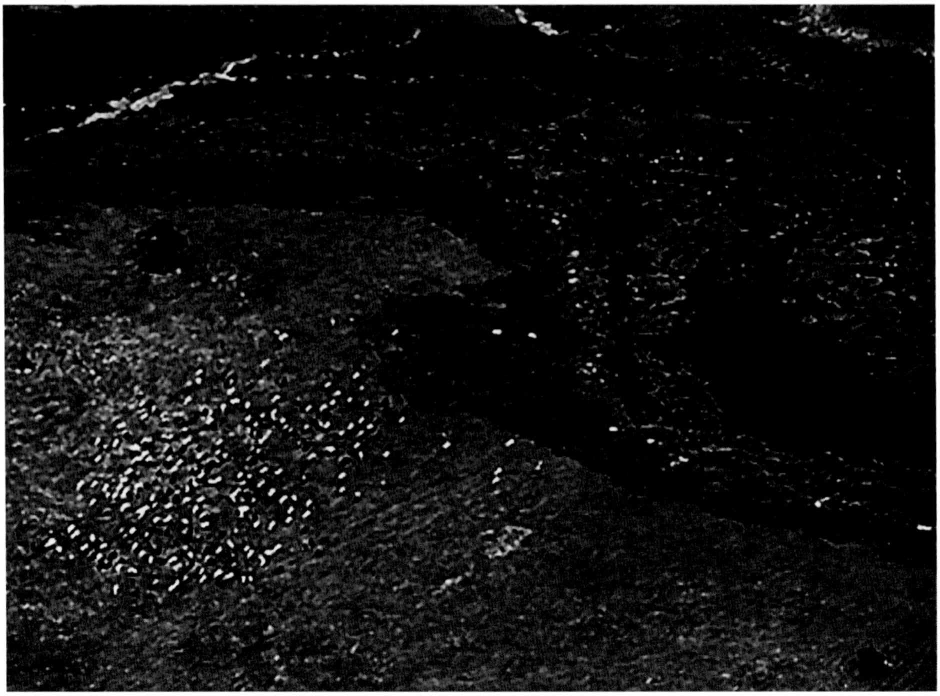

Figure 2.5 Aerial photography of Sudd wetland, South Sudan.

Figure 2.6 Aerial photography of Lake Victoria.

Egypt and Sudan's decision to withdraw from the Nile Basin Collaborative Mechanism contracted by the upstream countries may stymie such collaboration, which would have set a precedent for future infrastructure development. The competition to assert water rights through unilateral projects will continue. As a result, by synchronizing dam operations along the Nile, any damage caused by the initial refilling and administration of the GERD could also be minimized. On the other hand, Ethiopia credibly argues that the GERD would progressively improve downstream countries by stabilizing possible irregularities in river flow and boosting water availability during low-flow months. This would instantly improve Sudan's dams' unique ability to produce electricity. Ethiopia also claims that the embankment would minimize water damage and that Sudan would be able to typically purchase renewable energy produced by the dam [8, 15].

Egypt and Sudan want their 'historically gained rights' to be the starting point for any negotiations over the dam. Ethiopia, on the other hand, considers this unfair and illogical, and a ruin of the Statement of Moralities adopted by the three nations in 2015.

Ethiopia executed the third stage of reservoir refilling for its gigantic Blue Nile dam on August 12, 2022. A process that has continued to irritate Egypt and Sudan. The $4 billion GERD is crucial to Ethiopia's economic development, but Egypt and Sudan see it as a significant danger to their vital water resources. However, it does not appear that filling has caused a considerable impact on the downstream states since the initial filling in 2020 [12, 15, 16].

We will explore the Grand Ethiopian Renaissance Dam (GERD) and its significant impacts on riparian nations, climate change and the geology of the Nile River Basins. The subsequent sections and chapters will delve into these topics.

2.2 THE SIGNIFICANCE OF GERD

Currently, Ethiopia relies on sewage and biomass as major forms of energy. Unfortunately, over 60% of Ethiopians lack access to power. The dam can shape the trend of the country's growth. This would deliver clean and economical electricity to households, particularly those in rural Ethiopia. When GERD is completed, the dam's potential reservoir would ordinarily have a storage capacity of approximately 70 billion cubic meters. Its primary role is to generate electricity. GERD was inaugurated on 20 February 2022 and was 84% properly completed, with 18.5 billion cubic meters of flowing water in its local reservoir [14, 15, 17]. GERD was scheduled to generate about 750 megawatts of electrical power initially, but it has the potential to produce 6,000 megawatts. In this manner, the generated electricity has the potential to alter the agricultural sector, bring industry and manufacturing to impoverished areas, and minimize deforestation. It can revolutionize education and bring the internet to rural parts of the country [4, 8, 15].

Therefore, Ethiopia should export surplus power to neighbors like Sudan. As Ethiopia earnestly strives to typically emerge from a considerable period of economic and political turmoil, the completed dam has naturally become an iconic symbol of national pride. However, the impasse has been attributed to Ethiopia by Egypt because of treaties made during the colonial era [15, 17]. The author believes that the British colonial period wanted to control the flow of water to promote cotton cultivation in the Egyptian territory as the leading British-Egyptian collaboration industry [6, 14, 15].

2.3 THE GRAND ETHIOPIAN DAM (GERD)

The Grand Ethiopian Renaissance Dam (GERD), originally known as the Millennium Dam, was 85% complete in Ethiopia's Benishangul-Gumuz province (Figure 2.7) in 2022, on the Blue Nile River, about 40 km east of Sudan at latitude 11° 13′ 5.7432″ N and longitude 35° 5′ 36.9636″ E (Figure 2.8). The Benishangul-Gumuz province is an isolated, underdeveloped region that has a hot, and humid climate while being somewhat low-lying and receiving more than 1,000 mm of rain annually. The 1.03-million-person area is ruled by the little capital city of Asosa in the south.

From 1956 to 1962, the United States Bureau of Reclamation discovered the dam's location during a reconnaissance of the Blue Nile River. This tells us about the United States' early attention to establishing long-term conflict issues among the Nile basin countries, besides wanting diplomatic control of both Sudan and Egypt because of the Israeli-Arab conflict since 1948. This proves that GERD is the core reason for diplomatic stress between Egypt and Sudan and a great security battle for Israel.

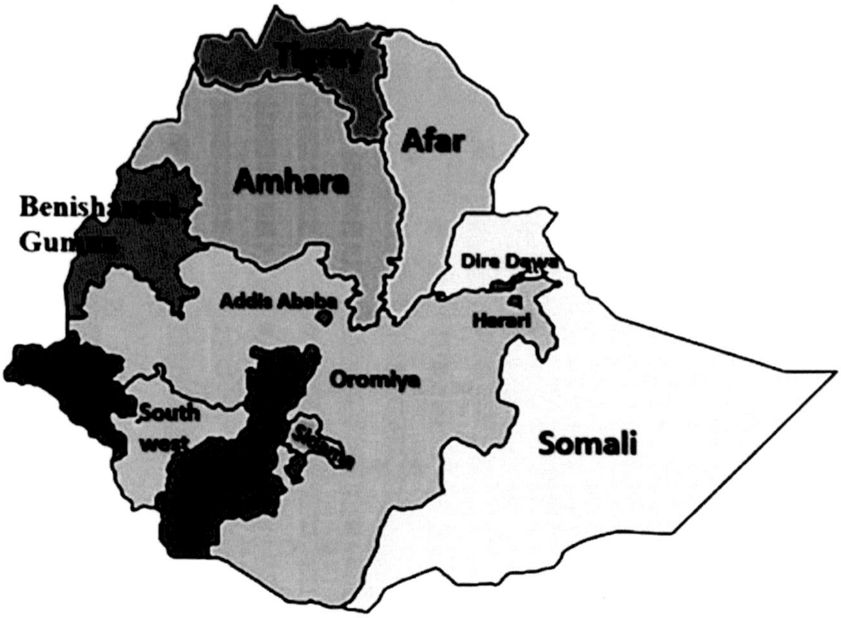

Figure 2.7 Benishangul-Gumuz province.

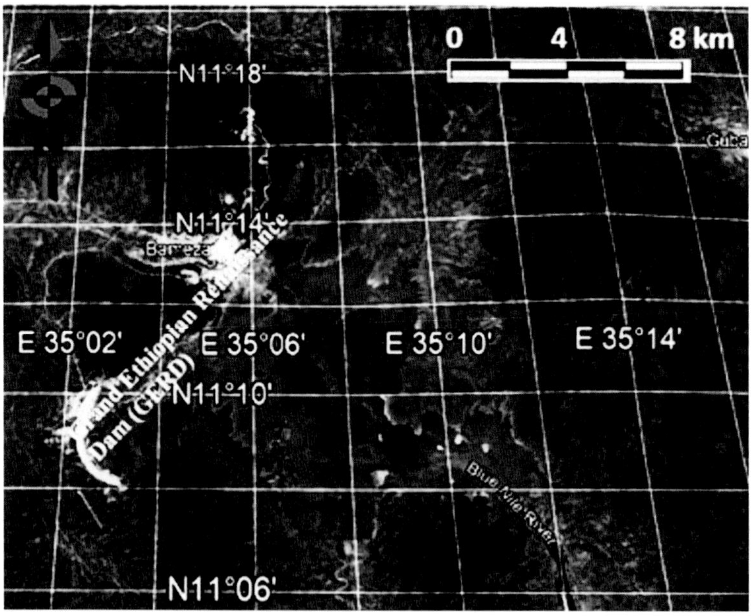

Figure 2.8 Location of GERD.

The Ethiopian Electric Power Corporation (EEPCO) owns this controversial project. The construction of the dam was initially expected to require five years. The new completion date was 2023. Ethiopia is expected to spend twice as much on the completed construction as originally estimated when work on the dam began in 2011. The acknowledged date of completed construction for the project was properly revealed in 2020. However, many Ethiopians continue supporting the dam, considering it the most significant public investment in the country's history. By handing surplus power to other Nile Basin governments, the country will probably be capable of recouping some of the cost overruns.

In July 2020, the first phase of filling the reservoir began and by August, the waterline had risen to 540 m. In other words, the water level is about 40 m higher than the bottom of the flowing river, which is 500 m above sea level. With no formal agreements with Egypt or Sudan, the second exploratory phase of recharge was typically completed on July 19, 2021, raising the water level to nearly 575 m. On August 12, 2022, the third fill was achieved at an elevation of 600 m, which is 25 m higher than the second fill in the previous year. However, depending on hydrological conditions at the time of filling, it will require 4 to 7 years to fill the lake with water [8–13].

The dam's main structure is made of roller-compact concrete (RCC). The dam is 145 m above the foundation level, with a crest length of 1780 m. At the full supply level of 640 m above sea level, the reservoir area would be 1874 km^2. The annual average energy production is estimated to be 15,692 GWh (Gigawatt hours). What is Gigawatt hours (GWh)? Gigawatt hours (GWh) is an energy measure that represents one billion (1,000,000,000) watt hours and is comparable to one million kilowatt hours. Gigawatt hours are frequently used to quantify the output of huge power energy stations [8, 11, 13].

The reservoir of the dam would be capable of storing approximately 70 billion cubic meters once construction is complete. Its primary purpose is to generate electricity. It was 84% finished and had 18.5 billion cubic meters in its reservoir as of February 20, 2022. The dam's initial projected power generated is 750 megawatts, although it has a maximum capacity of 6,000 megawatts.

2.4 GERD CONSTRUCTION CHARACTERISTICS

The Grand Renaissance dam is a gravity-type dam made of roller-compacted concrete (RCC) (section 1.3.4.1.2, chapter 1) (Figure 2.9), with two power plants, three spillways, and a saddle dam. Therefore, the major dam is 1,780 m long and 145 m high. Accordingly, RCC was preferred since the flows required to be abstracted are huge, and this kind of construction could resist corrosion. In other words, the gravity main dam is made of roller-compacted concrete and the saddle dam is made of rock (Figure 2.10). The spillway is a distinct building from the saddle dam and the main dam. Therefore, GERD consists of an embankment dam, a spillway on the right, and gravity dams on the main river channel. The natural river flow between two hills is where the gravity dam is built.

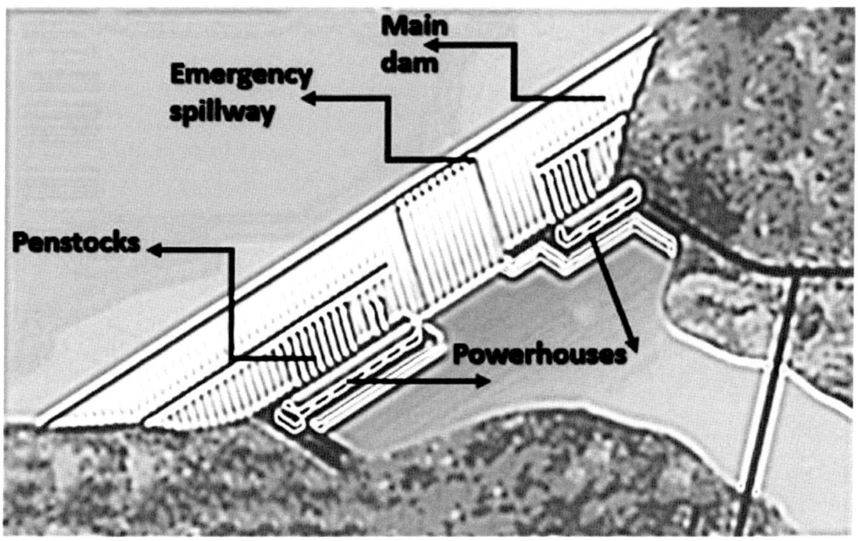

Figure 2.9 GERD layout plan.

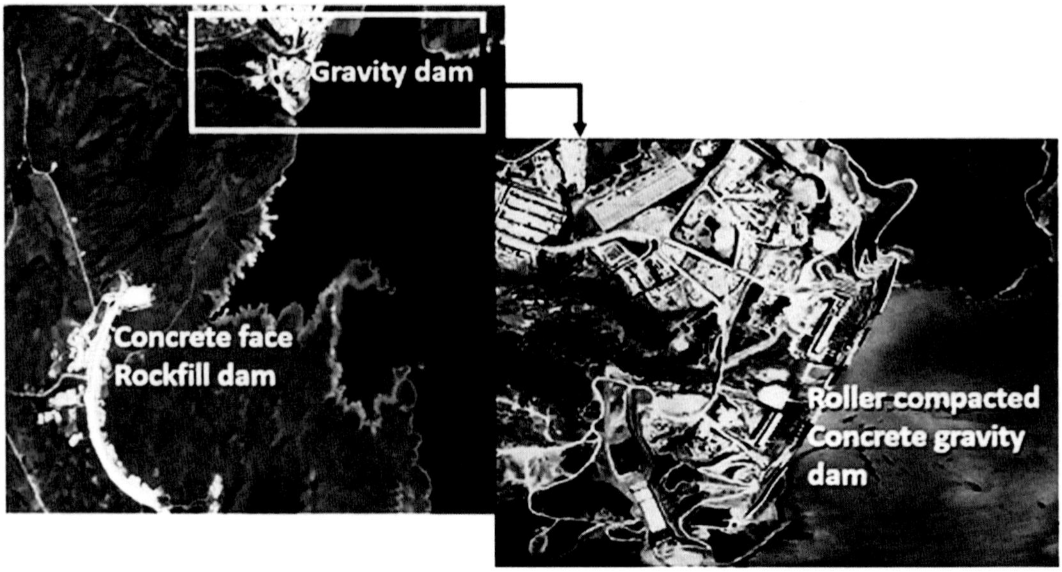

Figure 2.10 GERD dam- gravity dam and rockfill dam.

The GERD dam design is made up of two sections: (i) overflow section, and (ii) non-overflow section. The upstream edge is represented by the overflow section (stepped spillway), which has a horizontal-to-vertical ratio (H: V) of about 0.14:1. This shows the slope in the lower portion, which is below elevation 575 m (a.s.l.), and the vertical portion in the upper portion. The average slope of the stepped downstream face ranges from 0.77:1 to 0.95:1 (H: V). The upstream face of the non-overflow section, on the other hand, has a 0.10:1 (H: V) slope in the lower portion (below elevation 545 m a.s.l.) and is vertical in the upper portion. The mean slope of the stepped downstream face is 0.77:1 (H: V).

Both the upstream portion of GERD, where tensile strength under extreme seismic load and permeability criteria must be met, and the downstream toe, where compressive strength requirements must be met, use a larger cement concentration. To control the temperature rise and the ensuing risk of cracking mixes with low cement content are employed in the dam's middle zone [16–18].

Therefore, there are typically 85 monolith blocks in the constructed dam which are partitioned by cutting joints into the newly compacted RCC. In the upstream zone, the vertical contraction ties are adequately outfitted with uncommon double water ceases and accurately control proper drainage. Along the dam axis, the usual contraction joint spacing ranges from 18 to 27 meters. Thermal considerations and the unique attributes of the concrete electrical system structures (penstocks, drainage ditches, and bottom outflows) bridging the dam body properly regulate the joints' spacing. In this regard, two types of epoxy resin are exploited: (i) CEM I 42,5 LHHS (Portland Low Heat of Hydration); and (ii) High Sulfate Resistance cement for the first 300,000 m^3 of RCC production and CEM IV-A 32,5 R (Pozzolanic cement) cement for the remainder of the dam body (to date equal to about 7 Mm^3). Cement content varies from 75 to 125 kg m^{-3} for Portland cement and from 90 to 140 kg m^{-3} for pozzolanic cement through the dam's cross-section [18].

A systematic bedding mix is used in the corresponding region of the completed dam, for the first 3 m above the basement to properly sustain the impermeability criteria at lift joints. As a result, the bedding mix is only authorized for the warm and cold joints with immaturity larger than the initial curing time (8–12 hours) of the blend in the remaining portion of GERD. In this view, the grout-enriched RCC (GE-RCC) properly used on the impervious upstream face typically has a width range of 40 to 60. A GE-RCC transition is expected between the cement and the RCC mixes in the overflow section, where the downstream steps are covered by conventional concrete with silica fume. To improve remarkably the face appearance and excellent durability, a 40 cm wide GERCC layer is reliably used at the downstream slope of the non-overflow section [15–19].

The 3750 MW and 2250 MW installed capacity of the two outdoor power plants are on both banks of the river. The power units comprise 16 generator units with a total capacity of 375 MW. A 500 kV switchyard is also built in this accessible vicinity to transfer power from the control houses to the local grid. On December 2014, approximately half of the construction work was properly completed. In 2015, therefore, the dam produced 700 MW. This shows the first operation phase was efficiently achieved [14, 16, 18, 19].

2.5 THE RESERVOIR OF GERD

The ground level of the main dam, or zero level, is at a height of almost 500 m above sea level, or the level of the Blue Nile's riverbed. The main gravity dam, which is made of roller-compacted concrete, is roughly 1,780 m long and 145 m in height, taken from the base up. The dam's crest is 655 m above sea level. Because the outputs of the two powerhouses lie below ground level, the dam's overall height would be a little greater than its quoted height.

In other words, the main contractor building the dam estimates the height of the structure at 170 m, which could explain the additional depth of the dam below ground level that would require 15 m of excavation from the basement before the reservoir could be filled. The dam has a construction volume of 10,200,000 m^3. Therefore, the reservoir is built in the middle of the dam, between the left and right banks (Figure 2.11), and has a surface area of 1,680 m^2 at the full supply level (FSL). On the left side of the main dam; however, there are gated spillways with six radial gates. Each gate can discharge 2,450 m^{3}s^{-1} of water at the likely maximum flood event. Therefore, the main GERD saddle dam (Figure 2.12) is approximately 4,800 m long and 45 m high, respectively. It would have a side spillway that can be properly employed in a complex emergency to voluntarily evacuate flowing water into the potential Roseries reservoir.

Figure 2.11 Main GERD and reservoir.

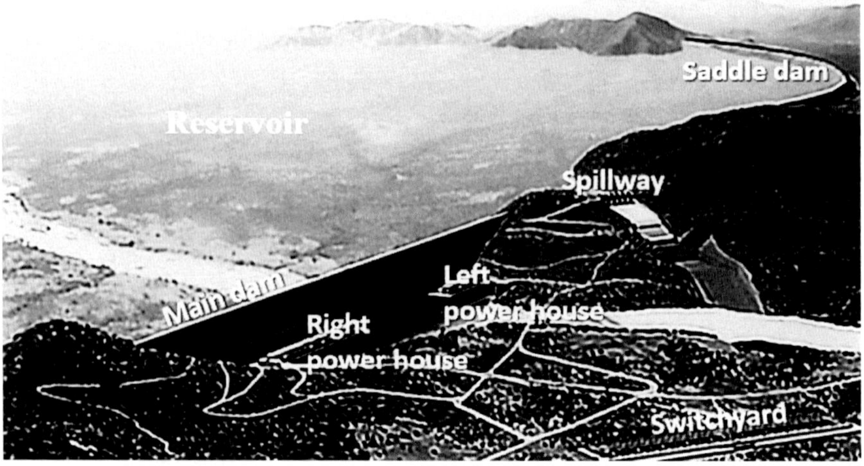

Figure 2.12 Main GERD and GERD saddle dam.

Therefore, a curved rock-fill saddle dam (Figure 2.13), 4.9 km long and 50 m high, adequately supports the main embankment and potential reservoir. The bottom level of the saddle wall is 600 m above sea level. The exterior of the saddle dam is properly covered with a distinct layer of bitumen to keep the interior dry. The saddle dam is closer to the Sudanese border than the main dam, with a distance of only 3.3–3.5 km.

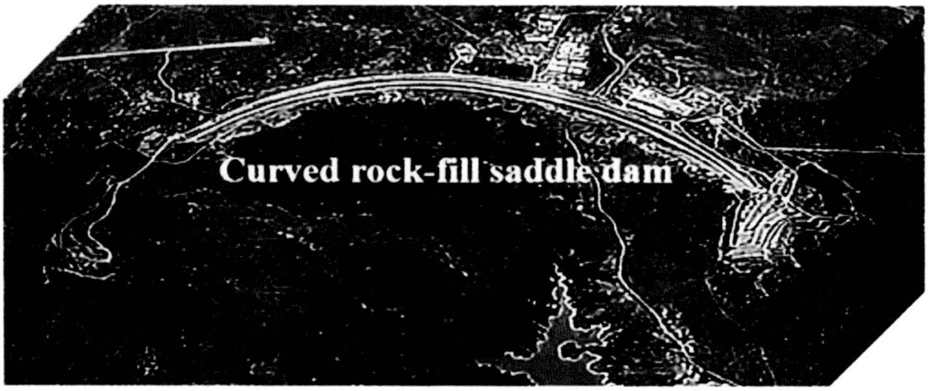

Figure 2.13 Curved rock-fill saddle dam.

With an area of 1874 km^2 and a storage capacity of 74 km^3, the reservoir behind the two dams has a full dam level of 640 m above sea level. Consequently, the ground level of the large dam is 140 m above the full supply level. Hydroelectric power can be generated between a reservoir level of 590 m, the minimum operating level, and 640 m, the full supply level. Between these two levels, 59.2 km^3 of reservoir space can be used for power generation. In this understanding, the living storage volume that can be depleted for power generation between the two levels is then 59.2 km^3. The first 90 m of the dam's height can be a dead height for the reservoir, resulting in a dead storage volume of the reservoir of 14.8 km^3.

In this sense, the appropriate level of the potential reservoir would rise sufficiently during the rainy season from June to September and fall in winter. To put it differently, GERD is now impounding water—containing 4.9 billion cubic meters of Blue Nile water after the season's rains since 2011.

2.6 GERD SPILLWAYS

There are three spillways on the dams in which 18,000 m^3 of concrete was used. Simultaneously, these spillways are built to withstand a flood of up to 38,500 m^3s^{-1}, which is the ostensible 'Probable Maximum Flood' and is not thought to occur ever. The Blue Nile should receive all the water from the three spillways before it reaches Sudanese territory. Therefore, the left of the main GERD involves the core and gated spillway, which are controlled by six floodgates (Figure 2.14). Moreover, they have a strategy discharge of 14,700 m^3s^{-1} altogether. The width of the spillway is 84 m at the outflow gates. The spillway, consequently, is designed with a base level of 624.9 m, which is well below the full supply level.

Figure 2.14 Gated spillways.

The auxiliary spillway, which has an open width of approximately 205 m is in the center of the main dam and is unhindered. This spillway's water table is exactly 640 meters below the reservoir's full level. On the left and right of the spillway, the dam's crest is 15 m higher. Only when the reservoir is full and the flow exceeds 14,700 m^3s^{-1}, a level expected to be exceeded only once every ten years, is the spillway projected to be deployed.

The third spillway, with a base elevation of 642 m, is an emergency spillway and is located to the right of the curved saddle dam. The edge of this spillway has an open area of approximately 1,200 m to sustain a catastrophic flood over 30,000 m^3s^{-1}, which corresponds to a flood that occurs only once in 10,000 years. This third spillway can precisely aid in transporting massive floodwater in such a scenario.

Accordingly, dual powerhouses, each with two 375 MW Francis turbine generators and 11 × 400 MW turbines, have been successfully constructed on either side of the unprotected auxiliary spillway in the middle of the dam. In this sense, a 5.150 MW would be the total installed capacity when entirely turbine generators are included. Therefore, the Blue Nile is expected to have an average annual flow of 1,547 m^3s^{-1}, which translates into an annual power generation of 16,153 GWh or an operation rate (or generation capacity) of 28.6%.

2.7 DRAINAGE SYSTEM AND GROUTING OPERATIONS

Grouting operations properly include curtain grouting and contact/deep foundation grouting. The efficient bulb grouting, which aims to tremendously improve the operational effectiveness of the effective shield near the upstream extremity of the completed dam where the hydraulic gradient is greatest, complements curtain grouting, which aims to properly establish an effective linear hydraulic barrier at the secure foundation of the upstream toe of the completed dam. Contact injections are aimed at improving contact at the dam/foundation interface and reduce the permeability of the surrounding rock mass, especially in the upstream foundation area. Deep foundation injections are used to consolidate the foundation rock and reduce the deformability of the rock where necessary. After excavation of the foundation, contact/deep foundation grouting was locally adjusted based on geological findings (voids, weak zones, etc.) [20–22].

Therefore, an extensive drainage system, comprising drainage arcades, basis drains, and embankment drainages at the upstream face has been installed correctly to precisely control the uplift pressure at the dam basis and carefully collect the flowing water eventually seeping through the proper foundation and the dam structure.

2.7.1 Operation Designing of Curtain Grouting and Bulb Grouting

Intentionally using the split spacing approach, which calls for sensibly diminishing the proper distance between each specific sequence of cavities from one series to the succeeding by half, curtain grouting has been accomplished starting at the primary hole, spaced at 12 m, and stretching along the entire dam upstream to the toe. In this context, a total length of approximately 39,600 meters is sufficient to effectively compensate for the systematic gaps. The systematic curtain's depth has been set at 30 m, which accurately represents in any exceptional circumstances half the proper height of the completed dam [21–23].

Consistent with the above perspective, every other primary hole is lengthened to 2/3 of the dam height for investigational purposes. Except for the gorge area, where the depth is increased to half of the dam height, secondary and tertiary holes are set to a depth of one-third of the dam height. The standard quaternary hole depth is 20 meters. Following the geo-structural zoning of the dam foundation, grout holes have been positioned (15° leftward and rightward at the left and right abutments, respectively, and vertically in the center). To close the area of transition between vertical and slanted holes, two overlapping portions are offered [19–22].

Consequently, to increase grout penetrability, the injections were performed using the GIN technique, which increases maximum pressure and grouting intensity with depth. For secondary and tertiary holes, a realistic value of 50 l/m of grout normally is considered and 40 l/m for quaternary cavities have been appropriately chosen as generation/extension criteria.

Without any additives, grout is accurately described by a w/c ratio of 0.7 (by weight). The actual values of the grout mix's density, viscosity, bleeding, and compressive strength generally fall within the parameters typical of an efficient GIN technique [17, 20, 23].

Along with the GIN technique, curtain injections are continuing, with approximately 90% of it filled as of now. With an average net grout consumption of about 10 kg/m, the results that are currently available indicate a low cement take. There have been localized high cement takes (200–400 kg/m) that correspond to rock mass problems (faults, highly fractured zones etc.) [24].

After the secondary hole series, there is a noticeable decrease in grout taken from one hole series to the next, indicating that the grout penetration distance is comparable to the distance between primary and secondary holes (6 m) [18, 25]. Bulb grouting consists of a series of holes drilled 10 meters deep in the rock, 15 and 30 degrees angled towards the reservoir alternately. The same parameters used for the first 10 meters of the curtain are used for the injections, which are carried out following the GIN procedure. The outcomes of the injections that have been performed so far agree with those of the curtain [18, 22, 25].

2.7.2 Contact/Consolidation Grouting

Now the question is: what is the tremendous role of contact grouting for GERD? The entire dam foundation is anticipated to use contact grouting. A maximum pressure of 5 bar is used to inject rock to refusal through holes spaced 6.0×6.0 m apart and 3.0 m deep.

Following the two powerhouses and ungated spillway, the grouting treatment has been deepened and intensified (consolidation grouting: keyholes spaced 6.0 × 6. m, 10. m in composed rock) at the embankment upstream extremity, over the entire dam, and at the dam downstream toe (deep foundation grouting: convenient holes spaced 6.0 × 4.5 m, 10. m in considered rock). With a maximum pressure of 10 bar, possible consolidation and deep foundation grouting were properly completed to consistent refusal. A particular region on the right side of the river gorge has some cavities that have been discovered during excavation. These cavities are the consequence of marble inside a schist band that is oriented east-west of GERD cement hydration [17, 18, 20, 22, 24].

A dedicated cavity treatment consisting of the following steps was carried out for the entire area located within the dam footprint: (i) cleaning of cavities and ducts with high-pressure water jets, (ii) filling cavities with concrete up to the foundation, (iii) the first stage of grouting through the cleaning pipes using a water-cement mixture with a w/c ratio of 0.5; and (iv) systematic grouting with primary and secondary holes spaced at 2.0 × 2.0 m and having depths of 25 m and 20 m in rock, respectively, injected at refusal with a maximum pressure of 25 bar [18–22].

2.7.3 Drainage System

One would wonder what main effects does contact grouting have on the drainage system.

A bottom gallery is positioned at the downstream toe and several transversal galleries adjoin the longitudinal one. The drainage system is composed of four longitudinal galleries at the upper side that are at various altitudes. In this perspective, water from the foundation drains and the lower section of the dam's upstream face drains are collected in the bottom gallery [18, 22, 25].

Consequently, the spacing between 150 mm-diameter drain holes bored into the upstream and the dam's foundation varies from 4 to 2 m. The spacing of the drainage hole is chosen based on the foundation's geological and geotechnical parameters, the dam body's permeability as determined by RCC monitoring and seepage information obtained during the first impounding [18, 23].

The drain holes of the foundation are angled in line with the curtain holes, whereas the upstream face drain holes are vertical. When the elevation rises above 518 meters above sea level, the water pooled from all the drainage galleries is partially discharged by gravity and partially by pumps housed in two drainage pits that are close to the powerhouses on the left and right banks, respectively [18, 20, 25].

Reasonably assuming the worst-case scenario and adequate redundancy, the peculiar capacity of drainage pumps has been dimensioned in a critical consideration of the controversial findings of the seepage analysis demonstrated hereunder.

The close spacing of drainages ensures high efficiency, substantially above the standard chosen for the stability of 67% [18]. This is achieved as well as the uplift pressure at the RCC/rock contact remains below the related reference case advised by USACE.

Therefore, under typical circumstances, the particular seepage flow for the entire length ranges between 0.15 and 0.30 l/s/m. In all cases, the calculated total discharges at the left and right drainage pits are significantly less than the combined 450 l/s pump capacity of the drainage pits. Along with the basic design plan, 100 l/s would be drained by gravity through the dam galleries and roughly 90 l/s would be discharged by the pumps located in the two drainage pits.

The predicted discharges are significantly increased, being more than three times those under normal circumstances, as a result of the increase in RCC permeability in a horizontal direction (defective RCC lifts) [15, 18, 24, 26].

On the other hand, assuming high RCC permeability values, results in a roughly 5-fold increase in the overall number of seepages. The two drainage pits release around 260 l/s/m through pumps, while the dam galleries discharge about 800 l/s through gravity. While studying the upper limit of seepage is suggested in case of abnormal permeability in the RCC, this scenario—where the permeability of the entire dam body exceeds the design values—is considered improbable. In the specific case of grouting work deficits, it has been reliably predicted that the seepages to be properly discharged from the left and right drainage pits would then increase by almost 50% (from 45 to 65 l/s) and approximately 140% (from 40 to 95 l/s), respectively, in a considerable stretch of the completed dam about 130 m long where the permeability of rock mass is higher [15–18].

2.8 FILLING SCENARIOS OF GERD

In the present circumstances, the issue arises: How long will it take for GERD to fill up? The completed dam is reasonably expected to fill with flowing water between 4 to 7 years. Water management began filling the created reservoir in 2020. This process could require several years to a decade, depending on the weather and how much of the Blue Nile's flow is held back by the dam managers. Ethiopia has the incentive to fast-fill the reservoir to start producing electricity and begin to pay for the $5 billion investment. However, as the Blue Nile undoubtedly contributes 60% of the reservoir that flows into the Nile, a quick filling might significantly reduce the water downstream.

The completed dam could, however, affect the countless millions of people who traditionally live and farm downstream in Egypt and active parts of Sudan as a specific result of intentionally altering the river's hydrology. Over 70 islands will be intentionally created behind GERD as a possible result of the GERD; forty of them will have more than 10 hectares of land, while the narrowest ones will occupy just five hectares.

The first phase of filling in 2020 (Figure 2.15) accumulated around 4.9 billion cubic meters of water, while the second phase of filling (Figure 2.16) added a further 6 billion cubic meters on July 19, 2021.

Figure 2.15 The first phase of filling.

Figure 2.16 The second phase of filling.

Consequently, at the end of the rainy season in September 2021, the reservoir was only 23% full, however as a result of evaporation and water releases, water levels gradually decreased. At a filling rate of approximately 11% per year for the first two years, it would take less than nine years to reach capacity at this rate.

The third filling took place between July 11th and August 11th, 2022, and only nine billion cubic meters of water were added, bringing the reservoir's total volume to 17 billion cubic meters (Figure 2.17). On August 12th, 2022, Ethiopia satisfactorily accomplished the third filling of the potential reservoir behind the Grand Ethiopian Renaissance Dam (GERD). Subsequently, the turbine became operational on August 11th, 2022; which indicates that Ethiopia has already exploited 270 megawatts of the second efficient turbine's renewable energy.

Figure 2.17 The third phase of filling.

Unit 10 of the completed dam's first turbine, which began efficiently generating electrical executive power on February 20, 2022, has been abundantly supplying 270 megawatts of hydropower, transferring the total power produced from GERD to date to well over 540 megawatts. The most massive hydroelectric dam in Africa is currently 83.3% constructed. The electromechanical projects are only 61% finished, whereas the civil work efficiently is 95% fully complete. The completed dam would therefore supply more than 6,000 megawatts of power to the African economies when it is completed in 2024. Although the number of turbines required for installation has decreased from 16 to 13, it is still expected to produce only 3,000 megawatts. The first turbine was installed at the beginning of this year, but no additional ones have been mounted subsequently. The dam's construction will continue for another 5 to 6 years since a rising middle wall is required for the reservoir's capacity to fulfil the original mission of 74 billion cubic meters. Needless to say, the fourth and fifth fillings could each be larger than 25 billion cubic meters if Ethiopia continues to fill the GERD in five years.

The fourth filling phase was concluded on September 10, 2023, with water levels reaching approximately 625 m. This phase added another 24 billion cubic meters to the reservoir, bringing cumulative storage to about 62.5 billion cubic meters. The ongoing developments have not only intensified regional disputes but have also highlighted the urgent need for cooperative water management strategies among the Nile Basin countries.

As of July 17, 2024, Ethiopia initiated the fifth filling phase, which is expected to conclude in early September 2024 (Table 2.1). The target for this phase is to raise the middle wall of the dam to approximately 640 m, allowing for an additional storage capacity of about 23 billion cubic meters, thereby increasing the total reservoir capacity to around 64 billion cubic meters. This latest filling has prompted renewed concerns from Egypt and Sudan regarding their water security as Ethiopia continues its unilateral actions concerning the dam's operation.

Table 2.1 Timeframe line of filling of GERD

Filling Phase	Start Date	Completion Date	Water Level Reached (m)	Volume of Water (billion m^3)
First Filling	July 2020	August 2020	540	4.9
Second Filling	July 2021	July 19, 2021	575	13.5
Third Filling	July 2022	August 12, 2022	600	22
Fourth Filling	July 2023	September 10, 2023	625	24
Fifth Filling	July 17, 2024	Early September 2024	640 (target)	Up to 23

The fifth filling of the Grand Ethiopian Renaissance Dam (GERD) represents a pivotal moment in Ethiopia's energy strategy, as it aligns with the country's ambitious goals for power generation while simultaneously raising significant concerns among downstream nations. Currently, the dam height has reached 640 m above sea level, a critical threshold necessary for the reservoir's functionality and capacity. This filling phase aims to increase the total water volume in the dam to approximately 63.7 billion cubic meters, a substantial rise from around 41 billion cubic meters prior to this filling.

As this fifth filling phase progresses, it is anticipated that Ethiopia will transition to regular energy generation, targeting an output of 1,308 GWh per month once the reservoir reaches its full capacity of 74 billion cubic meters. This operational capacity is not only essential for meeting domestic electricity needs but also ato enable Ethiopia to export power to neighboring countries, thereby enhancing regional energy security and economic development.

However, this ambitious plan has sparked controversy and heightened tensions with Egypt and Sudan, which rely heavily on the Nile River for their water supply. The unilateral actions taken by Ethiopia regarding the filling process have drawn criticism from these downstream nations that are concerned about potential reductions in their annual water shares due to the dam's operations.

These concerns underscore the pressing need for a legally binding agreement on the filling and operation of the dam to ensure equitable water management and mitigate regional conflicts.

Ultimately, while the current filling phase signifies progress toward Ethiopia's energy objectives, it simultaneously highlights the complexities of transboundary water resource management in the Nile Basin. The ongoing situation calls for constructive dialogue and cooperation among all stakeholders to address shared water security challenges and foster sustainable development throughout the region.

The critical question now is: What are the potential harmful effects on the planet from the filling of the Grand Ethiopian Renaissance Dam (GERD)? The potential harms of extended filling periods for the Grand Ethiopian Renaissance Dam (GERD) are multifaceted, impacting not only the immediate water management strategies but also the broader ecological and socio-economic dynamics in the Nile Basin.

One significant concern is the impact on downstream water availability. Prolonged filling can lead to reduced water flow to Egypt and Sudan, which depend heavily on the Nile for their agricultural and domestic water needs. This reduction could exacerbate water scarcity, particularly during dry seasons or drought years, leading to increased tensions between these countries and Ethiopia. The historical context of water sharing in the Nile Basin complicates this issue, as Egypt has long relied on a consistent flow from the river for its survival.

Moreover, there are environmental implications associated with extended filling. The GERD's reservoir may experience increased evaporation losses as its surface area expands, which could further decrease the amount of water available downstream. Studies suggest that evaporation rates could rise significantly, potentially offsetting some of the benefits derived from the dam's hydropower generation capabilities. Additionally, seepage from the reservoir has been noted as a concerning factor, with estimates indicating substantial losses that could affect overall water management strategies in the region.

From a socio-economic perspective, prolonged filling periods can disrupt agricultural cycles in downstream countries, particularly in Sudan and Egypt. As these nations rely on predictable water flows for irrigation, any significant alterations can lead to crop failures and food insecurity. This situation is compounded by the potential for increased competition over limited resources, which may escalate into conflicts or exacerbate existing tensions.

Finally, long-term health impacts cannot be overlooked. The stress and anxiety stemming from uncertainty about water availability can affect mental health in communities dependent on the Nile's resources. The fear of potential shortages may lead to increased psychological distress among populations that already face challenges related to food security and access to clean water [34–36].

In essence, while Ethiopia's ambitions for energy generation through the GERD are clear, the ramifications of extended filling periods necessitate careful consideration of regional dynamics. It is crucial for all stakeholders involved to engage in constructive dialogue to address these concerns collaboratively and ensure sustainable management of shared water resources in the Nile Basin.

2.9 INTENSE CRITICISM FROM SCIENTISTS OVER HEGGY'S STUDY: A CALL FOR SCIENTIFIC INTEGRITY AND RESPECTFUL DISCOURSE

In the previous chapter, the environmental impacts of dam filling were broadly covered (Section 1.7). A notable study by Heggy et al. [27] raised concerns about the potential effects of filling the Grand Ethiopian Renaissance Dam (GERD) on Egypt's environment and economy, specifically regarding its High Aswan Dam (HAD). This study sparked debate among experts, who argued that, according to Heggy et al. [27], Egypt faces an annual water deficit that could lead to severe agricultural and economic losses—estimated at $51 billion and affecting up to 72% of Egypt's farmland without new mitigation strategies.

In contrast, Eldardiry and Hossain [31] modeled the effects of GERD filling on HAD under different scenarios and found that a slower filling period or starting with a high HAD storage level could lessen downstream impact. Their simulation predicted that HAD's levels would remain above 170 meters even under a three-year GERD filling scenario.

Heggy et al.'s study has received substantial criticism from the scientific community. Wheeler et al. [28] argued that Heggy's analysis relies on faulty assumptions, including exaggerated water deficits and seepage losses that lack empirical grounding. They contend that the High Aswan Dam, currently at high water levels, would help offset deficits, and that Heggy's economic impact estimates are based on rudimentary calculations rather than robust economic models. Eladawy et al. [29] echoed these criticisms, arguing that Heggy's analysis overstates the impact of GERD, leading to unwarranted fear about the water future of Egypt and complicating diplomatic discussions.

Further studies provide additional context on GERD's impact. Mulat and Moges [33] found that GERD could harm Egypt's agricultural output if filled during a sequence of dry years, while Abdelhaleem and Helal [38] identified potential reductions in HAD's hydropower capacity. Other models, such as those by Kamara et al. [32], show varying impacts on Egypt's water resources and economy based on GERD's filling rate, with more gradual fills mitigating losses.

In recent studies, El-Rawy et al. [42] modeled declines in Nile flow to Egypt due to GERD, showing severe groundwater depletion scenarios during prolonged filling periods. The researchers suggest that operating GERD in a way that aligns with natural discharge cycles could minimize evaporation loss and help stabilize water levels downstream. Their findings, consistent with Heggy's concerns, support the need for careful management of Nile resources to address climate change and regional stability.

Political and scientific tensions over GERD emphasize the importance of rigorous research and respectful dialogue. As discussions continue, it remains crucial for all involved to prioritize fact-based analysis over adversarial positions, ensuring that the shared interests of all Nile Basin countries guide future water resource management.

2.10 GENERAL GERD IMPACTS ON THE ENVIRONMENT

Consistent with the above perspective, on a smaller scale, the environmental effects are substantially greater, and the majority of dams have significant environmental effects, which could have an impact on the construction of the Ethiopian project. The aquatic life upstream of the development will also be impacted. What was previously a river ecology will alter to resemble a lake, which will be very different for the local species. The barrier will also entirely separate the spawning areas by stopping fish species from migrating, which swim upstream to spawn. The amount of water that is being kept behind the dam will also cause changes in the water's temperature, chemical composition, and oxygen content, which will encourage non-native species to take over the new habitat.

Additionally, the shortage of water allotments to feed the Nile delta would cause an increase in seawater intrusion in the Egyptian north delta within the completed GERD construction. In this sense, navigation and Nile cruises would be impacted by the Nile's branches and canals having declining water levels. Because of a shortage of water, the Egyptian North Lakes would be experienced environmental degradation, an increase in pollution, and an imbalance in the natural system of life [37].

Another impact of climate change is felt in the Nile Delta, a sediment-rich landform at the northern end of the great river where it empties into the Mediterranean Sea. According to the Intergovernmental Panel on Climate Change, this region is one of the three areas of the world most affected by global warming, as the reduced flow makes it difficult for sea levels to rise. Since 1960, the Nile Delta has lost 35 to 75 meters of soil annually due to the erosion of the Mediterranean Sea. Therefore, 9 million people would be forced to relocate if the Nile rose by 1 meter and flooded 34% of the surrounding area in northern Egypt.

The problem worsens because of dwindling river reserves. The amount of Mediterranean water entering the Nile Delta will increase, while the Nile Valley's water supplies will decrease. This increases the risk of flooding and land erosion and changes the chemistry of the river. As freshwater inflows decrease, more and more saltwater from the Mediterranean Sea is found in the aquifer beneath the river delta. The salinity of the river water increases on its northern banks. Less than 1 billion cubic meters of drainage water from the river enters the Mediterranean Sea. That is absurdly little compared to what it was 40 or 50 years ago. In accordance with the United Nations Environment Program, the Mediterranean Sea will move 100 meters closer to the Nile Delta each year if temperatures continue to rise (UNEP). It is predicted that the Mediterranean Sea will eventually engulf 100,000 hectares of agricultural land that is less than 10 meters below sea level. The consequences for Egypt would be devastating, as 30 to 40% of the country's agricultural yield is produced in the north country. In this sense, vast areas of land have been damaged by Mediterranean salt, which weakens and damages crops. According to reports from farmers, the quality of vegetables has declined.

The Nile serves as both a source of water and energy in the ten countries through which it flows. More than half of Sudan's electricity production is generated by hydropower. In Uganda, the share rises to 80%. But the reliability of this energy source is declining. In Uganda, power outages are already frequent. Lower rainfall will result in lower water levels in Lake Victoria and the Nile River, which will reduce hydroelectric power generation. Despite the country's rapid economic progress, half of Ethiopia's 110 million people live without electricity every day. The country's authorities are counting on a massive dam to solve this problem, even if it means depriving neighboring countries of electricity.

Therefore, these changes and the lack of rainwater storage and treatment facilities make agriculture in Sudan extremely difficult and exacerbate a crisis in which one in four people suffer from severe hunger.

Sudan evaluates its resilience to climate change and ranks relatively low, similar to other countries along the Nile. However, the country's capacity to feed its people as well as provide electricity to power homes and industries would be impacted by rising temperatures [39, 41]. Flooding can be caused by short, heavy rains. Prolonged dry periods would cause water loss. The nations that form the Nile Basin also cannot exist without water, which is not considered in the research published by Wheeler et al. [28], Eladawy et al. [29], and Wheeler et al. [30].

2.11 GEOLOGICAL FEATURES OF ETHIOPIA

The region of Ethiopia looks back at about one billion years of geological history. The first event was the closure of the Mozambique Ocean between western and eastern Gondwana with the formation of the Ethiopian basement, which is between 880 and 550 Ma old. This folded and tilted Proterozoic basement underwent intense erosion that lasted a hundred million years and destroyed any relief of the Precambrian orogen. Fluviatile sediments of Ordovician to Silurian age and glacial deposits of Late Carboniferous to Early Permian age were deposited over an Early Paleozoic planing surface.

The initiation of Gondwana's breakup contributed to the Jurassic inundation of the Horn of Africa by a marine transgression from the Paleotethys and the emerging Indian/Madagascar Ocean. The Proterozoic terranes of Ethiopia are known as the East African Orogen [44], a megacollisional structure that extends from Palestine to Madagascar and was caused by the closure of the gap between western and eastern Gondwana, the ocean of Mozambique (Figure 2.18). The north-south orientations of the lithic components of the East African orogen, sometimes marked by ophiolite belts, stand out as prominent modern morphological expressions. Moreover, they essentially impact the trends of subsequent fragile structures. The East African Orogen forms the northern boundary.

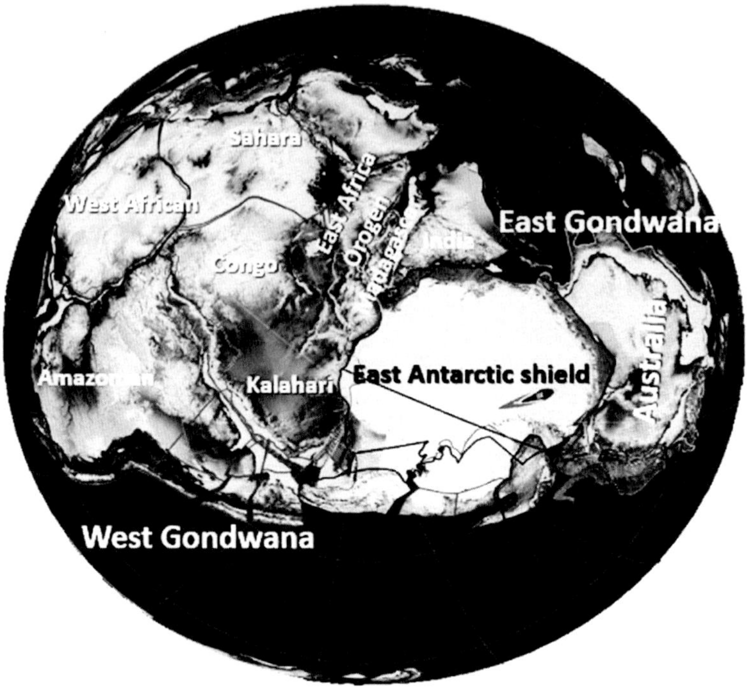

Figure 2.18 The East Africa orogen was sandwiched between western and eastern Gondwana.

Arabian-Nubian Shield, and Mozambique are to the south belt. The Nubian part of the shield lies in northern Ethiopia and consists of low-grade volcanic-sedimentary rocks overlain mainly by metasediments (stromatolitic carbonates and diamictites) associated with the 'Snowball Earth'. The Mozambique Belt, in southern Ethiopia, has a variety of amphibolites and granulitic metamorphic rocks and gneiss terranes [39–41]. Additionally, relics of less erodible rocks, such as granites and gneisses of various ages, are now inselbergs in the gently fissured basement landscape, e.g., near the border between Ethiopia and Sudan [44–46].

Following the Jurassic transgression and the deposition of Cretaceous continental sediments, the Ethiopian region was exposed to land for approximately seventy million years, during which period a new major peneplanation surface developed. Concurrent with the first phase of rifting of the Afro-Arabian plate, there was a strong outpouring of flood basalts, which occurred primarily during the Oligocene over a low-lying land surface. Therefore, the majority of the African continent (apart from its easternmost region) and the nearby oceanic crust to the west and south make up the African Plate, a sizable tectonic plate. Its western border is formed by the Mid-Atlantic Ridge, which also divides the South American and North American plates. Its eastern border is formed by the Arabian Plate and the Somali Plate, and its northern, eastern, and southern borders are formed by the Eurasian Plate, the Aegean Sea Plate, and the Anatolian Plate. Its southern border is formed by the Antarctic Plate (Figure 2.19) [45–47].

Along the East African Rift (Figure 2.20), the Somali Plate (Figure 2.21) started to separate from the African Plate 60 million years ago, and then again 10 million years later. The African plate is frequently referred to in literature as the Nubian plate to distinguish it from the continent as a whole because the African continent is made up of the crust of both the African and Somali

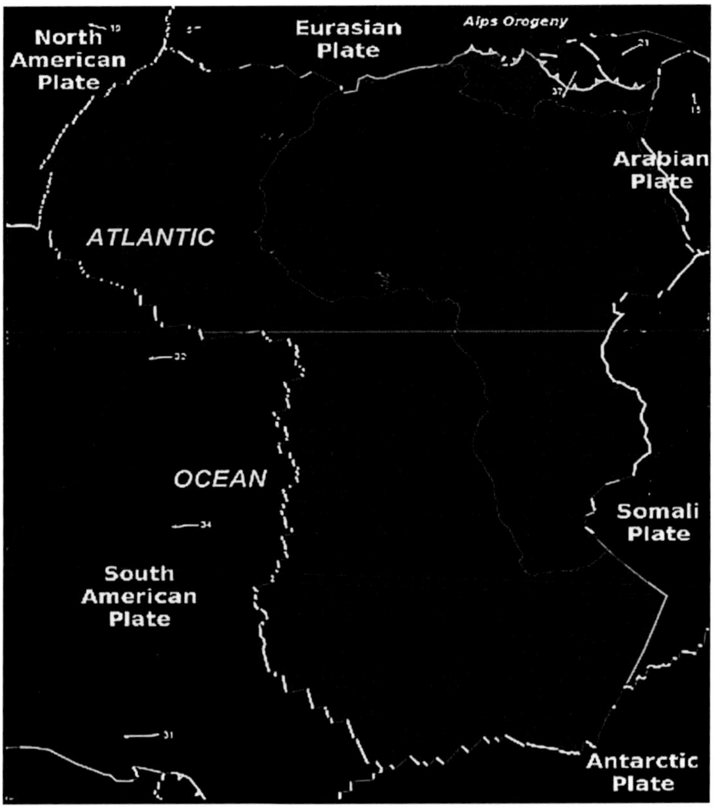

Figure 2.19 African plate.

plates. On the other hand, an active continental rift zone in East Africa is known as the East African Rift (EAR) or the East African Rift System (EARS) (Figure 2.20). At the start of the Miocene,

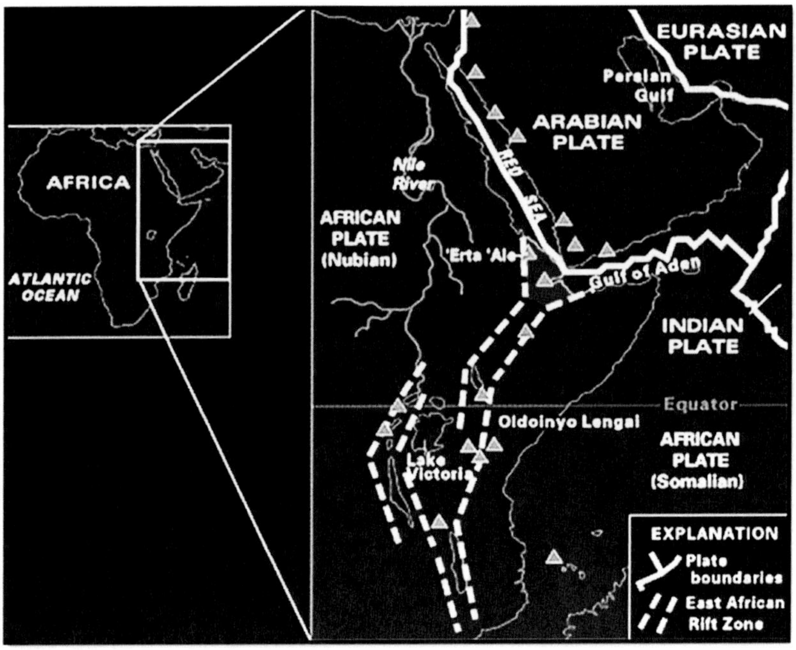

Figure 2.20 East African Rift.

22–25 million years ago, the EAR started to form. It was once thought to be a component of a broader Great Rift Valley that reached Asia Minor to the north. The African Plate breaks into the Somali Plate (Figure 2.21)and the Nubian Plate at a pace of 6–7 mm per year in the rift, a small zone that is an emerging divergent tectonic plate boundary. The Victoria microplate (Figure 2.22) is located to the north of the rift system, while the Rovuma and Lwandle microplates are located to the south. The Rovuma and Lwandle microplates (Figure 2.23) are to the south, and the Victoria Microplate is to the north. Compared to the African plate, the Victoria Microplate rotates counterclockwise. The EARS' arrangement of stronger and weaker lithospheric zones causes it to revolve (Figure 2.20) [44, 46, 48].

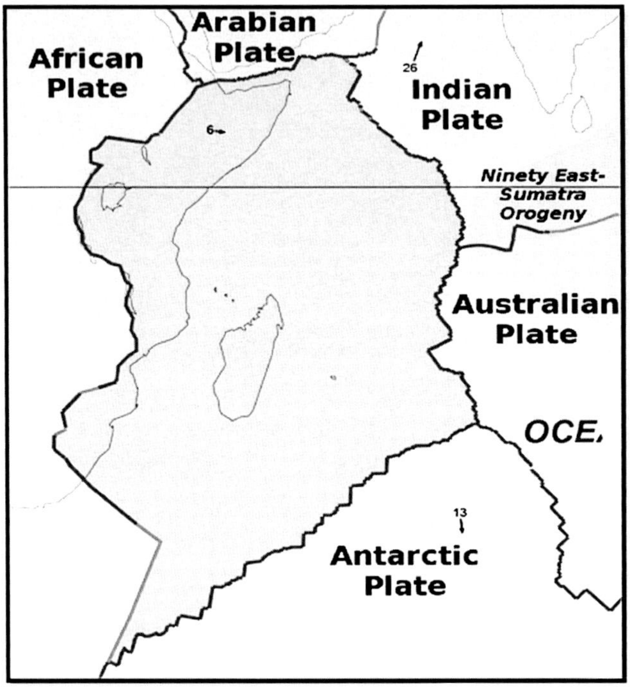

Figure 2.21 Somali plate.

Therefore, the Antarctic, Nubian, and Somali plates have a shared boundary with the Lwandle Plate, which is located between 30° and 50° east. Thus, the Lwandle Plate is oceanic and lies off the southeast coast of Africa. The southern part of Madagascar is commonly considered to be part of the Lwandle Plate because one of the plate boundaries cuts through the island. However, the majority of Lwandle's active plate boundaries remain unknown. What is generally understood is that they are closely associated with the following aspects [41, 45, 47, 49]:

Northern boundary: The Rovuma plate encircles the Lwandle plate on its northern edge [45].

Southern boundary: The Southwest Indian Ridge, an ultra-slow spreading ridge with a spread rate of about 12–18 mm/yr, serves as the southern boundary. Lwandle is separated from the Antarctic plate by this broad ridge [39, 41, 46, 48].

Eastern boundary: To the east, the margin could be more ambiguous. Current research is attempting to pinpoint the exact location of this frontier; it is conceived to typically cut across Madagascar, naturally making it Lwandle's only non-oceanic fringes. The primary constraint that accurately predicts actual plate boundaries in common is earthquake slip vectors. The predicted velocity at this arbitrary boundary efficiently is 1.3–1.4 mm/yr. This continuous line conveniently divides Lwandle from the Somali Plate [40, 45, 48].

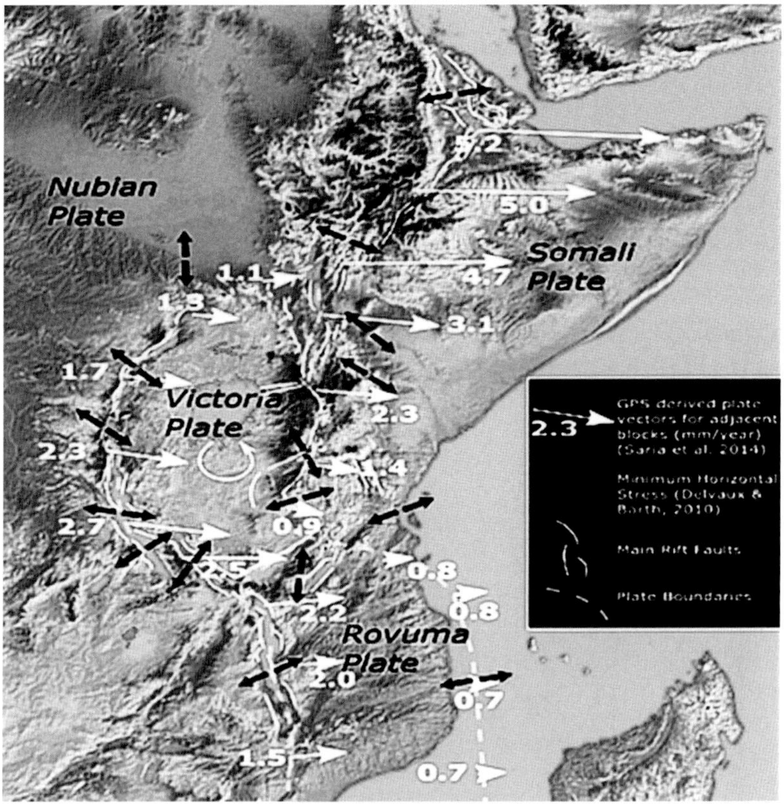

Figure 2.22 Victoria microplate.

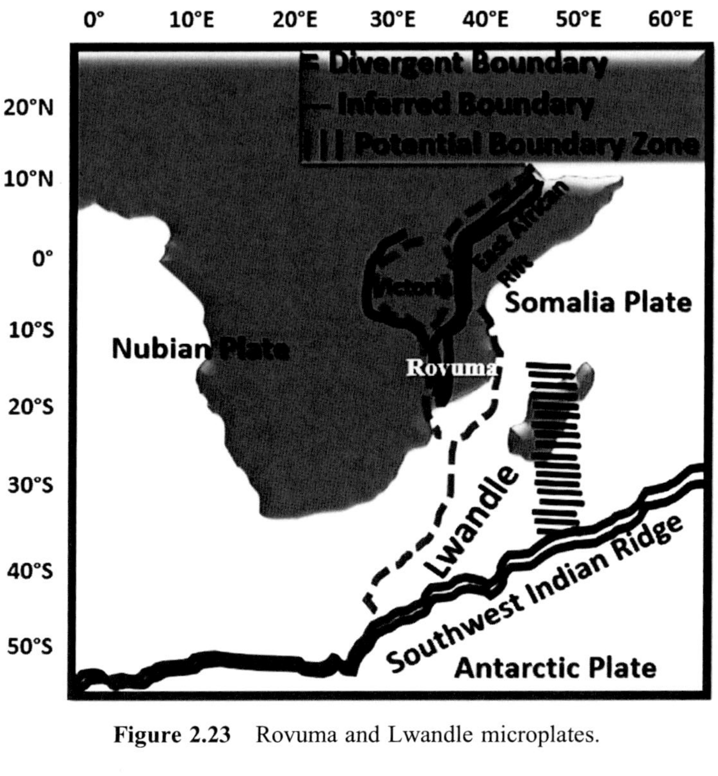

Figure 2.23 Rovuma and Lwandle microplates.

Western boundary: The western boundary, where there is relatively minor seismic activity, is distinctly imperfectly known. Nevertheless, a mismatch with a ground magnetic anomaly strongly suggests there is motion there. Graciously according to some accurate simulations, the precise boundary efficiently is a right-lateral strike with a strain rate of about 1 mm/year. Countless others go so far as to tentatively suggest that movement at this expanded boundary stopped about 11 million years ago, typically separating the Nubian Plate and Lwandle [44, 47].

Currently, the Lwandlean lithograph has been moving very slowly (1–2 mm/year) compared to the Nubian and Somali plates (Figure 2.24). It is predicted that there will be no vital seismic activity in this area, assuming the practical possibility of movement. This specific prediction is typically supported by empirical evidence. Although the expanding Lwandle Ridge is moving more swiftly relative to the Antarctic plate than the Nubian and Somali plates, it is still expanding more considerably than the sustainable growth of a human fingernail. It instantly appears that the Lwandle will remain steadfastly in its present position for a reasonable time.

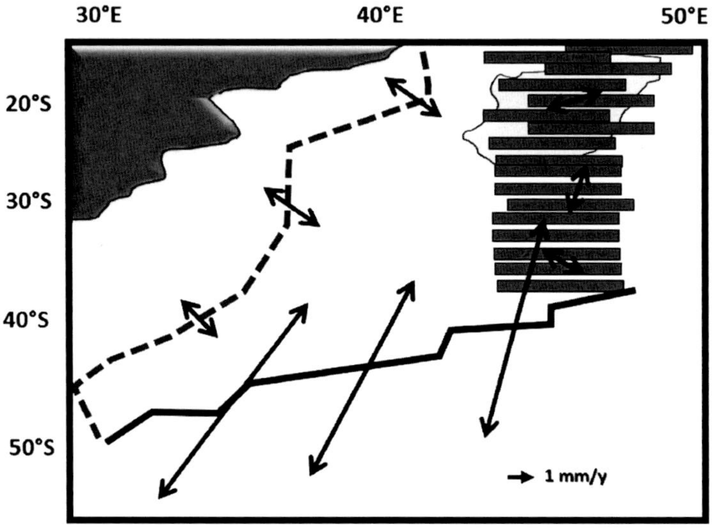

Figure 2.24 Modern movement along Lwandlean plate.

Therefore, the Afar Triangle (Figure 2.20) is a so-called triple junction (or triple point) where three plates are pulling apart from one another: the Arabian Plate and two parts of the African Plate—the Nubian and Somali Plates—splitting along the East African Rift Zone. Some of the historically active volcanoes are shown in Figure 2.21 of East Africa as red triangles, and the Afar Triangle is shaded in the center. Along with the African and Indian plates, the Arabian Plate (Figure 2.25) is one of three continental plates that have been advancing northward and slamming into the Eurasian Plate throughout geological time. As a result, plate fragments and mountain ranges that reach west from the Pyrenees across Southern Europe and Iran to form the Alborz and Zagros Mountains as well as ranges in Southeast Asia and the Himalayas are interacting [44–48].

Along the East African Rift, the African Plate is rifting in the eastern part of the continent. The Somali Plate is located in the east while the African Plate is located in the west across this rift zone. An alternate concept contends that rifting is essentially a zone of maximal weakness where the African Plate is deforming as plates to its east are moving quickly northward. One hypothesis contends that a mantle plume exists beneath the Afar region.

The African Plate is thought to be moving at a rate of 2.15 cm (0.85 inches) per year. Over the past 100 million years, it has moved toward the northeast. This causes subduction, where oceanic crust converges with continental crust and approaches the Eurasian Plate (e.g., parts of the central and eastern Mediterranean). The Azores-Gibraltar fault zone is where lateral pressures and pressures from the relative motions of the Eurasian and African plates are concentrated in the western Mediterranean. The Red Sea Trench, where the Arabian Plate separates from the African Plate, encloses the northeastern edge of the African Plate. Currently, the African Plate moves across the Earth's surface at a rate of 0.292° ± 0.007° per million years, relative to the "average" angular velocities of the Earth's crust (Figure 2.26) [46, 49].

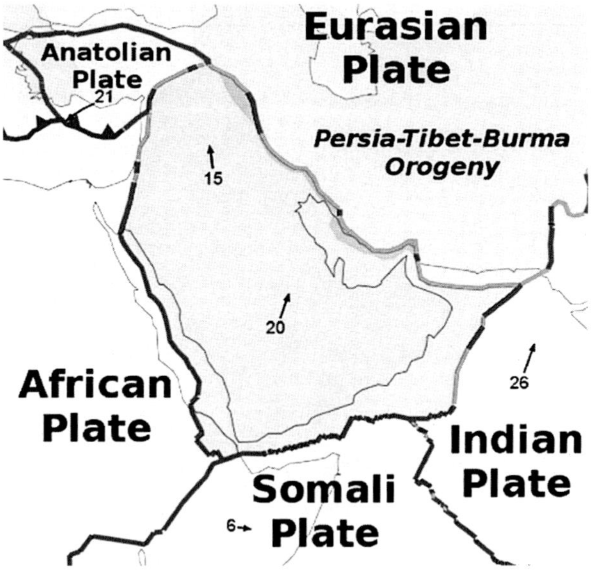

Figure 2.25 Arabian plate.

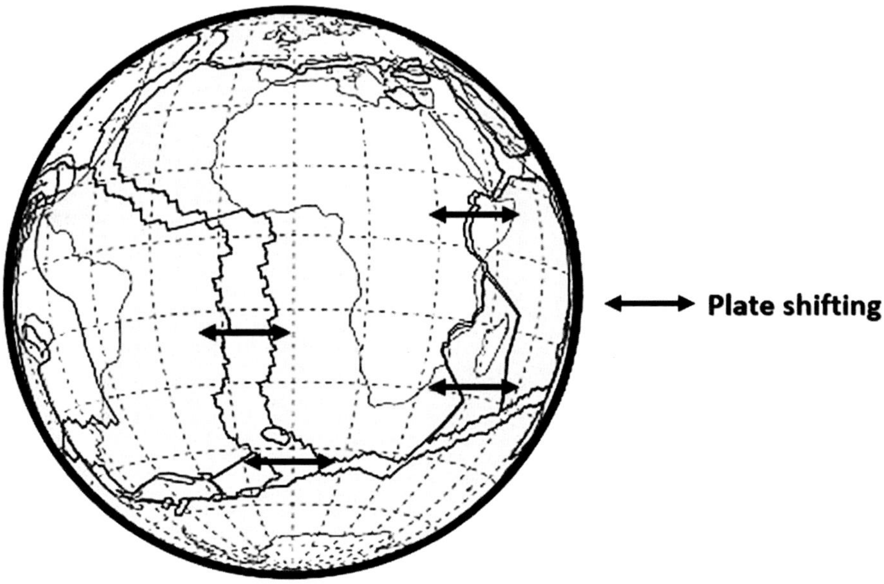

Figure 2.26 Modern African plate shifting or movement.

In other words, between the African (or Nubian) and Arabian plates realistically is the Red Sea Rift (Figure 2.27). From the Dead Sea to the Triple Junction at Afar, the trench extends the continuous length of the Red Sea. Numerous volcanoes, including Jabal al-Tair, are located within the Red Sea trench. This communicating trench progressively expands at a reasonable rate of 0.28 to 0.67 inches per year or about 7 to 17 mm/yr [44, 46, 48].

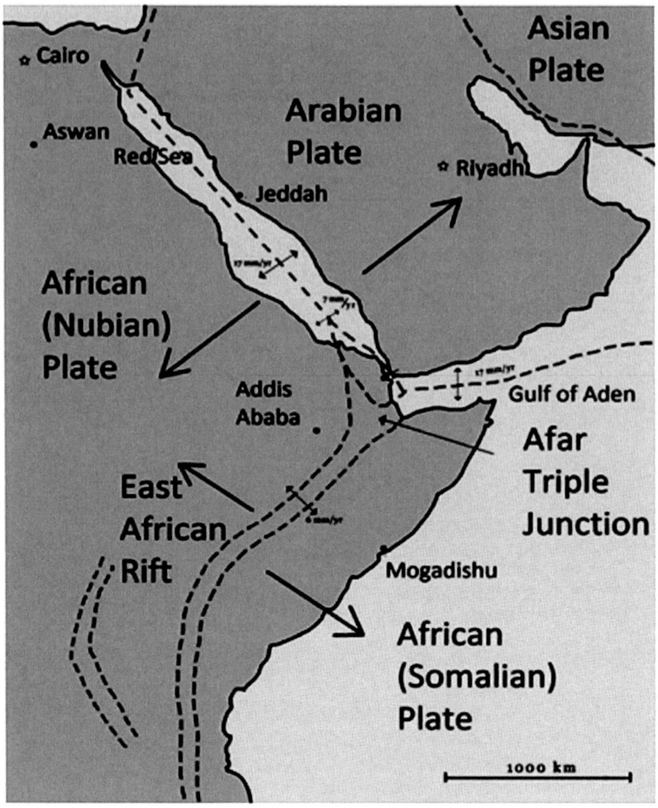

Figure 2.27 Red Sea Rift.

Therefore, the African (or Somali) and Arabian plates are typically separated by the Aden Ridge, a divergent plate boundary. The Owen Fracture Zone is precisely located where it naturally meets the Aden-Owen-Carlsberg triple junction of the African, Arabian, and Indo-Australian plates. It extends eastward from the triple junction. Near the Afar triple junction, the Aden Ridge is expanding tremendously at a reasonable rate of about 17 millimeters per year (0.67 inches/year) [40, 41, 45, 48, 50].

Therefore, the Afar Depression is a geological depression with an elevation of 3,280 to 390 feet or 1,000 to 120 meters. Numerous dome uplifts have occurred in the region, most notably the Afar Dome, which began 40 Mya. The massive crustal extension caused by this uplift resulted in trench and horst structures associated with conventional extensional faults. By 25 Mya, the uplift collapsed into the Afar Depression, which spread at a rate of 6 to 17 millimeters per year and covered more than 200,000 km^2 (77,000 mi^2) [44, 47, 49].

Needless to say, the movement of the three plates relative to each other is described as the ridge-ridge-ridge (RRR) type of Afar triple bifurcation. Relative to the surrounding plates, the Arabian, Somali, and Nubian plates all have divergent edges or ridges. According to the stability model proposed by Mackenzie and Morgan, the RRR geometry remains stable over time until the tectonic movement changes [47–50].

Compared to the other regions of the Afro-Arabian Rift System, the East African Rift System has a concentration of several volcanically active sites. Many of the projecting horsts have flood basalt layers that can be roughly dated using $^{40}Ar/^{39}Ar$-isotopic dating. Their age is estimated at 30 million years. The trap series was formed just before the onset of the main shift. The trap series surrounding the younger Neogene volcanics not only describes the crustal connection before rifting but also helps quantify the extent of crustal extension [44–49].

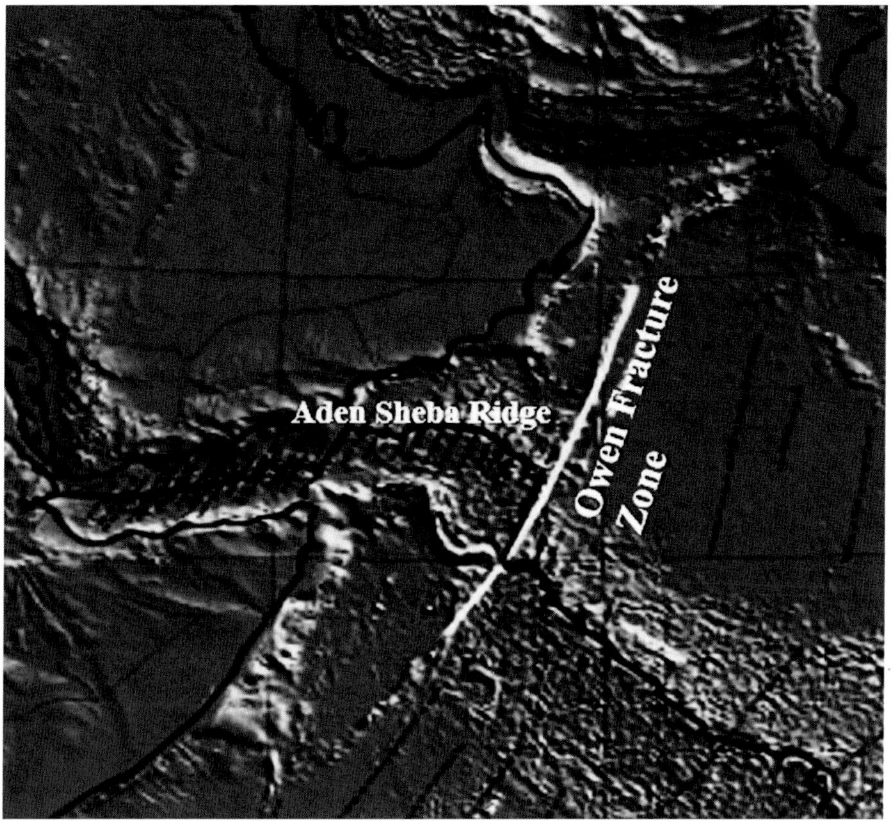

Figure 2.28 Aden ridge and Owen Fracture Zone.

According to the latest evidence, the Paleogene East African rift system is a typical complicated continental-continental rifting event. Both mantle uplift by several hot spots near EARS and far-field stress by the Arabian plate subducting beneath the Eurasian plate were responsible for its formation [40, 43, 47, 50].

Extensions, horst and graben structures, and even listric faults suggestive of a pre-oceanic basin have been produced as a result of this crustal uplift. The Nubian, Somali, and Arabian plates are anticipated to be separated ultimately by an ocean basin and a mid-ocean ridge if current tectonics remain unaltered. Though its future is undetermined, rifting is being researched by a variety of scientists using divergent speculative models [46–50].

2.12 IS A NEW OCEAN FORMING IN AFRICA?

Eastern Africa's entire Afar region is currently experiencing changes that could separate the continent and create a new ocean basin. Due to its resemblance to a mid-ocean ridge, the magmatism in Erta Ale might be the revealing evidence of this changeover. Concerning how close the Afar region is to this tectonic transition, there is disagreement. The petrology of the erupting lava could signify that we have not yet reached the stage of oceanic spreading, despite the geophysical properties of the magma storage at Erta Ale that could predict this hazard event.

In 5 million, 50 million, or 100 million years, the view from the top of Erta Ale will have significantly changed. The issue is whether the ongoing tectonic collisions in northern and eastern Africa will prevent the formation of a new ocean in the Erta Ale (Figure 2.29) and Afar region [51–53].

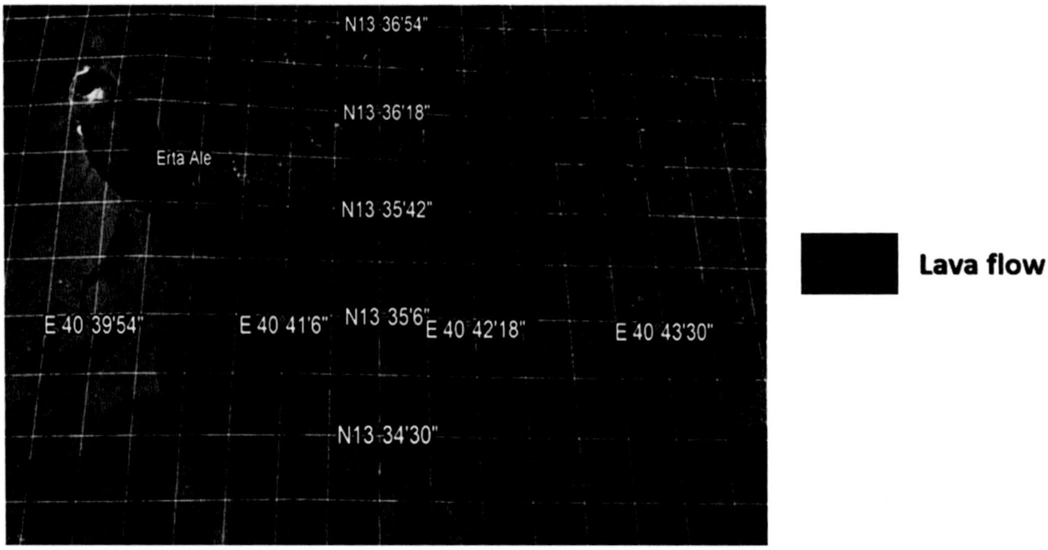

Figure 2.29 Location of Erta Ale.

The triple intersection of the Nubian, Somali, and Arabian tectonic plates is in the Afar region, as demonstrated in section 2.9. They all come together to form a giant Y on the Earth's surface around Djibouti and Eritrea. A southern opening extends more than 6,000 kilometers into the center of Africa, the Great Rift Valley. The Sinai Peninsula is where the Red Sea Rift meets the Northwest. The Aden Ridge (Figure 2.28), an oceanic spreading ridge, extends to the east. These boundaries are all expanding at a rate of up to 1.5 cm per year [51, 53].

In this sense, Africa is a continent ripping itself apart as a result of its complicated tectonic structure. The Somali plate, a portion of eastern Africa, has been separating from the rest of the continent for 30 million years. As a result, the Great Rift Valley was formed, which originates in Ethiopia and Eritrea, divides around the Kenya Dome, and then merges into the Malawi Rift. Looking at the string of lakes, including Tanganyika and Nyasa (Malawi), that line the bottom of the Great Rift Valley allows you to locate the rifting continent on a map. Some of the largest volcanoes on Earth were formed by the complicated tectonics of the Afar region. Because the eruptions of this volcanism are bimodal, they are at opposite ends of the compositional spectrum. Massive explosions of silica-rich lava coexist with massive basalt lava flows. The Great Rift Valley is lined by these volcanoes [52].

Massive basaltic shield volcanoes like Erta Ale in the Red Sea are to the north. Heading south, you'll come upon Ethiopia's enormous Corbetti Caldera, which might be one of world's most dangerous volcanoes. As you travel further into Africa, you'll come across oddities like the gigantic twin volcanoes Nyamuragira and Nyiragongo in the Democratic Republic of the Congo to the west, and the carbonatite-spewing Ol Doinyo Lengai in Tanzania to the east. These volcanoes all demonstrate rifting that is still active beneath eastern Africa [51, 53].

Why do rifting and spreading occur so frequently in this region? One theory is that the land was heated from below by the Afar plume. The African continent began to uplift and split as a result of this mantle plume, resulting in rift valleys. There seems to be a close connection between a huge mantle plume and the formation of new continental rifts and eventually two passive (continental) boundaries separated by seafloor spreading at ocean ridges, similar to how the North Atlantic opened 80 million years ago [52].

The significant question is: How can we predict when a continent breaks up and naturally forms a new ocean basin? The notable change from a continental rift to an oceanic spreading ridge may naturally take place in the active Afar region, conveniently located at the meeting point of tectonic plate boundaries. Attentively examining the specific area's crust raises one effective way to approach the subject of this remarkable transformation. The crust under Afar may be bare than 20 km thick, as opposed to the usual continental crust thickness of 35 km. This naturally contains the distinct layers of basalt from volcanic eruptions that have buried the distinct region for countless millions of years [52, 53].

The study of volcanoes in the region offers another opportunity to address the question of the possible change of the Afar from a continental rift valley to an oceanic spread. One of the most active volcanoes on earth is located in northwestern Ethiopia; its name is Erta Ale. The enormous shield volcano rises 600 m from the Danakil Depression, which is altogether 50 km long and about 100 km wide. Does this imply that Erta Ale is in the process of evolving into a mid-ocean ridge from a rift volcano?

The distinct Afar region is currently experiencing extensive continental rifting, but it is unclear if this rifting remains a primary cause of oceanic spreading. As a result of the Afar plume, melting beneath the Afar region instantly appears to arise from potential sources other than what one might reasonably anticipate at a freshly formed mid-ocean ridge. In this sense, across the Afar district, the crust is thin, but it is unclear if the lithospheric mantle beneath the crust has also shrunk. Additionally, the distinct Erta Ale region's rifting is moving northward significantly than southward as would be predicted if the Red Sea Rift typically constituted the possible extension's driving force. The active Afar region may currently be the closest known example of an alternative oceanic spreading center, despite everything [52–54].

The author; therefore, thinks that enormous changes would occur in Afar, similar to what has happened in the last five years when the continent has been cracking up under our feet. People are accustomed to understanding changes on earth on timelines of millions of years. However, in just ten days in 2005, a 60-kilometre-long section of the Earth expanded to an 8-meter breadth. In this way, the split is being produced by hot, molten rock that is slowly making its way to the surface from deep below the Earth. Because of the continuous subsurface eruptions, the horn of Africa could eventually slide away and a new ocean could arise [43, 51, 54].

As a result, the region has areas that are below sea level and is only separated from the ocean by a 20-meter-long section of land in Eritrea. Consistent with the author's understanding, this will ultimately drift apart, the sea would flood in, and a new ocean would begin to form. In this circumstance, a smaller Africa would form, along with a large island that drifts into the Indian Ocean as the land begins to split apart and sink further. Eventually, parts of southern Ethiopia and Somalia would drift off and create a new island [51–53].

What role would GERD serve in triggering this incident in this catastrophic scenario? The next question is also how GERD can remain stable in the context of the complicated geological structure.

In last chapter of this book, these questions would be answered by looking at how GERD interacts with different geological structures, the degree to which it is stable over time, and how it can contribute to seismic events. To better understand the role GERD plays in triggering seismic events, this book dives into how it interacts with different geological structures and how its stability varies across different contexts.

2.13 CONCLUSION

This chapter provides an in-depth examination of the Grand Ethiopian Renaissance Dam (GERD), focusing on its geographic setting, construction plans, and the resulting benefits and challenges, particularly for downstream nations like Sudan and Egypt. Situated within the tectonically active East African Rift System—home to active volcanoes like Erta Ale—GERD's geological environment is complex and presents unique stability considerations. Notably, the tectonic growth in this region may eventually lead to the formation of a new ocean beneath Africa.

The chapter also addresses the scientific debate surrounding GERD, specifically the criticism within the research community regarding Heggy's study. Concerns over methodological rigor and discourse style emphasize the importance of scientific integrity and respectful dialogue in advancing understanding of GERD's impacts. Such rigorous, constructive critique is essential in a topic as impactful as GERD, where scientific evaluations play a critical role in shaping discourse and policy.

Further, the chapter underscores GERD's significance within the broader Nile Basin dispute, noting its potential to alter water dynamics between Ethiopia, Egypt, and Sudan. Given the historical context of the 1959 Water Sharing Agreement, which heavily favors Egypt, GERD's operation could prompt resource challenges for downstream nations. By detailing the dam's potential benefits, like increased power generation and improved water management, alongside risks of unilateral filling practices, the chapter highlights the need for sustainable management and cooperative water strategies.

In upcoming chapters, advanced remote sensing technology will be explored to assess how the geological structure influences GERD's stability and regional impacts.

References

[1] Allan, J.A. 1999. The Nile basin: Evolving approaches to Nile waters management. Occasional Paper. 20: 1–1.

[2] Di Baldassarre, G., Elshamy, M., van Griensven, A., Soliman, E., Kigobe, M., Ndomba, P., et al. 2011. Future hydrology and climate in the River Nile basin: a review. Hydrological Sciences Journal. 56(2): 199–211.

[3] Yilma, A.D. and Awulachew, S.B. 2009. Characterization and Atlas of the Blue Nile Basin and Its Sub Basins. Intermediate Results Dissemination Workshop, Addis Ababa, 5-6 February 2009, 236 p.

[4] Abtew, W. and Melesse, A.M. 2014. The Nile River Basin. *In*: Nile River Basin. 7–21. Springer, Cham.

[5] Whittington, D., Wu, X. and Sadoff, C. 2005. Water resources management in the Nile basin: The economic value of cooperation. Water policy. 7(3): 227–252.

[6] Cascão, AE. 2008. Ethiopia–challenges to Egyptian hegemony in the Nile Basin. Water Policy. 10(S2): 13–28.

[7] Hamouda, M.A., Nour El-Din, M. and Moursy, F.I. 2009. Vulnerability assessment of water resources systems in the Eastern Nile Basin. Water Resources Management. 23(13): 2697–2725.

[8] Abtew, W. and Dessu, S.B. 2019. The Grand Ethiopian Renaissance Dam on the Blue Nile. Springer International Publishing.

[9] Kasimbazi, E. and Bamwine, F. 2021. Resolving the grand Ethiopian renaissance dam conflict through the African Union Nexus approach. pp. 61–78. *In*: Melesse, A.M., Abtew, W., Moges, S.A. (eds) Nile and Grand Ethiopian Renaissance Dam. Springer Geography. Springer, Cham.

[10] Pemunta, N.V., Ngo, N.V., Fani Djomo, C.R., Mutola, S., Seember, J.A., Mbong, G.A., et al. 2021. The Grand Ethiopian Renaissance Dam, Egyptian National Security, and human and food security in the Nile River Basin. Cogent Social Sciences. 7(1): 1875598.

[11] Tawfik, R. 2016. The grand Ethiopian renaissance dam: a benefit-sharing project in the Eastern Nile? Water International. 41(4): 574–592.

[12] Basheer, M., Nechifor, V., Calzadilla, A., Siddig, K., Etichia, M., Whittington, D., et al. 2021. Collaborative management of the Grand Ethiopian Renaissance Dam increases economic benefits and resilience. Nature Communications. 12(1): 1–2.

[13] Basheer, M. 2021. Cooperative operation of the Grand Ethiopian Renaissance Dam reduces Nile riverine floods. River Research and Applications. 37(6): 805–814.

[14] Hailu, A.D. 2022. Ethiopia hydropower development and Nile basin hydro politics. AIMS Energy. 10(1): 87–101.

[15] Mbaku, J.M. 2022. The Grand Ethiopian Renaissance Dam and the Nile Basin Conflict. Georgetown Journal of International Affairs. 23(1): 84–91.

[16] Turhan, Y. 2021. The hydro-political dilemma in Africa water geopolitics: The case of the Nile river basin. African Security Review. 30(1): 66–85.

[17] von Meding, R.E. 2022. The Grand Ethiopian Renaissance Dam: A large-scale energy project in violation of international law? LSU Journal of Energy Law and Resources. 10: 33.

[18] Pietrangeli, G., Bezzi, A., Rossini, C., Masciotta, A. and D'Alberti, D. 2017. Design of grand ethiopian renaissance RCC main dam (H = 175 m). Hydropower and Dams.

[19] Abtew, W. and Dessu, S.B. 2019. Grand Ethiopian Renaissance Dam analysis. *In*: The Grand Ethiopian Renaissance Dam on the Blue Nile. 79–96. Springer, Cham.

[20] Deere, Don U. 1982. Cement-bentonite grouting for dams. *In*: Mine Induced Subsidence: Effects on Engineered Structures. 279–300. ASCE.

[21] Kutzner, C. 2020. Grouting of Rock and Soil. CRC Press.

[22] Ewert, F.-K. 2012. Rock Grouting: With Emphasis on Dam Sites. Springer Science & Business Media.

[23] Weaver, Kenneth D., and Donald A. Bruce. 2007. Dam Foundation Grouting. Reston: ASCE Press.

[24] Li, S., Sha, F., Liu, R., Li, W., Li, Z. and Wang, G. 2017. Properties of cement-based grouts with high amounts of ground granulated blast-furnace slag and fly ash. Journal of Materials in Civil Engineering. 29(11): 04017219.

[25] Bezzi, A.M.A., Fontana, C., Fiorani, A., Masciotta, S., Pietrangeli, S. and Visconti, V.E. 2017. Evaluation of dam performance under seismic loads with linear time history analysis, case study Grand Ethiopian Renaissance RCC main dam.

[26] Rossini, C. and Schrader, E. 2013. Gibe III Dam: Project Summary, Mixes, Properties, Thermal Issues and Cores. Water Storage and Hydropower Development for Africa.

[27] Heggy, E., Sharkawy, Z. and Abotalib, A.Z. 2021. Egypt's water budget deficit and suggested mitigation policies for the Grand Ethiopian Renaissance Dam filling scenarios. Environmental Research Letters. 16(7): 074022.

[28] Wheeler, K., Jeuland, M., Strzepek, K., Hall, J., Zagona, E., Abdo, G., et al. 2022. Comment on 'Egypt's water budget deficit and suggested mitigation policies for the Grand Ethiopian Renaissance Dam filling scenarios'. Environmental Research Letters. 17(8): 088003.

[29] Eladawy, A., Asefa, T. and El Nour, S. 2022. Comment on 'Egypt's water budget deficit and suggested mitigation policies for the Grand Ethiopian Renaissance Dam filling scenarios'. Environmental Research Letters. 17(8): 088001.

[30] Wheeler, K.G., Jeuland, M., Hall, J.W., Zagona, E. and Whittington, D. 2020. Understanding and managing new risks on the Nile with the Grand Ethiopian Renaissance Dam. Nature Communications. 11(1): 1–9.

[31] Eldardiry, H. and Hossain, F. 2021. A blueprint for adapting high Aswan dam operation in Egypt to challenges of filling and operation of the Grand Ethiopian Renaissance dam. Journal of Hydrology. 598: 125708.

[32] Kamara, A., Ahmed, M. and Benavides, A. 2022. Environmental and Economic Impacts of the Grand Ethiopian Renaissance Dam in Africa. Water. 14(3): 312.

[33] Mulat, A.G. and Moges, S.A. 2014. Assessment of the impact of the Grand Ethiopian Renaissance Dam on the performance of the High Aswan Dam. Journal of Water Resource and Protection. 6(6): 583–598.

[34] Taye, M.T., Tadesse, T., Senay, G.B. and Block, P. 2016. The grand ethiopian renaissance dam: Source of cooperation or contention? Journal of Water Resources Planning and Management. 142: 02516001.

[35] Liersch, S., Koch, H. and Hattermann F.F. 2017. Management scenarios of the Grand Ethiopian Renaissance Dam and their impacts under recent and future climates. Water. 9(10): 728.

[36] Liersch, S., Tecklenburg, J., Rust, H., Dobler, A., Fischer, M., Kruschke, T., et al. 2018. Are we using the right fuel to drive hydrological models? A climate impact study in the Upper Blue Nile. Hydrology and Earth System Sciences. 22(4): 2163–2185.

[37] Ibrahim, A.I.R. 2017. Impact of Ethiopian Renaissance Dam and population on future Egypt water needs. American Journal of Engineering Research (AJER). 6: 160–171.

[38] Abdelhaleem, F.S. and Helal, E.Y. 2015. Impacts of Grand Ethiopian Renaissance Dam on different water usages in Upper Egypt. British Journal of Applied Science and Technology. 8(5): 461–483.

[39] Hamed, K.H. 2018. Stochastic investigation of the GERD-AHD interaction through first impoundment and beyond. Grand Ethiopian Renaissance Dam Versus Aswan High Dam. 95–117.

[40] Ramadan, S.M., Negm, A.M., Smanny, M. and Helmy, A.H. 2015. Quantifying the impacts of impounding grand Ethiopian renaissance dam reservoir on Nasser lake active storage. *In*: Proceedings of Eighteenth International Water Technology Conference (IWTC 18). Sharm ElSheikh. 12–14.

[41] Armanuos, A.M., Ibrahim, M.G., Mahmod, W.E., Negm, A., Yoshimura, C., Takemura, J., et al. 2017. Evaluation of the potential impact of Grand Ethiopian Renaissance Dam and pumping scenarios on groundwater level in the Nile Delta aquifer. Water Science and Technology: Water Supply. 17(5): 1356–1367.

[42] El-Rawy, M., Moghazy, H.E. and Eltarabily, M.G. 2021. Impacts of decreasing Nile flow on the Nile Valley Aquifer in EL-Minia Governorate, Egypt. Alexandria Engineering Journal. 60(2): 2179–2192.

[43] Abbate, E., Bruni, P. and Sagri, M. 2015. Geology of Ethiopia: A review and geomorphological perspectives. Landscapes and landforms of Ethiopia. 33–64.

[44] Varet, J. 2017. Geology of Afar (East Africa). Springer.

[45] Beyene, A. and Abdelsalam, M.G. 2005. Tectonics of the Afar Depression: A review and synthesis. Journal of African Earth Sciences. 41(1–2): 41–59.

[46] Chorowicz, J. 2005. The East African Rift system. Journal of African Earth Sciences. 43(1–3): 379–410.

[47] Ebinger, C.J. 1989. Tectonic development of the western branch of the East African Rift system. Geological Society of America Bulletin. 101(7): 885–903.

[48] Roberts, E.M., Stevens, N.J., O'Connor, P.M., Dirks, P.H., Gottfried, M.D., Clyde, W.C., et al. 2012. Initiation of the western branch of the East African Rift coeval with the eastern branch. Nature Geoscience. 5(4): 289–294.

[49] Mougenot, D., Recq, M., Virlogeux, P. and Lepvrier, C. 1986. Seaward extension of the East African rift. Nature. 321(6070): 599–603.

[50] Barnie, T.D., Oppenheimer, C. and Pagli, C. 2016. Does the lava lake of Erta 'Ale volcano respond to regional magmatic and tectonic events? An investigation using Earth Observation data. Geological Society, London, Special Publications. 420(1): 181–208.

[51] Klemetti, E. 2020. Are we seeing a new ocean starting to form in Africa? Eos, 2020 May 101, https://doi.org/10.1029/2020EO143740. Published on 08 May 2020.

[52] Bojanowski, A. 2020. Africa's New Ocean—A Continent Splits Apart. Online Report on the Website. Available online: http://www.spiegel.de/international/spiegel/africa-s-new-ocean-a-continent-splitsapart-a-405947.html (accessed on 20 December 2020).

[53] Dinka, M.O. 2017. Delineating the drainage structure and sources of groundwater flux for Lake Basaka, Central Rift Valley Region of Ethiopia. Water. 9(12): 797.

[54] Awulachew, M.T. 2020. Overview to potential sources, soil salinization and expansion level of Lake Basaka, Central Rift Valley Region of Ethiopia. Journal of Water Resources and Ocean Science. 9(4): 71–76.

Chapter 3

Utilizing Quantized Maxwell's Equations for Remote Sensing and Imaging of the Grand Ethiopian Renaissance Dam (GERD)

3.0 INTRODUCTION

This chapter focuses on understanding remote sensing mechanisms in imaging dam constructions, with a specific emphasis on the Grand Ethiopian Renaissance Dam (GERD). Most Egyptian research utilizing remote sensing for observing and monitoring the GERD has offered only a superficial understanding of the technology. These studies have been limited to basic techniques such as implementing vegetation index, simple equations or normalized water index and normalized different water index (MNDWI) using straightforward band ratio methods. Moreover, they have primarily relied on low-resolution satellite images, especially when employing interferometry synthetic aperture radar (InSAR) techniques. This chapter aims to establish a comprehensive concept of how advanced remote sensing technology can be utilized for imaging, monitoring, and forecasting the safety of the Grand Ethiopian Renaissance Dam (GERD).

3.1 ADVANCED REMOTE SENSING TECHNOLOGY

Quantum remote sensing is the advanced technology of remote sensing. What is quantum remote sensing and who is the pioneering scientist in this field? Quantum remote sensing involves the application of quantum mechanics principles to enhance the capabilities of remote sensing technologies. It leverages quantum phenomena such as superposition, entanglement, and quantum coherence to improve the accuracy, resolution, and sensitivity of remote sensing instruments. This approach can potentially revolutionize the field by enabling the detection and analysis of information that classical remote sensing methods cannot achieve.

The term 'quantum of remote sensing' refers to the fundamental quantum units or processes that underpin remote sensing techniques. These can include quantum bits (qubits) in quantum computing, quantum dots in imaging and the use of quantum entanglement for superior measurement precision and data acquisition.

DOI: 10.1201/9781003327325-3

Prof. Maged Marghany has been a pioneering figure in the field of quantum remote sensing. His work integrates advanced remote sensing technologies with quantum mechanics principles to develop innovative methodologies for environmental monitoring and disaster management. Marghany's contributions include the application of quantum image processing and quantum computing algorithms to enhance the interpretation of remote sensing data, thereby providing more accurate and detailed insights into various geospatial phenomena. His research has pushed the boundaries of traditional remote sensing, demonstrating the potential of quantum technologies to address complex challenges in environmental science and beyond [1].

Quantum remote sensing represents the next frontier in remote sensing technology, incorporating principles of quantum mechanics to enhance data acquisition, processing and interpretation. By leveraging quantum phenomena such as superposition, entanglement, and quantum coherence, quantum remote sensing aims to achieve unprecedented levels of accuracy, resolution and sensitivity that surpass the capabilities of classical remote sensing methods.

In quantum remote sensing, quantum bits (qubits) (Figure 3.1) are used to encode and process information, allowing for more complex and detailed analysis of remote sensing data. Quantum entanglement can improve the synchronization of sensors, leading to better spatial resolution and reduced noise [2]. Quantum coherence and quantum dots enhance imaging techniques, providing clearer and more precise observations of the Earth's surface and atmosphere.

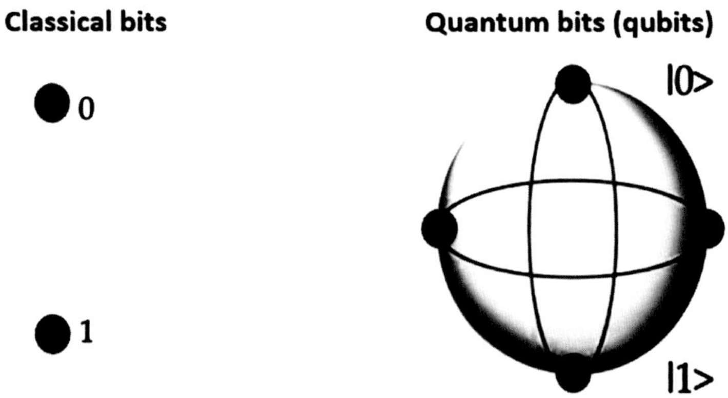

Figure 3.1 Example of quantum bits (qubits).

Prof. Maged Marghany's innovative research integrates quantum mechanics with remote sensing technologies, pushing the boundaries of what can be achieved in environmental monitoring and disaster management. Marghany's work includes the development of quantum image processing techniques and the application of quantum computing algorithms to improve the interpretation of remote sensing data. His contributions have significantly advanced the understanding and capabilities of remote sensing, demonstrating the transformative potential of quantum technologies in this field.

3.2 A NOVEL DEFINITION OF REMOTE SENSING

In recent advancements, Marghany proposed a novel definition of remote sensing, first published in 2019 and further supported by strong evidence in 2022. According to Marghany, remote sensing is defined as the quantum information compiled from any object, whether terrestrial or extraterrestrial, resulting from photon reflection or backscatter. This is directly influenced by the physical characteristics of the component atoms of the object.

Marghany's hypothesis posits that different elemental atoms interact uniquely with photons, absorbing and emitting them based on the photons' wavelengths and frequencies. This interaction forms the spectral signature, which serves as a photonic interaction signature reflecting the characteristics of elemental atoms. This interaction underpins quantum mechanics, a powerful tool for studying the composition and properties of matter. Quantum mechanics has broad applications across various scientific fields, including astronomy, chemistry, and physics, and has led to the development of transformative technologies such as lasers, transistors and MRI machines.

This innovative definition underscores the critical role of quantum mechanics in advancing remote sensing. By incorporating quantum principles, remote sensing can be revolutionized, offering new methodologies and applications in fields such as the monitoring of the Grand Ethiopian Renaissance Dam (GERD), environmental monitoring, disaster relief, and urban planning. Marghany's approach highlights the importance of integrating quantum mechanics into remote sensing studies, paving the way for significant technological and scientific advancements.

3.3 THE DIFFERENCE BETWEEN MARGHANY'S REMOTE SENSING DEFINITION AND THE CLASSICAL ONE

Classical remote sensing primarily involves the acquisition of information about an object or phenomenon without making physical contact with the object. This is typically achieved through the detection and measurement of radiation reflected or emitted from the surface of the object. It relies, therefore, on electromagnetic radiation, such as visible light, infrared, and radio waves (Figure 3.2), to capture images and data using sensors mounted on platforms like satellites and aircraft. On the contrary, Marghany's definition introduces a quantum mechanics perspective to remote sensing. It focuses on the quantum information derived from the interaction of photons with the component atoms of an object. According to Marghany, different elemental atoms interact with photons uniquely, absorbing and emitting them based on specific wavelengths and frequencies. This quantum interaction forms the basis of the spectral signature [1].

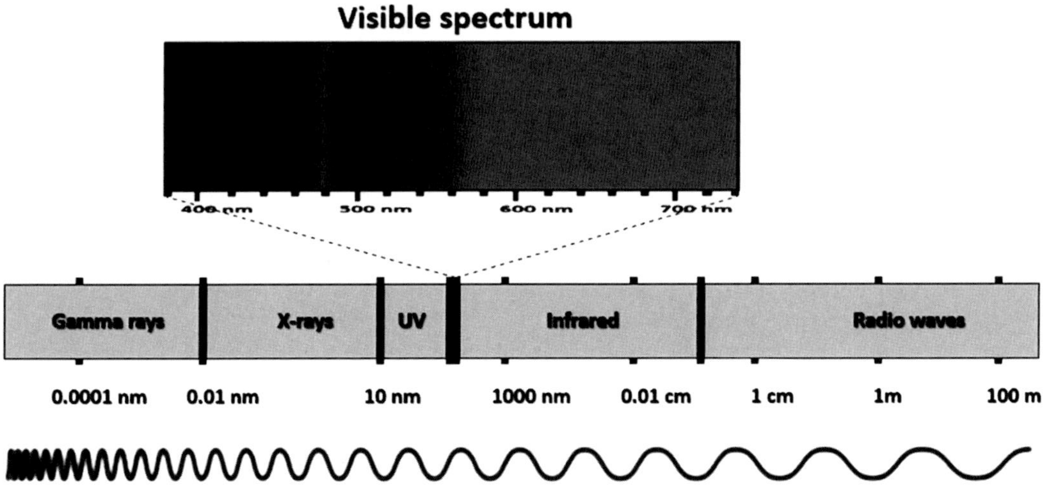

Figure 3.2 Electromagnetic wave spectra.

This approach emphasizes the need for advanced technologies and methodologies that can capture and interpret quantum-level information. It suggests a potential shift in remote sensing techniques towards more precise and detailed data acquisition and analysis. In this manner, the classical definition of remote sensing and Marghany's novel definition present distinct conceptual frameworks and methodologies, reflecting their underlying principles and technological implications.

Needless to say, while classical remote sensing remains a vital tool with broad applications, Marghany's novel definition introduces a transformative approach by integrating quantum mechanics. This new perspective not only enhances the detail and precision of data but also necessitates advanced technologies, promising significant advancements in remote sensing applications. Understanding these differences and leveraging the strengths of both approaches can lead to more comprehensive and accurate monitoring and analysis in various fields [1].

3.4 MECHANISM OF QUANTUM REMOTE SENSING IMAGING

Quantum remote sensing leverages the fundamental principles of quantum mechanics to describe the interactions between photons and atoms (Figure 3.3). These interactions can be mathematically represented using quantum mechanical equations. The energy E of a photon is related to its frequency ν and wavelength λ by the following equations:

$$E = h\nu = h\frac{c}{\lambda} \tag{3.1}$$

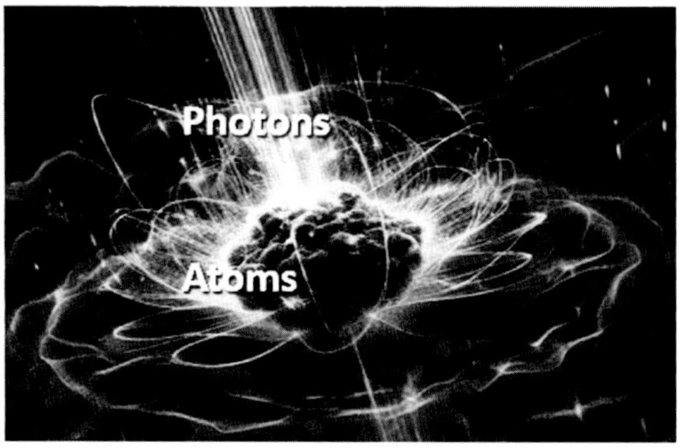

Figure 3.3 Photons interact with atoms.

where h is Planck's constant (6.626×10^{-34} Js), c is the speed of light in vacuum (3×10^8 m). Therefore, atoms have discrete energy levels and the transition between these levels involves the absorption or emission of photons (Figure 3.4). The energy difference ΔE between two energy levels E_i and E_f (Figure 3.5) in an atom is given by [1, 3]:

$$\Delta E = E_f - E_i = h\nu \tag{3.2}$$

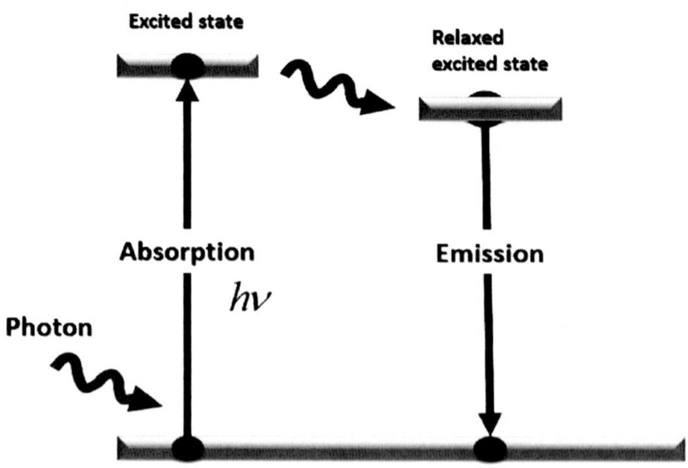

Figure 3.4 Photon absorption or emission.

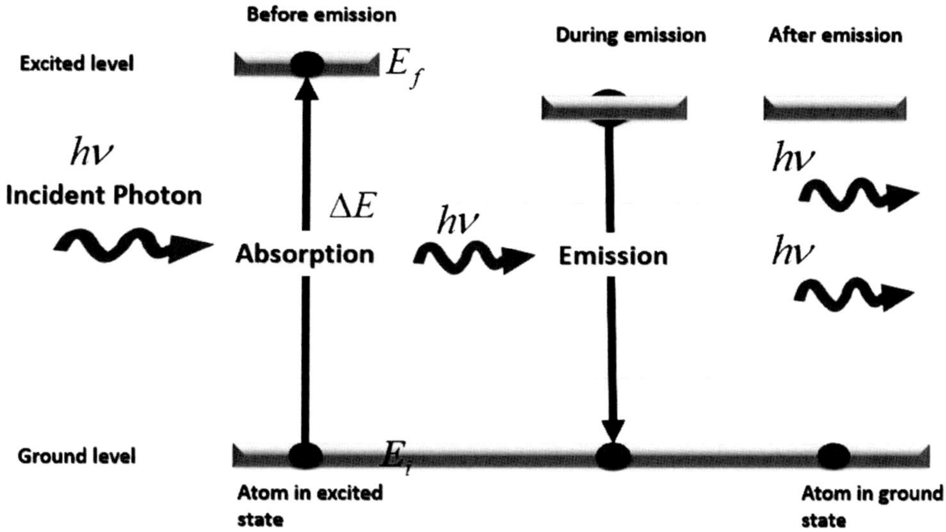

Figure 3.5 Energy difference due to emission.

The probability of a photon being absorbed or emitted by an atom can be described by the Einstein coefficients. For spontaneous emission (Figure 3.6), the probability per unit time A_{21} from a higher energy state:

$$A_{21} = 8\pi h\nu[c^{-3}][\mu_{21}]^2 \tag{3.3}$$

where μ_{21} is the dipole moment matrix element between states $|2\rangle$ and $|1\rangle$. Consequently, the absorption cross-section (ν) quantifies the likelihood of photon absorption by an atom:

$$\sigma(\nu) = \frac{\pi e^2}{m_e c} f_{12}\Phi(\nu) \tag{3.4}$$

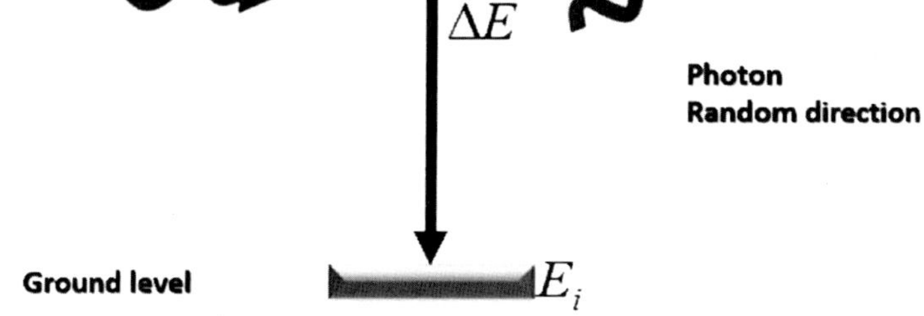

Figure 3.6 Spontaneous emission.

In equation 3.4, e is the electron charge (1.602×10^{-19}); m_e is the electron mass $(9.109 \times 10^{-31}\text{kg})$; f_{12} is the oscillator strength of the transition; and $\Phi(\nu)$ is the line profile function, which describes the distribution of absorption over frequencies [1, 4]. The rate equations; therefore, describe the population of atoms in different energy states. For an atom with two energy levels, the population N_1 in the lower state and N_2 in the upper state can be described as:

$$\frac{dN_1}{dt} = B_{12}\rho(\nu)N_1 + A_{21}N_2 + B_{21}\rho(\nu)N_2 \tag{3.5}$$

$$\frac{dN_2}{dt} = B_{12}\rho(\nu)N_1 - \left(A_{21}N_2 + B_{21}\rho(\nu)\right)N_2 \tag{3.6}$$

where B_{12} and B_{21} are the Einstein coefficients for absorption and stimulated emission, respectively. Therefore, (ν) is the spectral energy density of the incident radiation. Consequently, the spectral signature is formed by the collective absorption and emission spectra of the atoms. The intensity (ν) of the emitted or absorbed light at frequency ν can be expressed as:

$$I(\nu) = I_0 e^{-\sigma(\nu)NL} \tag{3.7}$$

here I_0 is the initial intensity, N is the number density of atoms, L is the path length through which the light travels. These equations illustrate the quantum mechanical framework that governs the interactions between photons and atoms. By comprehending and utilizing these principles, quantum remote sensing can achieve high-resolution, accurate imaging and analysis of various objects, enhancing the capabilities of conventional remote sensing techniques [3, 5, 7, 10].

3.5 QUANTIZED MAXWELL EQUATIONS

Maxwell's equations describe how electric and magnetic fields propagate and interact with matter:

$$\nabla \cdot \vec{E} = \rho\left[\varepsilon_0\right]^{-1}, \tag{3.8}$$

$$\nabla \cdot \vec{B} = 0, \tag{3.9}$$

$$\nabla \times \vec{E} = -\frac{\partial \vec{B}}{\partial t}, \tag{3.10}$$

$$\nabla \cdot \vec{B} = \mu_0 \vec{J} + \mu_0 \varepsilon_0 \frac{\partial \vec{E}}{\partial t}, \tag{3.11}$$

where $\vec{E}$ is the electric field, $\vec{B}$ is the magnetic field, ρ is the charge density, and $\vec{J}$ is the current density. In this view, equation 3.8 is the Gauss's law for electricity. This equation states that the divergence of the electric field $\vec{E}$ is equal to the charge density ρ divided by the permittivity of free space ε_0. It means that electric charges produce electric fields. The electric field lines diverge from positive charges and converge on negative charges [1, 3, 5]. The total flux of the electric field out of a closed surface is proportional to the charge enclosed by the surface $d\vec{A}$ (Figure 3.7).

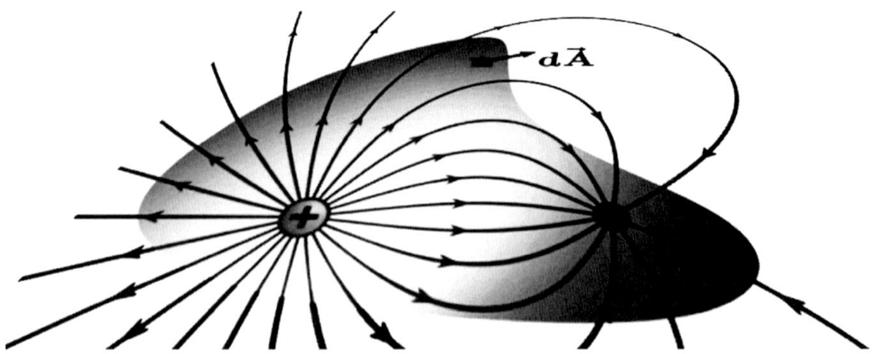

Figure 3.7 The Gauss's law for electricity.

Equation 3.9 presents Gauss's law for magnetism in which the divergence of the magnetic field $\vec{B}$ is zero. Consequently, Gauss's law for magnetism demonstrates that there are no 'magnetic charges' (also known as magnetic monopoles). In this perspective, magnetic field lines are continuous loops, and they do not begin or end at any point. The net magnetic flux through a closed surface is always zero (Figure 3.8).

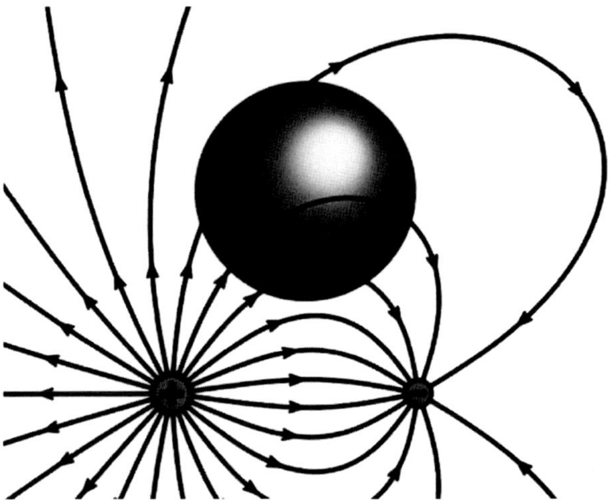

Figure 3.8 Zero electric charge through its surface.

Subsequently, Faraday's law of induction (equation 3.10) states that the curl of the electric field $\vec{E}$ is equal to the negative rate of change of the magnetic field $\vec{B}$ concerning time. A time-varying magnetic field induces an electric field [1, 7]. This is the principle behind electromagnetic induction, where changing magnetic fields can generate electric currents in conductors (Figure 3.9).

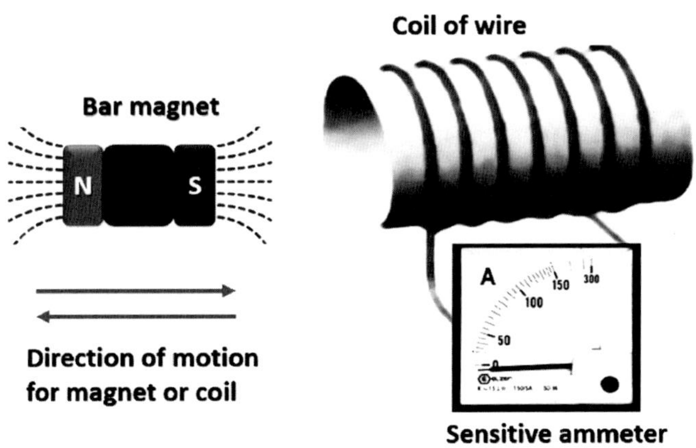

Figure 3.9 Faraday's law of induction.

These equations can be combined to derive the wave equations for the electric and magnetic fields. Therefore, equation 3.11 is Ampère's law (Figure 3.10). In this case, μ_0 is the permeability of free space (or the vacuum permeability), a constant that measures the ability of a vacuum to support the formation of a magnetic field. Its value is approximately $4\pi \times 10^{-7}$ H/m (Henries per meter). Additionally, ε_0 is the permittivity of free space (or the vacuum permittivity), a constant that measures the ability of a vacuum to permit electric field lines. Its value is approximately 8.854×10^{-12} F/m (Farads per meter).

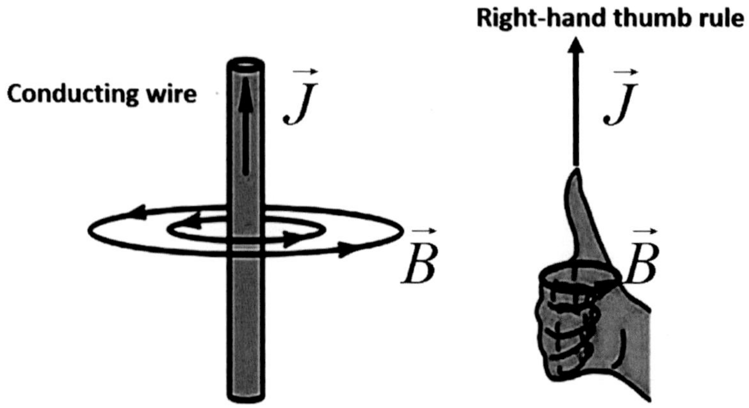

Figure 3.10 Ampère's law.

The complete equation 3.11 states that the curl of the magnetic field ($\nabla \cdot \vec{B}$) at any point in space is influenced both by the presence of electric currents ($\vec{J}$) and by the rate of change of the electric field $\left(\frac{\partial \vec{E}}{\partial t}\right)$. Thus, this equation is crucial for describing how electric and magnetic fields propagate and interact, particularly in the context of electromagnetic waves. Needless to say, Ampère's law with Maxwell's addition provides a complete description of how magnetic fields are generated by both electric currents and time-varying electric fields, forming a fundamental part of the framework for understanding classical electromagnetism [7–9].

The quantum electromagnetic field refers to a transition from classical electromagnetism to quantum mechanics. Both the fields $\vec{E}$ and $\vec{B}$ are considered operators that obey the theory of quantum mechanics. The quantization process begins by expressing the electromagnetic field in terms of vector potentials $\vec{A}$ and scalar potential ϕ, subject to the gauge conditions. This can be expressed mathematically as:

$$\vec{E} = -\nabla\phi - \frac{\partial \vec{A}}{\partial t} \tag{3.12}$$

$$\vec{B} = \nabla \cdot \vec{A} \tag{3.13}$$

To remove the redundancy in the potentials (since different potentials can give the same fields), we impose gauge conditions. A common choice is the Coulomb gauge (also known as the radiation gauge), where:

$$\nabla \cdot \vec{A} = 0 \tag{3.14}$$

$$\phi = 0 \tag{3.15}$$

Therefore, in quantum mechanics, the vector potential $\vec{A}$ and the scalar potential ϕ are promoted to operators. This means that they are no longer classical fields but quantum operators that act on the quantum state of the system [1, 7, 9]. Consequently, the operators $\vec{A}$ and the corresponding canonical momentum Π (related to the electric field) must satisfy the canonical commutation relations. For the electromagnetic field, the canonical momentum is given by:

$$\vec{\Pi} = -\varepsilon_0 \vec{E} \tag{3.16}$$

Thus, the commutation relations between the field operators are:

$$[\vec{A}_i(\vec{r}, t), \Pi j(\vec{r}\,', t)] = i\hbar\delta_{ij}\delta(\vec{r} - \vec{r}\,') \tag{3.17}$$

here i and j, denote the spatial components, while $\hbar$ is the reduced Planck constant, and δ is the Dirac delta function. In quantum electrodynamics (QED) (Figure 3.11), the electromagnetic field is quantized [1, 7, 10]. The electric and magnetic fields are expressed as operators in terms of creation and annihilation (Figure 3.12) operators [9, 11].

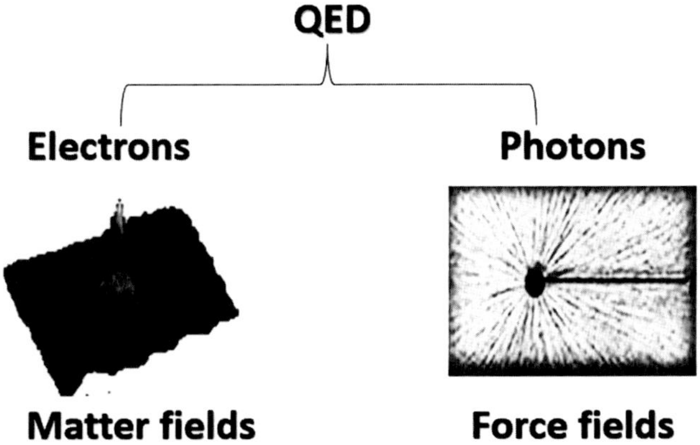

Figure 3.11 Concept of quantum electrodynamics (QED).

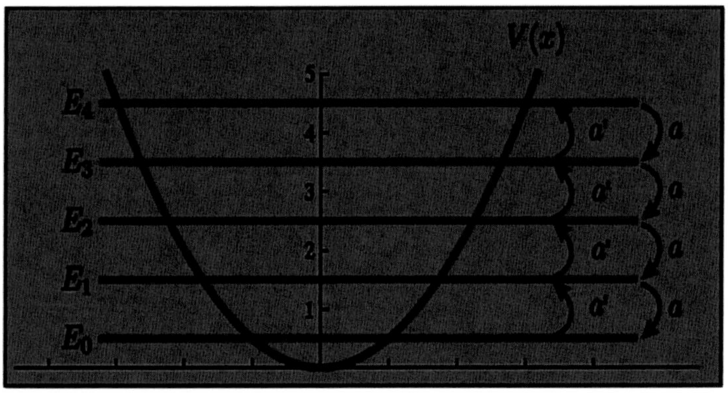

Figure 3.12 Generation of annihilation operators.

$$\vec{E}(\vec{r},t) = i\sum_{\vec{k},\lambda}\sqrt{\left[\frac{\hbar\omega_k}{2\varepsilon_0 V}\right]}\left(e_{\vec{k},\lambda}a_{\vec{k},\lambda}e^{i(\vec{k}\cdot\vec{r}-\omega_{\vec{k}}t)} - e^{*}_{\vec{k},\lambda}a^{\dagger}_{\vec{k},\lambda}e^{e^{i(\vec{k}\cdot\vec{r}-\omega_{\vec{k}}t)}}\right) \tag{3.18}$$

$$\vec{B}(\vec{r},t) = i\sum_{\vec{k},\lambda}\sqrt{\left[\frac{\hbar\omega_k}{2\varepsilon_0 V}\right]}\left(\left(\vec{k}\times e_{\vec{k},\lambda}\right)\left(a_{\vec{k},\lambda}e^{i(\vec{k}\cdot\vec{r}-\omega_{\vec{k}}t)} + a^{\dagger}_{\vec{k},\lambda}e^{e^{i(\vec{k}\cdot\vec{r}-\omega_{\vec{k}}t)}}\right)\right) \tag{3.19}$$

where $e_{\vec{k},\lambda}$ are the polarization vectors, $\vec{k}$ is the wave vector, λ is the polarization state, $a_{\vec{k},\lambda}$ and $a^{\dagger}_{\vec{k},\lambda}$ are the annihilation and creation operators respectively, V is the volume of quantization, and $\omega_{\vec{k}}$ is the angular frequency of the mode.

According to the above perspective, the quanta of the electromagnetic field are photons. In the quantized theory, the electromagnetic field can be described as a collection of photons, each with specific energy, momentum, and polarization. The quantum states can be described in terms of occupation numbers of photons in different modes of the field. The description of this statement can be addressed using Hilbert space [1, 9, 11].

3.6 HILBERT SPACE DESCRIPTION OF QUANTIZED MAXWELL'S EQUATIONS

The keystone question now is: how can quantized Maxwell's equations be derived by Hilbert space? Let us consider that the quantum states of the electromagnetic field can be represented as Fock states $|n_1, n_2, ...\rangle$, where n_i is the occupation number of photons in the i-th mode of the field. The Hilbert space $\mathcal{H}$ is the space of all possible quantum states [1, 12, 14]. The field operators include the creation operator $a^{\dagger}_{\vec{k}'}$ and the annihilation operator $a_{\vec{k}}$ for the k-th mode, which satisfies the following commutation relations:

$$\left[a_{\vec{k}} a^{\dagger}_{\vec{k}'}\right] = \partial_{\vec{k}\vec{k}'} \tag{3.20}$$

$$\left[a_{\vec{k}}, a_{\vec{k}'}\right] = \left[a^{\dagger}_{\vec{k}}, a^{\dagger}_{\vec{k}'}\right] = 0 \tag{3.21}$$

The quantized vector potential $\vec{A}_i(\vec{r}, t)$ and the electric field $\vec{E}(\vec{r}, t)$ can be written in terms of these operators. For simplicity, let us consider a single mode of the field:

$$\vec{A}(\vec{r}, t) = \sum_k \left(\vec{A}_k e^{i(\vec{k}\cdot\vec{r} - \omega_{\vec{k}} t)} a_k + \vec{A}^*_k e^{e^{i(\vec{k}\cdot\vec{r} - \omega_{\vec{k}} t)}} a^{\dagger}_k \right) \tag{3.22}$$

Therefore, the electric field operator is related to the time derivative of the vector potential:

$$\vec{E}(\vec{r}, t) = i \sum_k \omega_k \left(\vec{A}_k e^{i(\vec{k}\cdot\vec{r} - \omega_{\vec{k}} t)} a_k - \vec{A}^*_k e^{e^{-i(\vec{k}\cdot\vec{r} - \omega_{\vec{k}} t)}} a^{\dagger}_k \right) = \frac{\partial \vec{A}_k(\vec{r}, t)}{\partial t} \tag{3.23}$$

In this sense, the action of the creation and annihilation operators on the Fock states is given by:

$$a_k |n_1, n_2, ..., n_k, ...\rangle = \sqrt{n_k}\, |n_1, n_2, ..., n_k - 1, ...\rangle \tag{3.24}$$

$$a^{\dagger}_k |n_1, n_2, ..., n_k, ...\rangle = \sqrt{n_k + 1}\, |n_1, n_2, ..., n_k + 1, ...\rangle \tag{3.25}$$

Therefore, the number operator $\hat{N}_k = a^{\dagger}_k a_k$ gives the number of photons in the k-th mode:

$$\widehat{N}_k |n_1, n_2, ..., n_k, ...\rangle = n_k\, |n_1, n_2, ..., n_k, ...\rangle \tag{3.26}$$

From this perspective, altering the occupation number of photons in specific modes modifies the quantum states. The question that arises now is: how is the Hamiltonian of the quantized electromagnetic field written in terms of the creation and annihilation operators? The Hamiltonian of the quantized electromagnetic field can be written as:

$$\widehat{H} = \sum_k \hbar \omega_k (a^{\dagger}_k a_k + 0.5) \tag{3.27}$$

The field operators a_k and $a^{\dagger}_k$ act on quantum states in a Hilbert space, changing the occupation number of photons in specific modes. The electric and magnetic fields can be expressed in terms of these operators, and the entire field dynamics can be described using these quantized fields. The occupation number of each mode represents the number of photons in that mode, and the Hamiltonian describes the energy of the quantized field [1, 8, 11, 13, 15].

3.7 DERIVATIVE SPECTRAL SIGNATURE FROM QUANTIZED MAXWELL'S EQUATIONS

The spectral density is related to the expectation value of the field operators. For the electric field, the spectral density $S_E(\omega)$ at frequency (ω) is given by:

$$S_E(\omega) = \frac{1}{2\pi} \int_{-\infty}^{\infty} \langle E(t) \cdot E(t+\tau) \rangle e^{i\omega\tau} d\tau \tag{3.28}$$

Using equation 3.18, the spectral density $S_E(\omega)$ is then:

$$S_E(\omega) = \sum_{\vec{k},\lambda} \left[\frac{\hbar\omega_k}{2\varepsilon_0 V} \right] \left| \langle a_{\vec{k},\lambda} \rangle \right|^2 \partial\left(\omega - \omega_{\vec{k}}\right) \tag{3.29}$$

The spectral signature of the electromagnetic field can be derived by integrating the spectral density over the relevant modes and frequencies [16–18]. The spectral signature function $S(\omega)$ is given by:

$$S(\omega) = \sum_{\vec{k},\lambda} \left[\frac{\hbar\omega_k}{2\varepsilon_0 V} \right] \left| \langle a_{\vec{k},\lambda} \rangle \right|^2 \partial\left(\omega - \omega_{\vec{k}}\right) \tag{3.30}$$

This formula represents the distribution of energy in different modes of the electromagnetic field as a function of frequency, providing a detailed spectral signature. In this view, the sum $\sum_{\vec{k},\lambda}$ indicates that we are considering all possible modes of the electromagnetic field. Here, $\vec{k}$ represents the wavevector, which denotes the direction and wavelength of the mode, and λ represents the polarization state of the mode (e.g., vertical or horizontal polarization). Therefore, $\left[\frac{\hbar\omega_k}{2\varepsilon_0 V} \right]$ represents the energy contribution of a particular mode $\vec{k}$ and λ to the field. Additionally, the delta function ∂ ensures accurate detailed spectral distribution that is obtained, focusing on the energy present at specific frequencies. The overall result, $S(\omega)$, represents the spectral signature of the electromagnetic field. This signature tells us how the energy of the field is distributed across different frequencies, which is crucial for understanding the characteristics of the field in various applications, such as remote sensing, spectroscopy, and communication systems [18–20].

3.8 QUANTIZED BLACKBODY RADIATION

To derive the quantized black body radiation from the given formula for the spectral signature S(ω), which could be exploited with the Planck distribution in quantizing black body radiation (Figure 3.13) allowing for the calculation of energy levels and the analysis of radiation emitted by objects at different temperatures. Equation 3.30 shows how the energy is distributed among the different modes of the electromagnetic field in terms of their frequencies [1, 10, 21, 23]. Let us use the Planck distribution law for the average number of photons in each mode:

$$\left| \langle a_{\vec{k},\lambda} \rangle \right|^2 = \frac{1}{e^{\frac{\hbar\omega_k}{k_B T}} - 1} \tag{3.31}$$

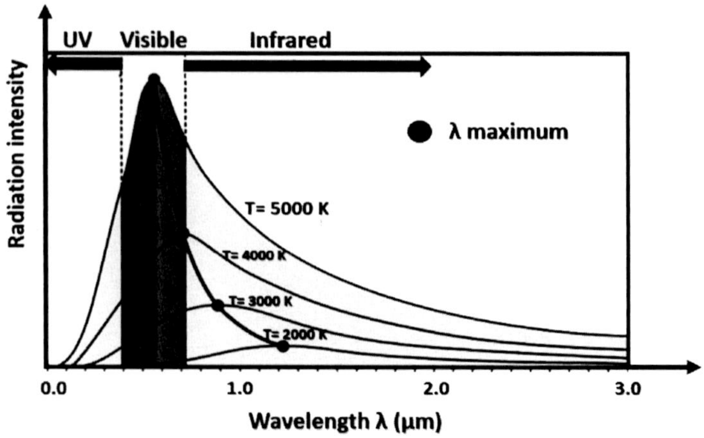

Figure 3.13 Blackbody spectra signature.

Substituting equation 3.31 into the spectral signature formula equation 3.30, one obtains:

$$S(\omega) = \sum_{\vec{k},\lambda}\left[\frac{\hbar\omega_k}{2\varepsilon_0 V}\right]\left(\frac{1}{e^{\frac{\hbar\omega_k}{k_B T}}-1}\right)\partial(\omega-\omega_{\vec{k}}) \tag{3.32}$$

Therefore, in a large volume V, the modes become continuous, and the sum can be replaced by an integral over the density of states (ω):

$$S(\omega) = \int\frac{\hbar\omega_k}{2\varepsilon_0 V}\left(\frac{1}{e^{\frac{\hbar\omega_k}{k_B T}}-1}\right)g\omega\, d\omega \tag{3.33}$$

The density of states (ω) for electromagnetic waves in a cavity (considering two polarization states) is:

$$g(\omega) = 2\omega^2 V[\pi^{-2}c^{-3}] \tag{3.33.1}$$

Substituting equation 3.33.1 into the integral of equation 3.33, one obtains:

$$S(\omega) = \frac{\hbar\omega_k}{2\varepsilon_0 V}\left(\frac{1}{e^{\frac{\hbar\omega_k}{k_B T}}-1}\right)2\omega^2 V[\pi^{-2}c^{-3}] \tag{3.34}$$

Thus, the spectral radiance $B(\omega)$ for black body radiation is:

$$B(\omega) = \hbar\omega^3[\varepsilon_0^{-1}\pi^{-2}c^{-3}]\left(\frac{1}{e^{\frac{\hbar\omega_k}{k_B T}}-1}\right) \tag{3.35}$$

here k_B is the Boltzmann constant, and T is the temperature. The spectral signature $S(\omega)$ derived from the quantized electromagnetic field indeed matches the form of the black body spectral radiance $B(\omega)$ up to a constant factor [21–23]. This demonstrates that black body radiation can be quantized using the spectral signature $S(\omega)$. The quantization procedure aligns with the Planck distribution, confirming that $S(\omega)$ effectively describes the energy distribution in black body radiation. Therefore, one can utilize equation 3.35 to interpret and analyze the spectral data obtained from multispectral and hyperspectral satellite sensors. This approach aids in extracting valuable physical information about the Earth's surface from the remote sensing data. The remain question is now how to derive the black body radiation energy from quantized spectral signature?

3.9 DERIVATIVE BLACKBODY RADIATION ENERGY FROM QUANTIZED SPECTRAL SIGNATURE

According to the above-mentioned perspective, given $B(\omega)$ describes the energy distribution of black body radiation, we can link it to the spectral signature $S(\omega)$. The spectral signature $S(\omega)$ could be interpreted as the observed radiance, which in an ideal black body scenario would match $B(\omega)$. Therefore, the total energy E radiated per unit area by a black body can be retrieved by integrating spectral radiance over all frequencies:

$$E = \int_0^\infty B(\omega) d\omega \tag{3.36}$$

Let $x = \frac{\hbar\omega}{k_B T}$, then $d\omega = \frac{k_B T}{\hbar} dx$. The limits of integration remain from 0 to ∞. Thus, the integral becomes:

$$E = \int_0^\infty \frac{\lambda \left(\frac{k_B T x}{\hbar} \right)^3}{\pi^2 c^2} [e^x - 1]^{-1} \left[\frac{k_B T}{\hbar} \right] dx \tag{3.37}$$

Simplifying the constants outside the integral:

$$E = \frac{\hbar (k_B T x)^4}{\pi^2 c^2 \hbar^3} \int_0^\infty x^3 [e^x - 1]^{-1} dx \tag{3.38}$$

In this sense, the integral term $\left(\int_0^\infty x^3 [e^x - 1]^{-1} dx \right)$ is known to equal $\frac{\pi^4}{15}$. Simplifying:

$$E = \frac{(k_B T x)^4}{c^2 \hbar^3} \frac{\pi^2}{15} \tag{3.39}$$

This result is consistent with the Stefan-Boltzmann law (Figure 3.14), which states the total energy radiated per unit surface area of a blackbody per unit time is proportional to the fourth power of the blackbody's temperature [1, 19, 23]. The Stefan-Boltzmann constant σ can be identified as:

$$\sigma = \frac{k_B{}^4}{c^2 \hbar^3} \frac{\pi^2}{15} \tag{3.40}$$

Using equation 3.40 into 3.39, one can obtain:

$$E = \sigma T^4 \tag{3.41}$$

Equation 3.41 illustrates the total radiation emitted per square meter (Watts m^{-2}), reflecting the Stefan-Boltzmann constant σ (5.67×10^{-8} Watts $m^{-2}K^{-4}$) and the absolute temperature $T(K)$. This equation shows how the Stefan-Boltzmann Law can be easily verified by comparing the integral values, or the area under the curves of the experimental black-body radiation distribution at various temperatures. Boltzmann derived this T^4 relationship from theory in 1884 by applying classical thermodynamic reasoning to a box filled with electromagnetic radiation and using Maxwell's equations to link pressure to energy density. Essentially, the minimum energy emitted from the cavity is inherently temperature-dependent, similar to the radiation intensity inside it [1, 22].

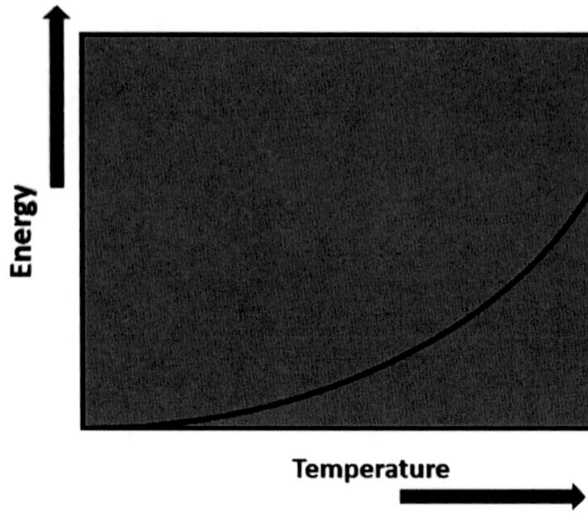

Figure 3.14 Simplification of Stefan-Boltzmann law.

Therefore, Boltzmann's classical thermodynamic approach involved: considering a cavity filled with electromagnetic radiation, and using Maxwell's equations to link pressure (P) to energy density (E_d).

$$P = \frac{1}{3} E_d \tag{3.42}$$

and using the relation between energy density and temperature:

$$E_d = aT^4 \tag{3.43}$$

where a is the radiation constant. Thus, the radiation intensity is temperature-dependent:

$$I = \sigma T^4 \tag{3.44}$$

In equation 3.44, T^4 dependence signifies that the energy emitted from a cavity (black body) depends on its temperature, following the Stefan-Boltzmann Law.

Consequently, in terms of the peak wavelength of light, the following is another illustration of Wien's Law (Figure 3.15):

$$\lambda_{\max} = bT^{-1} \tag{3.45}$$

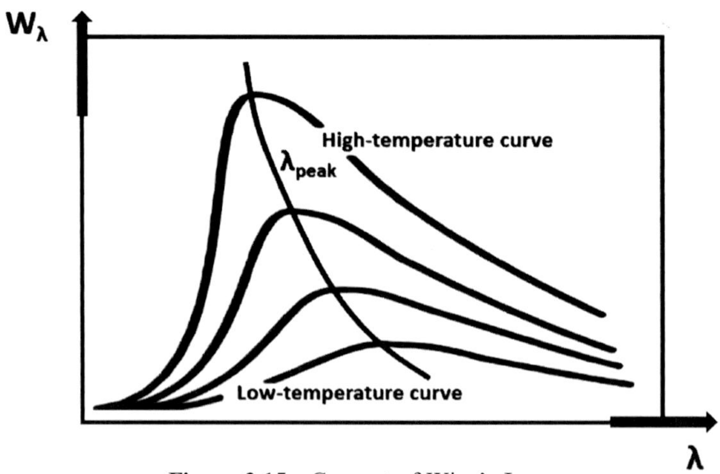

Figure 3.15 Concept of Wien's Law.

Equation 3.45 says T is the absolute temperature in kelvin and b is a proportionality constant known as the Wien's displacement constant, which is equal to 2.89×10^{-3} m K or, more conveniently, $b \approx 2{,}900\,\mu$m.K to obtain the wavelength in micrometers (μm). The relationship between wavelength and temperature is inverse. Therefore, the wavelength of thermal radiation will be shorter or smaller as temperature increases [1, 6, 10, 23]. The wavelength of thermal radiation is longer or bigger at lower temperatures. Hot things emit bluer light than cool ones do for visible radiation (Figure 3.16).

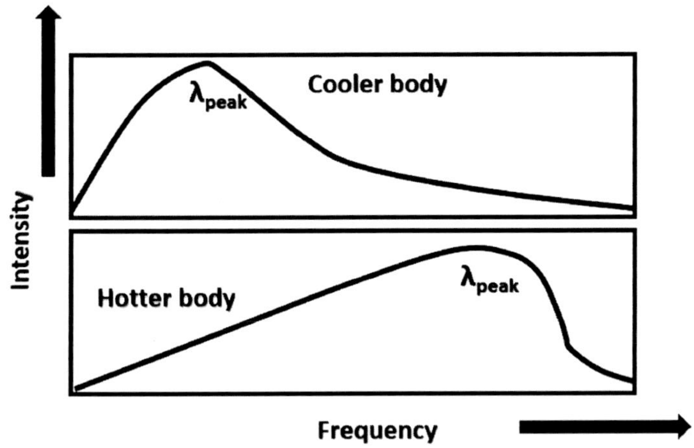

Figure 3.16 Peak wavelength is inversely proportional to the temperature of the blackbody.

3.10 QUANTIZED KIRCHHOFF'S LAWS

To describe quantized black body radiation in remote sensing satellite data using Kirchhoff's laws, we need to bridge classical thermodynamics with quantum mechanics, particularly Planck's law of black body radiation. Kirchhoff's laws, in the context of black body radiation, involve the principles of emission and absorption. Mathematically, this can be expressed as:

$$\varepsilon(\lambda, T) = \alpha(\lambda, T) \tag{3.46}$$

where $\varepsilon(\lambda,T)$ is the emissivity of the body at wavelength λ and temperature. Additionally, $\alpha(\lambda,T)$ is the absorptivity of the body at wavelength λ and temperature T. In other words, both $\varepsilon(\lambda,T)$, and, $\alpha(\lambda,T)$ are functions of wavelength λ and temperature T. In this sense, emissivity is a measure of a material's ability to emit radiation as a fraction of the radiation emitted by a perfect black body at the same temperature. It ranges from 0 to 1, where 1 indicates a perfect emitter (black body) and 0 indicates no emission. Instead, absorptivity is a measure of a material's ability to absorb incident radiation. Like emissivity, it ranges from 0 to 1, where 1 indicates complete absorption and 0 indicates no absorption [6, 25].

Alternatively, Planck's law describes the spectral radiance of blackbody radiation as a function of wavelength (λ) and temperature (T). Mathematically this can be shown as:

$$B(\lambda, T) = \frac{2hc^2}{\lambda^5}\left(\frac{1}{e^{\frac{\hbar\omega_k}{k_B T}} - 1}\right) \tag{3.47}$$

In remote sensing, we observe the radiation emitted by the Earth's surface and atmosphere, which can be approximated as black body radiation. By applying Kirchhoff's law, we understand that the observed radiation is a result of both emission and absorption processes (Figure 3.17). In this sense, satellites equipped with multispectral and hyperspectral sensors measure the radiance emitted and reflected by the Earth's surface and atmosphere. These measurements are taken at various wavelengths (or frequencies). Consequently, when describing the quantized black body radiation, we acknowledge that the energy levels of the emitting particles (e.g., molecules or atoms) are quantized [24–26]. Therefore, the emission spectrum is not continuous but consists of discrete lines corresponding to transitions between quantized energy levels. This is critical when analyzing high-resolution spectral data as it will be addressed later in section 3.11.3.

Figure 3.17 Simplification of blackbody Kirchhoff's law.

The spectral signature $S(\omega)$ of an object observed by a satellite is influenced by its temperature, material properties (emissivity (Figure 3.18) and absorptivity (Figure 3.19)), and atmospheric conditions. The radiative transfer equation that combines these effects is:

$$I(\lambda, T) = \varepsilon(\lambda, T)B(\lambda, T) + (1 - \varepsilon(\lambda, T))I_{\text{atm}}(\lambda) \tag{3.48}$$

Figure 3.18 Example of emissivity.

where $I(\lambda, T)$ is the intensity observed by the satellite, $\varepsilon(\lambda, T)$ is the emissivity of the surface, and $I_{\text{atm}}(\lambda)$ is the atmospheric contribution. Let us consider the observed spectral radiance $\mathcal{L}(\lambda,T)$ from a satellite sensor:

$$\mathcal{L}(\lambda,T) = \varepsilon(\lambda,T)\frac{2hc^2}{\lambda^5}\left(\frac{1}{e^{\frac{\hbar\omega_k}{k_BT}} - 1}\right) + (1 - \varepsilon(\lambda,T))I_{atm}(\lambda) \tag{3.49}$$

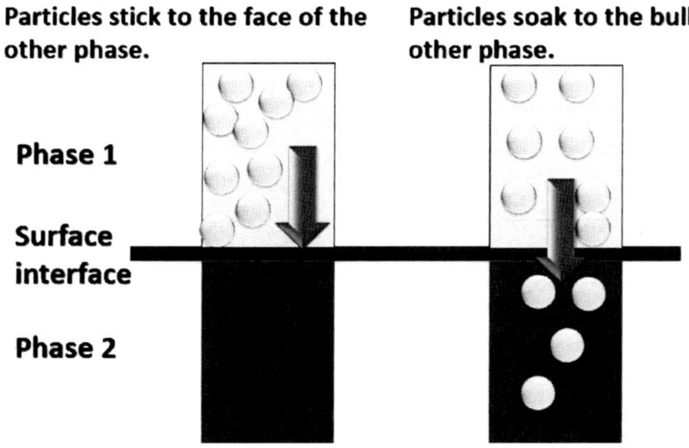

Figure 3.19 Concept of the absorptivity.

By solving this equation for different wavelengths and using known values for emissivity $\varepsilon(\lambda, T)$ and atmospheric radiance $I_{atm}(\lambda)$, one can extract the surface temperature T and further analyze the spectral features corresponding to quantized energy transitions. Needless to say, Kirchhoff's law establishes a direct relationship between a material's ability to emit and absorb thermal radiation. This principle is foundational in thermodynamics and quantum mechanics and is particularly useful in remote sensing for interpreting thermal radiation data from the Earth's surface and atmosphere [24, 26].

To effectively use Kirchhoff's law in investigating GERD dam construction from multispectral and hyperspectral data, it is essential to know the emissivity and absorptivity values of common construction materials like Roller Compacted Concrete (RCC) and rockfill. These values are critical for interpreting thermal infrared data collected from satellite sensors. Therefore, ASTER data that was acquired on November 6, 2020, and covered an area of 32 by 42.9 kilometers is examined (Figure 3.20). The Visible and Near-Infrared (VNIR) bands (Bands 1 to 3) are excluded as they are primarily utilized for detecting reflectance properties and are not suitable for emissivity measurements. Similarly, the Shortwave Infrared (SWIR) bands (Bands 4 to 9) are excluded because they are also used for reflectance measurements and are not typically employed for emissivity detection [5, 33, 35].

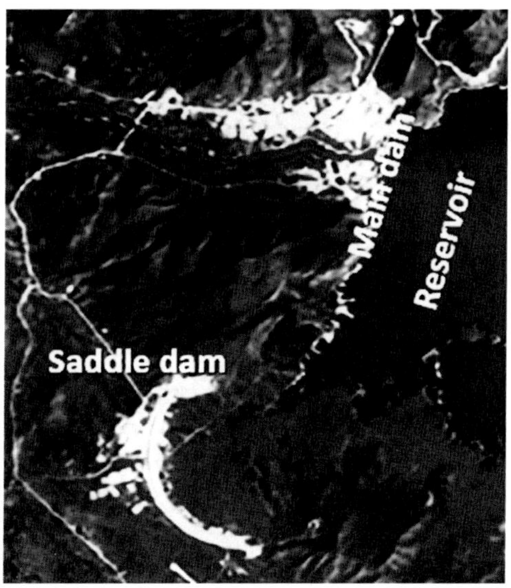

Figure 3.20 Tested ASTER data using Kirchhoff's law.

The spectral emissivity of concrete in ASTER bands, especially in the Thermal Infrared (TIR) range is generally higher than that of limestone and granite, as illustrated in Figure 3.21. Specifically, the maximum emissivity of concrete, reaching up to 0.95, occurs within the wavelength range of 8.925 to 9.275 µm. Conversely, granite exhibits the lowest emissivity values, notably within the wavelength range of 8.125 to 8.475 µm. This is primarily because granite is composed predominantly of minerals such as quartz, feldspar, and mica, which have inherently lower emissivity values compared to the materials found in concrete and limestone.

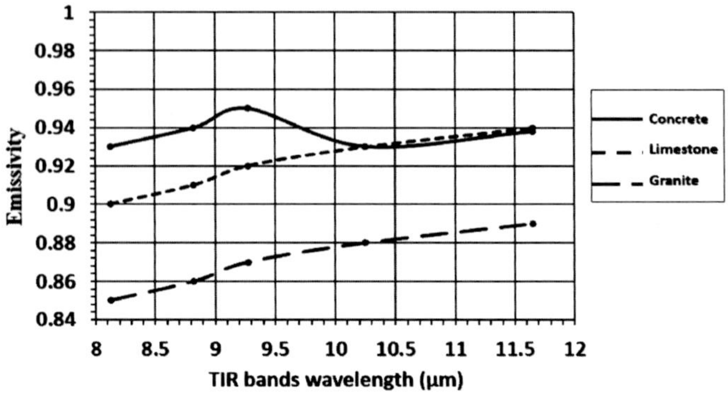

Figure 3.21 Emissivity spectra of GERD construction in TIR bands of ASTER image.

The minerals in granite reflect more thermal radiation, thereby reducing overall emissivity. Additionally, granite often possesses a smoother and more polished surface compared to concrete and limestone. Smoother surfaces are more reflective of thermal radiation and emit less, resulting in lower emissivity. Furthermore, the thermal conductivity and heat retention properties of granite differ significantly from those of concrete and limestone. Granite's lower thermal conductivity means it does not retain and emit heat as efficiently, contributing to its lower emissivity. Moreover, granite is denser and less porous than concrete and limestone. The lower porosity means fewer air pockets to trap and emit heat, leading to lower emissivity values [1].

The typical emissivity value for granite is approximately 0.90. According to Kirchhoff's law, for a body in thermal equilibrium, emissivity equals absorptivity at the same wavelength and temperature. Therefore, the emissivity and absorptivity of rockfill materials can vary depending on the type of rock and its surface characteristics. Common rock types used in rockfill include granite, basalt, and limestone, with emissivity and absorptivity values ranging from 0.85 to 0.95. Specifically, granite and limestone typically have values between 0.85 and 0.90, while concrete varies from 0.90 to 0.95.

Additionally, the surface condition of RCC (roller-compacted concrete) and rockfill, such as roughness, wetness, and the presence of vegetation or debris can influence emissivity. Moist surfaces, for instance, tend to have slightly higher emissivity. This variability underscores the importance of considering surface characteristics in remote sensing applications, particularly when interpreting thermal infrared data for materials such as concrete, granite, and limestone. Needless to say, only the TIR bands (Bands 10 to 14) can effectively detect the emissivity of granite, concrete and limestone in ASTER images.

3.11 THE ROLE OF SPATIAL DIMENSIONS IN SATELLITE REMOTE SENSING FOR GERD MONITORING AND ANALYSIS

The monitoring and analysis of large-scale infrastructure projects like the Grand Ethiopian Renaissance Dam (GERD) demand precise and reliable data. Satellite remote sensing, with its extensive spatial coverage and high-resolution imaging capabilities plays a crucial role in this context. Understanding the role of spatial dimensions in satellite remote sensing can significantly enhance our ability to monitor and analyze such projects effectively. Spatial dimensions in remote sensing refer to the resolution and extent of the data collected by satellite sensors (Figure 3.22). These dimensions are critical in determining the level of detail and the area covered by the imagery. In the context of the GERD, spatial resolution can influence how well we can observe and interpret various aspects of a dam's construction, reservoir levels and surrounding environmental impacts.

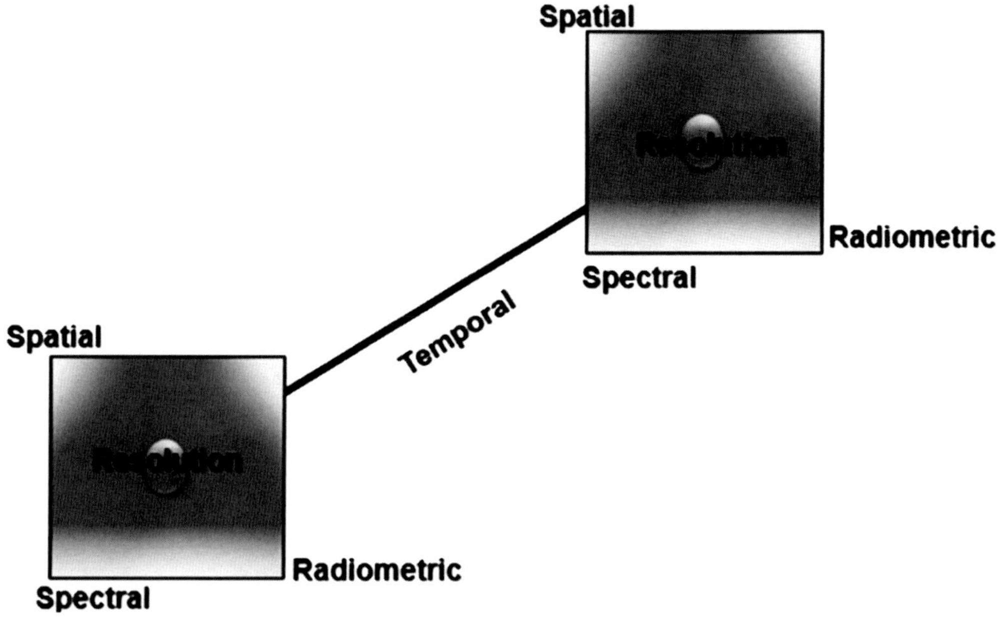

Figure 3.22 Various types of spatial dimensions can be utilized for monitoring the Grand Ethiopian Renaissance Dam (GERD) in Ethiopia.

3.11.1 Spatial Resolution

Spatial resolution refers to the size of the smallest object that can be resolved by the sensor. It is typically expressed in meters, indicating the ground area covered by one pixel of the image. In this view, high spatial resolution and medium spatial resolution perhaps can be implemented in monitoring and analysis such as GERD over time. Therefore, high-resolution satellite imagery, such as that provided by systems like WorldView-3 or GeoEye-1, offers detailed views of the dam structure, construction progress and associated infrastructure [5, 28]. This level of detail is essential for monitoring specific features and detecting any structural anomalies or changes over time [27–30]. On the contrary, satellites like Landsat-8 and Sentinel-2 provide medium to low-resolution images but cover larger areas (Table 3.1). These images are valuable for broader environmental monitoring, such as assessing changes in land use, vegetation and hydrological patterns in the vicinity of the GERD.

Therefore, Table 3.1 provides a clear overview of the satellite options available for comprehensive monitoring and analysis of the GERD dam, facilitating the selection of appropriate sensors for different monitoring needs.

In light of the aforementioned viewpoint, high-resolution imagery allows for the detailed observation of the construction progress of the GERD. This includes monitoring the placement of concrete, the building of spillways and the installation of turbines. Any deviations from planned construction activities can be quickly identified and addressed.

Table 3.1 Spatial resolution satellite data is used in monitoring GERD

Category	Satellite	Spatial-resolution	Revisit Frequency	Primary Uses
High-Spatial Resolution	WorldView-3	31 cm (panchromatic)	1–2 days	Detailed structural monitoring, construction progress.
	GeoEye-1	41 cm (panchromatic)	3 days	High-detail mapping, infrastructure inspection.
	Pleiades	50 cm (panchromatic)	1–2 days	Detailed site analysis, and environmental impact assessment.
	QuickBird	61 cm (panchromatic)	1–3.5 days	Construction site monitoring, and change detection.
	IKONOS	82 cm (panchromatic)	3 days	Detailed observation, high-precision mapping
Medium to Low-Spatial Resolution	Landsat-8	15 m (panchromatic), 30 m (multispectral)	16 days	Broad area environmental monitoring, land use change detection.
	Sentinel-2	10 m (multispectral)	5 days	Vegetation analysis, water quality monitoring, environmental changes.
	ASTER	15 m (VNIR), 30 m (SWIR), 90 m (TIR)	16 days	Detailed terrain mapping, thermal analysis, and vegetation studies.
	MODIS	250 m to 1 km	Daily	Large-scale environmental monitoring, climate studies.

3.11.2 Temporal Resolution

Temporal resolution refers to the frequency with which a sensor can revisit the same location. Higher temporal resolution means more frequent data collection. This means the temporal resolution, or revisit frequency of satellites is crucial for continuous monitoring. Systems like Sentinel-1 and Sentinel-2, with their relatively high revisit times, enable frequent observation of the dam and its surroundings, allowing for timely detection of changes and potential issues [27, 30, 32].

3.11.3 Spectral Resolution

The spectral resolution, or the ability to capture data in multiple wavelength bands, enhances the analysis capabilities of remote sensing [28, 31, 33, 35]. Hyperspectral sensors, which capture data in hundreds of narrow bands, can provide detailed information on the materials and conditions at the dam site. This includes monitoring water quality, and sediment levels, and identifying different types of vegetation [29–35].

Given the perspective mentioned earlier, multispectral and hyperspectral data help in assessing the environmental impact of the dam. Changes in vegetation cover, land use patterns, and water quality can be monitored over time. This information is critical for ensuring that the dam's construction and operation do not adversely affect the surrounding ecosystem.

Unquestionably, the spatial dimensions of satellite remote sensing—encompassing spatial, temporal, and spectral resolutions—are vital for the comprehensive monitoring and analysis of the Grand Ethiopian Renaissance Dam. By exploiting these capabilities, stakeholders can ensure the efficient construction, operation and management of the dam while mitigating potential environmental impacts and risks. This multifaceted approach underscores the indispensable role of advanced remote sensing technologies in modern infrastructure monitoring and management.

3.12 CONCLUSION

This chapter presents the concept of quantized remote sensing, building on the novel definition established by Marghany. It underscores the importance of understanding spectral signatures through the quantized blackbody model. The application of Kirchhoff's law to the Thermal Infrared (TIR) bands in ASTER data has been explored to analyze the material compositions used in the construction of the Grand Ethiopian Renaissance Dam (GERD).

The concept of a quantized blackbody is pivotal in understanding spectral signatures. By applying Kirchhoff's law, the chapter investigates the emissivity spectra of various materials used in GERD. Granite, a primary component of GERD, is noted for its lower emissivity compared to concrete and limestone. This distinction in emissivity is critical for remote sensing applications and the accurate monitoring of the dam's structural integrity.

The chapter delves into the role of spatial dimensions in monitoring the GERD over time. The ability to observe and analyze spatial changes is essential for assessing the dam's status and ensuring its stability. However, a more detailed examination of the spectral signature of GERD is necessary to fully understand its composition and the surrounding environment.

To address this need, the next chapter will focus on implementing the quantum spectral signature. This approach aims to provide a comprehensive analysis of GERD's composition, including the materials used in construction and the dam reservoirs. By leveraging the quantum spectral signature, the study will offer a clearer and more detailed understanding of GERD and its environmental context.

In conclusion, this chapter has laid the groundwork for advanced remote sensing analysis of GERD by integrating the concepts of quantized blackbody and Kirchhoff's law. The insights gained from examining material emissivity and spatial dimensions will inform future research and monitoring efforts. The subsequent chapter will build on these findings, utilizing quantum spectral signatures to enhance our understanding of GERD's structure and surroundings.

References

[1] Marghany, M. 2022. Remote Sensing and Image Processing in Mineralogy. CRC Press.

[2] Gruska, J. 1999. Quantum Computing (Vol. 2005). London: McGraw-Hill.

[3] Pirandola, S., Bardhan, B.R., Gehring, T., Weedbrook, C. and Lloyd, S. 2018. Advances in photonic quantum sensing. Nature Photonics. 12(12): 724–733.

[4] Degen, C.L., Reinhard, F. and Cappellaro, P. 2017. Quantum sensing. Reviews of Modern Physics. 89(3): 035002.

[5] Aggarwal, S. 2004. Principles of remote sensing. Satellite Remote Sensing and GIS Applications in Agricultural Meteorology. 23(2): 23–28.

[6] Arbab, A.I. 2017. Quantized Maxwell's Equations. Optik. 136: 64–70.

[7] Minasyan, V.N. 2009. Quantized Equations of Maxwell. arXiv preprint arXiv:0902.3057.

[8] Moore, G.T. 1970. Quantum theory of the electromagnetic field in a variable-length one-dimensional cavity. Journal of Mathematical Physics. 11(9): 2679–2691.

[9] Gersten, A. and Moalem, A. 2015. Consistent quantization of massless fields of any spin and the generalized Maxwell's equations. Journal of Physics: Conference Series. 615(1): 012011. IOP Publishing.

[10] Strocchi, F. 1970. Gauge problem in quantum field theory. III. Quantization of Maxwell equations and weak local commutativity. Physical Review D. 2(10): 2334.

[11] Chew, W.C., Na, D.Y., Bermel, P., Roth, T.E., Ryu, C.J. and Kudeki, E. 2020. Quantum Maxwell's equations made simple: Employing scalar and vector potential formulation. IEEE Antennas and Propagation Magazine. 63(1): 14–26.

[12] Gessner, W. and Ernst, V. 1980. The rosette of rosettes of Hilbert spaces in the indefinite metric state space of the quantized Maxwell field. Journal of Mathematical Physics. 21(1): 93–111.

[13] Hawton, M. 2021. Photon quantum mechanics in real Hilbert space. Physical Review A. 104(5): 052211.

[14] Hawton, M. 2019. Maxwell quantum mechanics. Physical Review A. 100(1): 012122.

[15] Bennett, R., Barlow, T.M. and Beige, A. 2015. A physically motivated quantization of the electromagnetic field. European Journal of Physics. 37(1): 014001.

[16] Rozina, C., Asma, A., Poedts, S., Ali, S. and Shah, H.A. 2023. Signatures of quantized magnetic field on the nonlinear Landau damping of transverse electromagnetic waves. Physica Scripta. 98(9): 095601.

[17] Tomassetti, L. 2020. Physics of RS. TORUS 2–Toward an Open Resource Using Services: Cloud Computing for Environmental Data. 23–46.

[18] Darrigol, O. 1986. The origin of quantized matter waves. Historical Studies in the Physical and Biological Sciences. 16(2): 197–253.

[19] Aurilia, A. and Takahashi, Y. 1981. Generalized Maxwell equations and the gauge mixing mechanism of mass generation. Progress of Theoretical Physics. 66(2): 693–712.

[20] Maurer, P., Gonzalez-Ballestero, C. and Romero-Isart, O. 2023. Quantum electrodynamics with a nonmoving dielectric sphere: quantizing Lorenz–Mie scattering. JOSA B. 40(12): 3137–3155.

[21] Theimer, O. and Peterson, P.R. 1974. Statistics of classical blackbody radiation with ground state. Physical Review D. 10(12): 3962.

[22] Ostilli, M. and Presilla, C. 2017. Thermalization of noninteracting quantum systems coupled to blackbody radiation: A Lindblad-based analysis. Physical Review A. 95(6): 062112.

[23] Boyer, T.H. 1969. Derivation of the blackbody radiation spectrum without quantum assumptions. Physical Review. 182(5): 1374.

[24] Agassi, J. 1967. The Kirchhoff-Planck Radiation Law: Considering Kirchhoff's law as it was initially meant may help us understand the rise of quantum theory. Science. 156(3771): 30–37.

[25] Hollandt, J. 2012. The "Black Body" and the Quantization of the World. PTB-Mitteilungen. 122(2).

[26] Zhao, B., Shi, Y., Wang, J., Zhao, Z., Zhao, N. and Fan, S. 2019. Near-complete violation of Kirchhoff's law of thermal radiation with a 0.3 T magnetic field. Optics Letters. 44(17): 4203–4206.

[27] Chuvieco, E. 2020. Fundamentals of satellite remote sensing: An environmental approach. CRC press.

[28] Zhou, G. 2020. Urban High-Resolution Remote Sensing: Algorithms and Modeling. CRC Press.

[29] Oka, H. 2017. Generation of broadband ultraviolet frequency-entangled photons using cavity quantum plasmonics. Scientific Reports, 7(1): 1–0.

[30] de J, León-Montiel, R., Svozilik, J., Salazar-Serrano, L.J. and Torres, J.P. 2013. Role of the spectral shape of quantum correlations in two-photon virtual-state spectroscopy. New Journal of Physics. 15(5): 053023.

[31] Richards, J.A. and Jia, X. 2006. Remote sensing digital image analysis: An introduction. Springer Science & Business Media.

[32] Zhang, W., Zhang, D., Qiu, X. and Chen, L. 2019. Quantum remote sensing of the angular rotation of structured objects. Physical Review A. 100(4): 043832.

[32] De Jong, S.M., Van der Meer, F.D. and Clevers, J.G.P.W. 2004. Basics of remote sensing. Remote sensing image analysis: Including the spatial domain. 1–15.

[33] Elachi, C. and Van Zyl, J.J. 2021. Introduction to the Physics and Techniques of Remote Sensing. John Wiley & Sons.

[34] Sabins Jr, F.F. and Ellis, J.M. 2020. Remote sensing: Principles, interpretation, and applications. Waveland Press.

[35] Schanda, E. 2012. Physical Fundamentals of Remote Sensing. Springer Science & Business Media.

Chapter 4

Quantum Spectral Signature of the Grand Ethiopian Renaissance Dam Construction and its Surrounding Geological Features

4.0 INTRODUCTION

In a mountainous region, the Grand Ethiopian Renaissance Dam (GERD) is being assembled. However, inadequate information has been compiled on strategic ecological aspects. There is a significant lack of comprehensive data regarding site conditions, potential hazards and environmental impacts. This scarcity of critical information poses substantial challenges to the project's planning, execution and long-term sustainability.

Given the complexity of the mountainous terrain thorough geological and environmental investigations are imperative. These investigations are crucial to understand the stability of the foundation, assess seismic risks and identify suitable construction materials. Without this data, the risk of unforeseen geological challenges increases, potentially compromising the structural integrity of the dam.

Further, the absence of detailed environmental impact assessments can lead to unforeseen consequences on local ecosystems, water quality, and sedimentation patterns. The potential for adverse effects on biodiversity, water resources and the livelihood of those living downstream underscores the necessity for comprehensive environmental studies.

Addressing these gaps in knowledge is essential for the safe and sustainable development of GERD. It is imperative to implement a robust framework for continuous monitoring and data collection throughout the construction and operational phases of the dam. This approach will help mitigate risks, ensure compliance with environmental standards, and promote the project's long-term success.

By prioritizing detailed environmental and geological assessments, stakeholders can make informed decisions that balance developmental goals with environmental stewardship, thereby securing the benefits of GERD for future generations.

This chapter is dedicated to comprehensively understanding the critical geological features of the Grand Ethiopian Renaissance Dam (GERD) and its surrounding environment. This

DOI: 10.1201/9781003327325-4

understanding is achieved through the use of advanced remote sensing sensors and sophisticated image processing algorithms, primarily grounded in quantum computing principles.

Advanced remote sensing sensors, such as multispectral and hyperspectral sensors, provide detailed spectral information across a wide range of wavelengths. These sensors can capture subtle differences in the reflectance and emissivity properties of various geological materials. By leveraging these capabilities, we can identify and characterize the diverse geological formations in the GERD region with high precision.

The application of quantum computing to image processing algorithms represents a significant leap forward in data analysis. Quantum algorithms can handle vast amounts of data more efficiently than classical algorithms, enabling faster and more accurate processing of remote-sensing images. This efficiency is crucial for extracting meaningful insights from the complex datasets generated by advanced sensors.

In this chapter, we explore how these cutting-edge technologies are utilized to map and analyze the geological features of the GERD site. This includes identifying rock types, assessing soil properties and detecting potential geological hazards. The integration of remote sensing data with quantum computing techniques allows for a more detailed and nuanced understanding of the site conditions, which is essential for informed decision-making in the construction and management of the dam.

By advancing our knowledge of the GERD's geological context, this chapter underscores the importance of combining state-of-the-art remote sensing technology with innovative computational methods. This approach not only enhances our ability to monitor and analyze complex environments but also paves the way for future applications in environmental and geological studies.

4.1 THE IMPORTANCE OF GEOLOGICAL INVESTIGATIONS FOR DAM SITE SELECTION

Geological investigations are critically important for selecting suitable dam sites due to several key reasons, each contributing to the overall stability, safety and sustainability of the dam. First, foundation stability is paramount. Comprehensive geological assessments ensure that the dam is built on a solid and stable foundation, preventing issues such as settlement, sliding, or collapse, which could lead to catastrophic failures. The comparison here between a well-investigated site and one that isn't thoroughly examined underscores the critical nature of these investigations.

Second, seepage control is another crucial factor. Geological investigations help identify the permeability characteristics of the rock and soil at the proposed site. By understanding these properties, engineers can design effective seepage control measures, preventing water from leaking through or under the dam. This contrasts with sites lacking thorough investigations, where unforeseen seepage issues can lead to significant structural and environmental problems [1–3].

Third, the availability of construction materials is also essential. Geological investigations provide valuable information about the availability of suitable construction materials, such as rock and soil, in the vicinity of the dam site. This can significantly reduce construction costs and logistical challenges compared to sites where materials need to be transported from distant locations.

Furthermore, seismic risk assessment is vital. By examining the geological history and fault lines in the region, geologists can assess the seismic risks associated with the site. This information is crucial for designing dams that can withstand earthquakes and other seismic activities, ensuring long-term structural integrity. Comparing dams built in seismically active areas with and without proper geological assessments highlights the importance of this step [1, 4, 6].

Additionally, environmental impact assessments are necessary. Geological studies help predict potential environmental impacts of dam construction, such as alterations in groundwater flow, erosion and sedimentation. Understanding these impacts is crucial for implementing sustainable and environmentally responsible construction practices. This stands in contrast to projects that proceed without thorough environmental assessments, often resulting in significant ecological damage.

Moreover, hydrological considerations play a key role. Geological investigations offer insights into the hydrological characteristics of the site, including groundwater flow and storage. This knowledge is vital for effective water resource management and for ensuring the dam's operational efficiency over its lifespan. Comparatively, dams built without this understanding can face operational challenges and inefficiencies [7–10].

Hence, geological investigations are important for dam site selection. They ensure the dam's safety, cost-effectiveness, and environmental sustainability by providing essential data and analysis for informed decision-making. By comparing the outcomes of projects with and without these investigations, the critical nature of geological assessments becomes abundantly clear, highlighting their role in the successful completion and operation of dam projects.

4.2 HOW GEOLOGICAL INVESTIGATIONS IMPACT DAM SAFETY?

Geological investigations are essential for enhancing emergency preparedness for dams. These studies provide critical information that aids in planning, designing and implementing effective emergency response strategies. Geological investigations identify fault lines and assess seismic activity in the region. This information is crucial to understand the potential impact of earthquakes on the dam and to develop seismic emergency response plans. By analyzing slope stability and identifying areas prone to landslides, geological studies help in preparing to mitigate the effects of landslides that could threaten the dam [4, 7, 9].

Understanding the geological conditions ensures that the dam is built on a stable foundation. This reduces the risk of foundation failure, which is critical for emergency preparedness. Identifying suitable construction materials ensures the dam's structural integrity, reducing the likelihood of structural failures that could necessitate emergency responses.

According to the above-mentioned perspective, geological investigations assess the permeability of the foundation and surrounding materials helping to design effective seepage control measures. This prevents water from undermining the dam which is vital for emergency preparedness. Additionally, understanding erosion patterns eases in designing measures to prevent erosion. Erosion can compromise the dam's safety and require emergency interventions. Moreover, geological investigations provide insights into groundwater flow and storage, which is essential to manage water levels and prevent overflow or flooding during emergencies. Assessing the environmental impact of the dam helps in developing strategies to mitigate adverse effects, ensuring that emergency plans consider environmental factors [3, 7, 10].

4.3 EMERGENCY RESPONSE PLANNING

Geological data is used to model various emergency scenarios, such as earthquakes, landslides and floods. This helps in developing comprehensive emergency response plans that address potential geological hazards. Understanding the geological conditions helps in allocating resources effectively for emergency preparedness, such as positioning emergency equipment and personnel in strategic locations.

Geological investigations help in the development of training programs for dam operators and emergency responders. These programs include drills and simulations based on realistic geological scenarios. In this sense, educating the local community about geological risks and emergency procedures enhances overall preparedness and ensures a coordinated response during emergencies.

Needless to say, geological investigations provide the foundational knowledge necessary for effective emergency preparedness for dams. They help in identifying and assessing risks, ensuring structural integrity, controlling seepage and erosion, understanding hydrological impacts, planning emergency responses and conducting training and public awareness programs. These efforts collectively enhance the safety and resilience of dam structures [6–10].

4.4 SOME FAMOUS DAMS BUILT ON CHALLENGING GEOLOGICAL SITES

Several renowned dams have been constructed on geologically challenging sites, showcasing exceptional engineering prowess.

4.4.1 Hoover Dam (USA)

The Hoover Dam was constructed in the Black Canyon of the Colorado River, it (Figure 4.1) faced significant challenges due to the hard rock formations and the necessity to divert the river during construction. Engineers employed innovative techniques, such as constructing diversion tunnels and utilizing large-scale concrete pouring methods to overcome these obstacles [11].

Figure 4.1 Hoover Dam.

4.4.2 Three Gorges Dam (China)

The Three Gorges Dam is situated on the Yangtze River (Figure 4.2). This dam had to contend with complex geological conditions, including fault lines and the risk of landslides. Extensive geological surveys and advanced construction techniques were implemented to ensure the dam's stability and safety [12].

Figure 4.2 Three Gorges Dam.

4.4.3 Jinping-I Dam (China)

The Jinping-I Dam (Figure 4.3), that is located in a deep gorge in Sichuan province, had to face geological challenges. It had to deal with steep slopes and high seismic activity. The design incorporated deep foundations and extensive rock reinforcement to address these geological challenges [5, 7, 10, 13].

Figure 4.3 Jinping-I Dam

4.4.4 Itaipu Dam (Brazil/Paraguay)

The Itaipu Dam was built on the Paraná River. It (Figure 4.4) faced challenges related to the river's high flow rate and the need to excavate large volumes of rock. Engineers constructed massive cofferdams and employed advanced excavation techniques to manage the river flow and stabilize the foundation [14].

Figure 4.4 Itaipu Dam.

4.4.5 Kariba Dam (Zambia/Zimbabwe)

On the other hand, the Kariba Dam is situated in a seismically active zone on the Zambezi River (Figure 4.5). It had to address the risk of earthquakes and the presence of weak rock formations. The dam was designed with a flexible structure to withstand seismic activity, and extensive grouting was used to strengthen the foundation [15].

Figure 4.5 Kariba Dam.

The above mentioned dams exemplify how innovative engineering solutions can overcome significant geological challenges, ensuring the safety and functionality of these critical structures [13–15].

4.5 THE ROLE OF INSTRUMENTATION IN MONITORING DAM BEHAVIOR

Instrumentation plays a crucial role in monitoring the behavior of dams, ensuring their safety, stability and functionality. Strain gauge devices measure the strain on various parts of the dam structure, helping to detect any deformation or stress that could indicate potential issues. These sensors track the movement of the dam structure, including any shifts or tilts, which are critical for identifying structural instability. Consequently, Piezometers (Figure 4.6) are installed within the dam and its foundation. Piezometers measure the water pressure and help detect seepage paths. This is vital for preventing internal erosion and potential dam failure. Subsequently, seepage meter instruments (Figure 4.7) measure the rate of water seepage through the dam body and foundation, providing data to assess the effectiveness of seepage control measures [16].

Figure 4.6 Piezometer Monitoring Systems for tailings storage facilities in dam construction.

Figure 4.7 Seepage meter device.

Additionally, water level sensors monitor water levels (Figure 4.8) in the reservoir and downstream, ensuring that the dam operates within safe limits and help in flood management. In this context, flow meters (Figure 4.9) are also installed in spillways and outlets. Flow meters measure the discharge rates, which is essential for managing water release and maintaining reservoir balance.

Figure 4.8 Water level sensor for monitoring water levels in dam construction.

Figure 4.9 Example of flow meter device.

Weather stations collect data on precipitation, temperature, wind speed and other meteorological factors that can affect the dam and reservoir. Further, sediment sensors measure sediment accumulation in the reservoir, which can impact water storage capacity and dam operation. Additionally, seismographs (Figure 4.10) can be implanted at the dam site; seismographs detect and record seismic activity. This data is crucial for assessing the dam's response to earthquakes and ensuring its structural integrity [17].

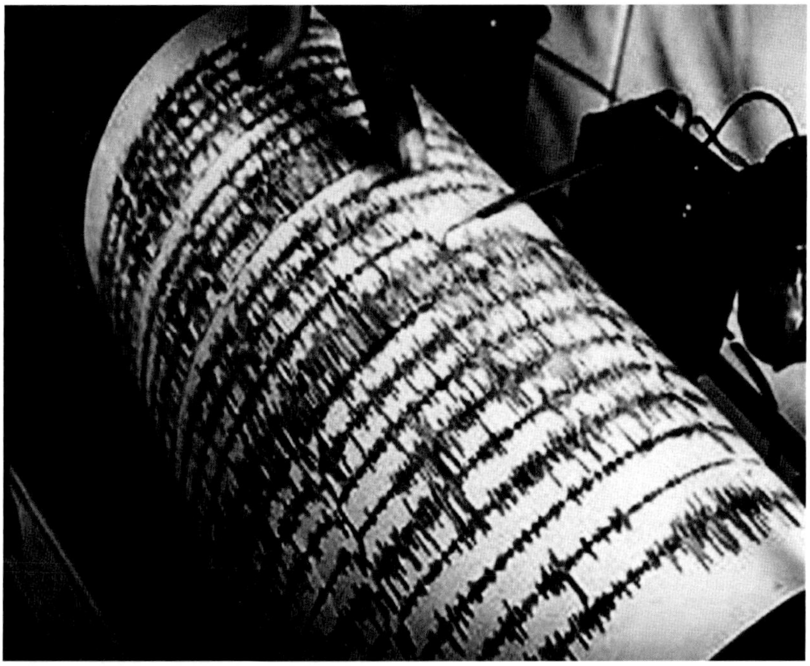

Figure 4.10 Seismograph device for detecting seismic activity.

Accelerometer devices measure the acceleration of the dam structure during seismic events, providing insights into the dynamic behavior of the dam. Inclinometer instruments can measure the inclination or tilt of the dam and its abutments, helping to detect any movement that could indicate instability [16–18].

The question arising now is: what is the role of remote sensing in geotechnical dam monitoring?

4.6 REMOTE SENSING MULTISPECTRAL AND HYPERSPECTRAL MONITORING OF DAM BEHAVIOR

Remote sensing technologies, particularly multispectral and hyperspectral imaging, have revolutionized the monitoring and management of dam behavior. These advanced techniques offer significant advantages in terms of spatial coverage, frequency of data acquisition and the ability to detect subtle changes in the dam and its surrounding environment.

The inquiry into the nature of multispectral and hyperspectral remote sensing technologies reveals significant insights into their functionalities and applications. Multispectral imagery is produced by sensors that measure reflected energy within several specific sections, known as bands, of the electromagnetic spectrum (Figure 4.11). Hyperspectral sensors, on the other hand, measure energy in much narrower and more numerous bands than multispectral sensors (Figure 4.12).

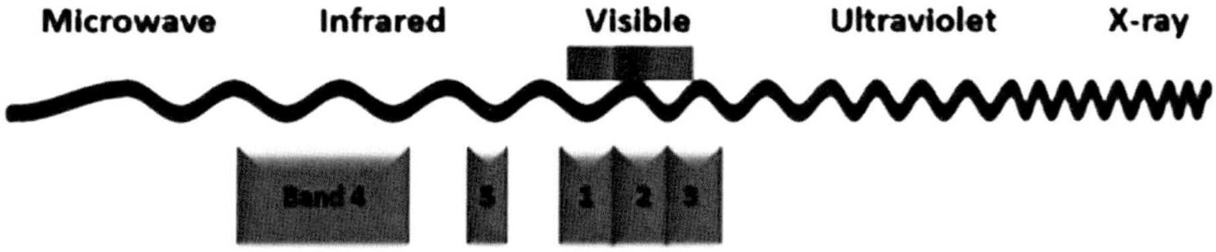

Figure 4.11 Multispectral spectral bands.

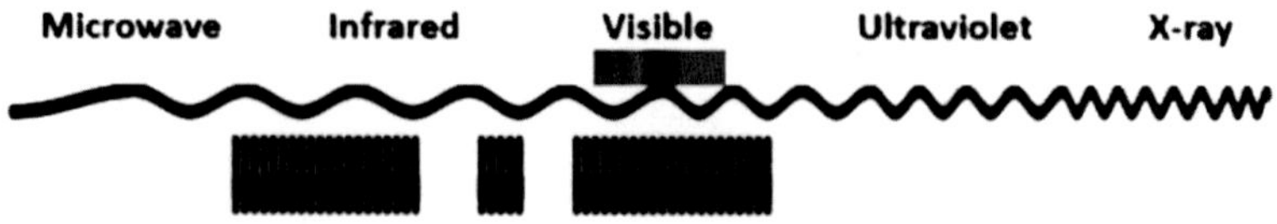

Figure 4.12 Hyperspectral spectral bands.

4.6.1 Key Differences Between Multispectral and Hyperspectral Sensors

The primary distinction between multispectral and hyperspectral sensors lies in the number and width of the bands they capture. Multispectral imagery (MSI) typically encompasses 3 to 10 bands, whereas hyperspectral imagery (HSI) consists of bands that are much narrower (10–20 nm) and far more numerous, often totaling to hundreds or even thousands (Figure 4.12). This extensive spectral detail in hyperspectral images enhances the capability to detect subtle differences in material composition and surface characteristics [19–21].

While hyperspectral remote sensing can distinguish between three closely related minerals due to their high spectral resolution, the multispectral Landsat Thematic Mapper may fail to do so. However, the complexity of working with hyperspectral data, such as managing redundancy across 200 narrow bands poses significant challenges.

Both multispectral and hyperspectral remote sensing have myriad real-world applications. Hyperspectral imagery, for example, is invaluable in mapping invasive species and aids in mineral exploration. Additionally, these technologies are extensively used in agriculture, ecology, oil and gas exploration, atmospheric studies and more.

The output from a hyperspectral sensor, often referred to as a 'data cube,' can be visualized as a stack of hundreds of images, each representing a specific wavelength. This data cube provides a unique spectral signature or fingerprint of an object, greatly enhancing its identification and classification capabilities.

4.6.2 Comparing MSI and HSI Systems

An MSI array system is analogous to an HSI array system, with key differences. While HSI arrays capture continuous wavelengths, MSI focuses on discrete wavebands selected based on specific requirements. Common RGB sensors illustrate the differences between HSI and MSI systems well. In RGB sensors, a Bayer pattern—comprising red, green, and blue filters—covers the pixels. These filters allow wavelengths from specific color bands to be absorbed while attenuating the rest of the light. Therefore, these bandpass filters have transmission bands in the range of 100 to 150 nm and have a slight spectral overlap (Figure 4.13). In other words, the RGB camera quantum efficiency curve views the overlap between red, green, and blue [20–23].

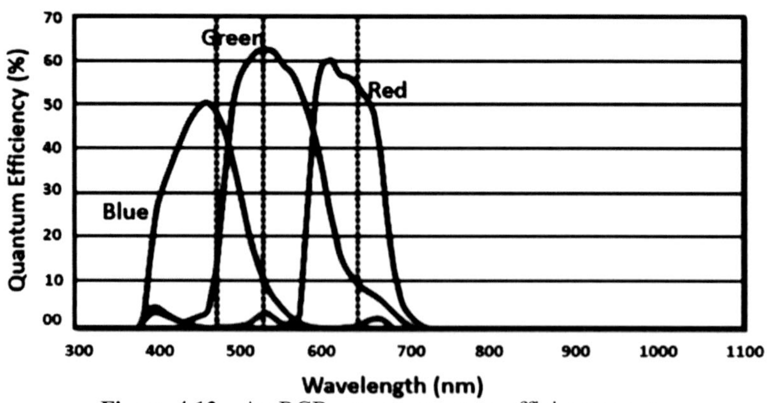

Figure 4.13 An RGB camera quantum efficiency curve.

In most MSI applications, the wavelength bands are significantly narrower and more numerous than those in RGB sensors. MSI typically captures wavebands in the tens of nanometers range and can include UV, NIR, and thermal wavelengths, depending on the application.

4.6.3 Practical Implications

In particular, multispectral imaging (MSI) can be considered a less sophisticated version of hyperspectral imaging (HSI) due to its reduced spectral resolution. Hyperspectral imaging, in reality, is optimal for applications that require sensitivity to subtle variations in signals along a continuous spectrum. These small signals may be overlooked by a system that samples larger wavebands [21–23]. However, certain systems require significant portions of the electromagnetic spectrum to be blocked, allowing only specific segments of the electromagnetic radiation to be selectively captured (Figure 4.14).

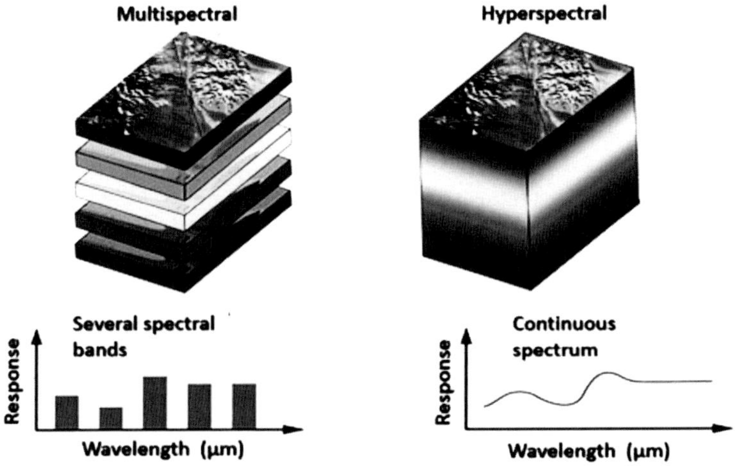

Figure 4.14 Multi- and hyperspectral stacks of images.

It can be said that the comparison between MSI and HSI highlights that while MSI captures several images across discrete wavelength regions, HSI images cover a larger continuous range of wavelength bands. This difference underscores the broader and more detailed spectral information that HSI provides, making it superior for applications sensitive to slight signal variations. Both technologies, however, play critical roles in remote sensing, offering unique advantages tailored to specific needs and challenges [24, 26].

4.7 QUANTIZATION TECHNIQUES FOR MULTISPECTRAL AND HYPERSPECTRAL DATA

In the 1980s, Richard Feynman proposed the concept of the quantum computer, which offers potential applications, such as quantum image processing. However, there has been limited research conducted on quantum image processing for mineral mapping using multispectral and hyperspectral data. Previous chapters established the theories of remote sensing mineral quantization and the basics of quantum image processing. Before applying quantum image processing for mineral exploration in remote sensing data, it is necessary to transform the image into a qubit format [35].

To comprehend this transformation, it is crucial to understand the differences between bits and qubits. A bit is the fundamental unit used to store spectral information (either 0 or 1) in conventional computers or classical devices and to represent a pixel in an image. Conversely, the minimum unit to store spectral information in a quantum computer is the qubit.

A conventional bit can exist in one of two possible states: $|0\rangle$ or $|1\rangle$, corresponding to the binary values 0 and 1 [32]. In contrast, a qubit can exist in a superposition of these states. Mathematically, a qubit state $|\psi\rangle$ can be expressed as:

$$|\Psi\rangle = \alpha|0\rangle + \beta|1\rangle \tag{4.1}$$

where α and β are complex probability amplitudes that satisfy the normalization condition:

$$|\alpha|^2 + |\beta|^2 = 1 \tag{4.2}$$

Equation 4.2 states that the wave function $|\psi\rangle$ associated with a pixel is related to the probability of retrieving the pixel in various gray level states, such as gray level 0 (gl_0), gray level 1 (gl_1), and so on. This relationship effectively represents the quantum pixel image, including the values of the pixels.

To clarify, in quantum image processing, each pixel in an image can be represented by a quantum state or wave function $|\psi\rangle$. This quantum state encodes the probability of the pixel being in specific gray level states. For instance, gray level 0 (gl_0) and gray level 1 (gl_1) correspond to different intensity values of the pixel. The quantum representation allows for the superposition of these states, providing a richer and more complex encoding of image information than classical methods [27–30].

Mathematically, if a pixel's quantum state is given by:

$$|\psi\rangle = \alpha_0|gl_0\rangle + \alpha_1|gl_1\rangle + \alpha_2|gl_2\rangle + \cdots + \alpha_n|gl_n\rangle \tag{4.3}$$

where α_i are complex coefficients representing the probability amplitudes for each gray gl_i, then the probability of the pixel being in a specific gray level gl_i is given by $|\alpha_i|^2$. These probability amplitudes must satisfy the normalization condition:

$$\sum_{i=0}^{n} |\alpha_i|^2 = 1 \tag{4.4}$$

This equation illustrates how the wave function $|\psi\rangle$ encapsulates the likelihood of the pixel's intensity values across all possible gray levels, thus forming the quantum pixel image. Pointless to say, Equation 4.4 conjoins the wave function $|\psi\rangle$ of a pixel to the probabilities of various gray level states, providing a quantum framework for representing and processing pixel values in an image [29–32].

This superposition allows a qubit to represent a much richer set of information than a classical bit. When applied to image processing, each pixel, represented as a qubit, can simultaneously encode multiple spectral features, enhancing the capability of spectral construction mapping for GERD in multispectral and hyperspectral data. Therefore, this process mimics the visual observation of an image, where the image is primarily stored in quantum memory in a precise 'quantum digital' format (Figure 4.15).

In this quantum digital format, the intensity values of each pixel are represented by quantum states, allowing for the storage and manipulation of the image data in a quantum computer. The use of quantum memory enables the efficient encoding, retrieval, and processing of complex image information, exploiting quantum properties such as superposition and entanglement:

$$|\Psi_{\Sigma}^{D}\rangle = (N)^{-0.5} \sum_{DN} |\Sigma(DN)\rangle |DN\rangle, \tag{4.5}$$

the index register $|DN\rangle$ is given by:

$$|DN\rangle = \otimes_{l=1}^{\log_2(N)} |DN_l\rangle \tag{4.6}$$

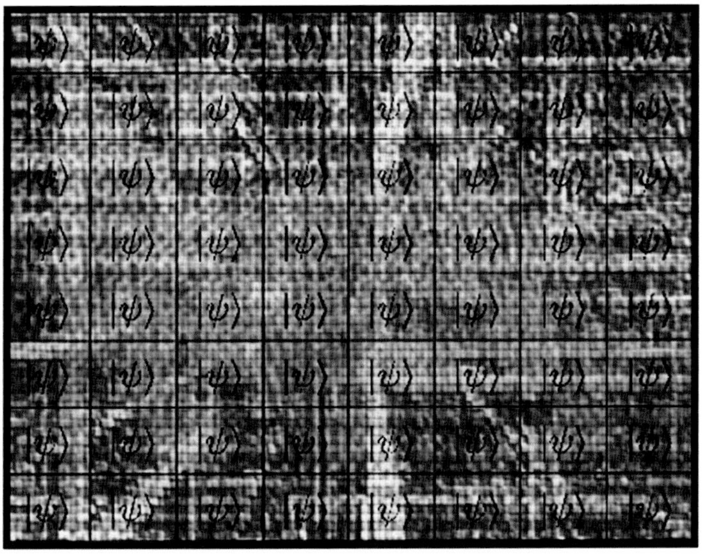

Figure 4.15 Quantum image digital format.

Equation 4.6 reveals that the quantum state of an image pixel is represented by qubits. The state of two qubits, for instance, requires four complex amplitudes, which can be stored in a 4D vector as given by:

$$\psi|DN\rangle = DN_{00}|01\rangle + DN_{01}|10\rangle + DN_{11}|11\rangle \tag{4.7}$$

Here, $|\psi\rangle$ represents the quantum state of the two qubits, and DN_{00}, DN_{01}, DN_{10}, and DN_{11}, are complex probability amplitudes associated with each possible state of the two qubits. These amplitudes must satisfy the normalization condition:

$$|DN_{00}|^2 + |DN_{01}|^2 + |DN_{10}|^2 + |DN_{11}|^2 = 1 \tag{4.8}$$

This 4D vector (DN_{00}, DN_{01}, DN_{10}, and DN_{11}) encapsulates all the information needed to describe the state of the two qubits. Expanding this concept to an image, where each pixel is represented by multiple qubits, the quantum state of the entire image can be described by a high-dimensional vector encompassing the amplitudes of all possible states of the qubits representing the image pixels. This approach allows for sophisticated quantum image processing techniques that leverage the inherent parallelism and complexity of quantum computing [31, 33, 35].

Equation 4.8 illustrates that the quantum state of an image pixel is represented by qubits, with the state of two qubits requiring a 4D vector of complex amplitudes to describe all possible states. This foundational concept enables the advanced manipulation and analysis of image data in the quantum domain [35].

For (n) qubits, it is required to sustain the track of (2^n) complex amplitudes. Consequently, these vectors grow exponentially with the number of qubits. This exponential growth is a key reason why quantum computers with large numbers of qubits are extremely complicated to simulate.

For instance, a modern laptop can easily simulate a general quantum state of approximately 20 qubits. This is because the computational resources required for such a simulation are manageable, given that (2^{20}) is approximately 1 million complex amplitudes. However, simulating 100 qubits is vastly more complex, as it involves tracking (2^{100}) (approximately (1.27×10^{30}) complex amplitudes. This immense number of amplitudes far exceeds the capacity of even the largest supercomputers, which is why simulating quantum systems with such a high number of qubits remains a significant challenge in computational science [27, 35].

In condensation, while simulating quantum states with a relatively small number of qubits is feasible with current technology, the exponential increase in the number of required complex amplitudes makes it extraordinarily difficult to simulate systems with a large number of qubits. This limitation underscores the need for actual quantum hardware to explore and harness the full potential of quantum computing [27–29].

Figures 4.16 and 4.17 demonstrate two sources of raw hyperspectral and multispectral data represented in quantum image digital format. The data are derived from Landsat-8 (Table 4.1) and ASTER (Table 4.2), both of which illustrate low-quality pixel variations in their raw form.

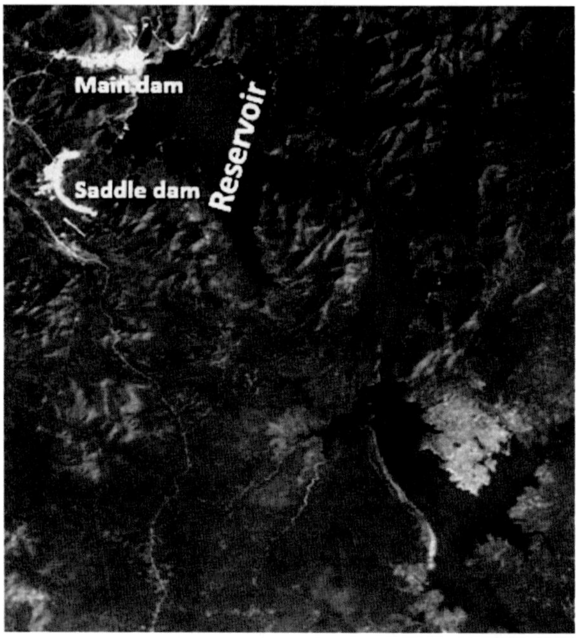

Figure 4.16 Raw true hyperspectral ASTER image color of GERD.

Figure 4.17 Raw true multispectral Landsat-8 image color of GERD.

Landsat-8's Operational Land Imager (OLI) and ASTER (Advanced Spaceborne Thermal Emission and Reflection Radiometer) provide higher spatial resolutions of 15 meters and 30 meters, respectively, as detailed in Table 4.1. However, this higher spatial resolution comes at the cost of less frequent data collection.

Data from ASTER and Landsat-8 are accessible for free through the EROS (Earth Resources Observation and Science) Center, managed by the U.S. Geological Survey (USGS). Specifically, ASTER data were acquired from the U.S. Geological and Science Center (EROS) on November 6, 2020, via the EarthExplorer portal (https://earthexplorer.usgs.gov/) and Landsat-8 data were obtained from the USGS Global Visualization Viewer (GloVis) (https://glovis.usgs.gov/) on February 14, 2022, respectively [35].

True color Landsat-8 imagery (Table 4.1), similar to the interpretation of aerial photography and a natural color image from ASTER provide valuable visual information. Natural color imagery attempts to replicate the overall appearance of a true color image using green, blue, and grayish tones. However, ASTER lacks the specific red, green, and blue (RGB) visible wavelength bands required for a true "true color" image, resulting in some limitations in color representation (Table 4.2).

Table 4.1 Landsat-8 band spectral characteristics

Subsystem	Band Number	Spectral Range (μm)	Ground Resolution (m)
	1	0.433–0.453	
	2	0.450–0.515	
VNIR	3	0.525–0.600	30
	4	0.630–0.680	
	5	0.845–0.885	
	Pan	0.500–0.680	15
SWIR	6	1.560–1.660	15
	7	2.100–2.300	
TIR	9	1.360–1.390	
	10	10.30–11.30	100
	11	11.50–12.50	

Table 4.2 ASTER band spectral characteristics

Subsystem	Band Number	Spectral Range (μm)	Ground Resolution (m)
	1	0.520–0.600	
VNIR	2	0.630–0.690	15
	3	0.780–0.860	
	4	1.600–1.700	
	5	2.145–2.185	
SWIR	6	2.185–2.225	30
	7	2.235–2.285	
	8	2.295–2.365	
	9	2.360–2430	
	10	8.125–8.475	
	11	8.475–8.825	
TIR	12	8.925–9.275	90
	13	10.25–10.95	
	14	10.95–11.65	

When the raw image pixels are transformed into a quantum digital format, the information of the images is stored in qubits. This transformation enhances the information quality of the images. In the quantum pixel image, based on the superposition principle in quantum mechanics (Figures 4.18 and 4.19), the quantum states of a qubit are used to represent the complete gray level of each pixel $|DN\rangle$.

In this approach, a qubit/pixel is used to encode every pixel of the images, capturing the intensity value across all gray scales. This method effectively converts multispectral and hyperspectral raw images into a format represented by a series of quantum superposition states $|DN_{00}\rangle$, $|DN_{01}\rangle$, ..., $|DN_{NN}\rangle$. It reduces the impact of variations in gray levels $|DN\rangle$ from one bit plane to another [35–37].

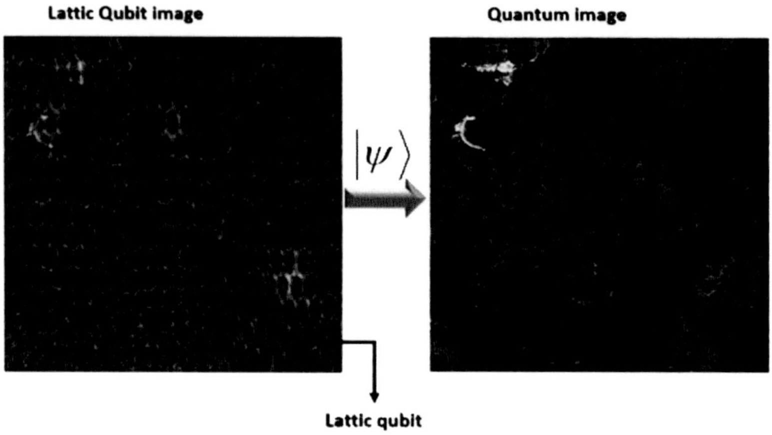

Figure 4.18 Encoded ASTER image in the qubit.

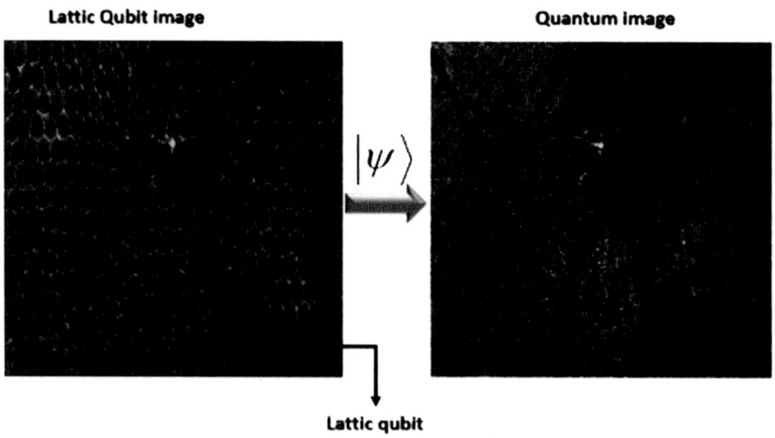

Figure 4.19 Encoded Landsat-8 image in the qubit.

4.8 SPECTRAL REFLECTANCE QUANTUM IMAGE FORMATION (SRQIF)

The spectra signature is simulated from the USGS Spectral Library version 7.0. To this end, a flexible representation for quantum images is implemented for encoding the matched real spectral reflectance in remote sensing data [32,36]. In this context, the similarity between the spectral library $|\psi\rangle_r$ and the spectral reflectance of images $|\psi\rangle_{smr}$ is mathematically articulated as:

$$Sim\left(|\psi\rangle_{smr}, |\psi\rangle_r\right) = f(DN^0_{i,j}, DN^1_{i,j}, \ldots, DN^{2^{2n-1}}_{i,j}), \tag{4.9}$$

here $DN_{i,j}$ is the pixel spectral information along raw i and column j; respectively and equation 4.9 reveals that $Sim(|\psi\rangle_{smr}, |\psi\rangle_r) \in [0, 1]$ [32, 35, 37]. In this circumstance, the matched spectral library $|\psi\rangle_r$ automatically overprint $|\psi\rangle_{smr}$ and determine the specifically requested mineral as a function of the following criteria:

In the circumstance of $\forall i, DN^n_{i,j} = \frac{\pi}{2}$, $Sim(|\psi\rangle_{smr}, |\psi\rangle_r) = 0$ and both $|\psi\rangle_r$ and $|\psi\rangle_{smr}$ are signified as being absolutely dissimilar. However, the specifically matched spectra signature of GERD and its surrounding environment overprint automatically in remote sensing if $\forall i, DN^n_{i,j} = 0$, then $Sim(|\psi\rangle_{smr}, |\psi\rangle_r) = 1$, where $n = 0, 1, 2, \ldots, 2^{2n} - 1DN^n_{i,j}$.

Let us consider that cluster of different Quantum spectral construction mapping for GERD and its surrounding environment $Sim(|\psi\rangle_{smr}, |\psi\rangle_r)$ derived post-implementation of equation 4.9 for remote sensing data as $C|\psi\rangle_{smr} \in (Sim(|\psi\rangle_{smr}, |\psi\rangle_r)$; which signify by

$$C|\psi\rangle_N = \left[C|\psi\rangle_1, C|\psi\rangle_2, \ldots, C|\psi\rangle_N\right] \tag{4.10}$$

The centroid of every mineral cluster is given by:

$$C|\psi\rangle_{c_j} = \frac{1}{M_j} \sum_{k \in C|\psi\rangle_j} C\left(Sim\left(|\psi\rangle_{smr}, |\psi\rangle_r\right)\right)_k \tag{4.11}$$

Equation 4.8 reveals that each $C(Sim(|\psi\rangle_{smr}, |\psi\rangle_r))$ is the data point of the cluster $C|\psi\rangle$ and M_j is represents the number of data points of that cluster [28, 32, 35].

4.9 ENTANGLEMENT SHAPES THE FORMATION OF SPECTRAL LIBRARIES FOR GERD

Entanglement, in the context of spectral library formation for projects like the Grand Ethiopian Renaissance Dam (GERD), refers to the intricate interconnection of various factors influencing the composition and development of spectral data repositories. These libraries are crucial in remote sensing and monitoring efforts, providing essential data for understanding environmental dynamics and project impacts.

In the case of GERD, the formation of spectral libraries involves the systematic collection and categorization of spectral data from the dam's vicinity and surrounding areas. This data encompasses a wide range of electromagnetic wavelengths captured by remote sensing instruments, reflecting the diverse characteristics of land cover, water bodies and infrastructure. However, the process of forming these libraries is influenced by multiple intertwined factors, such as geographical variability, seasonal changes and anthropogenic activities. These elements contribute to the complexity and richness of the spectral data, which are organized into libraries to serve as reference resources for ongoing monitoring and analysis.

Understanding how entanglement shapes the formation of these spectral libraries is essential for effectively utilizing remote sensing data in assessing environmental impacts, managing resources, and supporting sustainable development initiatives associated with large-scale infrastructure projects like GERD.

Developing a novel theory to understand the quantization of the spectral library involves leveraging the principles of quantum entanglement. Let us assume an entangled pair of electrons, where one electron represents the spin state of a sensor (such as those used in spectroscopy and remote sensing) and the other represents the spin state of a spectra signature of GERD and its surrounding environment. This pair can be described in a particular Bell state, which is one of the fundamental forms of quantum entanglement. Specifically, the Bell state can be expressed as:

$$|\Phi_0\rangle = \frac{1}{\sqrt{2}}\left(|\uparrow\rangle_s \otimes |\uparrow\rangle_m + |\downarrow\rangle_s \otimes |\downarrow\rangle_m\right). \quad (4.12)$$

Equation 4.12 indicates that entangled electrons are present between the sensors and the minerals. In this context, the sensor measures the 'Bell-state' of the system, which entangles the quantum state of the sensor with that of the spectra signature of GERD. Therefore, the sensor transmits the result of its GERD spectra signature through the specific conventional technique of photon-electron interaction. Consequently, the Pauli matrices are used to theorize Bell states, which form an orthonormal basis of entangled states for the tensor product space of spin- $\frac{1}{2}$ particles:

$$|\Phi_0\rangle = I \otimes \sigma_0 |\Phi_0\rangle = \frac{1}{\sqrt{2}}\left(|\uparrow\rangle_s \otimes |\uparrow\rangle_m + |\downarrow\rangle_s \otimes |\downarrow\rangle_m\right). \quad (4.13)$$

$$|\Phi_1\rangle = I \otimes \sigma_1 |\Phi_0\rangle = \frac{1}{\sqrt{2}}\left(|\uparrow\rangle_s \otimes |\uparrow\rangle_m + |\downarrow\rangle_s \otimes |\uparrow\rangle_m\right). \quad (4.14)$$

$$|\Phi_2\rangle = I \otimes \sigma_2 |\Phi_0\rangle = \frac{i}{\sqrt{2}}\left(|\uparrow\rangle_s \otimes |\downarrow\rangle_m - |\downarrow\rangle_s \otimes |\uparrow\rangle_m\right). \quad (4.15)$$

$$|\Phi_3\rangle = I \otimes \sigma_3 |\Phi_0\rangle = \frac{1}{\sqrt{2}}\left(|\uparrow\rangle_s \otimes |\uparrow\rangle_m - |\downarrow\rangle_s \otimes |\downarrow\rangle_m\right). \quad (4.16)$$

Mathematically and based on equations 4.13 to 4.16, Pauli matrices can be articulated as:

$$\sigma_0 = I = \begin{pmatrix} 1 & 0 \\ 0 & 1 \end{pmatrix}, \ \sigma_1 = \begin{pmatrix} 0 & 1 \\ 1 & 0 \end{pmatrix}, \ \sigma_2 = \begin{pmatrix} 0 & -i \\ i & 0 \end{pmatrix}, \ \sigma_3 = \begin{pmatrix} 1 & 0 \\ 0 & -1 \end{pmatrix}. \quad (4.17)$$

Accordingly, the spin operators laterally each axis are delineated as $\frac{\hbar}{2}$ σ_0, σ_1, σ_2, and σ_3 for the x, y, and z, axes; respectively.

Remote sensing satellites or spectroscopy measure the entanglement quantity between photons and the materials used in the construction of the Grand Ethiopian Renaissance Dam (GERD). The rate of entanglement varies across different spectrum bands due to interactions with these construction materials.

In this context, the concept of quantum teleportation is speculative, identifying how remote sensing sensors measure the spectral components of GERD construction materials across distinct bands. Essentially, spectral measurements disentangle the sensors from GERD and entangle the reflectance in separate bands. Depending on the specific entangled state determined by the sensor, each spectral band precisely reflects how the materials used in the construction of GERD interact (mr), enabling manipulation of GERD's spectral signature to match the original reflectance spectra. Thus, the state of different spectral channels or bands is 'teleported' from sensors to GERD, aligning GERD's state to mirror the original spectral appearance. It is crucial to note that quantum mechanics' principles, such as the no-cloning and no-deletion theorems, ensure that the state variations in stored binary spectral signatures are preserved in these processes [33–37].

Therefore, the state of spectral signatures $|\psi\rangle_{sm_r}$, determined in separate individual binary bands or channels can be expressed as:

$$|\Psi\rangle_{sm_r} = \sum_{i=0}^{n} 0.5 |\Phi\rangle_{sm_r} \otimes \begin{pmatrix} 0 & 1 \\ 1 & 0 \end{pmatrix} |\Psi\rangle_m \quad (4.18)$$

Consequently, constituent atoms of GERD materials have changed the spin state of particle owing to photon-atom interaction; which is expressed as:

$$|\Psi\rangle_m = c_1 |\uparrow\rangle_m + c_2 |\downarrow\rangle_m, \quad (4.19)$$

Using equation 4.18 into 4.19, then we obtain the full scenario of how the remote sensing operates to identify constituent atoms of GERD materials as:

$$|\Psi\rangle_{sm_r} = \sum_{i=0}^{n} 0.5|\Phi\rangle_{sm_r} \otimes \begin{pmatrix} 0 & 1 \\ 1 & 0 \end{pmatrix} c_1|\uparrow\rangle_m + c_2|\downarrow\rangle_m. \tag{4.20}$$

Equation 4.20 reveals that the state to be teleported and the state of one of the entangled particles. The state to be teleported is the quantum state of constituent atoms of GERD materials spectral reflections and state to entangled is photon-molecule interactions (Figure 4.20). In this understanding, the final spin state of photon-atom interaction looks similar to constituent atoms of GERD materials spectral reflectance state.

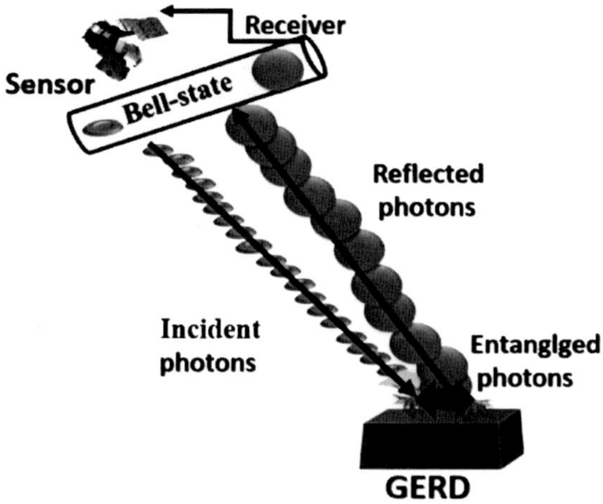

Figure 4.20 Remote sensing quantum teleportation and entanglement of GERD materials.

4.10 MULTISPECTRAL AND HYPERSPECTRAL DATA STORAGE IN QUANTUM COMPUTING

Storing images in quantum computing involves representing the pixel values of an image as quantum states. This process leverages the principles of quantum superposition and entanglement to encode information efficiently. Each pixel value in a grayscale image can be represented as a quantum state. For simplicity, let us assume the image has $N \times N$ pixels, and each pixel value ranges from 0 to 255, which can be represented using 8 bits. Therefore, we can encode each pixel value into a qubit or a group of qubits. For an 8-bit grayscale image, we need 8 qubits to represent each pixel value [33–36]. The state of a single pixel p_{ij} at position (i, j) can be written as:

$$|p_{ij}\rangle = |p_{ij,7}p_{ij,6}\cdots p_{ij,1}p_{ij,0}\rangle \tag{4.21}$$

where p_{ijk} is the k-th bit of the pixel value p_{ij}. To encode the entire $N \times N$ image, we need N_2 such qubit states. The overall quantum state of the image can be represented as a tensor product of the individual pixel states:

$$|\Psi\rangle = \otimes_{i=0}^{N-1} \otimes_{j=0}^{N-1} |p_{ij}\rangle \tag{4.22}$$

In quantum computing, we often work with normalized states. The superposition principle allows us to represent a combination of all possible pixel values. For a single pixel, the superposition state can be written as:

$$|p_{ij}\rangle = \sum_{k}^{255} \alpha_k |k\rangle \tag{4.23}$$

Here α_k is complex coefficients such that $\left|p_{ij}\right\rangle = \sum_k^{255} \left|\alpha_k\right|^2 - 1$. In this perspective, quantum images can be represented in various models, such as the Flexible Representation of Quantum Images (FRQI). In the FRQI model, a grayscale image is represented by a single qubit rotation angle to encode pixel intensity and a set of qubits to encode the pixel positions.

$$|I\rangle = \frac{1}{\sqrt{2^n}} \sum_{i=0}^{2^n-1} (cos(\theta_i)|0\rangle + sin(\theta_i)|1\rangle)|i\rangle \tag{4.24}$$

where $n = \log_2 (N \times N)$ is the number of qubits needed to encode the positions of the pixels, and θ_i is the angle representing the intensity of the i-th pixel [33, 38].

4.11 MODELING QUANTUM SPECTRA SIGNATURE OF GERD CONSTRUCTION AND ITS SURROUNDING ENVIRONMENT

To simulate the quantum spectral signature of the GERD construction and its surrounding environment from multispectral and hyperspectral data, we assume the system behaves like a quantum linear system. The optical remote sensing data can be represented using a Hermitian matrix. In this context, the Hermitian matrix captures the interactions and properties of the spectral data as a Hermitian matrix I^H and $DN \in I^H$. In other words, in quantum mechanics, a Hermitian matrix (or Hermitian operator) is a complex square matrix that is equal to its own conjugate transpose. Hermitian matrices are used to represent observable quantities, such as the spectral radiance in this case [28–35]. Mathematically, this can be expressed as:

$$|\Psi\rangle_{sm_r} = \left(\frac{|L(\lambda)\rangle_{\max}}{254} - \frac{|L(\lambda)\rangle_{\min}}{255} \right) \otimes |DN\rangle\langle DN| + |L(\lambda)\rangle_{\min} \tag{4.25}$$

Equation 4.25 reveals the quantum state of the spectral reflectance in optical remote sensing image; in which the quantum state of $|DN\rangle\langle DN| \in I^H$ is entangled the quantum state of spectral reflectance $|\Psi\rangle_{sm_r}$ as demonstrated previously in section 4.9.

Moreover, the quantum gain state is presented in $\left(\frac{|L(\lambda)\rangle_{\max}}{254} - \frac{|L(\lambda)\rangle_{\min}}{255} \right)$ and quantum state of minimum spectral radiance is presented by $|L(\lambda)\rangle_{\min}$. Consequently, the spectral reflectance of a single band of the optical DNs image can be formulated as:

$$|\Psi\rangle_{sm_r} = \left(\frac{|L(\lambda)\rangle_{\max} - |L(\lambda)\rangle_{\min}}{|DN\rangle_{\max} - |DN\rangle_{\min}} \right) \otimes \left(|DN\rangle - |DN\rangle_{\min}\right) + |L(\lambda)\rangle_{\min} \pm |\varepsilon\rangle \tag{4.26}$$

In equation 4.26, the gate complexity is the number of 2-qubit gates; which is exploited. An algorithm is gate-efficient if its gate complexity is larger than its query complexity only by a logarithmic factor. Formally, an algorithm with query complexity Q is gate-efficient if its gate complexity is $Q(Q \cdot \text{poly}(\log(qN)))$. Subsequently, the runtime of the quantum algorithm for obtaining precise spectral reflectance can be demonstrated implementing $O(\kappa^2 \log N\varepsilon^{-1})$, where $\varepsilon > 0$ is the error and κ is condition number. Therefore $\kappa = \frac{\max_i |\psi\rangle_{smr}}{\min_i |\psi\rangle_{smr}}$ in which, If the condition number is not too much larger than one, the estimated spectral reflectance of the mineral is well-conditioned, and it means that its inverse can be computed with good accuracy [35].

In such a technique, the gray level of each individual pixel could be encoded as a qubit per pixel, where the qubit represents the pixel's value within the full grayscale range, effectively encoding a spectral image. In the quantum form, this could be, therefore, understood by exploitation polynomials to designate the code in a conventional approach. Nevertheless, the entire technique is effective, as in the quantum world, both $|0\rangle$ and $|1\rangle$ states are present, but the 0s are suppressed. Then the equivalence of polynomial zeros is interpreted as the coefficient of the base state |0, and the equivalence of polynomial ones as the coefficient of the base state $|1\rangle$.

4.12 QUANTUM SPECTRA SIGNATURE OF THE SURROUNDING ENVIRONMENT OF GERD

The analyzed ASTER image of the GERD surrounding environment site demonstrates that the spectral bands in ASTER (Advanced Spaceborne Thermal Emission and Reflection Radiometer) cover three main regions: visible and near-infrared (VNIR), shortwave infrared (SWIR), and thermal infrared (TIR). Consequently, the exact values may vary depending on the specific composition and environmental conditions of the rocks. The plot above illustrates the synthetic quantum spectral reflectance curves for various metamorphic and sedimentary rocks across the VNIR and SWIR bands of the ASTER sensor. The rocks included are: metamorphic rocks: chlorite, amphibolite, talc schist, biotitic schist, quartz graphite; and sedimentary rocks: conglomerate, sandstone (Figures 3.21 and 3.22).

In the ASTER image, chlorite shows higher quantum spectra reflectance, about 0.3 at 0.56 μm, with significant features peaking at 0.5 around 2.2 μm (Figure 4.21). Conversely, chlorite exhibits a quantum spectra reflectance of approximately 0.2 at 0.5 μm, increasing to around 0.3 at 2.2 μm in Landsat-8 data. In Landsat-8, amphibolite quantum spectra reflectance values are relatively low, starting at around 0.15 at 0.5 μm and gradually increasing to 0.25 at 2.2 μm (Figure 4.22). However, amphibolite's quantum spectra reflectance in ASTER is more pronounced, starting at 0.2 at 0.56 μm and showing distinct absorption features, peaking at 0.35 at 2.2 μm.

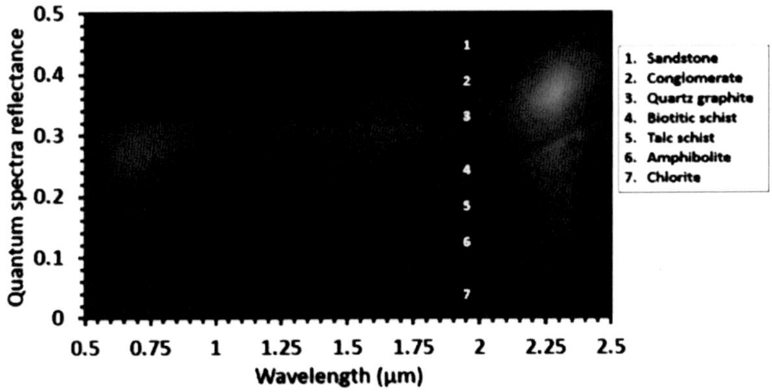

Figure 4.21 Quantum spectra reflectance for GERD surrounding environment in ASTER data.

Similarly, talc schist in Landsat-8 images exhibits moderate quantum spectra reflectance values, beginning at 0.2 at 0.5 μm and increasing to 0.3 at 2.2 μm. This gradual increase indicates a consistent response across the quantum spectra reflectance range, useful for general identification. In contrast, ASTER data is dominated by the highest quantum spectra reflectance in the VNIR region, starting at 0.25 at 0.56 μm, with peaks around 0.4 at 2.2 μm. ASTER's higher quantum spectra reflectance resolution captures more detailed features, making it better suited for identifying subtle mineral variations in talc schist.

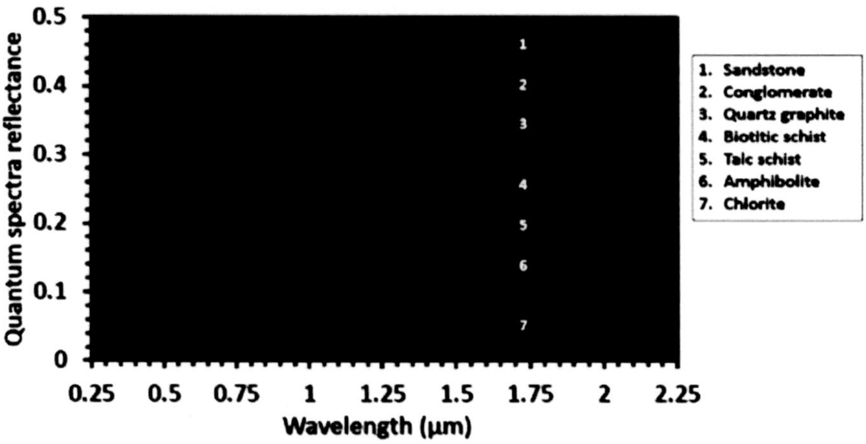

Figure 4.22 Quantum spectra reflectance for GERD surrounding environment in Landsat-8 data.

Additionally, biotitic schist shows moderate quantum spectra reflectance, approximately 0.2 at 0.5 μm, increasing to 0.3 at 2.2 μm in Landsat-8 data. The pattern is similar to that of talc schist, providing a baseline for comparison across metamorphic rocks. In ASTER images, however, biotitic schist displays higher quantum spectra reflectance, about 0.25 at 0.56 μm, with distinct features peaking at 0.4 at 2.2 μm. These distinct quantum spectra reflectance features enhance the ability to identify biotitic schist more precisely [22, 26, 35].

Furthermore, quartz graphite in Landsat-8 demonstrates low to moderate quantum spectra reflectance, starting at 0.15 at 0.5 μm and slightly increasing to 0.25 at 2.2 μm. The relatively flat quantum spectra reflectance indicates less variability, which can be a challenge for detailed analysis. ASTER, nevertheless, captures higher quantum spectra reflectance, beginning at 0.2 at 0.56 μm, with notable features around 0.35 at 2.2 μm (Figure 4.21). The enhanced spectral resolution of ASTER highlights subtle variations in quartz graphite that Landsat-8 might miss.

Moreover, sedimentary rocks such as conglomerate in Landsat-8 exhibit relatively high quantum spectra reflectance, about 0.25 at 0.5 μm, gradually increasing to 0.35 at 2.2 μm. The broad quantum spectra reflectance response is useful for distinguishing conglomerate from other rock types. In ASTER images, on the other hand, conglomerate shows higher quantum spectra reflectance in the VNIR bands, starting at 0.3 at 0.56 μm and peaking at 0.45 at 2.2 μm. This higher resolution helps capture the complex composition of conglomerate more effectively.

Lastly, sandstone in Landsat-8 images has high quantum spectra reflectance values around 0.3 at 0.5 μm, increasing to 0.4 at 2.2 μm. The distinct quantum spectra reflectance response is indicative of sandstone's mineral content, making it easier to identify in multispectral images. In ASTER images, sandstone exhibits even higher quantum spectra reflectance, beginning at 0.35 at 0.56 μm, with peaks around 0.5 at 2.2 μm. The detailed quantum spectra reflectance data provided by ASTER facilitates precise identification and analysis of sandstone.

The comparison of quantum spectra reflectance between Landsat-8 and ASTER images reveals significant differences in their capabilities. ASTER, with its higher spectral resolution, is more adept at capturing detailed spectral features, particularly in the VNIR and SWIR regions. This enhanced capability makes ASTER more suitable for identifying subtle mineralogical variations in both metamorphic and sedimentary rocks.

While Landsat-8 provides a general overview of reflectance patterns useful for broad geological assessments, ASTER's ability to capture fine spectral details facilitates more precise identification of rock types. This makes ASTER a valuable tool for detailed geological studies and applications, particularly where subtle variations in mineral composition are critical.

4.13 COMPARISON BETWEEN QUANTUM SPECTRA SIGNATURE OF GERD CONSTRUCTION IN LANDSAT-8 AND ASTER DATA

In comparing the quantum spectral reflectance of construction materials as observed by Landsat-8 and ASTER, several key trends emerge (Figures 4.23 and 4.24). For concrete, both sensors show a steady increase in reflectance values; however, Landsat-8 starts measuring at 0.44 μm while ASTER begins at 0.56 μm. Consequently, the trend for cement is similar to that of concrete in both datasets, but cement exhibits slightly higher reflectance values in both sensors.

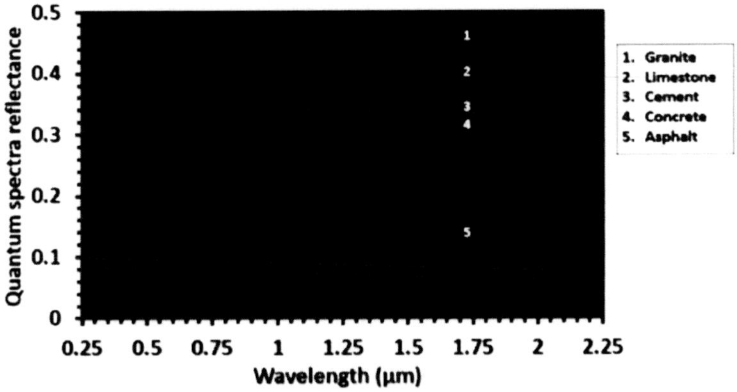

Figure 4.23 Quantum spectra reflectance for GERD construction in Landsat-8 data.

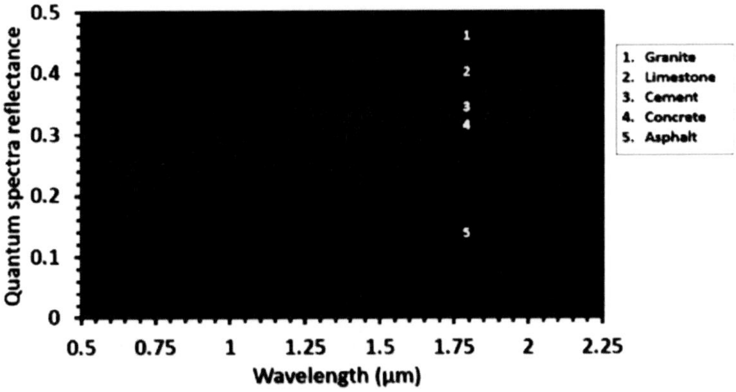

Figure 4.24 Quantum spectra reflectance for GERD construction environment in ASTER Landsat-8 data.

Asphalt, on the other hand, demonstrates the lowest reflectance across the spectrum in both Landsat-8 and ASTER images, though it shows a slight increase in reflectance across the spectrum in the ASTER data. Limestone's reflectance increases moderately in both sensors, peaking at 2.2 μm. Notably, granite exhibits the highest reflectance values in both Landsat-8 and ASTER datasets, with a consistent increase across all measured wavelengths (Figures 4.23 and 4.24).

In comparing these values, concrete in both Landsat-8 and ASTER images shows a consistent increase in reflectance, though ASTER starts at a higher initial reflectance (0.25 vs. 0.2). Cement follows a similar pattern in both datasets, with ASTER displaying higher values overall (0.3 to 0.45) compared to Landsat-8 (0.25 to 0.4) (Figures 4.23 and 4.24).

Asphalt consistently shows the lowest reflectance across both sensors, with a slightly higher initial reflectance in ASTER (0.15 vs. 0.1) but similar increasing trends. Limestone's reflectance values are moderately higher in ASTER, peaking at 0.45 compared to Landsat-8's peak of 0.4.

Therefore, granite has the highest reflectance values in both datasets, with ASTER showing a higher starting point (0.35 vs. 0.3) and reaching a higher peak (0.5 vs. 0.45). Thus, while both sensors provide a similar trend for the spectral reflectance of construction materials, ASTER starts its measurements at a slightly higher wavelength and generally shows higher reflectance values for cement and asphalt. These observations indicate the higher spectral resolution of ASTER, allowing for more detailed analysis.

Pointless to declare, while both Landsat-8 and ASTER provide consistent trends in spectral reflectance for construction materials, ASTER's higher initial values and higher peaks across various materials indicate its superior spectral resolution. This enhanced capability makes ASTER more adept at capturing detailed spectral features, particularly useful for identifying and analyzing construction materials with greater precision [18, 20, 35].

4.14 QUANTUM SPECTRAL SIGNATURES ACCURATELY IDENTIFIES GERD CONSTRUCTION AND ITS SURROUNDING ENVIRONMENT

It is worth mentioning that both Landsat-8 and ASTER deliver accurate information in dissimilar quantum spectra signature of both GERD construction and its surroundings differentiations In practice, Figure 4.25 provides an insight into the usefulness of quantum spectral identification, which is validated by Pearson correlation to predict the occurrence percentage of spatial mineral variations along with their magnitude. In the quantum spectral identification algorithms indicate a high correlation with quantum spectral library computing algorithm of 90% and 95% confidence levels for Landsat-8 and ASTER images, respectively. This work agrees with that of Marghany. [35].

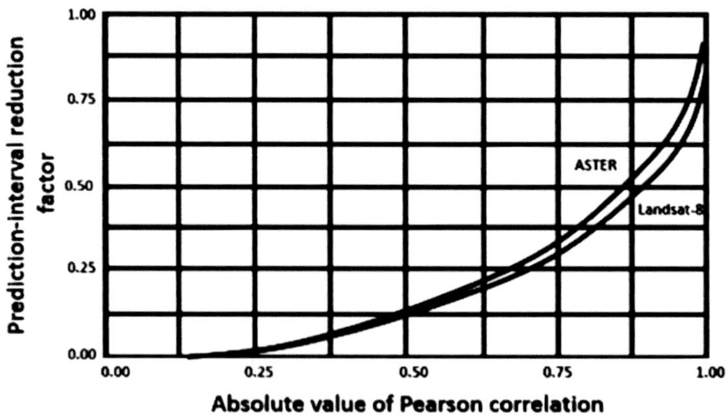

Figure 4.25 Pearson correlation for predicting the percentage of occurrence of spatial quantum spectra signature variations in ASTER and Landsat-8 images.

The hyperspectral data from ASTER, which has more bands than Landsat-8, enhances the application of the quantum spectral algorithm in identifying a greater number of minerals compared to Landsat-8 data. The novelty of quantum spectral mineral percentage algorithm is the first quantum computing algorithm for mineral identifications. Conventional image band ratio procedures just identify the possible occurrence of specific minerals. However, the band ratio algorithm is just a sort of image enhancement without delivering an exact quantity of quantum spectra signature of both GERD construction and its surroundings through optical remote sensing data.

Further confirmations can be also delivered by quantum cosine similarity between quantum spectral library and quantum spectra signature of both GERD construction and its surroundings images. This can be mathematically expressed by [35]:

$$\Psi|\cos(\theta)\rangle = \frac{\rho(|\psi\rangle, C|\psi\rangle).|\Psi\rangle_{sm_r}}{\|\rho(|\psi\rangle, C|\psi\rangle)\| \| |\Psi\rangle_{sm_r} \|} \tag{4.27}$$

Cosine similarity displays the cosine similarity scores between the quantum spectral reflectance curves of different rock types in both ASTER and Landsat-8 images. A value of 1 indicates perfect similarity, while 0 indicates no similarity. The rows and columns represent different rock types. The color scale ranges from blue (low similarity) to red (high similarity), showing how closely related the spectral signatures of different rocks are.

According to the above perspective, rocks with high similarity scores have similar spectral reflectance properties across both ASTER and Landsat-8 images. Rocks with low similarity; however, scores have distinct spectral reflectance properties.

Both Landsat-8 and ASTER data demonstrate fluctuations in the spatial similarity distribution of quantum spectral signatures for rock identifications within the GERD zone. Specifically, ASTER data show a remarkably high similarity level of 0.995 compared to Landsat-8 data (Figure 4.26). This significant similarity can be attributed to the advanced physical characteristics inherent in hyperspectral data, which offer greater detail and accuracy than multispectral data. Hyperspectral sensors, like those in ASTER, capture continuous spectral information across a wide range of wavelengths, allowing for a more precise identification of rock types based on their unique spectral signatures.

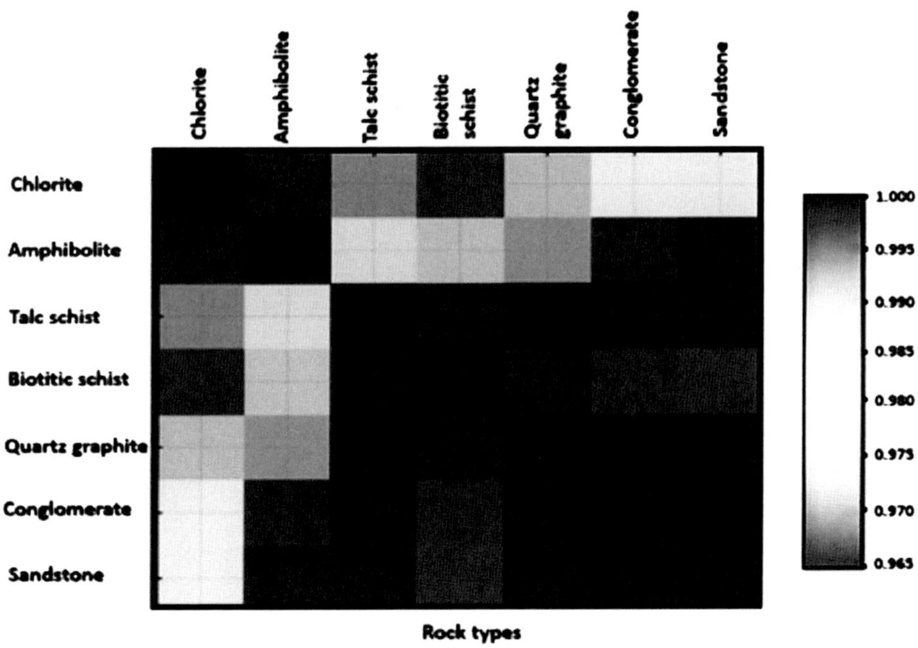

Figure 4.26 Quantum cosine similarity between quantum spectra signature for Landsat-8 and ASTER data.

In contrast, multispectral sensors, like those in Landsat-8, sample fewer and broader wavelength bands, which can result in a loss of detailed spectral information. This discrepancy highlights the importance of utilizing hyperspectral data for applications requiring fine spectral resolution, such as geological mapping and mineral exploration.

The application of quantum entanglement theory in the analysis of satellite data underscores the necessity of exercising caution when assigning physical reality to individual quantum spectral properties. Quantum entanglement suggests that the properties of quantum particles are interconnected, such that the state of one particle instantaneously influences the state of another, regardless of the distance between them. This phenomenon necessitates a holistic approach to interpreting quantum spectral signatures in satellite imagery.

In this context, the quantum spectral signature serves as a powerful tool for clustering rock types prevalent in the GERD vicinity as depicted in the ASTER image (Figure 4.27). The northern part of the dam site is predominantly composed of post-tectonic granitoids, such as granite and diorite. These rocks are typically formed from the cooling and solidification of magma and are known for their coarse-grained texture.

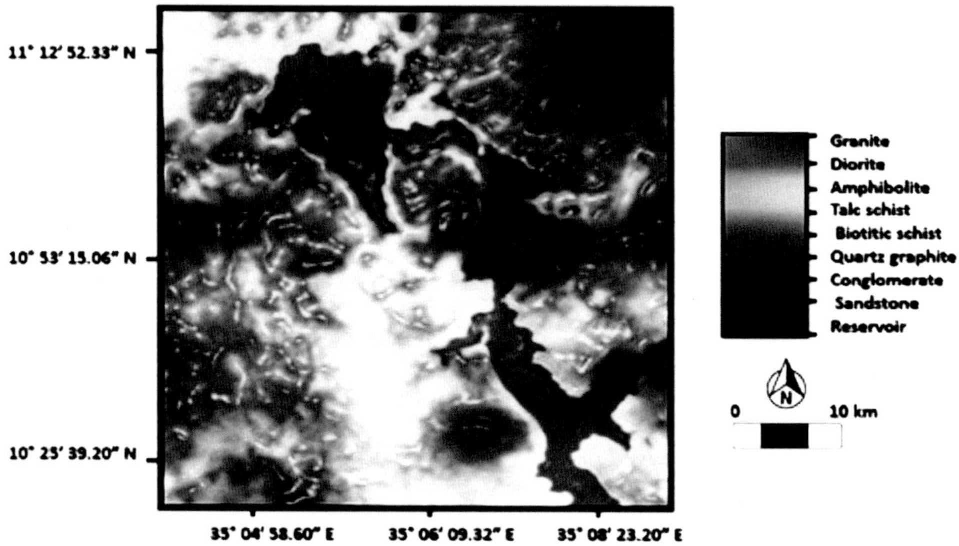

Figure 4.27 Quantum spectra cluster map of rock types along the GERD surrounding.

On the other hand, the southern part of the dam site is characterized by a variety of metamorphic and sedimentary rocks. Metamorphic rocks present include chlorite, amphibolite, talc schist, biotitic schist, and quartz graphite. These rocks have transformed due to high pressure and temperature conditions, resulting in distinct mineral compositions and structures. Sedimentary rocks in the area, such as conglomerates and sandstone, are formed from the accumulation and cementation of mineral and organic particles over time. Their layered structure and composition provide valuable information about the geological history of the region. This finding agrees with the work of Mohamed and Elmahdy [39].

Unquestionably, the comparison of Landsat-8 and ASTER data highlights the superior capability of hyperspectral imaging for detailed geological analysis. The advanced spectral resolution of ASTER allows for more accurate identification and clustering of rock types based on their quantum spectral signatures, offering valuable insights into the geological composition of the GERD zone. This detailed understanding is crucial for effective resource management and planning in the region.

The quantum spectra signature algorithms demonstrated in this chapter determine the level of the quantum entanglement properties like the angular momentum of rock-type electrons interactions with photon wavelength encoded in the DN of different bands. In other words, two quantum systems are entangled: photon quantum energy and rock-type quantum electron energies owing to photon-electron interactions. In this circumstance, shapes or colors in pixels using pseudo-color tools cannot be a good index for rock-type occurrences [35–39].

4.15 CONCLUSION

In this chapter, we focused on demonstrating quantum spectral modeling. The key focus was to introduce a new approach to obtaining spectra signature reflectance of the GERD and its surroundings using quantum spectral computing. We also explored practical theories of quantum entanglement and quantum teleportation. Our findings reveal that the quantum spectra signature algorithm of the GERD construction and its surroundings exhibits high efficiency in mapping the rock-types of the GERD area and its construction components. Therefore, the next chapter will discuss the sedimentation issue in the GERD reservoir, which is crucial for the efficiency of dam operations.

References

[1] Jehlik, D.G. 1964. Geology and dam construction. Transactions of the Kansas Academy of Science 1903. 67(2): 296–300.

[2] Barjasteh, A. 2019. Influence of geological structure on dam behavior and case studies. *In*: Dam Engineering. IntechOpen.

[3] Du Toit, A.L. 1922. Geology in dam construction. Civil Engineering—Siviele Ingenieurswese. 1922(1): 25–56.

[4] Adamo, N., Al-Ansari, N., Sissakian, V., Laue, J. and Knutsson, S. 2020. Dams safety and geology. Journal of Earth Sciences and Geotechnical Engineering. 10(6): 133–189.

[5] Rezaei, A., Karimi, H. and Zhan, H. 2017. The importance of understanding the hydrogeology and geochemistry of karst terrains for safely siting dams. Journal of Cave and Karst Studies. 79(1).

[6] Sissakian, V.K., Adamo, N. and Al-Ansari, N. 2020. The role of geological investigations for dam siting: Mosul dam a case study. Geotechnical and Geological Engineering. 38(2): 2085–2096.

[7] World Commission on Dams. 2000. Dams and development: A new framework for decision-making: The report of the world commission on dams. Earthscan.

[8] Poff, N.L. and Hart, D.D. 2002. How dams vary and why it matters for the emerging science of dam removal: an ecological classification of dams is needed to characterize how the tremendous variation in the size, operational mode, age, and number of dams in a river basin influences the potential for restoring regulated rivers via dam removal. BioScience. 52(8): 659–668.

[9] Ansar, A., Flyvbjerg, B., Budzier, A. and Lunn, D. 2014. Should we build more large dams? The actual costs of hydropower megaproject development. Energy policy. 69: 43–56.

[10] Ertunç, A. 1999. The geological problems of the large dams constructed on the Euphrates River (Turkey). Engineering Geology. 51(3): 167–182.

[11] Stevens, J.E. 1990. Hoover Dam: An American Adventure. University of Oklahoma Press.

[12] Dam, T.G. 2019. Three Gorges Dam. The Sanxia. Encyclopaedia Britannica. Disponível em: https://www.britannica.com/topic/Three-Gorges-Dam. Acesso em, 20.

[13] Tao, Z., Liu, Y., Cheng, L. and Yang, Q. 2019. Failure and stability analysis of Jinping-I arch dam based on geomechanical model test and nonlinear numerical analysis. Rock Mechanics and Rock Engineering. 52: 2245–2260.

[14] Blanc, J. 2019. Before the Flood: The Itaipu Dam and the Visibility of Rural Brazil. Duke University Press.

[15] Bertoni, F., Castelletti, A., Giuliani, M. and Reed, P.M. 2019. Discovering dependencies, trade-offs, and robustness in joint dam design and operation: An ex-post assessment of the Kariba Dam. Earth's Future. 7(12): 1367–1390.

[16] Adamo, N., Al-Ansari, N., Sissakian, V., Laue, J. and Knutsson, S. 2021. Dam safety: Use of instrumentation in dams. Journal of Earth Sciences and Geotechnical Engineering. 11(1): 145–202.

[17] Adamo, N., Al-Ansari, N., Sissakian, V., Laue, J. and Knutsson, S. 2021. Dam safety: Monitoring of tailings dams and safety reviews. Journal of Earth Sciences and Geotechnical Engineering. 11(1): 249–289.

[18] Shijun, W.A.N.G., Yanchang, G.U. and Congbing, G.E. 2019. Technical framework for dam safety monitoring system evaluation. Hydro-Science and Engineering. (4): 63–67.

[19] Adão, T., Hruška, J., Pádua, L., Bessa, J., Peres, E., Morais, R., et al. 2017. Hyperspectral imaging: A review on UAV-based sensors, data processing and applications for agriculture and forestry. Remote Sensing. 9(11): 1110.

[20] Vasefi, F., MacKinnon, N. and Farkas, D.L. 2016. Hyperspectral and multispectral imaging in dermatology. *In*: Imaging in Dermatology. 187–201. Academic Press.

[21] Hecker, C., van Ruitenbeek, F.J., van der Werff, H.M., Bakker, W.H., Hewson, R.D. and van der Meer F.D. 2019. Spectral Absorption Feature Analysis for Finding Ore: A Tutorial on Using the Method in Geological Remote Sensing. IEEE Geoscience and Remote Sensing Magazine. 17(2): 51–71.

[22] Hunt, G.R. 1977. Spectral signatures of particulate minerals in the visible and near infrared. Geophysics. 42(3): 501–513.

[23] Salisbury, J.W., Walter, L.S., Vergo, N. 1989. Availability of a library of infrared (2.1–25.0 μm) mineral spectra. American Mineralogist. 74(7–8): 938–939.

[24] Goetz, A.F., Vane, G., Solomon, J.E. and Rock, B.N. 1985. Imaging spectrometry for earth remote sensing. Science. 228(4704): 1147–1153.

[25] Pearlman, J.S., Barry, P.S., Segal, C.C., Shepanski, J., Beiso, D. and Carman, S.L. 2003. Hyperion, a space-based imaging spectrometer. IEEE Transactions on Geoscience and Remote Sensing. 41(6): 1160–1173.

[26] Guanter, L., Kaufmann, H., Segl, K., Foerster, S., Rogass, C., Chabrillat, S., et al. 2015. The EnMAP spaceborne imaging spectroscopy mission for earth observation. Remote Sensing. 7(7): 8830–8857.

[27] Yan, F., Iliyasu, A.M. and Le, P.Q. 2017. Quantum image processing: A review of advances in its security technologies. International Journal of Quantum Information. 15(03): 1730001.

[28] Markham, B.L. and Barker, J.L. 1985. Spectral characterization of the Landsat Thematic Mapper sensors. International Journal of Remote Sensing. 6(5): 697–716.

[29] Mastriani, M. 2017. Quantum image processing? Quantum Information Processing. 16(1): 27.

[30] Wang, Z., Xu, M. and Zhang, Y. 2022. Review of quantum image processing. Archives of Computational Methods in Engineering. 29(2): 737–761.

[31] Ruan, Y., Xue, X. and Shen, Y. 2021. Quantum image processing: opportunities and challenges. Mathematical Problems in Engineering. 2021(1): 6671613.

[32] Steane, A. 1988. Quantum computing. Reports on Progress in Physics. 61(2): 117.

[33] Cai, X.D., Weedbrook, C., Su, Z.E., Chen, M.C., Gu, M., Zhu, M.J., et al. 2013. Experimental quantum computing to solve systems of linear equations. Physical Review Letters. 110(23): 230501.

[34] Yuan, S., Mao, X., Chen, L. and Xue, Y. 2013. Quantum digital image processing algorithms based on quantum measurement. Optik. 124(23): 6386–6390.

[35] Marghany, M. 2022. Remote Sensing and Image Processing in Mineralogy. CRC Press.

[36] Hey T. 1999. Quantum computing: An introduction. Computing and Control Engineering Journal. 10(3): 105–112.

[37] Preskill, J. 1998. Quantum computing: Pro and con. *In*: Proceedings of the Royal Society of London. Series A: Mathematical, Physical and Engineering Sciences. 454(1969): 469–486.

[38] Montanaro, A. 2016. Quantum algorithms: An overview. NPJ Quantum Information. 2(1): 1–8.

[39] Mohamed, M.M. and Elmahdy, S.I. 2017. Remote sensing of the Grand Ethiopian Renaissance Dam: A hazard and environmental impacts assessment. Geomatics, Natural Hazards and Risk. 8(2): 1225–1240.

Chapter 5

Quantum Support Vector Machine for Simulating Sedimentation Concentration in the GERD Reservoir

5.0 INTRODUCTION

Despite extensive research on the Grand Ethiopian Renaissance Dam (GERD), there remains a critical gap in understanding the sediment concentration within the reservoir. To date, studies have largely focused on water level fluctuations, discharge variations, and the reservoir's changing dimensions over time. However, sedimentation—a crucial factor impacting dam stability and operational efficiency—has received comparatively little attention. High sediment concentration can have severe implications for GERD's long-term functionality, posing risks to both its structural integrity and water discharge capacity. When excessive sediments accumulate behind the dam, they not only reduce storage volume but can also obstruct turbine operation, potentially leading to significant power generation losses, structural stress and even dam failure.

This chapter addresses this gap by presenting a novel investigation into sediment concentration levels within the GERD reservoir. By employing advanced techniques in quantum computing, we aim to develop a new algorithm based on Quantum Support Vector Machines (QSVM) to accurately model and predict sedimentation patterns. Leveraging the computational power of QSVM offers several advantages, including enhanced data processing speed, precision, and adaptability in handling the complex dynamics of sediment behavior within the reservoir.

This chapter will first explore the current state of sediment accumulation and its potential threats to GERD. Next, it will outline the theoretical foundation of QSVM and its relevance in modeling sediment concentration. By examining the intricacies of sediment deposition through quantum-based algorithms, this study seeks to provide critical insights into GERD's future operational sustainability and to offer practical strategies for sediment management.

DOI: 10.1201/9781003327325-5

5.1 UNDERSTANDING THE IMPORTANCE OF RESERVOIR SEDIMENTATION

Reservoir sedimentation plays a crucial role in the management of dams and hydropower facilities. Firstly, sedimentation significantly impacts hydropower production. As sediment accumulates in the reservoir, it reduces the available storage capacity of water. Consequently, this diminished storage capacity leads to decreased energy generation, as less water is available to pass through the turbines for electricity production. Secondly, sediment deposition in reservoirs raises safety concerns [1]. The accumulation of sediments can increase the load on the structure and gates of the dam, potentially damaging mechanical components. This added pressure can compromise the structural integrity of the dam, posing significant risks to its operation and the surrounding areas.

Moreover, improper sediment management can have detrimental environmental effects. Sediment-laden water released downstream can harm aquatic ecosystems, disrupt habitats and alter riverbed morphology. These environmental impacts underscore the importance of managing sediments effectively to preserve the health and balance of downstream ecosystems. In addition to production and safety concerns, effective sediment management is crucial for ensuring the longevity of reservoirs. By maintaining the storage capacity and preventing excessive silt buildup, proper sediment management practices help extend the operational life of reservoirs. This, in turn, supports sustainable water resource management and long-term hydropower production [1, 3].

Therefore, addressing sedimentation is essential for sustainable dam operation, safety and environmental preservation. Proper management techniques include diverting sediment, removing deposits and minimizing the inflow of sediments into reservoirs. These strategies help maintain the efficiency and safety of dam operations, protect the environment and ensure the longevity of hydropower facilities [1–3].

5.2 RESERVOIR SEDIMENTATION MECHANISM

The reduction in storage capacity due to sedimentation in dam reservoirs currently exceeds the gains from constructing additional reservoirs for hydropower, irrigation or residential use. Dominik et al. [4] highlighted this trend, projecting a significant decline in the storage capacity of large reservoirs. Chinese reservoirs, with an average lifespan of twenty-two years due to high sedimentation rates, have the shortest longevity globally, whereas European and North American reservoirs can last up to 250 years. The global average annual loss in reservoir volume capacity due to sedimentation is 1–2%, indicating a widespread and substantial impact [5, 14].

Sediment mass is produced by the erosive action of water streams in watersheds, which subsequently drain into rivers [6]. A river transports suspended sediment particles effectively functioning as conduits for both water and sediment [14]. When these rivers enter the still waters of a reservoir, their flow velocity decreases while depth and area increase. This reduction in flow velocity results in a decreased sediment carrying capacity (SCC), leading to sediment deposition within the reservoir [7].

Schleiss et al. [8] stated that sediment deposition begins near the mouth of the reservoir, where coarser particles settle first. Finer particles travel further upstream within the reservoir and deposit closer to the dam body. The finest sediment particles remain suspended due to the upward currents in the reservoir. This process follows a deltaic depositional pattern. Therefore, Shen [9] and Julien [10] identified four main parts of deltaic sediment deposition: the front reach, front set, top set, and tail reach (Figure 5.1).

Continuous reservoir operations lead to the migration of smaller sediment particles closer to the dam body. As more sediment enters the reservoir, the depositional pattern may evolve into a wedge shape. The topset of the sediment delta is composed of coarser materials, while finer particles form the bottomset deposition. Fan and Morris [11] have identified several key characteristics of reservoir deltas:

(i) An abrupt change in slope between the topset and tail reach deposits.
(ii) The topset layer contains coarser particles compared to the frontset bed, with a noticeable change in particle diameter between these layers.
(iii) The elevation of the zone between the topset and frontset beds is influenced by the reservoir's operational rules and water surface levels.

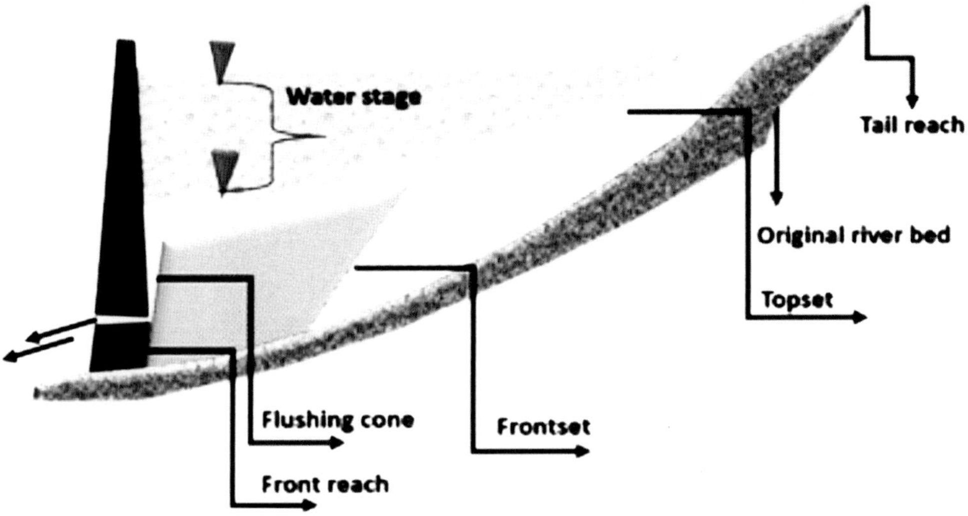

Figure 5.1 The sedimentation pattern in reservoirs.

Without doubt, comprehending the complex processes of reservoir sedimentation is essential for effective management and operational longevity. Implementing appropriate sediment management strategies can mitigate storage capacity loss and ensure the safety and efficiency of dam operations.

Reservoir sedimentation processes are complex and vary depending on a range of intricate conditions. These conditions encompass entire basin characteristics, such as watershed sediment yield, which are influenced by land use and anthropogenic factors. Additionally, Idrees [14] reported that the sediment transport capacity of the river channel, the frequency of floods and the river's cross-sectional and longitudinal geometry play crucial roles in sedimentation in reservoirs. The properties of the sediment and the dam's operational rule curves also determine the shape and consolidation of the reservoir delta.

Morris and Fan [12] and Idrees [14] classified longitudinal reservoir sediment depositional patterns into four general types, as illustrated in Figure 5.2. Typically, coarser fractions of sediment deposits or a larger quantity of finer sediment mass, such as silt, shape the deltaic reservoir sedimentation pattern. The tapering and uniform patterns mostly appear in longer reservoirs with fine sediment mass inflow and narrow reservoirs with variable discharge and a small sediment load inflow, respectively. When water inflow, composed largely of fine sediment mass, enters a smaller reservoir, the sedimentation patterns follow a wedge-shaped deposition [14].

In reservoirs with regular water release, the delta tip moves toward the dam body and the deltaic deposition transforms into a wedge-shaped deposition, considered the long-term equilibrium state of the reservoir [13]. The sediment depositional patterns are inherently a function of the interactions among water discharge, sediment grain size, channel morphology and reservoir operational rules.

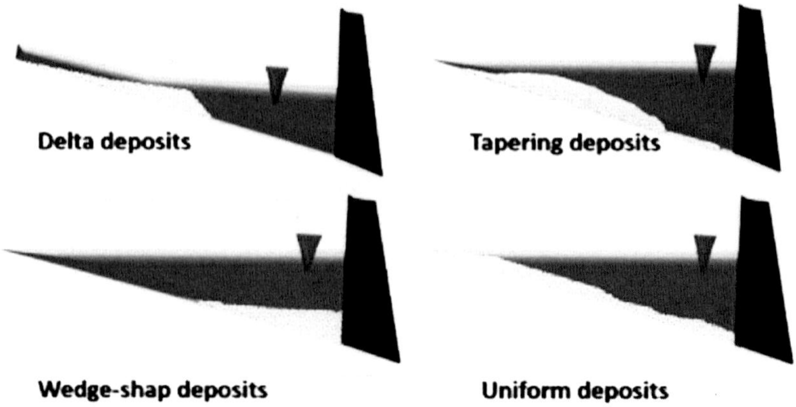

Figure 5.2 Classification pattern of reservoir sediment depositions.

5.3 SEDIMENT TYPES IN GERD RESERVOIR

According to the aforementioned perspective, sedimentation in dam reservoirs is a critical factor that affects the long-term operational efficiency and safety of hydropower facilities. The Grand Ethiopian Renaissance Dam (GERD), situated on the Blue Nile River, is particularly susceptible to sedimentation due to the high sediment load carried by the river. This sedimentation process is influenced by various factors, including watershed characteristics, river morphology and dam operations.

In this context, it is essential to understand the types and patterns of sedimentation occurring in the GERD reservoir. Given the GERD's location and the characteristics of the Blue Nile River, the sedimentation is likely to exhibit the following types: (i) deltaic sedimentation (ii) wedge-shaped sedimentation and (iii) layered sedimentation [13, 15].

As the Blue Nile River enters the still waters of the GERD reservoir, a delta forms at the reservoir's head where the coarser sediments settle first. This pattern is typical of many reservoirs where the inflow river carries a significant sediment load. The deltaic deposition will be characterized by a progression from coarser materials at the delta front to finer sediments further into the reservoir. Therefore, with continuous sediment inflow and regular water releases from the dam, the sediment deposition can transform into a wedge shape, moving towards the dam body. This is because finer sediments remain in suspension longer and travel further into the reservoir before settling. Over time, the sedimentation will exhibit stratification, with layers of varying sediment sizes deposited during different flow conditions. Subsequently, high-flow events may deposit coarser materials, while finer sediments will settle during lower-flow periods [11–14].

The characteristics of the sediments, including grain size and cohesiveness, influence their deposition behavior. Coarser sediments settle quickly near the reservoir mouth, while finer sediments travel further before settling. Therefore, the frequency and magnitude of floods significantly impact sediment deposition. High-magnitude floods can transport large quantities of sediments into the reservoir, altering the sedimentation pattern [11, 14].

Understanding the sedimentation patterns in the GERD reservoir is crucial for effective reservoir management and dam operation. Deltaic, wedge-shaped, and layered sedimentation patterns are expected to develop in the GERD reservoir due to the interplay of various factors such as watershed characteristics, river morphology, flood frequency, sediment properties and dam operation rules. These patterns have significant implications for reservoir storage capacity, dam safety and the long-term sustainability of the GERD hydropower facility. In other words, the accumulation of sediments near the dam structure can pose safety risks. Increased sediment loads can add stress to the dam and its gates, potentially leading to structural damage. Additionally, sediment buildup can affect the functionality of spillways and other mechanical components, compromising the overall safety of the dam.

The Grand Ethiopian Renaissance Dam (GERD) is a pivotal hydropower project on the Blue Nile River. Designed to trap sediment inflow for up to 100 years, GERD faces significant sedimentation challenges that could impact its operational efficiency and longevity. The key question, therefore, is: how does GERD address these challenges? Understanding and implementing effective sediment management strategies is essential for maintaining GERD's stability and extending the reservoir's lifespan.

5.4 THE SOURCES OF SEDIMENTATION IN THE GERD RESERVOIR

Based on the perspective presented, the Blue Nile plays a crucial role as a primary source of sedimentation. This aspect will be further elaborated upon in the following discussion to provide a comprehensive understanding.

5.4.1 Watershed Characteristics

The Grand Ethiopian Renaissance Dam (GERD) is strategically located in the Ethiopian highlands at the origin of the Blue Nile River (Figure 5.3). This location is significant because the Blue Nile contributes a significant amount of sediment to the GERD reservoir. The watershed surrounding GERD contains various geological sources that influence sediment influx: (i) Wadi Sediments; (ii) shales; (iii) Nubian sandstones [15].

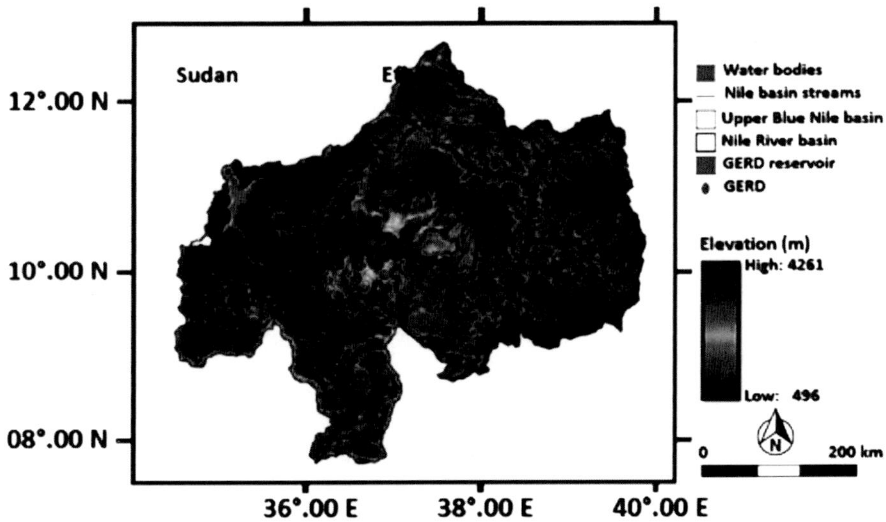

Figure 5.3 GERD location in the Nile River Basin.

The sediments consist of fine particles such as clay and silt and they are transported by seasonal rivers called wadis when there is heavy rainfall. This process leads to the accumulation of significant sediment within the reservoir. Shale is a type of sedimentary rock made up of fine-grained mud (Figure 5.4). It consists of tiny fragments of minerals like quartz and calcite, along with flakes of clay minerals (Figure 5.5). Shale has thin layers and can easily break along these layers, which are less than one centimeter thick. Mudstones (Figure 5.6) have a similar composition to shale but do not have the same tendency to break along thin layers. In addition to shales, the Nubian sandstones present in the regional geological formations also play a role in adding to the overall sediment entering the reservoir [16].

Figure 5.4 Sedimentary rock of shale.

Figure 5.5 Clay minerals.

Figure 5.6 Mudstone mineral.

5.4.2 Soil Composition of the Blue Nile Basin, Ethiopia

The Blue Nile basin in Ethiopia is predominantly covered with clay soils, significantly influencing its agricultural productivity and sedimentation processes. Here is an overview of the soil composition in the basin: (i) vertisols (Figure 5.7); and (ii) red soils (Figure 5.8). Vertisols are clay-rich soils that are prone to swelling and shrinking and cover approximately 15% of the basin. They are primarily found in poorly drained areas. Therefore, red soils are located on well-drained sites with moderate slopes, particularly in regions with high rainfall or long-term stability (such as the lowlands), where red soils dominate. These include (i) alisols (ii) acrisols and (iii) nitosols (Figure 5.9). Collectively, these red soils cover about 42% of the basin [15, 17].

Figure 5.7 Vertisols soil pattern.

Figure 5.8 Red soil pattern.

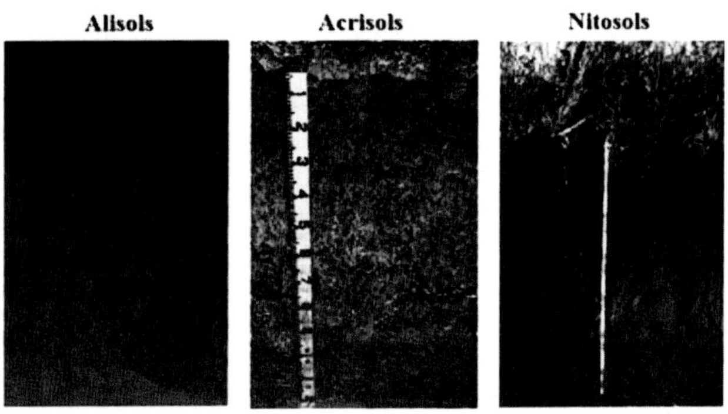

Figure 5.9 Types of red soil: alisols, acrisols, and nitosols.

In addition to these, luvisols, known for their high fertility and productivity, occupy 9% of the basin. Furthermore, about 32% of the basin is covered by shallow and moderately deep soils. This category comprises leptosols, which are found on steep slopes and cover 21.5% of the basin, and cambisols, young and moderately deep soils accounting for 9.5% of the basin. Additionally, regosols and arenosols (Figure 5.10), typically found in less stable areas, together make up 1.3% of the basin.

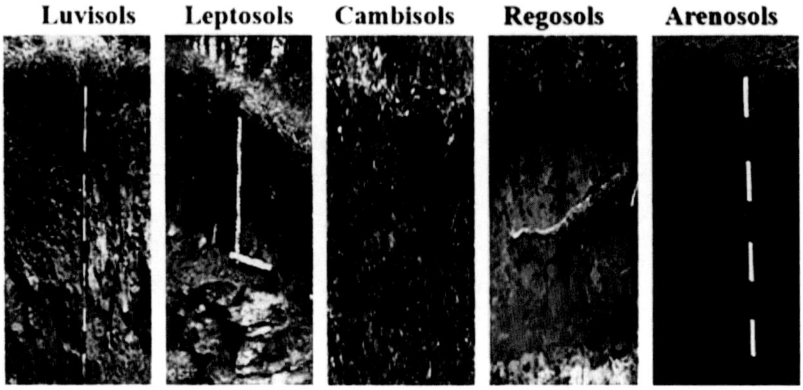

Figure 5.10 Various soil compositions of the Blue Nile Basin.

The diverse soil composition in the Blue Nile basin requires tailored land management and agricultural planning to optimize productivity and mitigate sedimentation issues. The unique properties of each soil type play a crucial role in the region's overall environmental and agricultural dynamics.

Understanding the distribution and characteristics of these soil types is crucial for effective land management and agricultural planning in the Blue Nile basin. The varying soil properties also play a significant role in influencing sediment transport and deposition patterns in the region.

5.4.3 Sediment Yield

The sediment yield in the GERD reservoir is influenced by several factors, notably the characteristics of the Lake Tana sub-basin. The average sediment yield in the Lake Tana sub-basin, where GERD is located, is around 2,500 tons per square kilometer per year (t/km^2/yr). In regions within the watershed that exhibit high erosion potential, therefore, sediment yields can soar to 6,500 t/km^2/yr. This elevated sediment yield significantly impacts the rate of sedimentation in the reservoir.

These factors collectively underline the complex dynamics of sedimentation at GERD. Understanding the sources and rates of sediment influx is crucial for developing effective sediment management strategies to ensure the dam's operational efficiency and longevity [15–17].

Needless to say, the Grand Ethiopian Renaissance Dam (GERD) faces considerable sedimentation challenges due to the characteristics of its watershed and high sediment yields. Through comprehensive watershed management, sediment trapping and removal and flood management strategies, GERD aims to mitigate these challenges. These efforts are essential to ensure the dam's long-term stability, efficiency, and contribution to Ethiopia's sustainable development.

5.5 OPTICAL REMOTE SENSING OF SOIL INVESTIGATIONS

The precise spectral libraries are not consistent across all the regions of the Earth. Clay minerals are frequently used as climatic indicators because their formation is closely influenced by weather conditions, such as temperature fluctuations and precipitation levels during soil formation processes. Consequently, the dominant clay minerals transition from chlorite/illite to vermiculite, then to montmorillonite, and finally to kaolinite, depending on the prevailing climate conditions. These transitions reflect changes from cool and dry to warm and moist climates.

In this framework, the primary clays in soils exhibit diagnostic absorptions in the Short-Wave Infrared (SWIR) spectrum, characterized by vibrational transitions that result in sharp and narrow spectral features. The key analytical bands are primarily located around ~1.4 μm (attributed to OH), ~1.9 μm (attributed to molecular water), and ~2.2 μm (attributed to Al-OH combination tones). Additionally, weaker absorption bands in the 2.3–2.5 μm region are associated with the presence of Fe-OH and/or Mg-OH in the clay minerals. These spectral features provide valuable insights into the mineralogical composition of soils and their climatic history.

Chlorites are a group of clay minerals that include specific octahedral positive ions such as Fe, Mg, and Al. Their reflectance spectra exhibit a weak absorption band around 1.4 μm and three absorption features near 2.3 μm. The absorption bands at 2.25 μm and 2.35 μm are associated with Fe-OH and Mg-OH, respectively. Illite; therefore, is characterized by three prominent absorption bands at approximately 1.4 μm, 1.9 μm, and 2.2 μm. Additionally, two secondary Al-OH absorption peaks near 2.3 μm and 2.4 μm are enhanced by the substitution of Fe and Mg cations. Consequently, vermiculite displays two broad absorption bands at 1.4 μm and 1.9 μm, along with two weaker absorptions around 2.2 μm and 2.3 μm. Montmorillonite features three strong and sharp absorption bands at approximately 1.4 μm, 1.9 μm, and 2.2 μm, which are similar to but generally stronger than those of illite. On the other hand, kaolinite is distinguished by two spectral doublets: one around 1.4 μm and the other near 2.2 μm. In this regard, these spectral characteristics provide essential insights into the mineralogical composition and structural differences among these clay minerals.

At a specific depth within the clay, carbonates leach from the surface and gradually accumulate in the soil over time. The presence of carbonates is extensively utilized as a fundamental indicator of soil characteristics, aiding in the identification of soil types and the assessment of soil erosion. In the VNIR domain, multiple absorptions of carbonates are caused by the overtones and combinations of fundamental vibrations of the CO_3^{2-} ion. Specifically, in the NIR region, carbonates exhibit a strong absorption band around 2.35 μm. Additionally, carbonates display three weaker absorption bands at approximately 1.9 μm, 2.0 μm, and 2.16 μm; respectively [18]. Furthermore, carbonates exhibit a double-band structure near the 2.35 μm absorption. These spectral features are critical for understanding the mineralogical composition and environmental processes affecting soil formation and stability [16–20].

In general, the fundamental nutrients found in water, soil and various biochemical compounds on Earth are nitrogen (N), phosphorus (P), and potassium (K). Nitrogen plays a crucial role in ecological and biochemical processes, with its absorption bands primarily located in the visible, short-wave near-infrared (SWNIR), and near-infrared (NIR) regions of the spectrum at wavelengths such as 2.35, 2.3, 2.18, 2.06, and 1.51 μm. These absorption features have a significant impact on the spectral reflectance properties of various biochemical compounds, including chlorophylls (Chls), proteins, and plant leaves, thereby serving as a "fingerprint" for nitrogen. Similarly, phosphorus absorption features are spread across the visible, SWNIR, and NIR regions, indirectly influencing the spectral characteristics of diverse biochemical compounds.

Potassium concentration in leaves has an insignificant effect on needle morphology. The presence of K affects the NIR reflectance. It thickens the sclerenchyma cell walls in leaves and the high concentration of K increases the NIR reflectance. The evaluation of the nutrient concentration is very much guided by the image data and spectral measurements. Several analysis protocols and methods such as derivative analysis, spectral matching, and spectral index analysis are developed for accurate assessment of nutrient contents.

5.6 METHODS FOR CALCULATING CATION EXCHANGE CAPACITY IN THE LABORATORY

The Cation Exchange Capacity (CEC) refers to the soil's ability to retain exchangeable cations. Soils with higher CEC values can retain more cations. The CEC of soil varies based on the type and quantity of clay and organic matter present. High-clay soils possess a greater capacity to hold exchangeable cations compared to low-clay soils. Additionally, CEC increases with the organic matter content. Clay minerals typically exhibit CEC values ranging from 10 to 150 meq/100g, while organic matter shows higher CEC values, ranging from 200 to 400 meq/100 g. Therefore, both the type and amount of clay and organic matter significantly influence the soil's CEC. This cation exchange process is crucial for soil fertility as it impacts soil acidity and basicity.

The Cation Exchange Capacity (CEC) is the capacity of soil to hold exchangeable cations. The higher the CEC of soil, the more cations it can retain. Soil differs in its capacity to hold exchangeable K^+ and other cations. The cations exchange capacity depends on the amount and kind of clay and organic matter present. High clay soil can hold more exchangeable cations than low clay soils. CEC also increases as organic matter increases. Clay mineral usually ranges from 10 to 150 meq/100 g. In CEC values organic matter ranges from 200 to 400 meq/100 g. Therefore, the kind and amount of clay and organic matter content greatly influence the CEC of soil. Cation exchange is an important reaction in soil fertility in causing and correcting soil acidity and basicity [20–22].

The cation exchange capacity (CEC) measurement involves determining the total amount of negative charges per unit weight of soil that are neutralized by exchangeable cations. It is defined as the soil's capacity to absorb exchangeable cations, expressed as milliequivalents per 100 grams of soil or centimoles of positive charge per kilogram (C mol (P^+)/kg) on an oven-dry basis. In other words, the total cations consist of base cations plus acid cations (Ca + Mg + K + Na) + (H + Al). For example, if the concentration of basic cations is low and the concentration of acidic cations is high, it indicates a high degree of weathering and leaching and the soil is likely to be acidic [20].

Let us define the total exchangeable base (TEB) as the sum of exchangeable cations: calcium (Ca), magnesium (Mg), sodium (Na), and potassium (K). Thus, TEB can be expressed as TEB = Ca + Mg + Na + K (mmolc/kg). Similarly, the total exchangeable acids (TEA) are defined as the sum of exchangeable hydrogen (H) and aluminum (Al). Therefore, the cation exchange capacity (CEC) can be calculated using the formula:

$$\text{Percentage Base Saturation (\%BS)} = \text{TEB/CEC} \times 100 \tag{5.1}$$

$$\text{Percentage Acid Saturation (\%AS)} = \text{TEA/CEC} \times 100 \tag{5.2}$$

The percentage saturation of clay ion $\%C_I$ is mathematically computed as:

$$\%C_I = \frac{C}{\text{CEC} \times 100} \tag{5.3}$$

here CEC is the sum of exchangeable base cations and exchangeable acid cations, representing the total capacity of the soil to hold and exchange cations [20, 23].

Under certain conditions, particularly in extremely acidic environments with a pH less than 4.5, hydrous oxides of iron (Fe) and aluminum (Al), as well as some aluminosilicate clays can develop a positive charge. This phenomenon occurs due to the protonation of the hydroxyl (OH) groups at the exposed crystal edges. For instance, the process can be represented as Al—OH + H^+ ---------- Al—OH_2^+.

When clays develop positive charges through this protonation, they can adsorb anions, through a process known as anion adsorption. Consequently, anions such as chloride Cl^-, sulfate (SO_4^{2-}), and dihydrogen phosphate ($H_2PO_4^-$) can be held by the positively charged clay surfaces. This mechanism plays a crucial role in soil chemistry, particularly in influencing nutrient availability and mobility within acidic soils.

Clays are frequently negatively charged and thus adsorb cations in solution, referred to as clay colloids or micelles. A colloid forms when one phase is finely dispersed within another phase, such as solid clay within a liquid phase (soil solution). Clays are small and finely dispersed in the soil solution. Figure 5.11 illustrates that different cations are adsorbed around a clay colloid.

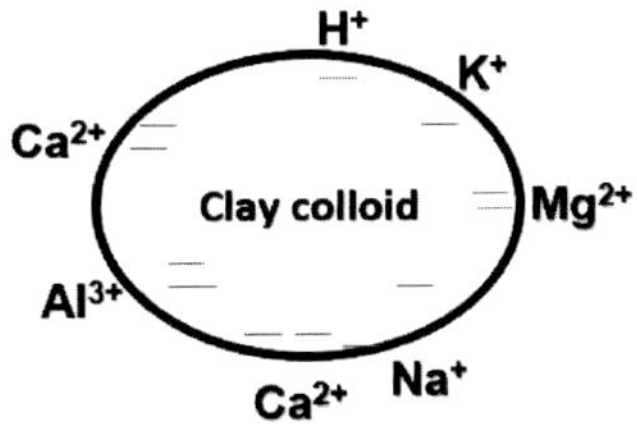

Figure 5.11 The notion of clay colloid.

Cations in the soil originate from rock weathering, minerals, fertilizers and the breakdown of organic matter. The preferential order of adsorption follows the sequence: $Al^{3+} > Ca^{2+} > Mg^{2+} > K^{+} > Na^{+} > H^{+}$. Thus, clay will preferentially adsorb Al^{3+} over Ca^{2+} if both cations are present in equal concentrations. In this context, cations are adsorbed onto clay surfaces, meaning they are loosely held and can be exchanged with other cations in the soil solution. This exchangeability is a crucial aspect of soil chemistry, influencing nutrient availability and soil fertility [19, 22].

When a clay micelle with calcium ions is exposed to high concentrations of potassium ions, the calcium ions will be exchanged for the potassium ions, a process known as cation exchange (see Figure 5.12). This reaction maintains a balance to satisfy all the negative charges on the clay micelle.

Moreover, the effect of adding lime can also be observed through a cation exchange reaction (see Figure 5.13). Adding lime, which typically contains calcium carbonate, increases the concentration of calcium ions in the soil. Consequently, calcium ions can replace other cations, such as potassium or magnesium, on the clay micelle, thereby improving soil structure and fertility. This process is essential for maintaining soil health and optimizing nutrient availability for plants.

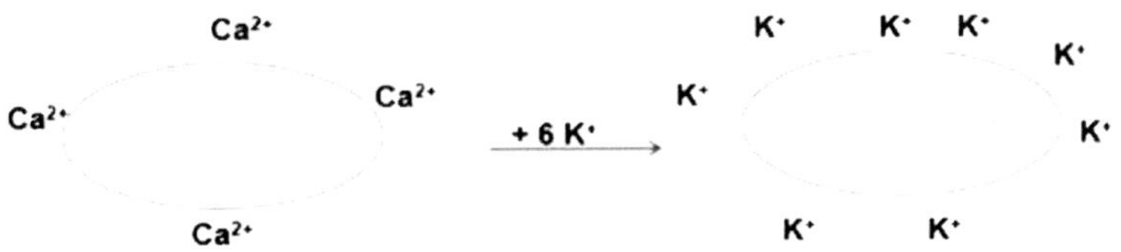

Figure 5.12 Concept of cation exchange.

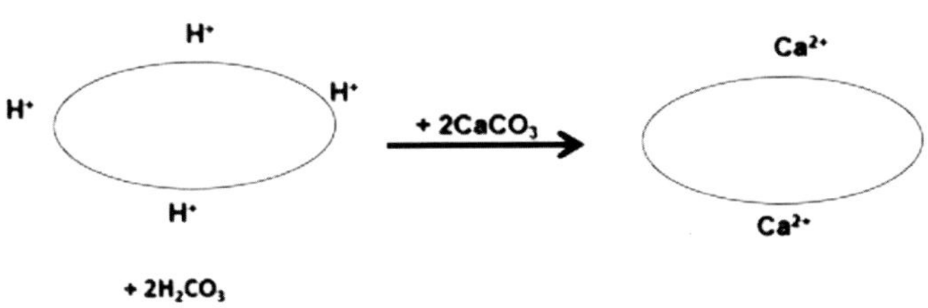

Figure 5.13 Lime was added to the cation reaction.

Consistent with Marghany [20], clay particles and organic matter carry negative charges on their surfaces, which enable them to adsorb positively charged particles, known as cations. Various cations, such as H^{+}, Ca^{++}, K^{+}, NH^{+}_{4}, Mg^{++} and others can replace each other depending on their concentration (mass action) and their replacing power. This process, known as cation exchange, allows cations to swap places based on their relative affinities for the negatively charged surfaces.

The order of replacing power of some cations is given as $H^+ > Ca^{++}$. $Mg^{++} > K^+ > NH^+_4 > Na^+ > Si^{++++}$ [20–23].

5.7 SENTINEL-2 SATELLITE DATA ACQUISITION

The data used in modeling clay percentage along the GERD reservoir is Sentinel-2 (S2). Sentinel-2 (S2) Multi-Spectral Instrument (MSI) collects multispectral data with 13 bands in the visible/near-infrared (VNIR) and shortwave infrared (SWIR) regions of the spectrum at intervals of five days or less. These data are freely accessible from [USGS Earth Explorer] (https://glovis.usgs.gov/app). Unlike traditional multispectral satellite sensors such as Landsat, Sentinel-2 includes three vegetation red edge bands. Specifically, it captures Blue (0.490 μm), Green (0.560 μm), Red (0.665 μm), and NIR (0.842 μm) bands at a spatial resolution of 10 meters. Additionally, the three Vegetation Red Edge bands (0.705 μm, 0.740 μm, and 0.783 μm), the Narrow NIR band (0.865 μm), and two shortwave infrared bands (1.610 μm and 2.190 μm) are captured at a spatial resolution of 20 meters.

In this study, we used S2 MSI Level 1C data, which represents top-of-atmosphere (TOA) reflectance scaled by 10,000 and is available on Google Earth Engine (GEE). The choice of S2-TOA imagery was due to its higher data accessibility compared to the processed S2-Surface Reflectance (SR) data. To maintain consistency, all 10-meter bands of the S2 images were resampled to 20 meters to align with the 20-meter bands of Sentinel-2 images [20, 24].

The Multi-Spectral Instrument (MSI) on Sentinel-2 measures reflected radiance across thirteen spectral bands, with spatial resolution varying according to the specific band. Firstly, there are four bands with a 10-meter resolution: blue (0.490 μm), green (0.560 μm), red (0.665 μm), and near-infrared (0.842 μm) (Table 5.1). These high-resolution bands are particularly useful for detailed monitoring of vegetation, water bodies and urban areas.

Table 5.1 Characteristics of Sentinel-2 satellite data

Band Number	Central Wavelength (μm)	Bandwidth (μm)	Spatial Resolution (m)
1	0.433–0.453	0.020	60
2	0.458–0.523	0.065	10
3	0.543–0.578	0.035	10
4	0.650–0.680	0.030	10
5	0.698–0.713	0.015	20
6	0.733–0.748	0.015	20
7	0.773–0.793	0.020	20
8	0.785–0.900	0.115	10
8a	0.855–0.875	0.020	20
9	0.935–0.955	0.020	60
10	1.365–1.385	0.030	60
11 (SWIR-1)	1.565–1.655	0.090	20
12 (SWIR-2)	2.100–2.280	0.180	20
TCI*	RGB	Composite	10

Secondly, the MSI includes six bands at a 20-meter resolution. Four of these are narrow bands designed for vegetation characterization (0.705 μm, 0.740 μm, 0.783 μm, and 0.865 μm). Additionally, there are two broader shortwave infrared (SWIR) bands (1.610 μm and 2.190 μm), which are employed for applications such as snow/ice/cloud detection and assessing vegetation moisture stress.

Lastly, three bands operate at a 60-meter resolution, and are primarily used for cloud screening and atmospheric corrections. These include bands at 443 nm for aerosol detection, 0.945 μm to monitor water vapor and 1.375 μm to detect cirrus clouds *i*. This combination of spectral and spatial resolutions enables the MSI to support a wide range of environmental monitoring and management applications, providing detailed and comprehensive data for various scientific and practical uses. In this study, Sentinel-2 data were utilized, specifically those acquired from July 2023 to July 2024. Some data, could not be used due to heavy cloud cover, especially during August 2023.

5.8 HOW CLAY POTENTIAL PERCENTAGE IN REMOTE SENSING DATA CAN BE RETRIEVED

This section introduces a novel algorithm for determining clay saturation using optical remote sensing data. Both multispectral and hyperspectral imagery include dual short-wave infrared (SWIR) bands: SWIR-1 and SWIR-2. For Sentinel-2, SWIR-1 spans from 1.565 to 1.655 μm, while SWIR-2 covers 2.100 to 2.280 μm. To develop the new algorithm, it is essential to retrieve the clay saturation percentage from these SWIR bands.

Assuming that the clay saturation percentage is embedded within both SWIR-1 and SWIR-2 bands, we denote this percentage as C_I. From a quantum mechanical perspective [24–27], let $\Psi|C_I\rangle\alpha\Psi|SWIR_1\rangle|SWIR_2\rangle^{-1}$. Therefore, the spectral reflectance in SWIR-1 and SWIR-2 is described by the quantum state $C_I \in SWIR_{1,2}$ and probability $p \leq 0.5$. Consequently, the probability distribution of clay saturation $\Psi|C_I\rangle$ is proportional to the quantum state of the spectral reflectance $|S\rangle_r$. Thus, the quantum state of clay saturation $\Psi|C_I\rangle$ and the spectral reflectance $|S\rangle_r$ are inherently linked as $\Psi|C_I\rangle\alpha\Psi|S\rangle_r$ and mathematically is expressed as:

$$\Psi|C_I\rangle\alpha\,\Psi|SWIR_1\rangle|SWIR_2\rangle^{-1} \quad \text{or} \quad \Psi|C_I\rangle\alpha\,\Psi|S_r\rangle \tag{5.4}$$

Equation 5.4 can be solved in a linear form by incorporating constant values $|\hat{b}\rangle$ on the right side of the equation, as outlined below:

$$\Psi|C_I\rangle = |\hat{b}\rangle + \sum_{i=1}^{k}|S_r\rangle_i \otimes \left[\frac{\partial|C_I\rangle_i}{\partial|S_r\rangle_i}\right] \tag{5.5}$$

Given a Hermitian $K \times k$ matrix H and a unit vector $\vec{h}$, one can obtain a vector $\vec{x}$ that satisfies a specific condition. To solve this problem, we need to know the specific condition or equation that $\vec{x}$ should satisfy. If the goal is to find $\vec{x}$ that satisfies an equation of the form:

$$\mathcal{H}\hat{b} = \vec{h} \tag{5.6}$$

Where

$$\mathcal{H} = \begin{bmatrix} n & \sum S_{2j} & .... & \sum S_{kj} \\ \sum S_{2j} & \sum S_{2j}^2 & .... & \sum S_{2j}^2 S_{kj} \\ .... & .... & .... & .... \\ \sum S_{kj} & \sum S_{kj} S_{2j} & .... & \sum S_{kj}^2 \end{bmatrix} \tag{5.6.1}$$

$$\hat{b} = \begin{bmatrix} \hat{b}_1 \\ \hat{b}_2 \\ \vdots \\ \vdots \\ \hat{b}_k \end{bmatrix} \tag{5.6.2}$$

and

$$\vec{h} = \begin{bmatrix} \sum \Psi |C_I\rangle_j \\ \sum \Psi |C_I\rangle_j |S_r\rangle_{2j} \\ \vdots \\ \sum \Psi |C_I\rangle_j |S_r\rangle_{kj} \end{bmatrix} \tag{5.6.3}$$

Accordingly, $\mathcal{H}$ is a $k \times k$ estimated matrix with both spectral radiance $|S_r\rangle$ of SWIR-1 and SWIR-2; respectively that are exploited to model the percentage saturation of clay C_I. Consequently, both $\hat{b}$ and $\vec{h}$ are $k \times 1$ column vectors. The solution to the least-squares normal equation is derived from minimizing the sum of the squared differences between observed values and predicted values in a linear regression problem.

$$\hat{b} = \mathcal{H}^{-1}\vec{h} \tag{5.7}$$

Given a solution to the least-squares normal equations, the retrieval is computed using the fitted multiple regression model as follows:

$$\Psi|C_I\rangle = \left[\sum_{i=1}^{k} |S_r\rangle_i \otimes \left[\frac{\sum_{i=1}^{k} \left(|S_r\rangle_i - |\bar{S}_r\rangle\right) \otimes \left(|C_I\rangle_i - |\bar{C}_I\rangle\right)}{\sum_{i=1}^{k} \left(|S_r\rangle_i - |\bar{S}_r\rangle\right)^2} \right] \right] \oplus \left[|\bar{C}_I\rangle - |\bar{S}_r\rangle \otimes \left(\frac{\sum_{i=1}^{k} \left(|S_r\rangle_i - |\bar{S}_r\rangle\right) \otimes \left(|C_I\rangle_i - |\bar{C}_I\rangle\right)}{\sum_{i=1}^{k} \left(|S_r\rangle_i - |\bar{S}_r\rangle\right)^2} \right) \right] \pm \left[\frac{\sum_{i=1}^{k} \sqrt{\left[\Psi|C_I\rangle - \Psi|C_{lab}\rangle\right]^2}}{k} \right] \tag{5.8}$$

where $|S_r\rangle$ is computed by:

$$|S_r\rangle = \frac{\sum_{i=1}^{k} \left(\frac{|S(\lambda_i)\rangle_{\max}}{254} - \frac{|S(\lambda_i)\rangle_{\min}}{255} \right) \otimes |DN\rangle_i + |S(\lambda_i)\rangle_{\min}}{\sum_{j=1}^{k} \left(\frac{|S(\lambda_j)\rangle_{\max}}{254} - \frac{|S(\lambda_j)\rangle_{\min}}{255} \right) \otimes |DN\rangle_j + |S(\lambda_j)\rangle_{\min}} \tag{5.8.1}$$

Equation 5.8 is the general derived equation to estimate the percentage saturation of clay C_I in any multispectral or hyperspectral data. It is considered a linear hyperbolic quantum algorithm to determine the variability of percentage saturation of clay C_I in optical remote sensing data. This quantum algorithm aims at uncertainty reduction in computing the percentage saturation of clay C_I. This algorithm can be titled 'the quantized Marghany clay saturation algorithm'.

Equation 5.8 represents a general formula for estimating the percentage saturation of clay C_I from multispectral or hyperspectral data. This equation is derived using a linear hyperbolic quantum algorithm designed to address the variability in clay saturation estimates from optical remote sensing data. The goal of this quantum algorithm is to reduce uncertainty in the computation of clay saturation percentages C_I.

This approach was developed by Marghany [20] and named the 'Quantized Marghany Clay Saturation Algorithm'. This title reflects the incorporation of quantum mechanics principles into the algorithm distinguishing it from traditional methods. The algorithm leverages quantum mechanics to improve the accuracy and reliability of clay saturation C_I measurements by minimizing uncertainties inherent in remote sensing data. This framework underscores the integration of advanced quantum techniques to improve remote sensing analysis and estimation precision.

5.9 QUANTIZED MARGHANY CLAY SATURATION ALGORITHM FOR RETRIEVING SEDIMENTATION OCCURRENCE IN GERD RESERVOIR

To date, no comprehensive study has examined sediment concentration across the GERD (Grand Ethiopian Renaissance Dam) reservoir. The images analyzed in this study were selected based on minimal cloud cover over the GERD reservoir and occurrences of sedimentation floods. It is important to note that not all images or months exhibit sedimentation events in the GERD reservoir. Sentinel-2 data spanning from March 2023 to July 2024 were chosen to simulate clay concentration. Clay is a key component of sedimentation and suspended sediments are detectable in remote sensing data. Sedimentation patterns observed in ASTER data or other multispectral or hyperspectral data serve as excellent indicators of the upstream flow dynamics.

The quantum spectral signature plot for different soil types (vertisols, red soils, nitosols, alisols, and acrisols) using Sentinel-2 data provides a visual representation of how each soil type reflects electromagnetic radiation across various wavelengths (Figure 5.14). The general trends indicate that all soil types exhibit an increase in reflectance from the visible region (0.490 μm) to the near-infrared (NIR) region (0.842 μm). Conversely, the quantum spectral signature decreases in the short-wave infrared (SWIR) regions (1.610 μm and 2.190 μm), which is consistent with typical suspended sediment spectral behavior. For instance, vertisols show peak reflectance in the NIR region (0.842 μm and 0.865 μm) before decreasing significantly in the SWIR bands (Figure 5.14). The quantum spectral signature of vertisols is relatively high in the blue region (0.490 μm) and steadily increases towards the NIR region.

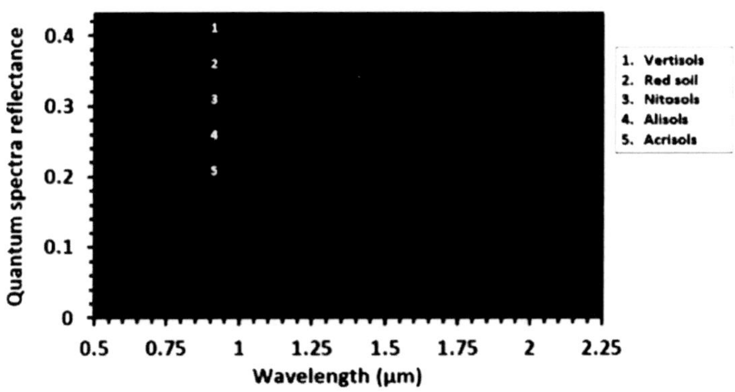

Figure 5.14 Quantum spectra signature of clay soils contains in suspended sediment in Sentinel 2 data.

However, the quantum spectral reflectance values of red soils are generally lower than those of vertisols but follow a similar increasing trend toward the NIR region. Red soils exhibit distinctive absorption features, particularly in the SWIR region, indicating the presence of iron oxides which impart their characteristic red color. Nitosols also display a trend similar to other soil types but with generally lower reflectance values. Their quantum spectral reflectance increases toward the NIR region but not as sharply as vertisols and red soils, suggesting different mineral compositions and possibly higher clay content.

Alisols, on the other hand, have reflectance values between those of vertisols and nitosols. They show a steady increase in reflectance from the visible to the NIR region, with notable absorption features in the SWIR region. In contrast, acrisols have the lowest reflectance values among the soil types in the plot. Their quantum spectral signature increases toward the NIR region but remains lower than the other soils, indicating higher organic matter and moisture content. The sharp decline in acrisols' reflectance in the SWIR bands indicates significant absorption features typical of soils with high clay content.

Indeed, quantum spectral signatures in the SWIR bands (1.610 μm and 2.190 μm) are sensitive to soil moisture and clay content. This sensitivity is evident in acrisols and nitosols, which show lower SWIR quantum spectral signatures due to higher clay content and moisture levels. Further evidence of clay sedimentation in Sentinel-2 data is provided by the presence of regosols and leptosols, both of which exhibit lower SWIR reflectance (Figure 5.15).

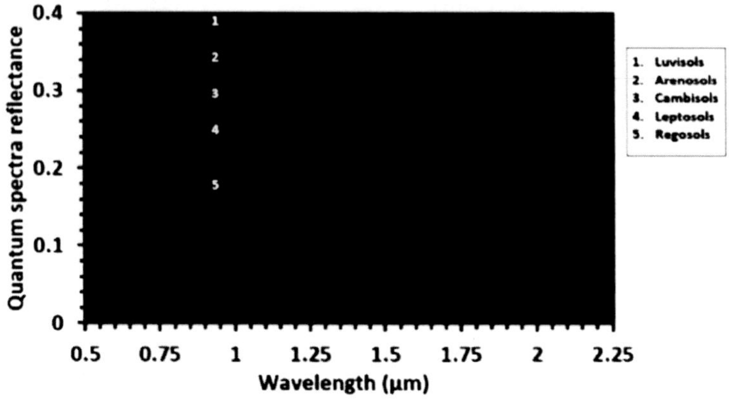

Figure 5.15 Evidence of the existence of clay sediment in GERD reservoir in Sentinel-2 data.

The central question is: what types of clay are contained in the suspended sediments across the GERD reservoir as seen in Sentinel-2 data? Lower reflectance values in the visible bands (0.490 μm, 0.560 μm, 0.665 μm) indicate higher organic matter and darker mineral colors. For instance, kaolinite, with its lower reflectance, suggests higher organic matter and darker hues compared to vermiculite and chlorite. Conversely, higher reflectance in the NIR region (0.842 μm, 0.865 μm) is typically associated with higher mineral brightness and lower organic content. In this context, vermiculite and chlorite exhibit the highest reflectance in the NIR, reflecting their unique mineral compositions and lower moisture content.

In the SWIR bands (1.610 to 2.190 μm), the sharp decrease in reflectance for all minerals highlights absorption due to hydroxyl and water content (Figure 5.16). Montmorillonite and kaolinite which show lower SWIR reflectance indicate higher moisture levels and hydroxyl content. The quantum spectral signature profile thus illustrates the differences in reflectance characteristics across various clay minerals. These differences can be attributed to variations in suspended sediment composition and clay structure.

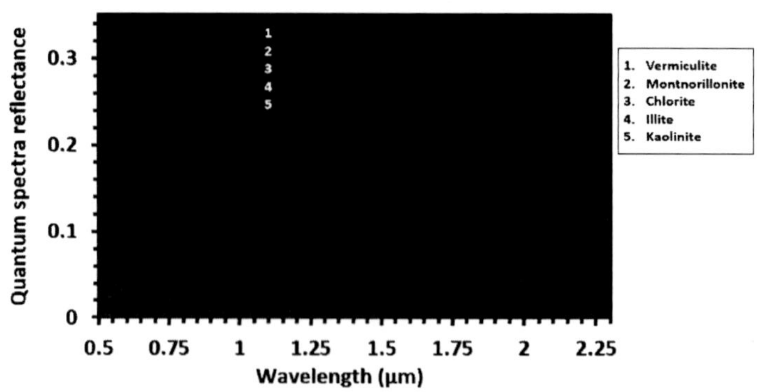

Figure 5.16 Analysis of quantum spectra to identify various clay minerals using Sentinel-2 data.

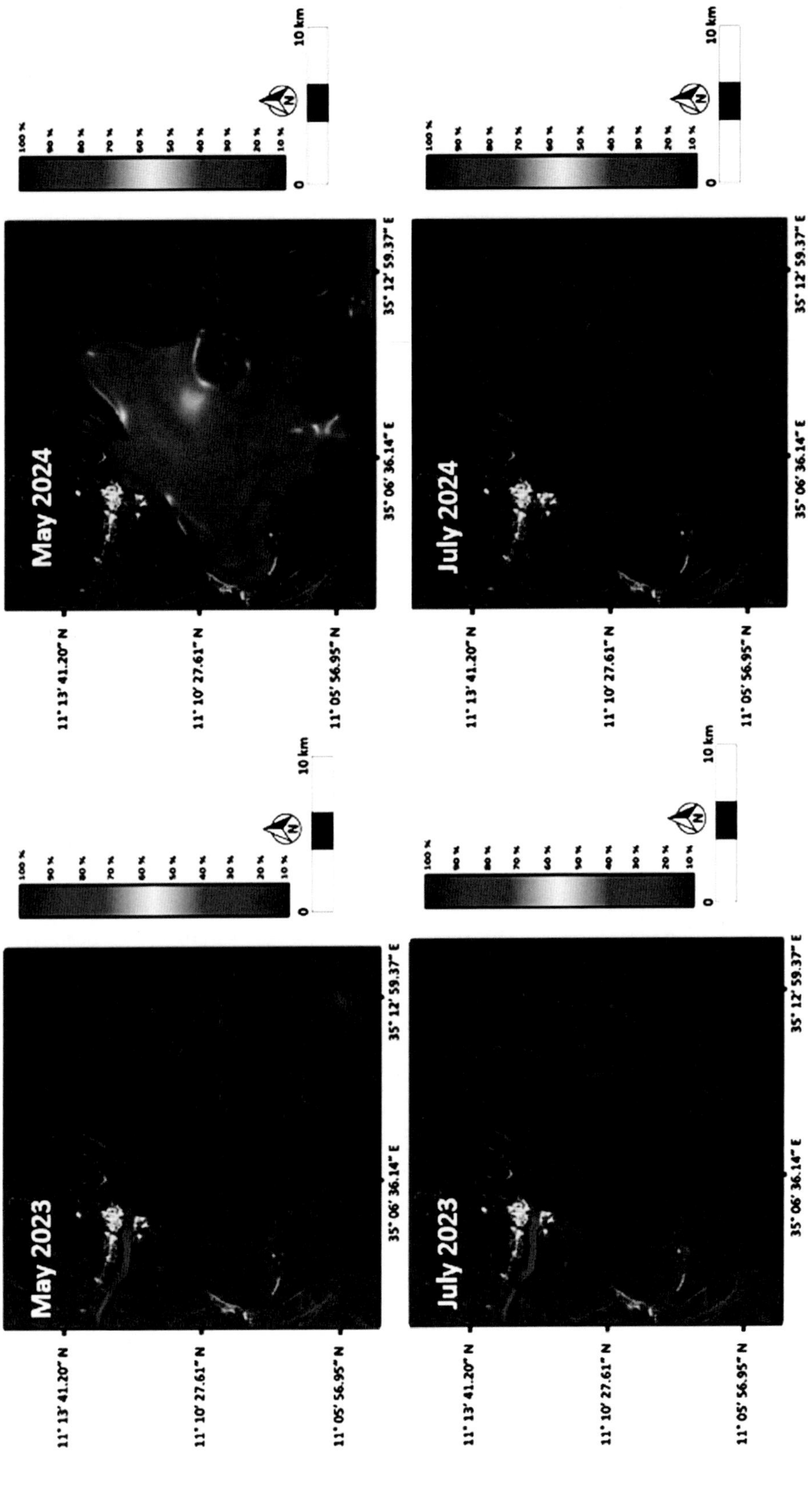

Figure 5.17 The percentage rate of the saturation of clay in the GERD reservoir in May, and July 2023 and 2024, respectively.

Sentinel-2 data, with its extensive spectral coverage proves useful in distinguishing between these clay sediment compositions, based on their unique spectral signatures. This capability underscores the potential of remote sensing in analyzing and monitoring sediment types and their properties in the GERD reservoir.

According to the above perspective, both May and July of 2023 and 2024 exhibit high rates of clay saturation, reaching 100% along the main GERD dam and Saddle dam, as well as in the GERD reservoir (Figure 5.17). In contrast, in April 2024, the rate of clay saturation decreased to 65% in the center of the reservoir but remained high along the Saddle Dam, with the highest rate recorded at 85% (Figure 5.18).

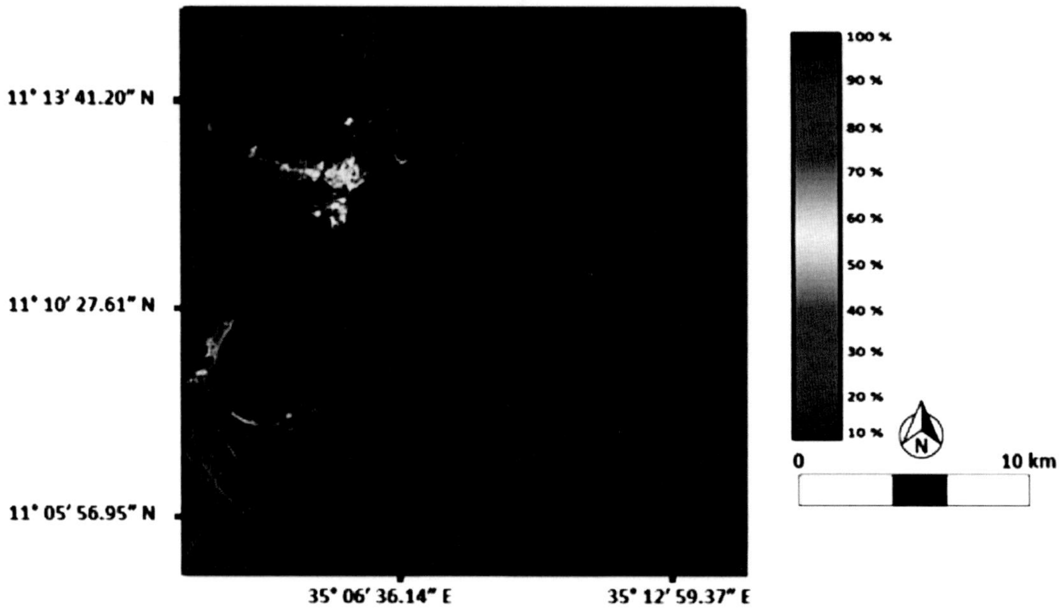

Figure 5.18 The percentage rate of the saturation of clay in the GERD reservoir in April, 2024.

However, from January 2024 to March 2024 (Figure 5.19), the clay saturation percentage experienced a decrease ranging from 10 to 50%. The lowest rate was observed in January 2024 affecting the entire reservoir. Similarly, November and December 2023 (Figure 5.20) exhibited a low clay saturation rate of 10%. In contrast, from July 2023 to October 2023 (Figure 5.21), the clay saturation rate increased to 100% across the entire reservoir.

The variation in clay saturation levels through the GERD reservoir can be attributed to several hydrological, climatic and environmental factors that influence sediment transport and deposition. Here are some key reasons for the observed increments and decrements.

The Blue Nile River, which feeds the GERD reservoir, experiences significant seasonal variations in flow due to rainfall patterns. The rainy season, typically from June to September, leads to higher river discharge, bringing a large volume of sediment, including clay, into the reservoir. This influx is evident in the high clay saturation rates from July to October 2023, reaching up to 100%. Conversely, during the dry season, from November to March, river discharge decreases reducing the amount of incoming sediment. This reduction is reflected in the lower clay saturation rates observed in January to March 2024 and the low rates in November and December 2023.

According to Mellander et al. [28], and Woldesenbet and Elagib [29], the Blue Nile basin receives significant annual rainfall, which plays a crucial role in the water supply for the Nile River. The spatially averaged annual rainfall in the Blue Nile basin is approximately 1490 mm. In other words, the spatially averaged monthly rainfall over the upper Blue Nile Basin exhibits

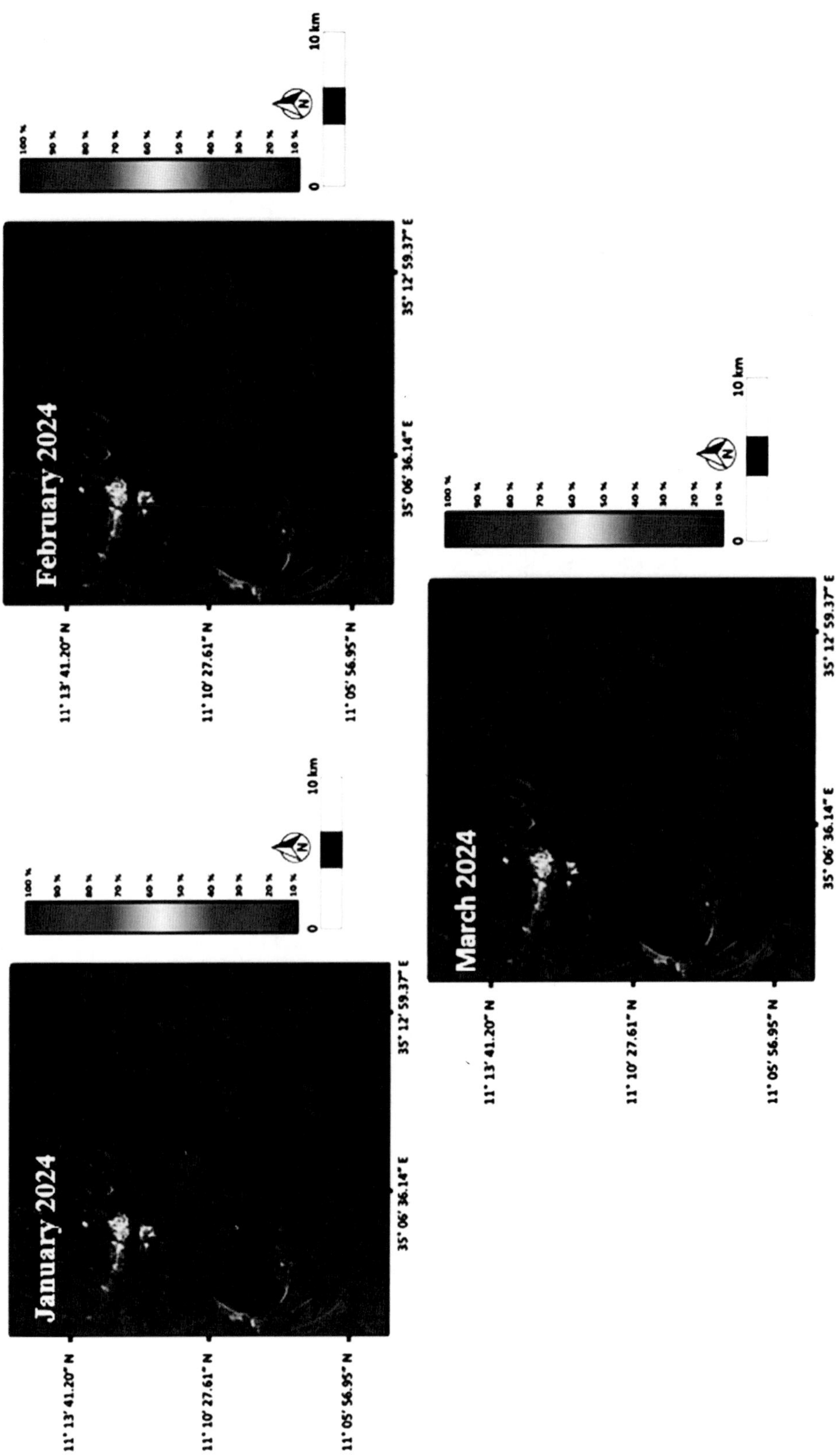

Figure 5.19 The percentage rate of the saturation of clay in the GERD reservoir from January to March 2024.

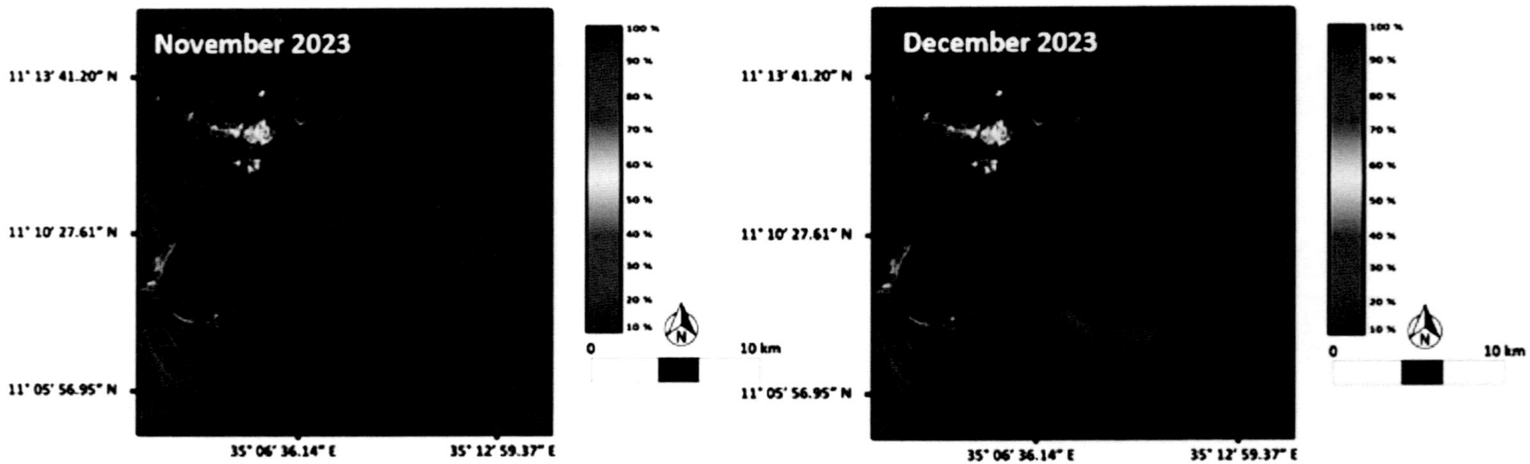

Figure 5.20 The percentage rate of the saturation of clay in the GERD reservoir in November and December 2023, respectively.

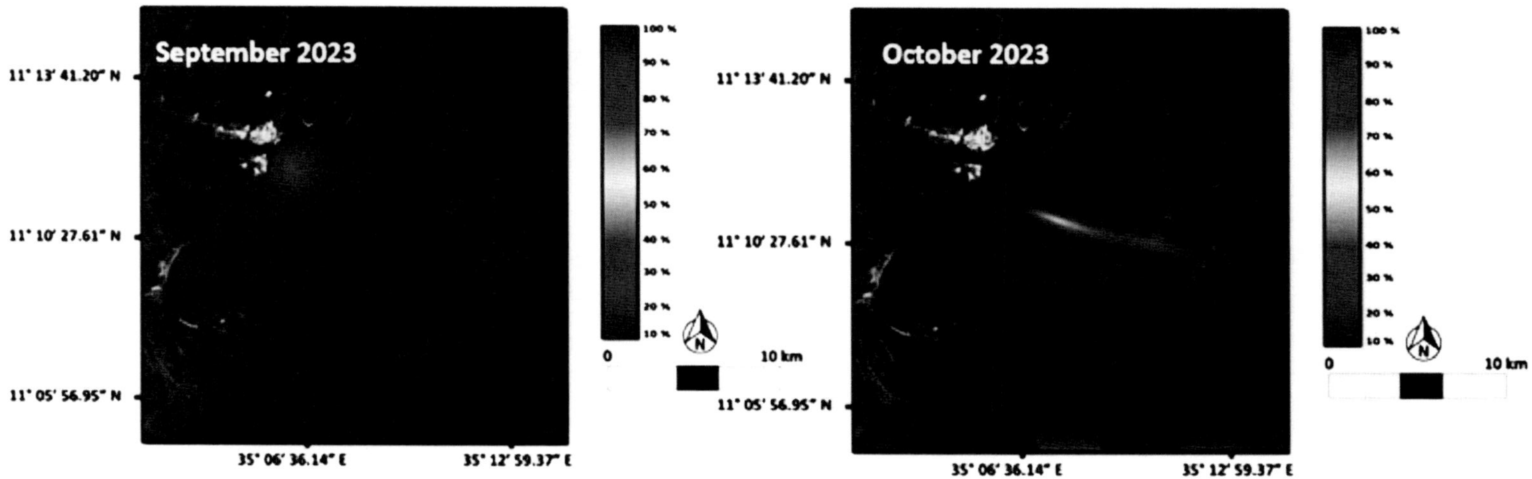

Figure 5.21 The percentage rate of the saturation of clay in the GERD reservoir in September and October 2023, respectively.

significant variation throughout the year. During December, rainfall is minimal, averaging just 9 mm. In contrast, July sees the highest average rainfall, reaching up to 290 mm, with considerable variability (Figure 5.22).

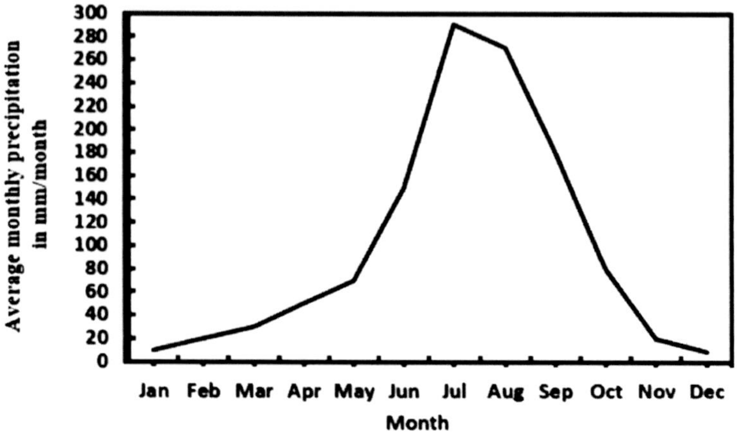

Figure 5.22 Average monthly precipitation in mm/month.

Therefore, the validation of the quantized Marghany clay saturation algorithm is achieved using regression correlation with average monthly precipitation. Notably, the correlation coefficient (R^2) is 0.913, with a p-value less than 0.00005 (Figure 5.23). This high correlation indicates a strong relationship between clay saturation levels and monthly precipitation, validating the efficacy of the quantized Marghany clay saturation algorithm in predicting sedimentation patterns in the GERD reservoir. During the wet season (June to September) increased rainfall leads to higher river discharge and sediment load, resulting in higher clay saturation levels in the reservoir. This is consistent with the observed 100% clay saturation rates from July to October 2023 (Figures 5.17 and 5.21).

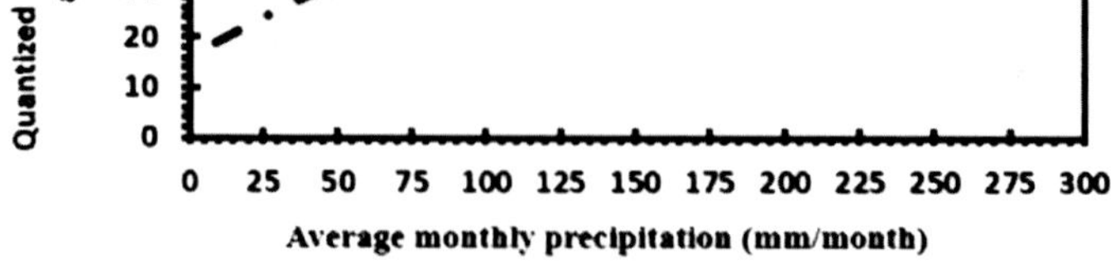

Figure 5.23 Validation of quantized Marghany clay saturation algorithm.

5.10 SUPPORT VECTOR MACHINES

What is meant by Support Vector Machines (SVM)? A support vector machine (SVM) is a type of deep learning algorithm that performs supervised learning for the classification or regression of data clusters. An SVM constructs a learning model that assigns new examples to one cluster or another.

In this process, SVMs are recognized as non-probabilistic, binary linear classifiers. Consequently, SVMs can use methods such as Platt Scaling to operate within probabilistic classification settings [30–32]. The question now is: what are the key features and operations of SVMs? Like other supervised learning algorithms, a support vector machine (SVM) requires labeled data for training. It categorizes the training data into distinct clusters, organizing them into clearly separated groups in space. This organization allows the SVM to classify training data by finding the optimal hyperplane that best separates the clusters. This hyperplane is chosen to maximize the margin between the two closest data points from each cluster, known as support vectors.

Moreover, SVMs are inherently non-probabilistic, meaning they do not output probabilities for classifications but rather provide a clear decision boundary. They are also binary classifiers, typically distinguishing between two classes. Therefore, the hyperplane that separates the dataset with the maximum margin is considered the best because it enhances the model's generalization ability to new datasets. The margin is the distance between the hyperplane and the nearest data points from each class.

However, SVMs can be sensitive to noise, such as mislabeled data, which can reduce their accuracy. This sensitivity arises because the algorithm strives to find the hyperplane that maximizes the margin, and mislabeled data points can distort this boundary. As a result, it is crucial to ensure data quality when using SVMs to maintain their effectiveness and reliability. The separating hyperplane is defined as:

$$w^T \phi(x) + b = 0 \tag{5.9}$$

where w is the decision hyperplane normal vector, X_i is a vector of the dataset mapped to a high dimensional space with $i = 1, 2, \ldots, n$, yi is the class label of data point i (+1 or −1), ξ is the classification error [Figure 5.24], ρ is the margin separator called the width of separation between support vectors of classes given by $\rho = \frac{2}{\|w\|}$.

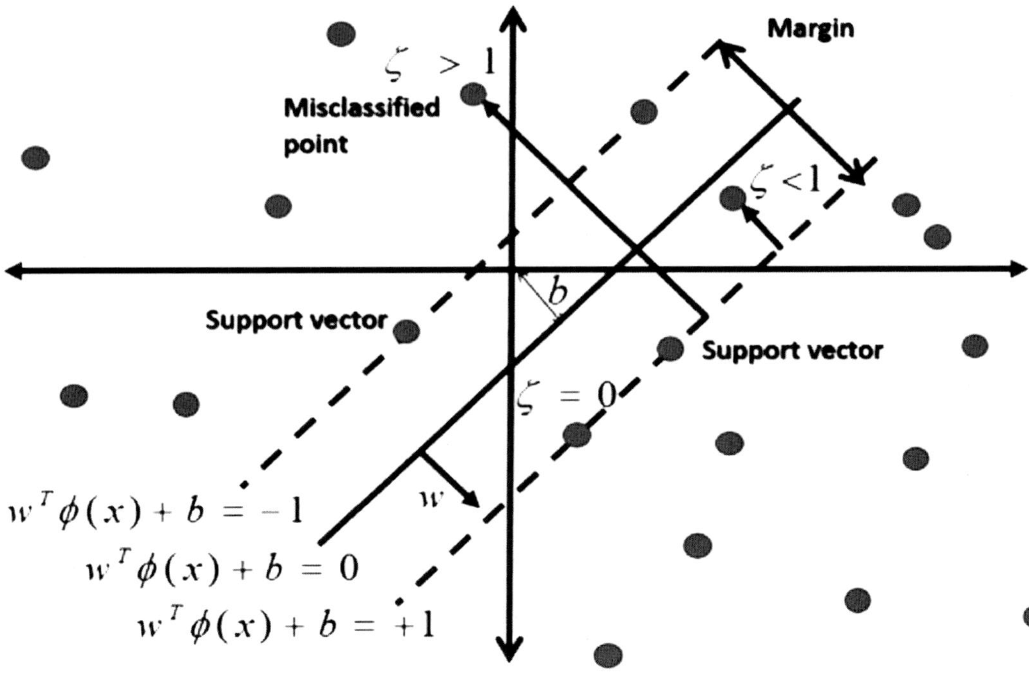

Figure 5.24 SVM classifications involve support vectors critical points that are close to the decision boundary.

How does SVM Work? A support vector machine (SVM) employs a procedure known as the kernel trick to transform input training data. Based on these transformations, it identifies an optimal boundary between the potential outputs. This method allows SVMs to handle non-linear data by mapping it into a higher-dimensional space where a linear separation is achievable. Consequently, SVMs are also referred to as support vector networks (SVNs).

This method of image classification is both accurate and efficient, particularly in remote sensing applications. One of the significant advantages of SVMs is that they require only small amounts of training data, specifically data points located near the interclass boundaries in the feature space. Although SVM classification needs only a single sample per class, it is crucial that this sample is close to the class boundary to ensure accurate classification.

In contrast, maximum likelihood (ML) classification necessitates a large training sample size, especially when dealing with data in a high-dimensional feature space. This requirement makes SVMs more efficient and practical in many scenarios where obtaining extensive training data is challenging. Needless to say, SVMs exploit the kernel trick to transform data and determine optimal decision boundaries, requiring minimal training data near interclass boundaries, thereby offering a highly accurate and efficient classification method, especially useful in remote sensing [30, 32].

Figure 5.25 demonstrates the percentage rate of clay saturations delivered by SVM from Sentinel-2 satellite data from July to October. Like the quantized Marghany clay saturation algorithm, the spatial variation of the percentage rate of clay saturations shows high concentration during wintertime (Figures 5.18 and 5.21). On the other hand, SVM in comparison with the quantized Marghany clay saturation algorithm, shows a lower r^2 value of 0.743 and root mean square error RMSE of ±7.78 (Figure 5.26). Indeed, the SVM requires the specification of a kernel function and involves the slow development of multiclass relationships. In contrast, the quantized Marghany algorithm is able to reduce standard errors due to the increased number of training samples fed into the SVM. Unlike the quantized Marghany algorithm, SVM cannot reduce standard errors due to the increase in training samples fed into the algorithm. However, the major advantage of SVM is that it will not require too much time to collect data during the training.

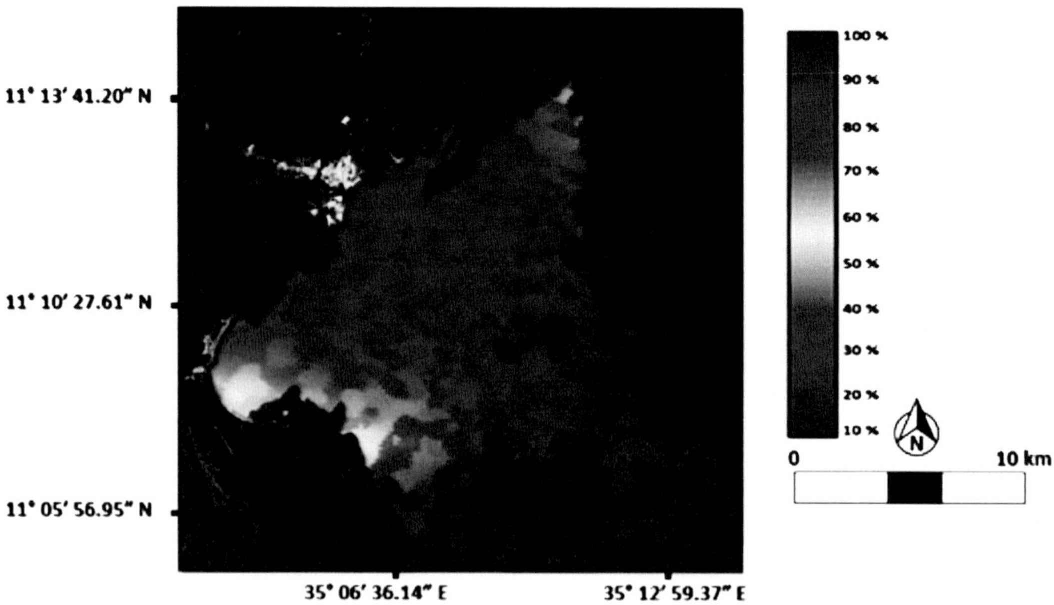

Figure 5.25 SVM for computing percentage rate of the saturation of clay in Sentinel-2 data from July to October.

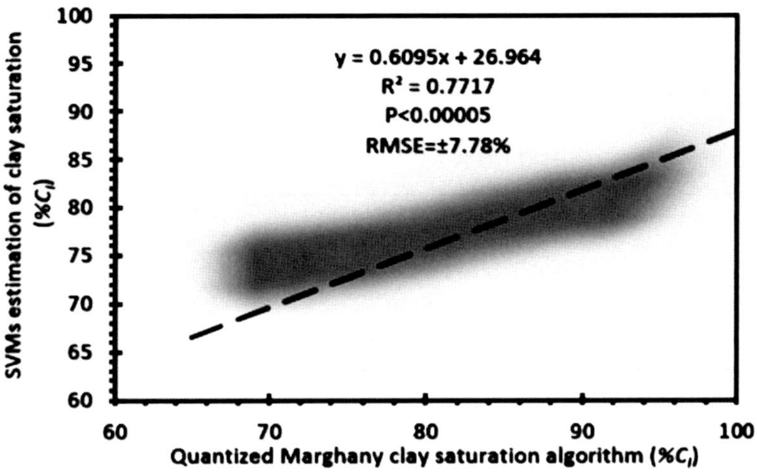

Figure 5.26 Correlation between SVM and quantized Marghany clay saturation algorithm.

5.11 QUANTUM SUPPORT VECTOR MACHINES

Following Marghany [20], let us assume that the training data for the percentage rate of quantized Marghany's clay saturation algorithm is

$\Psi|\%C_I\rangle_{\text{Marghany}_i},\left\{\left(\%\vec{C}_{I_j},\Psi|\%C_I\rangle_{\text{Marghany}_j}\right):\%\vec{C}_{I_j}\in\mathbb{R}^d,\Psi|\%C_I\rangle_{\text{Marghany}_j}=\pm 1\right\}$, and a kernel k.

In this context, the classification of new data based on quantized Marghany's clay saturation algorithm can be formulated as:

$$\Psi|\%\vec{C}_I\rangle_{\text{Marghany}_i}=\text{sgn}\left(\sum_{j=1}^{m}\alpha_j k\left(|\%\vec{C}_I\rangle_{\text{Marghany}_j},|\%\vec{C}_I\rangle_{\text{Marghany}_i}\right)+b\right). \tag{5.10}$$

In the foundation of training set labels of $\Psi|\%\vec{C}_I\rangle_{\text{Marghany}_i}$, the anticipated SVM parameters are encoded in the amplitudes of the ultimate state-run:

$$|b,\vec{\alpha}\rangle=\left[b^2+\sum_{k=1}^{m}\alpha_k^2\right]^{-0.5}\left(b|0\rangle+\sum_{k=1}^{m}\alpha_k|k\rangle\right), \tag{5.11}$$

Then there exists a quantum algorithm that solves the training problem of the least-square support vector machine within the accuracy ε in the time $\tilde{O}(\log(md)\kappa^2\varepsilon^{-3})$. Determination of the new classified clay features can be expressed mathematically as:

$$\begin{aligned}\Psi|\%\vec{C}_I\rangle_{\text{Marghany}_i}|0\rangle|0\rangle &= \sum_{j=1}^{m+1}\langle u_j|\%C_{\text{Marghany}_i}\rangle|0\rangle|0\rangle\\ &\xrightarrow[\text{result}=1]{\text{Compute}}\sum_{j=1}^{m+1}\frac{\langle u_j|\%C_{\text{Marghany}_i}\rangle}{\tilde{\lambda}_j}|u_j\rangle\end{aligned} \tag{5.12}$$

Equation 5.12 says that $\tilde{\lambda}_j$ is the eigenstate of the Hamiltonian matrix, and unitary u_j can be given by:

$$|u\rangle=\left[b^2+\sum_{k=1}^{m}\alpha_k^2\left|\%C_{I_k}\right|^2\right]^{-0.5}\left(b|0\rangle|0\rangle+\sum_{k=1}^{m}\alpha_k\left|\%\vec{C}_I\right||k\rangle|\%\vec{C}_{I_k}\rangle\right) \tag{5.13}$$

$$\left|\%C_{\text{Marghany}_i}\right\rangle = \left[\mathcal{H}\left|\%C_{\text{Marghany}_i}\right|^{-2}\right]^{-0.5}\left(|0\rangle|0\rangle + \sum_{k=1}^{m} \left|\%\vec{C}_I\right|\left|k\right\rangle\left|\%\vec{C}_I\right\rangle\right) \tag{5.14}$$

In equation 5.14, the Hermitian matrix $\mathcal{H}$ can be expressed as $\left\langle \%C_{\text{Marghany}_i}\right|\mathcal{H}\left|\%C_{\text{Marghany}_i}\right\rangle$. If we calculate only the kernel matrix by quantum means, we have a complexity of $O(\mathcal{H}^2(\mathcal{H}+\varepsilon^{-1}\log N))$. There is more to gain: exponential speedup in the number of training examples is possible when using the least-squares formulation of support vector machines. In this circumstance, obtaining the normalized kernel matrix can assist in an exponential speedup of QSVM as:

$$\hat{k} = \frac{1}{tr(k)}\sum_{i,j=1}^{m}\left\langle \%C_{\text{Marghany}_i}\middle|\%C_{\text{Marghany}_j}\right\rangle \left\|\%C_{\text{Marghany}_i}\right\|\left\|\%C_{\text{Marghany}_j}\right\| |i\rangle\langle j|. \tag{5.15}$$

$\hat{k}$ is a normalized Hermitian matrix $\mathcal{H}$, which makes it a prime candidate for quantum self-analysis. The exponentiation; therefore, is completed in $O(\log N)$ the procedure.

In Figure 5.27, the quantized Marghany's clay saturation percentage rate involving QSVM is demonstrated. The addition of QSVM has significantly improved the identification of cluster spatial variations within the quantized Marghany algorithms, particularly in the Sentinel-2 SWIR-1 and SWIR-2 spectral domains. This improvement results in the extremely clear identification of boundaries between different clusters of clay saturation percentage rate. QSVM can effectively determine the composition of clay-rich sediment, including shale, sandstone, silt, and clay. It is important to note that clay sedimentation dominates in GERD reservoirs, with the highest rate occurring from July to October. Similar to the quantized Marghany's clay saturation algorithm, QSVM indicates a maximum clay saturation rate of 100% in the entire GERD reservoir from July to October. However, QSVM further enhances the quantized Marghany's clay saturation algorithm by revealing a clear pattern of clay concentration composition.

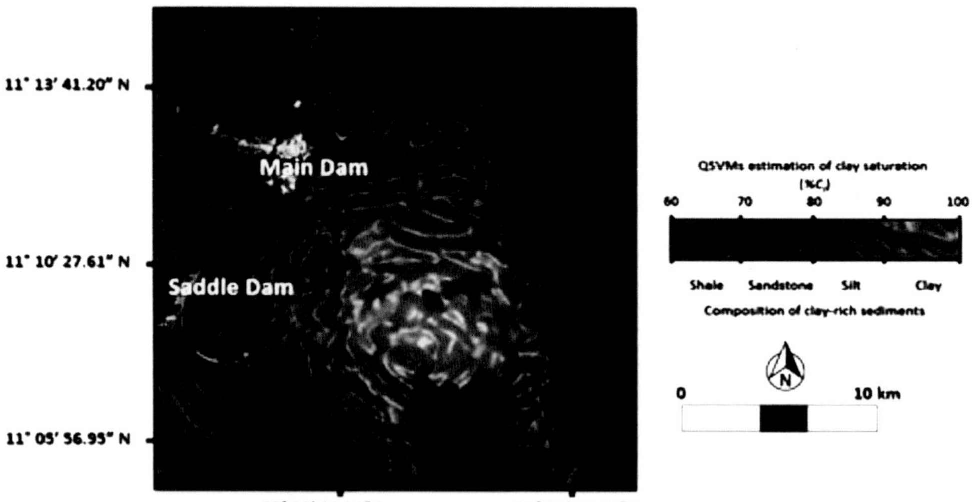

Figure 5.27 QSVM clay saturation percentage with identifying cluster of clay-rich-sedimentation sorts.

5.12 QSVM ENTANGLES QUANTIZED MARGHANY'S CLAY SATURATION ALGORITHM

QSVM demonstrates higher accuracy than SVM, achieving a lower RMSE of ±6.92% (Figure 5.28). On the other hand, QSVM improves the accuracy of the quantized Marghany clay saturation algorithm by identifying clear cluster features, resulting in an error reduction difference of ±0.86%. In this context, QSVM shows strong agreement with the quantized Marghany clay saturation algorithm, with an R^2 value of 0.97.

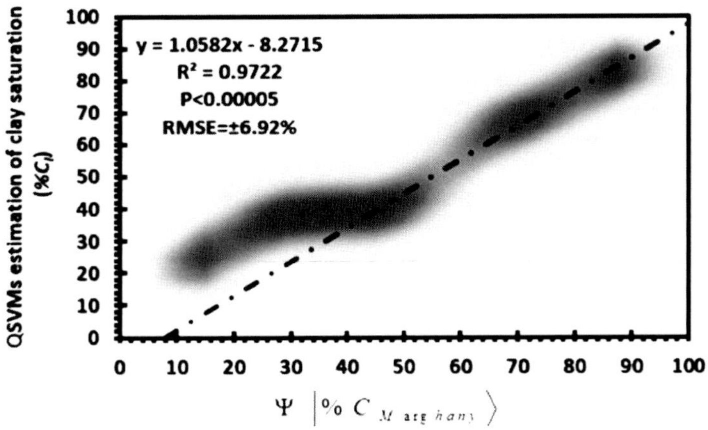

Figure 5.28 Correlation between QSVM and the quantized Marghany's clay saturation algorithm $\Psi\left|\%C_{\text{Marghany}}\right\rangle$.

Consequently, QSVM employs an optimization search to replace sequential minimum optimization in a discretized search space. By incorporating least-squares support vector machines, it translates the optimization problem into a set of linear equations, necessitating the rapid computation of the kernel matrix—one key source of speedup in the quantum version. Based on this formulation and assuming quantum input and output space, quantum support vector machines can achieve an exponential speedup over their classical counterparts [33–36].

Specifically, QSVM is integrated with the quantized Marghany clay saturation algorithm to classify new clay saturation percentage rates. This effort demonstrates the feasibility of using QSVM in conjunction with the quantized Marghany clay saturation algorithm on near-term quantum computers. However, further advancements in quantum hardware and error correction are necessary to facilitate large-scale testing on satellite data.

It is worth noting that little to no effort has been made to investigate sedimentation patterns in the GERD reservoirs, as most studies have concentrated on monitoring water level fluctuations and assessing the dam's stability. In this regard, the crucial question now is whether the actual concentration of suspended sediment in the GERD reservoirs can be accurately determined.

5.13 QUANTUM INSPIRED ALGORITHM FOR RETRIEVING SUSPENDED SEDIMENT CONCENTRATION FROM SENTINEL-2 DATA

To model Suspended Sediment Concentration (SSC) using Sentinel-2 satellite images with a quantum mechanics-inspired approach, we can draw on principles of spectral reflectance and photon interactions with particles. While the actual phenomenon is not quantum mechanical, we can develop a reflectance-based model that mimics certain probabilistic aspects of quantum mechanics.

Let us assume reflectance $R(\lambda)$ is the observed spectral reflectance at wavelength λ. In this, the Sentinel-2 bands in the visible to near-infrared range (e.g., Bands 4, 5, and 6 corresponding to wavelengths ~665 nm, 705 nm, and 740 nm) are used. The reflectance in these bands is related to SSC, as suspended sediments increase scattering and absorption, especially in the red and red-edge portions of the spectrum.

Therefore, let us define SSC as a function of the reflectance values $R(\lambda)$ at the specific bands. A common empirical form for SSC estimation is:

$$SSC = a_0 + a_1 R(\lambda_1) + a_2 R(\lambda_2) + \cdots + a_n R(\lambda_n) + \varepsilon \tag{5.16}$$

where a_0, a_1,..., a_n are empirical coefficients determined through calibration with in-situ SSC measurements. In this sense, $R(\lambda_1)$, $R(\lambda_2)$, ..., $R(\lambda_n)$ are the reflectance values at specific Sentinel-2 bands and ϵ represents the error term.

To give it a probabilistic or 'quantum-inspired' aspect, we could model the interaction of photons with suspended particles by treating each reflectance value as a probability amplitude for sediment presence. In quantum mechanics, the probability amplitude is squared to get the actual probability of a state. For reflectance, we can introduce an 'effective reflectance' for each band:

$$\tilde{R}(\lambda_i) = | R(\lambda_i) |^2 \tag{5.17}$$

The final model for SSC as a function of Sentinel-2 reflectance values becomes:

$$SSC = b_0 + b_1 | R(\lambda_4) |^2 + b_2 |R(\lambda_5)|^2 + b_3 |R(\lambda_6)|^2 + \varepsilon \tag{5.18}$$

where $R(\lambda_4)$, $R(\lambda_5)$, and $R(\lambda_6)$ are the reflectance values for Sentinel-2 Bands 4, 5, and 6. Additionally, b_0; b_1; b_2; and b_3 are coefficients that can be determined empirically by correlating satellite reflectance data with in-situ SSC measurements. In this perspective, the quantum-inspired model for SSC is:

$$\Psi|SSC\rangle = b_0 + b_1\Psi\left||R(665)|^2\right\rangle + b_2\Psi\left||R(705)|^2\right\rangle + b_3\Psi\left||R(740)|^2\right\rangle + \varepsilon \tag{5.19}$$

Equation 5.19 encapsulates the probabilistic nature of interactions between sediment and light by representing reflectance as a probability amplitude. In this model, the squared values are interpreted as 'probability densities' for the concentration of sediment. Figure 5.29 illustrates that the quantum-inspired algorithm estimates the total suspended sediment in the Grand Ethiopian Renaissance Dam (GERD) reservoirs to range between 30,000 mg/L and 150,000 mg/L.

The suspended sediment concentration in the Grand Ethiopian Renaissance Dam (GERD) reservoirs, reaching values between 30,000 mg/L and 150,000 mg/L, is a critical finding. These high concentrations indicate a substantial sediment load, which carries significant implications for the dam's operational life, water quality and regional hydrodynamics.

At concentrations of this magnitude, sediment can accumulate rapidly, reducing the reservoir's storage capacity and impacting the dam's efficiency. Over time, high sedimentation rates may require costly maintenance, such as dredging or other sediment management interventions, to maintain reservoir volume and prolong the dam's lifespan. Furthermore, sediment-laden water can damage turbines, erode infrastructure, and disrupt aquatic ecosystems. Thus, monitoring and managing sediment in reservoirs like GERD is essential for sustainable energy production and water resource management.

To validate these findings, a comparative analysis using a retrieval method based on the Quantum Support Vector Machine (QSVM) algorithm was conducted. Figure 5.30 illustrates the accuracy of this simulation, as evidenced by a high R^2 value of 0.95 and a Root Mean Square Error (RMSE) of ±5,228 mg/L. The highest recorded concentration of suspended sediment, approximately 150,000 mg/L, accumulates behind the main dam, indicating significant sediment saturation in the reservoir. This high sediment load suggests reduced water storage efficiency and a likely decrease in discharge flow to downstream regions, an impact that will be further examined in subsequent chapters.

Additionally, it is observed that the suspended sediment concentration near the Saddle dam ranges from 30,000 to 60,000 mg/L. This distribution suggests that sedimentation patterns may vary within the reservoir, with sediment accumulation possibly influenced by water flow dynamics and dam design.

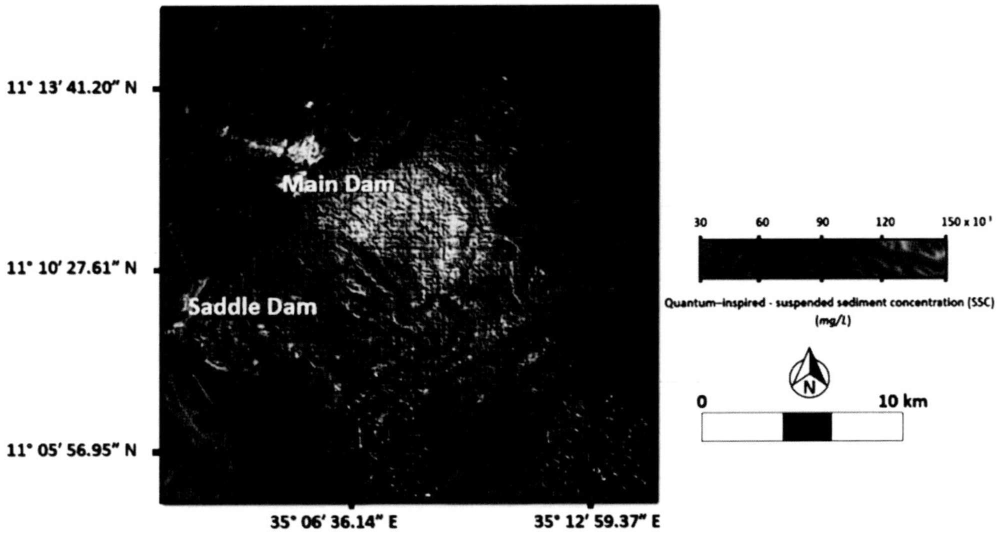

Figure 5.29 Total suspended sediment retrieved using quantum-inspired algorithm from sentieal-2 data in GERD reservoir.

Therefore, the substantial accumulation of sediment in the GERD reservoir underscores the importance of advanced sediment monitoring and management techniques. The quantum-inspired approach was employed, through QSVM, provides a highly accurate framework for tracking sediment dynamics, offering valuable insights that can aid in optimizing reservoir operations. This level of precision enables us to assess sediment load and distribution within the reservoir, directly informing strategies to mitigate sedimentation impacts on water storage, hydropower efficiency, and downstream flow.

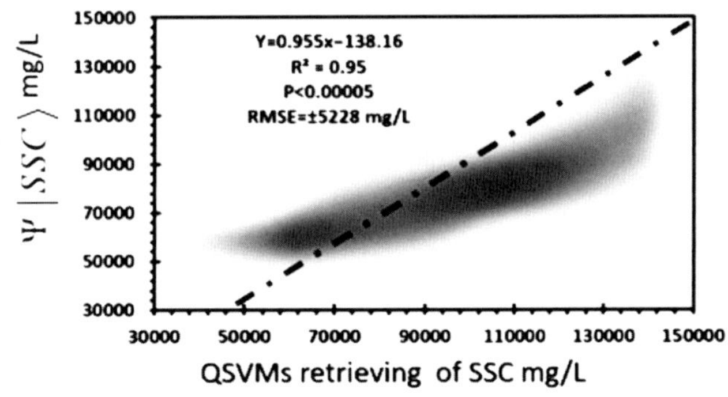

Figure 5.30 Correlation accuracy between quantum-inspired algorithm and QSVM.

From a scientific standpoint, the use of a quantum-inspired algorithm to estimate these concentrations represents a leap in our capabilities for environmental monitoring. Quantum Support Vector Machines (QSVM) allow us to process complex, probabilistic data with higher precision, capturing variability and subtle interactions that traditional methods might miss. This innovative approach opens new avenues for environmental science, providing tools to monitor large-scale projects with the granularity and accuracy required for informed decision-making.

The quantum-inspired techniques applied here represent not only a breakthrough in sediment monitoring but also an opportunity for proactive reservoir management. By precisely understanding sediment dynamics, we can mitigate risks associated with reservoir sedimentation, enhance the dam's operational lifespan, and contribute to the sustainable development of hydropower in the Nile Basin.

5.14 CONCLUSION

This chapter introduced a groundbreaking quantum algorithm, the 'quantized Marghany clay saturation algorithm', developed to identify clay saturation percentage and suspended sediment concentration rates based on the SWIR-1 and SWIR-2 bands of Sentinel-2 data in the GERD reservoir. The algorithm's accuracy is demonstrated by an exceptionally low RMSE of ±0.86%, and when combined with the Quantum Support Vector Machine (QSVM), it achieves an R^2 value of 0.97, underscoring its high precision.

The findings reveal that the GERD reservoir has reached full sediment saturation, with clay and other sediments now making up 100% of the concentration. Specifically, suspended sediment concentrations range between 30,000 and 150,000 mg/L, aligning with a full saturation of clay content. This accumulation represents a maximum level of fine particles, including clay, silt, and organic material primarily transported from upstream sources.

Such high levels of sediment pose critical challenges for the reservoir's longevity, structural integrity and storage capacity. Clay and fine sediment particles, known for their low permeability, can compact densely over time, reducing water flow and increasing pressure on the dam structure. This accumulation is likely to diminish the reservoir's effective storage capacity, impacting both water availability and hydropower production. Additionally, the dense sediment layer may require frequent dredging or other costly measures to maintain the reservoir's operability.

These findings underscore the urgent need for proactive sediment management strategies, such as upstream erosion control or sediment flushing techniques. Without intervention, the high sediment load could jeopardize the long-term sustainability of the Grand Ethiopian Renaissance Dam, diminishing its anticipated benefits and requiring extensive, complex maintenance efforts to ensure continued functionality.

References

[1] World Bank. 2019. Sediment Management for Hydropower Reservoirs: A Guide for Practitioners. Retrieved from https://openknowledge.worldbank.org/handle/10986/31676

[2] United States Society on Dams. 2010. Guidelines for Dam Safety: Selecting and Accommodating Inflow Design Floods for Dams. Retrieved from https://www.ussdams.org/wp-content/uploads/2017/01/2010-Guidelines-for-Dam-Safety.pdf

[3] International Commission on Large Dams (ICOLD). 2019. Bulletin 164: Sedimentation of Reservoirs. Retrieved from https://www.icold-cigb.org/GB/publications/Bulletins/GB_164_Sedimentation_of_reservoirs.pdf

[4] Wisser, D., Frolking, S., Hagen, S. and Bierkens, M.F. 2013. Beyond peak reservoir storage? A global estimate of declining water storage capacity in large reservoirs. Water Resources Research. 49(9): 5732–5739.

[5] Kondolf, G.M. 2013. Sustainable sediment management in reservoirs and regulated rivers: Experiences from five continents, Earth's Future, 2, doi: 10.1002/2013EF000184.

[6] Samad, N., Ch, M.H., Ashraf, M., Saleem, M., Hamid, Q., Babar, U., et al. 2016. Sediment yield assessment and identification of check dam sites for Rawal Dam catchment. Arabian Journal of Geosciences. 9(6): 466.

[7] Wang, Z. and Hu, C. 2009. Strategies for managing reservoir sedimentation. International Journal of Sediment Research. 24(4): 369–384.

[8] Schleiss, A.J., Franca, M.J., Juez, C. and De Cesare, G. 2016. Reservoir sedimentation. Journal of Hydraulic Research. 54(6): 595–614.

[9] Shen, H.W. 1999. Flushing sediment through reservoirs. Journal of Hydraulic Research. 37(6): 743–757.

[10] Julien, P.Y. 2010. Erosion and Sedimentation. Cambridge University Press.

[11] Fan, J. and Morris, G.L. 1992. Reservoir sedimentation. I: Delta and density current deposits. Journal of Hydraulic Engineering. 118(3): 354–369.

[12] Morris, G.L. and Fan, J. 1998. Reservoir Sedimentation Handbook, Design and Management of Dams, Reservoirs, and Watersheds for Sustainable Use. McGraw-Hill Book Co., New York.

[13] Shen, H.W. and Lai, J.S. 1996. Sustain reservoir useful life by flushing sediment. International Journal of Sediment Research. 11(3): 10–17.

[14] Idrees, M.B. 2023. Storage depletion of surface water reservoirs due to sediment deposition and possible management options. *In*: Water and Environment for Sustainability: Case Studies from Developing Countries. 131–144. Cham: Springer International Publishing.

[15] Li, P., He, Z., Cai, J., Zhang, J., Belete, M., Deng, J., et al. 2022. Identify the impacts of the Grand Ethiopian Renaissance Dam on watershed sediment and water yields dynamics. Sustainability. 14(13): 7590.

[16] Borji, T.T. 2013. Sedimentation and sustainability of hydropower reservoirs: cases of Grand Ethiopian Renaissance Dam on the Blue Nile River in Ethiopia (Master's thesis, Institutt for vann-og miljøteknikk).

[17] Luna, L., Awange, J., Song, Y., Bui, L.K., Zerihun, A. and Kuhn, M. 2024. Surface water and geomorphological changes of the Blue Nile dynamics associated with the Grand Ethiopian Renaissance Dam (GERD): a multi-temporal analysis. GIScience and Remote Sensing. 61(1): 2346383.

[18] Fang, Q., Hong, H., Zhao, L., Kukolich, S., Yin, K. and Wang, C. 2018. Visible and near-infrared reflectance spectroscopy for investigating soil mineralogy: A review. Journal of Spectroscopy. 2018(1): 3168974.

[19] Hunt, G.R. and Ashley, R.P. 1979. Spectra of altered rocks in the visible and near infrared. Economic Geology. 74(7): 1613–1629.

[20] Marghany, M. 2022. Remote Sensing and Image Processing in Mineralogy. CRC Press.

[21] Aprile, F. and Lorandi, R. 2012. Evaluation of cation exchange capacity (CEC) in tropical soils using four different analytical methods. Journal of Agricultural Science. 4(6): 278.

[22] Hendershot, W.H., Lalande, H. and Duquette, M. 1993. Ion exchange and exchangeable cations. Soil sampling and methods of analysis. 19: 167–176.

[23] Meier, L.P. and Kahr, G. 1999. Determination of the cation exchange capacity (CEC) of clay minerals using the complexes of copper (II) ion with triethylenetetramine and tetraethylenepentamine. Clays and Clay Minerals. 47(3): 386–388.

[24] Liu, X., Fatoyinbo, T.E., Thomas, N.M., Guan, W.W., Zhan, Y., Mondal, P., et al. 2021. Large-scale high-resolution coastal mangrove forests mapping across West Africa with machine learning ensemble and satellite big data. Frontiers in Earth Science. 8: 560933.

[25] Wang, G. 2017. Quantum algorithm for linear regression. Physical Review A. 96(1): 012335.

[26] Schuld, M., Sinayskiy, I. and Petruccione, F. 2016. Prediction by linear regression on a quantum computer. Physical Review A. 94(2): 022342.

[27] Gilyén, A., Song, Z. and Tang, E. 2022. An improved quantum-inspired algorithm for linear regression. Quantum. 6: 754.

[28] Mellander, P.E., Gebrehiwot, S.G., Gärdenäs, A.I., Bewket, W. and Bishop, K. 2013. Summer rains and dry seasons in the Upper Blue Nile Basin: the predictability of half a century of past and future spatiotemporal patterns. PloS one. 8(7): e68461.

[29] Woldesenbet, T.A. and Elagib, N.A. 2021. Analysis of climatic trends in the upper Blue Nile basin based on homogenized data. Theoretical and Applied Climatology. 146(1): 767–780.

[30] Balabin, R.M. and Lomakina, E.I. 2011. Support vector machine regression (LS-SVM)—an alternative to artificial neural networks (ANNs) for the analysis of quantum chemistry data? Physical Chemistry Chemical Physics. 13(24): 11710–11718.

[31] Noble, W.S. 2006. What is a support vector machine? Nature Biotechnology. 24(12): 1565–1567.

[32] Wang, L. (Ed.). 2005. Support Vector Machines: Theory and Applications, (Vol. 177). Springer Science & Business Media.

[33] Ding, S.F., Qi, B.J. and Tan, H.Y. 2011. An overview on theory and algorithm of support vector machines. Journal of University of Electronic Science and Technology of China. 40(1): 2–10.

[34] Ezawa, M. 2022. Variational quantum support vector machine based on Γ matrix expansion and variational universal-quantum-state generator. Scientific Reports. 12(1): 6758.

[35] Zhu, X., Xiong, J. and Liang, Q. 2018. Fault diagnosis of rotation machinery based on support vector machine optimized by quantum genetic algorithm. IEEE Access. 6: 33583–33588.

[36] Ding, C., Bao, T.Y. and Huang, H.L. 2021. Quantum-inspired support vector machine. IEEE Transactions on Neural Networks and Learning Systems. 33(12): 7210–7222.

Chapter 6

Advanced Quantum Image Processing for Automatic Lineament and Fault Detection along GERD using Multispectral Remote Sensing Data

6.0 INTRODUCTION

In our prior analysis, we identified significant clay concentrations within the GERD reservoir. This finding is critical because clay-rich areas can have substantial implications for the structural integrity and safety of the dam. In the context of the GERD, faults can act as pathways for water seepage, potentially undermining the dam's foundation and leading to structural instability. Therefore, the presence of lineaments suggests areas of weakness in the geological framework, which can be exacerbated by the presence of clay.

6.1 LINEAMENT

This intriguing question leads to a comprehensive understanding of economic geology, encompassing both petroleum and mining. In literature, there are numerous definitions of the term 'lineament'. Lineament attributes are often associated with terms such as geologic lineament (Figure 6.1), tectonic lineament (Figure 6.2), photo lineament, and geophysical lineament. These descriptors typically refer to the presumed origin of the linear feature or the data source from which it has been derived. Some researchers also use the terms 'fracture trace' or 'photo linear' as alternatives [1–3]. A lineament is a crucial geological structure often associated with faults.

According to the above-mentioned perspective, each lineament definition has a slightly different perspective. From a geological standpoint, a lineament is a linear (Figure 6.3) or curvilinear (Figure 6.4) feature of a surface, where segments align in a straight or significantly curved arrangement. A lineament appears as a linear feature in a landscape, reflecting an underlying geological structure such as a fault (Figure 6.5). Geologically, a fault is defined as a planar fracture or discontinuity within a volume of rock, across which there has been significant displacement due to rock-mass movements [1]. Lineaments are often visible on geological or topographic maps and can also be detected as linear features on aerial or satellite imagery [2].

DOI: 10.1201/9781003327325-6

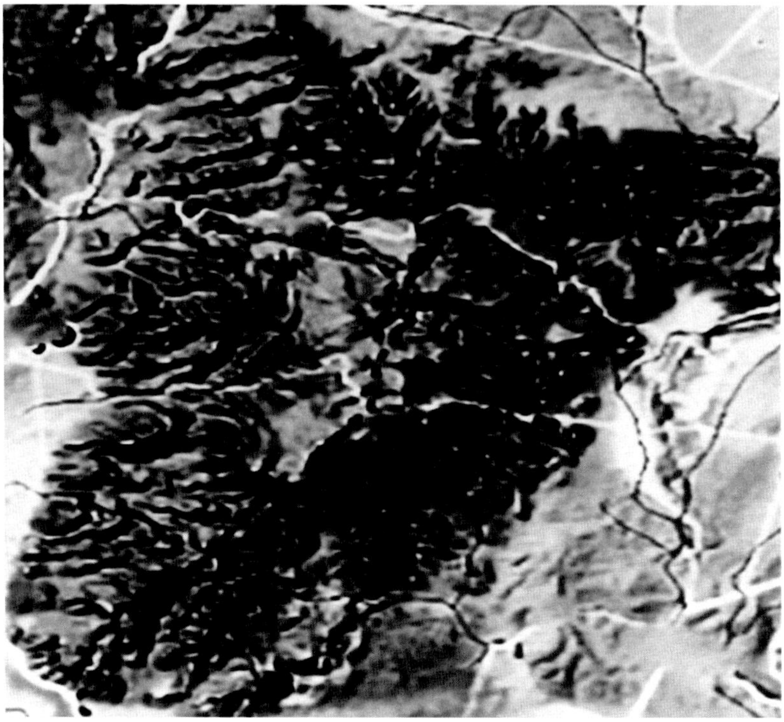

Figure 6.1 Geological lineament.

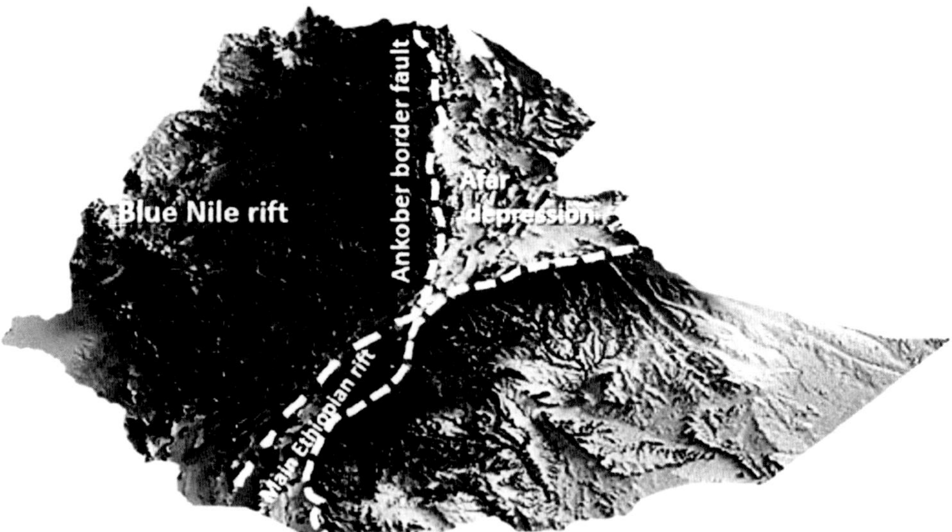

Figure 6.2 Tectonic lineament.

Inherently, lineaments are significant geological structures often aligned with faults, folds or other linear features in the landscape. These features can take various forms. Firstly, fault-aligned valleys run parallel to fault lines, formed by the movement and displacement of the Earth's crust along the fault (Figure 6.6). Secondly, fault or fold-aligned hills are aligned with fault lines or folds in the rock layers, indicating the underlying geological structure (Figure 6.7). Lastly, straight coastlines can occur where geological faults intersect the coastline creating a linear feature.

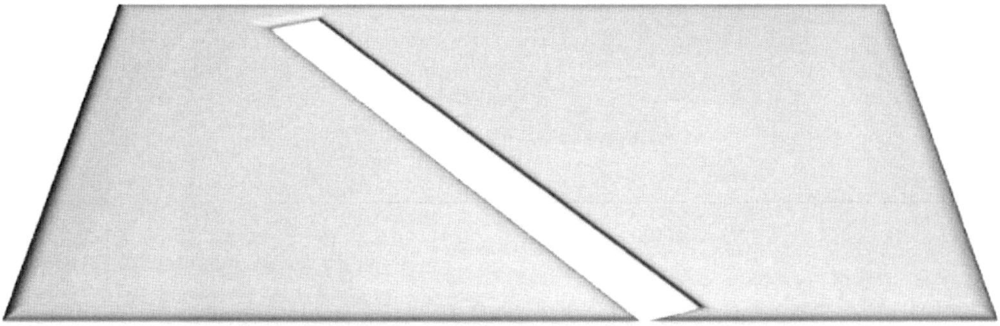

Figure 6.3 Lineament linear feature.

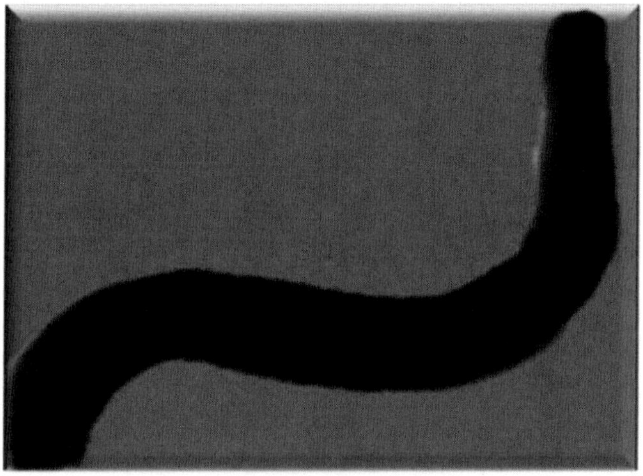

Figure 6.4 Curvilinear feature of lineament.

Figure 6.5 Landscape lineaments (faults).

Additionally, other geological phenomena can also create lineaments. Fracture zones are formed where rock has been broken and displaced by tectonic forces resulting in linear features. Shear zones occur where rocks have been deformed by shear stress, often forming elongated, linear structures. Moreover, igneous intrusions (Figure 6.8), such as dykes (Figure 6.9) are formed when magma intrudes into cracks in the rock, solidifies and creates linear features visible on the surface.

Figure 6.6 Fault-aligned valley.

Figure 6.7 Fold-aligned hills.

Figure 6.8 Igneous intrusions.

Figure 6.9 Visible dykes.

Lineaments are typically identified at the provincial scale. They are not considered mesoscopic or microscopic features. Structural lineaments, which result from the localization of recognized geological features are characterized by precise alignments of topographic features such as ridges, depressions or escarpments [4]. Additionally, the term 'megalineament' is used to describe such features on a continental scale. For example, the trace of the San Andreas Fault can be considered a megalineament (Figure 6.10) [2].

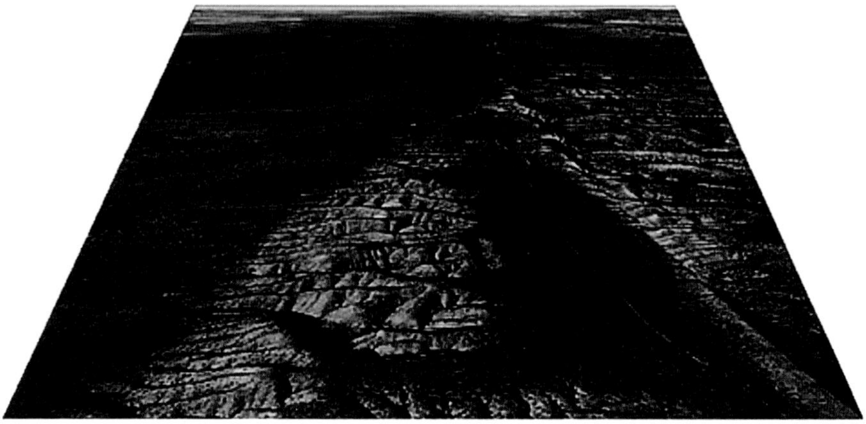

Figure 6.10 San Andreas Fault.

The Great Glen Fault in Scotland extends for about 100 kilometers and is a significant geological feature in the Scottish Highlands. Similarly, the Trans-Brazilian Lineament is one of the longest shear zones on Earth, spanning approximately 4,000 kilometers across Brazil. Furthermore, the Trans-Saharan Belt, an extensive shear zone, stretches across the Sahara Desert and is considered one of the longest coherent shear zones. Additionally, the Coast Range Megalineament in southeastern Alaska extends southeastward for about 550 kilometers from its junction with the Chatham Strait-Lynn Canal fault. These megalineaments are critical geological structures that provide valuable insights into the tectonic processes shaping our planet.

6.2 THE DIFFERENT TYPES OF LINEAMENTS

Lineaments are categorized into two principal types: positive and negative. Positive lineaments are characterized by ridge trends which manifest as elevated geological features. These can be observed as a series of hills or mountains, elongated elevations on the ocean floor or linear crests. An illustrative example of such a ridge is the mountainous strip in the southeastern region of Mount Everest in Nepal (Figure 6.11). Ocean floor features are elongated elevations on the ocean floor such as submarine ridges. Therefore, linear crests present a linear series of crests, which may be visible on land (Figure 6.12) or underwater. These features are typically the result of tectonic uplift or volcanic activity and are often associated with areas of high relief.

Figure 6.11 A strip of mountains.

Figure 6.12 Linear crests are visible on land.

Conversely, negative lineaments are represented by river valleys, also known as fluvial valleys. These are typically V-shaped and are sculpted by the action of flowing water (Figure 6.13). The specific morphology of a river valley is contingent upon the characteristics of the stream that carves it. In regions with steep gradients, such as mountain ranges, rivers tend to create valleys with steep walls and narrow bottoms. In contrast, areas with shallower slopes may give rise to broader and more gently sloping valleys. As a river reaches its lowest stretch, approaching its base level, it begins to deposit sediment, transforming the valley bottom into a floodplain.

Figure 6.13 River valleys are associated with lineaments and mountains.

Therefore, the keystone question now is: what role does water play in the creation of negative lineaments? Water is a fundamental agent in the creation of negative lineaments through its erosional power. By carving through rock and soil, transporting sediment and facilitating weathering, water shapes the landscape forming valleys and depressions that are characteristic of negative lineaments [5–7]. This process is essential for understanding the geological history and dynamics of a region.

Flowing water is a powerful erosional agent that can carve through rock and soil creating valleys and depressions. Over time, rivers and streams can significantly alter a landscape, forming distinctive V-shaped valleys. Therefore, the velocity of flowing water and the erodibility of the underlying materials are key factors in the creation of negative lineaments. Faster-flowing water tends to erode materials more quickly, resulting in steeper and narrower valleys. Conversely, slower-flowing water may create broader and more gently sloping valleys. Consequently, water not only erodes materials but also transports sediment downstream. This sediment can be deposited in lower-energy environments, such as floodplains, contributing to the formation and evolution of negative lineaments. Water also facilitates chemical and physical weathering which weakens rock and makes it more susceptible to erosion. This process is particularly important in the initial stages of valley formation. Subsequently, changes in the base level of a river, which is the lowest point to which it can erode, can also influence the formation of negative lineaments. For example, if the base level drops, the river can erode deeper, creating a deeper valley.

6.3 THE FUNDAMENTAL DIFFERENCES BETWEEN POSITIVE AND NEGATIVE LINEAMENTS IN TERMS OF THEIR TOPOGRAPHIC EXPRESSION AND FORMATION PROCESSES

Positive and negative lineaments are two primary categories used by geologists to classify linear geological features. These categories are distinguished by their contrasting geological characteristics and the landscapes they form. The fundamental differences between positive and negative lineaments are related to their topographic expression and the processes that form them. Therefore, positive lineaments are associated with the uplift and the creation of elevated features, while negative lineaments are associated with erosion and the formation of valleys and depressions. This classification aids geologists in interpreting the geological history and dynamics of a region. Table 6.1 clearly outlines the fundamental differences between positive and negative lineaments, highlighting their distinct topographic expressions, formation processes, associated landscapes and geological significance.

Table 6.1 Key differences between positive and negative lineaments

Aspect	Positive Lineaments	Negative Lineaments
Topographic Expression	Elevated features such as ridges and mountains	Depressions or valleys, typically V-shaped
Formation Processes	Tectonic uplift or volcanic activity	Erosion, particularly by flowing water
Associated Landscapes	Areas of high relief, such as mountain ranges	River valleys and floodplains
Geological Significance	Indicate areas of tectonic activity or past volcanism	Provide insights into erosional processes and water flow

6.4 SATELLITE REMOTE SENSING AND IMAGE PROCESSING TO DETECT LINEAMENT FEATURES

Analyzing and interpreting lineaments in imagery is crucial for researchers to uncover their true nature and origin. Using additional data sources like geological maps or field surveys is instrumental in confirming the presence of geological lineaments and distinguishing them from other linear features. Therefore, it is important to carefully analyze and verify the nature of lineaments observed in satellite imagery before coming to any conclusions about their origin. Additionally, advanced image processing techniques can assist in distinguishing between natural geological lineaments and artificial features.

In the early days of Landsat, one of the most common applications of space imagery in geology was the detection of linear features that appeared as tonal discontinuities. These features were often suspected to be geological, primarily attributed to faults or fracture systems controlled by joints. This means, lineaments are linear features that can be identified as lines in aerial or space imagery, often due to contrasts in terrain or ground cover on either side. Geologically, these features typically represent faults, joints, or boundaries between stratigraphic formations. However, lineaments can also be caused by non-geological factors such as roads, railroads, fence lines, or even 'false alarms' due to unknown factors [11–13].

Optical remote sensing techniques have shown great promise for mapping geological feature variations over wide areas. According to Koike et al. [18] and Walsh et al. [19], lineament information extraction in satellite images can be broadly divided into three categories: (i) lineament enhancement and extraction (ii) image classification and (iii) data integration. Primarily, lineament enhancement and extraction involve identifying and highlighting features such as faults, folds and fractures to better understand the subsurface geology. By extracting these lineaments, geologists can analyze the structural controls on mineralization or groundwater flow. Therefore, the image classification technique utilizes remote sensing data to identify and classify different geological features or mineral deposits based on their unique spectral signatures [6, 10, 14, 17]. It is commonly used in exploration and resource management in the mining industry to efficiently target areas with high mineral potential. Lastly, data integration presents a superposition of satellite images and multiple data sources (geological, geochemical, geophysical) in a geographical information system. This process allows for a comprehensive analysis of the Earth's surface and subsurface, aiding in the identification of potential mineral deposits or geological features. By combining various datasets, researchers can gain a more holistic understanding of the area being studied, leading to more informed decision-making in resource exploration and management. Therefore, real-time data assimilation of both passive and active remotely sensed data is also a powerful tool for extracting and mapping significant geological features such as lineaments, curvilinear features, and faults [4, 10, 17, 29].

According to the above-mentioned perspective, various image processing tools have been used for lineament feature detection, including: (i) image enhancement techniques [20–22], and (ii) edge detection and segmentation [15–23].

Researchers often prefer spatial domain filtering techniques to remove artificial lineaments and verify disjoint lineament pixels in satellite data. Techniques such as band-rationing, and linear and Gaussian nonlinear stretching enhancement have been implemented to determine lineament populations. The High Pass Filter enhancement technique, for example, has been found to provide an accurate geological map by selectively enhancing small-scale features while maintaining larger-scale features [11, 24].

Therefore, edge detection algorithms such as the Haar and Hough transforms have been used for lineament detection in Landsat-TM satellite data. The Haar transform is effective in extracting subtle features with finer details, concentrating variance energy in localized zones. It can be used for image enhancement due to its low and high-frequency components. The Hough transform, on the other hand, is suitable for fault feature mapping, as it is robust against noise and can extract lines even in areas with pixel gaps. It works by mapping collinear sets of edge pixels into a parameter space (Hough space), where collinear sets of pixels in the image correspond to peaks in the Hough space [13–15].

However, the major algorithms for lineament detection are Laplacian, Sobel, and Canny. These are the major algorithms used for lineament feature detection in remotely sensed data. Additionally, Marghany [40] invented a novel algorithm for automatic lineament detection named 'Marghany Adaptive Modification Algorithm for lineament automatic detection'.

Needless to say, understanding and detecting lineaments is crucial in geological studies. Remote sensing and image processing techniques provide valuable tools for identifying and mapping these features, whether they are geological or caused by other factors. These techniques enable geologists to gain insights into the Earth's crust, assist in resource exploration and contribute to hazard assessment and environmental management.

6.5 CHALLENGES IN EXTRACTING GEOLOGICAL FEATURES FROM REMOTE SENSING DATA

Geological investigations necessitate the use of standardized techniques and processes to acquire precise information. Nevertheless, the intricate geography of the planet may provide challenges when using conventional techniques. The main benefit of using satellite remote sensing in geology is its extensive coverage of the region of interest. This allows for the extraction of precise and valuable information, such as structural patterns and spectral characteristics from the data.

Yet, the richness of geological characteristics is not well understood. The extraction of lineaments plays a crucial role in the application of remote sensing to geology. However, the true meaning of lineament is still uncertain. Lineaments must be distinguished from other line characteristics that are not caused by geological formations. Lineament extraction should be carefully evaluated by geologists [24–27].

The identification of lineaments in remote sensing data is often hindered by the presence of a dense vegetation canopy, extensive weathering and recent unconsolidated deposits such as fluvial terraces, pediments or alluvial sediments. Dense vegetation can obscure the ground surface, making it difficult to discern lineaments that may be present. The canopy can create a uniform appearance and mask the subtle variations in terrain that indicate lineaments. Vegetation has specific reflectance properties that can dominate the spectral signature of the area, making it harder to identify the underlying geological features. Additionally, vegetation can also create shadows that obscure the ground surface, further complicating the identification of lineaments.

Weathering processes can lead to erosion and deposition of materials which can obscure or alter original lineament features. This can make it difficult to distinguish between natural variations in terrain and actual lineaments. Extensive weathering can smooth out the surface, reducing the contrast between lineaments and the surrounding terrain. This makes it harder to detect lineaments using remote sensing techniques. Moreover, weathering can lead to the development of soil which can cover and obscure underlying geological features, including lineaments.

Both dense vegetation canopies and extensive weathering can reduce the visibility of lineaments in remote sensing imagery. This makes it challenging to accurately identify and map these features. Hence, the presence of vegetation and weathering can introduce ambiguity in the interpretation of remote sensing data. Features that appear as lineaments may be vegetation patterns or weathered surfaces leading to false positives.

Not all faults are expressed topographically, and therefore, the poor visibility of a lineament in remote-sensed imagery does not necessarily indicate the absence of joints or faults in the area. Remote sensing satellite imagery, however, is continually improving in its ability to recognize these linear or slightly curvilinear features compared to field investigations. Consequently, the continuity of a fault may also be obscured in satellite imagery due to 'noise' caused by factors such as thermal noise, shadowing, and dense cloud cover. Additionally, some 'lineaments' reported in literature are artifacts resulting from the interaction of sunlight with the ground surface and thus do not possess geological significance.

The foremost challenge with automatic lineament extraction methods is that, in typical scenarios, the filtering algorithm produces segmented images containing many false lineament pixels. These false pixels must be removed using complex edge-linking algorithms. Additionally, the current lineament extraction processes only consider the extraction of edge pixels without taking into account the topographic information present in remotely sensed data. Moreover, these procedures are not able to effectively extract lineaments from low-contrast pixels and in mountain shadows. This results in short, dense lineaments that are difficult to associate with tectonically significant structures [25, 28].

The Segment Tracing Algorithm (STA) is a non-filtering technique implemented to address these challenges. The core principle of STA involves identifying a line of pixels as a vector component by examining the internal variation of the grey level in the digital image and connecting these line segments along their anticipated paths. In this context, the threshold values for the extraction and connection of line segments are path-dependent. Consequently, STA outperforms traditional filtering methods by effectively identifying continuous valleys and extracting additional lineaments, particularly those aligned with the sun's azimuth and those located in shadow areas.

By combining STA with the Hough Transform, it is possible to scan a line-element image to detect continuous grey-level edges and generate what is known as the Segment Tracing and Rotation Transformation (START) [30]. This approach enhances the ability to identify and extract lineaments from remote sensing data, providing a more comprehensive and accurate analysis of geological features.

6.6 HOW MULTISPECTRAL REMOTE SENSING DATA IDENTIFY LINEAMENTS?

Lineament extraction from Sentinel-2 data involves leveraging the multispectral capabilities of the Sentinel-2 satellite to enhance the visibility of geological features. Sentinel-2, part of the Copernicus Program led by the European Space Agency (ESA), provides high-resolution multispectral imagery that is valuable for a wide range of applications, including geological mapping and lineament extraction. According to Marghany [34], Sentinel-2 acquires images in thirteen spectral bands ranging from the visible to the shortwave infrared (SWIR) spectrum, with spatial resolutions varying from 10 to 60 m depending on the band.

Before selecting bands for lineament identification, atmospheric correction was applied to the raw Sentinel-2 imagery obtained from the Earth Explorer platform (www.earthexplorer.usgs.gov). The imagery, acquired on May 8, 2024, over the GERD region in Ethiopia, underwent this process to eliminate atmospheric effects and enhance image quality (Figure 6.14). This step was essential for accurately interpreting spectral signatures. Additionally, orthorectification was performed to correct distortions caused by the sensor's perspective and the Earth's curvature, ensuring precise geospatial alignment.

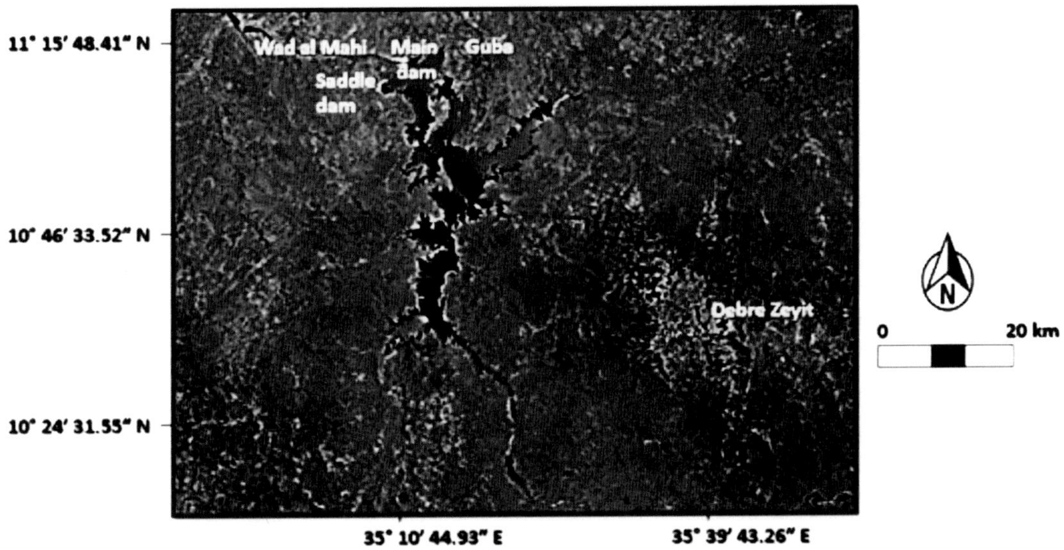

Figure 6.14 Raw Sentinel-2 imagery was acquired on May 8, 2024.

Appropriate bands for lineament extraction were selected. Common band combinations include Bands 4 (Red), 3 (Green), and 2 (Blue) create a natural color composite. False color (NIR-R-G): Bands 8 (NIR), 4 (Red), and 3 (Green) enhance vegetation and soil features. False-color composites were created using these band combinations to enhance the visibility of lineaments. For instance, combining the Near-Infrared (NIR) band with the Red and Green bands can help distinguish between vegetation and geological features (Figure 6.15). Consequently, bands in the SWIR range (e.g., Bands 11 and 12) can be useful for highlighting mineral and moisture content.

The band ratio of the Red band (Red), and Near-infrared (NIR) has the potential role of revealing the lineament features in such multispectral Sentinel-2 imagery. In other words, the normalized difference vegetation index (NDVI) has a background theory where the index is based on the difference in reflectance in the near-infrared (NIR) and red bands of the electromagnetic spectrum (EMS). Its valid results fall between 0 and +1. This mathematically Normalized Difference Vegetation Index (NDVI) can be expressed as:

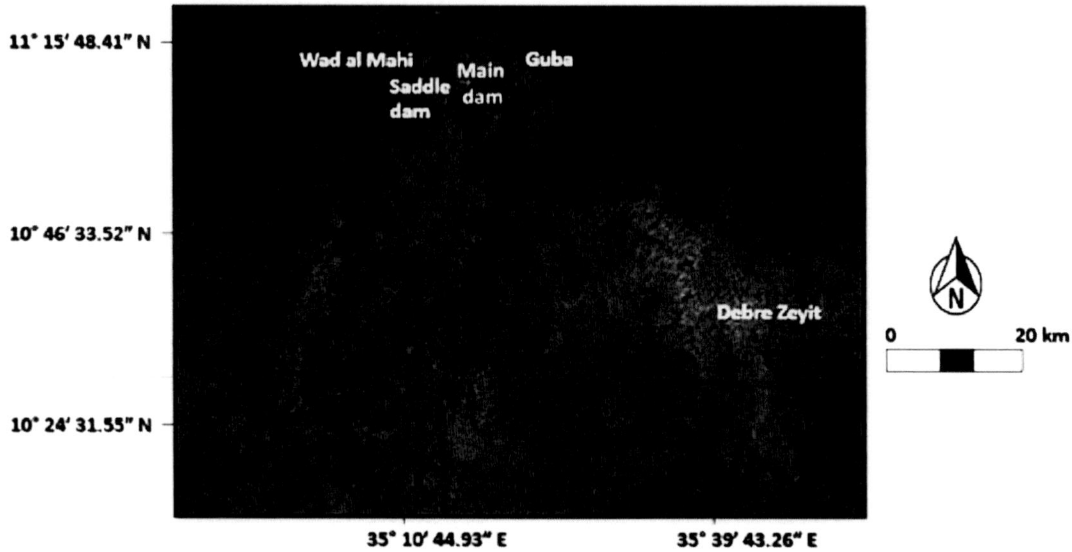

Figure 6.15 False color image combination of NIR, Red, and green bands.

$$NDVI = \frac{(\text{Band 8} - \text{Band 4})}{(\text{Band 8} + \text{Band 4})} \tag{6.1}$$

Higher NDVI values indicate denser green vegetation. The highest NDVI values, approaching 1, are observed in the elevated areas surrounding the main dam and saddle dam. This suggests that these regions have the densest and healthiest vegetation. Conversely, lower NDVI values, ranging from 0.2 to 0.4, are found in the lowland areas, particularly to the south of the dam reservoir. These lower values indicate sparser vegetation, which could be due to various factors such as soil moisture, land use or vegetation type.

The NDVI (Normalized Difference Vegetation Index) is a widely used remote sensing metric that measures vegetation health and density. Higher NDVI values (close to 1) typically represent lush, green vegetation, while lower values (closer to 0) indicate less vegetation or bare soil. In this context, the elevated areas around the dams likely benefit from favorable conditions for vegetation growth, such as better drainage and less human disturbance (Figure 6.16). In contrast, the lowland areas may experience conditions less conducive to dense vegetation, such as waterlogging or higher human activity.

Equation 6.1 can be developed in the quantum spectral reflectance as follows:

$$|\Psi(\text{Re}\,d)\rangle_{sr} = \left(\frac{|L(\lambda)\rangle_{\max}}{254} - \frac{|L(\lambda)\rangle_{\min}}{255}\right) \otimes |DN\rangle\langle DN| + |L(\lambda)\rangle_{\min} \tag{6.2}$$

$$|\Psi(NIR)\rangle_{sr} = \left(\frac{|L(\lambda)\rangle_{\max}}{254} - \frac{|L(\lambda)\rangle_{\min}}{255}\right) \otimes |DN\rangle\langle DN| + |L(\lambda)\rangle_{\min} \tag{6.3}$$

In the circumstance of $|\Psi(\text{Re}\,d)\rangle_{sr} \notin |\Psi(NIR)\rangle_{sr}$ and $|\Psi(\text{Re}\,d)\rangle_{sr} \neq |\Psi(NIR)\rangle_{sr}$. The quantum vegetation index can be expressed as:

$$|\Psi(\text{R})\rangle_{sr} = \frac{|\Psi(NIR)\rangle_{sr} - |\Psi(\text{Re}\,d)\rangle_{sr}}{|\Psi(NIR)\rangle_{sr} \oplus |\Psi(\text{Re}\,d)\rangle_{sr}} \tag{6.4}$$

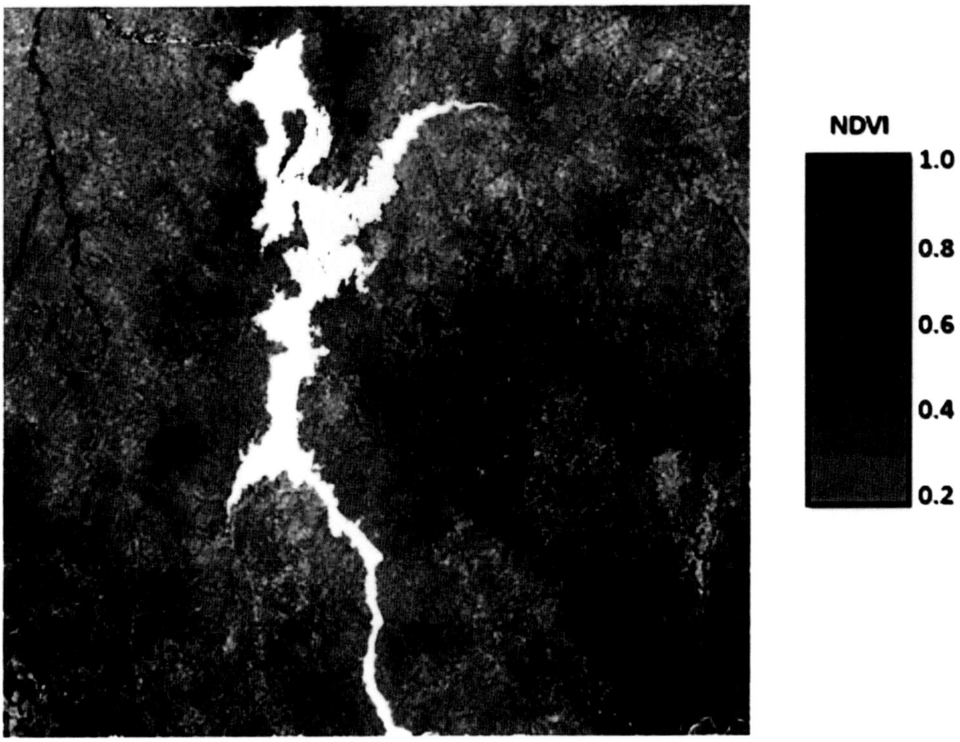

Figure 6.16 Normalized difference vegetation index across the GERD.

Equation 6.4 demonstrates the entanglement of electromagnetic photons of NIR and Red quantum spectral reflectance with the variation of the vegetation density variation as a function of the entanglement spectral reflectance ratio of Red and NIR photon energies $|\Psi(\mathrm{R})\rangle_{sr}$. The vegetation index can be represented as the wave function of the quantum photons entanglement of spectra reflectance of vegetation signature in both NIR and Red quantum spectral constrained range. The wave function $|\Psi(\mathrm{R})\rangle_{sr}$ collapses in the absence of vegetation density cover when $|\Psi(\mathrm{R})\rangle_{sr} = 0$. A wave function higher than zero indicates the existence of the vegetation covers. Therefore, the wave function $|\Psi(\mathrm{R})\rangle_{sr} = 1$ reveals healthy vegetation covers. In this sense, the entanglement absorption in the Red spectrum occurs through chlorophyll while the entanglement reflectance in the NIR spectrum occurs by mesophyll tissue. In other words, the strong entanglement reflectance induces varieties of continuous illuminations in remote sensing sensors as a function of the particle of mesophyll tissue in vegetation leaves.

Equation 6.4 has been applied to Sentinel-2 imagery obtained from EarthExplorer on May 8, 2024, covering the Grand Ethiopian Renaissance Dam (GERD) area in Ethiopia (Figure 6.17). It is noteworthy that the wave function collapses to a value of zero, indicating non-vegetation zones, which correspond to the dam and reservoir areas. Conversely, the wave function approaches a value of one, signifying dense vegetation cover, which is depicted in dark grey. These dense vegetation zones are predominantly found in the highest mountainous regions surrounding the GERD (Figure 6.17).

Additionally, the areas with the highest vegetation index are associated with visible geological structures such as faults and lineaments, particularly in Debre Zeyit. This observation aligns with the findings of Haeruddin et al. [31] and Marghany [34]. The application of Equation 6.4 to Sentinel-2 imagery allows for a detailed analysis of vegetation distribution and its correlation with geological features. The zero values in the wave function effectively highlight non-vegetated areas, such as the dam and reservoir, while values approaching one indicate regions of dense vegetation.

This method provides valuable insights into the ecological and geological characteristics of the GERD area, aiding in environmental monitoring and management efforts.

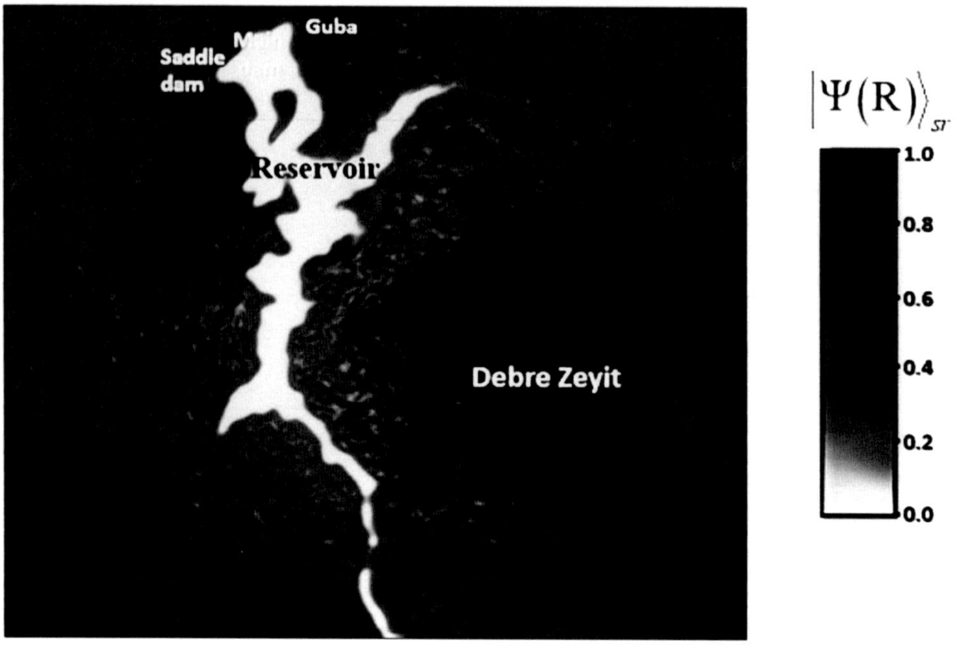

Figure 6.17 The wave function of the normalized difference vegetation index.

Equations 6.1 and 6.4 have been exploited to detect and map geologic linear features (lineaments) in hard-rock terrains as a function of tone, color, and the identified textural patterns. Besides, the presence of lineaments explain their influence on the diversity of land cover through an analysis of the vegetation index. In other words, the existence of the lineament would incessantly impact the existence of vegetation patterns around the geological structure zone. Thus, the spatial variation of vegetation is anticipated to be associated with the existence of lineaments.

Boyer and McQueen [8] proved the usefulness of NDVI in detecting fractures and faults which can be associated with the occurrence of vegetation alignment. The mapping of such structural features will ease groundwater accretion in a rocky terrain where permeability and porosity are of negligible value [9]. The Sobel algorithm is implemented on the wave function image $|\Psi(R)\rangle_{sr}$ to delineate the variation of lineaments as a function of NDVI as:

$$\left|\Psi\left(\frac{\partial}{\partial x}\right)\right\rangle = \left[|\Psi(R)\rangle_{sr_i}\left\langle|\Psi(R)\rangle_{sr_j}\right|\right] * \left(\left|\frac{\partial}{\partial x}G\right\rangle\right) \tag{6.5}$$

The Sobel operator relies on central variances in which it is regarded as a guess of the first Gaussian derivative. In this sense, the Sobel operator corresponds to the first derivate of Gaussian blurring image convolutes by $\left(\frac{\partial}{\partial x}G\right)$ and is attained by instigating a 3 × 3 mask to the image. In this context, Figure 6.18 illustrates the varying intensity of lineament features as a function of the NDVI wave function. It is noteworthy that the lineament density, highlighted in dark grey, overlays areas of dense vegetation where the wave function exceeds 0.5 and reaches up to 0.85. This indicates that regions with higher NDVI values, which signify denser vegetation, also exhibit a higher density of lineaments.

The direct proportionality between lineament density and high NDVI wave function values suggest a strong correlation between geological structures and vegetation cover. This observation is consistent with the findings of Marghany [34], who also reported a similar relationship.

The presence of dense lineaments in areas with high NDVI values could be attributed to the influence of geological structures on vegetation growth. Faults and lineaments often create microenvironments that can support diverse and dense vegetation due to factors such as increased soil moisture and nutrient availability. Additionally, these geological features can influence water flow and soil stability, further affecting vegetation patterns.

Understanding this relationship is crucial for geological and ecological studies, as it provides insights into how geological structures impact vegetation distribution. This information can be valuable for environmental monitoring, land use planning, and conservation efforts, particularly in regions with significant geological and ecological interactions.

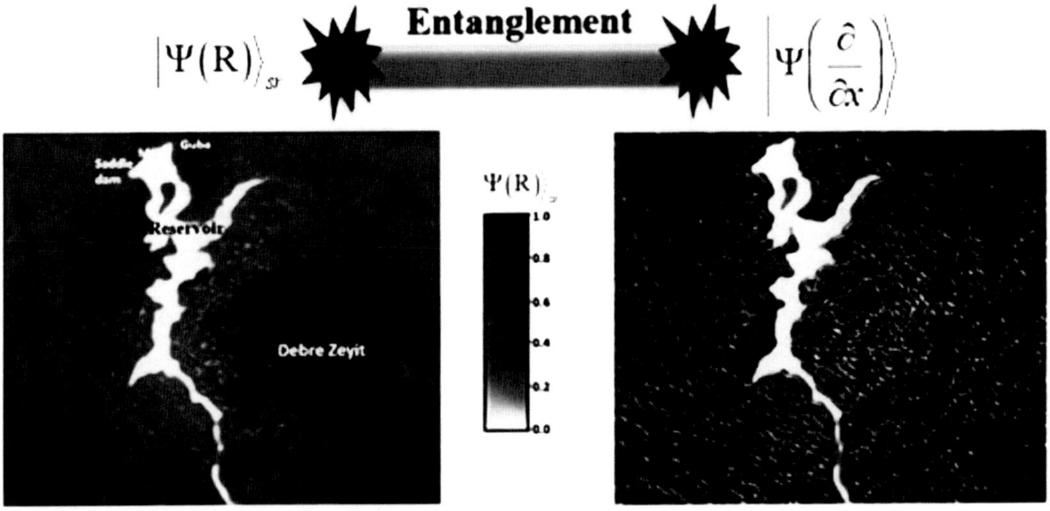

Figure 6.18 Lineament density variations as a function of the wave function of the NDVI.

It is intriguing to propose a novel explanation for why dense vegetation covers are associated with the highest variations in lineament density. This phenomenon can be understood through the concept of quantum entanglement between the quantum states of vegetation cover and lineament variation. When these quantum states are entangled, the highest lineament densities are observed beneath areas of dense vegetation. This implies that the wave functions of both quantum states are interconnected, leading to a direct correlation between lineament density and vegetation cover.

The use of Sentinel-2 imagery, which is freely available from sources such as EarthExplorer, is particularly advantageous for lineament mapping due to its cost-effectiveness and widespread availability. Sentinel-2 imagery is extensively used in geological studies because it provides high-resolution data that can reveal detailed features of the Earth's surface. However, it is important to note that older archival imagery may highlight features that are no longer visible in more recent images due to factors such as urbanization and environmental changes.

Despite the benefits of using satellite imagery, it is crucial to validate the identified lineaments through ground-truthing. Ground-truthing involves verifying the satellite data with on-the-ground observations to ensure the accuracy of the mapped features. This step is essential to confirm the presence and characteristics of lineaments, as well as to understand their geological significance.

Needless to say, the correlation between dense vegetation and high lineament density can be explained through quantum entanglement, and the use of Sentinel-2 imagery provides a valuable tool for mapping these features. This approach aligns with the principles of rigorous scientific investigation and contributes to our understanding of the complex interactions between geological structures and vegetation cover in such GERD zones.

6.7 CAN DIGITAL ELEVATION MODELS BE USED FOR LINEAMENT DELINEATION?

Digital Elevation Models (DEMs) can be derived from various sources, including topographical maps, Shuttle Radar Topography Mission (SRTM) data or aerial photographs. DEMs are extensively used in lineament mapping due to their ability to provide a three-dimensional representation of the terrain, which enhances the interpretation of topographical features crucial for understanding lineaments.

By overlaying satellite imagery onto a DEM, a stereo view is created, which can significantly improve the ability to identify and interpret lineaments. Typically, SRTM DEMs are employed to detect lineaments. These DEMs are converted into multi-shaded relief images, which highlight the terrain's features from different angles. Lineaments can then be extracted using various detection algorithms, such as the Hough Transform, the Canny Edge Detection Algorithm, or the combination of the Statistical Texture Analysis (STA) and Hough Transform.

Another effective technique for delineating lineaments is analytical hill shading in DEMs. This method involves creating shaded topographic images based on the reflection of artificial electromagnetic beams emitted from remote sensing sensors. DEMs can be artificially illuminated from specific directions, which is not possible with optical satellite imagery. The beams need to be perpendicular to the topography to enhance the visibility of lineaments. Consequently, lineaments that are not visible in aerial photographs or satellite imagery can be detected in shaded relief images.

Shaded-relief images are particularly useful as they reveal bare-ground surfaces and can identify features obscured by vegetation or land use. This method allows for the detection of lineaments that might otherwise remain hidden, providing a more comprehensive understanding of the geological structures.

Consistent with the above perspective, the utilization of hill shading in Digital Elevation Models (DEMs) has proven to be successful in mapping lineaments. In this section, ASTER GDEM data is combined with Landsat-8 imagery to conduct a lineament density analysis in the study area. The data, which can be downloaded from EarthExplorer, has a resolution of 30 m by 30 m (1 arc-second) (Figure 6.19).

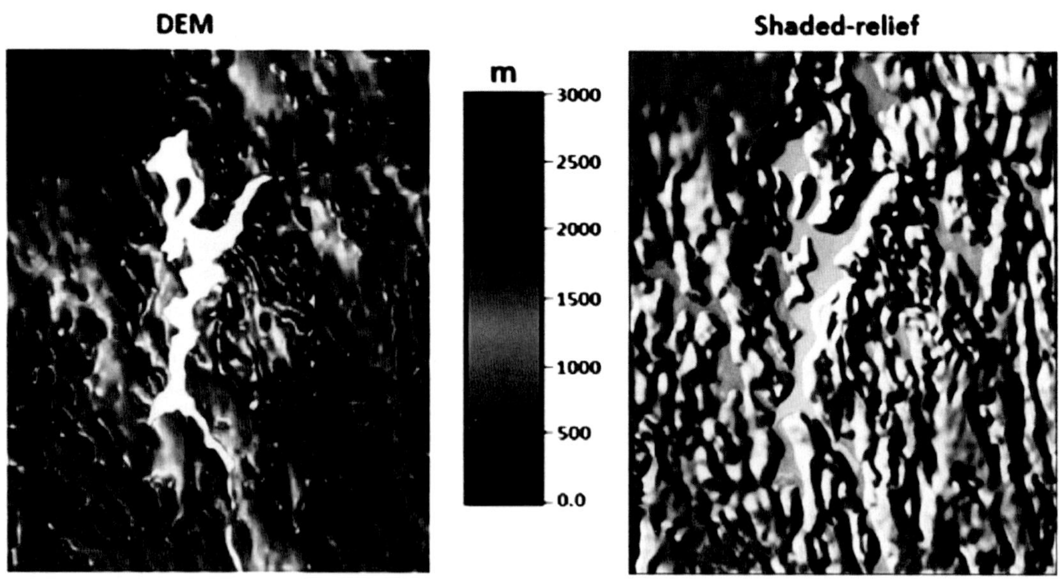

Figure 6.19 DEM and shaded-relief image of Landsat-8.

The integration of ASTER GDEM with Landsat-8 imagery enhances the ability to detect and analyze lineaments by providing detailed topographical and spectral information. Hill shading techniques applied to DEMs create shaded relief images that highlight the terrain's features from various angles, making it easier to identify geological structures such as faults and lineaments. This method is particularly effective in areas with complex topography, where traditional optical imagery might not reveal all the geological features.

Figure 6.20 illustrates the extraction of lineament density using shaded-relief images derived from Digital Elevation Models (DEMs). In this context, the Sobel operator's wave function effectively integrates the DEM with shaded-relief imagery and variations in vegetation cover. This integration enhances the detection of lineaments by highlighting changes in elevation and vegetation. The high density of lineament variations is particularly prominent in areas of high elevation, such as the mountains of Debre Zeyit and Guba. These regions exhibit a significant concentration of lineaments, which are geological features like faults and fractures. The elevated terrain provides a clearer view of these structures, as the steep slopes and varied topography accentuate the lineaments.

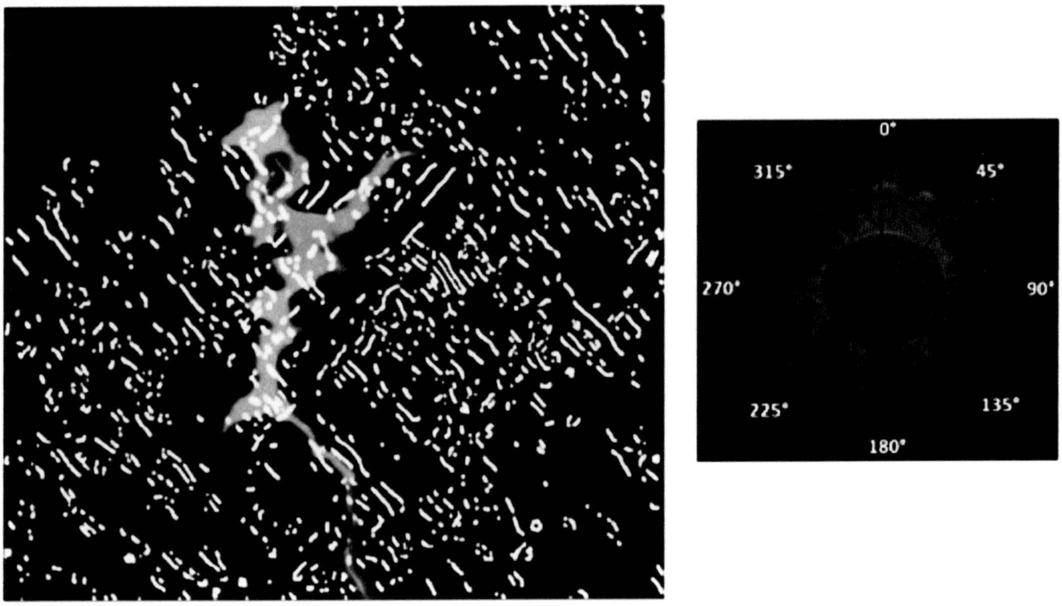

Figure 6.20 Lineament density variation derived based on the shaded-relief image.

Lineaments commonly exhibit trends in the NNE–SSW, NE–SW, and NW–SE directions. These orientations align with the broader African tectonic trends that intersect the Nubian Block. To elaborate, lineaments are linear features on the Earth's surface that represent zones of structural weakness, such as faults, fractures, or joints. The NNE–SSW (North-Northeast to South-Southwest), NE–SW (Northeast to Southwest), and NW–SE (Northwest to Southeast) orientations are significant because they reflect the regional stress fields and tectonic forces that have shaped the geological landscape over time.

These directional trends are consistent with the African tectonic framework, which is characterized by complex interactions between various tectonic plates and blocks. The Nubian Block, a major tectonic unit within the African Plate, is cross-cut by these lineaments, indicating a history of tectonic activity and deformation. The alignment of lineaments with the African trend suggests that the forces driving the tectonic movements in this region have influenced the development and orientation of these structural features. Understanding the orientation and distribution of lineaments is crucial for geological studies, as they can provide insights into the tectonic history, stress regimes and potential areas of mineralization or groundwater flow. In regions like the Nubian Block, where tectonic activity has played a significant role in shaping the landscape, analyzing lineament patterns can reveal important information about the geological processes at work. This work proves the study done by Mohamed and Elmahdy [35].

The common trends of lineaments in the NNE–SSW, NE–SW, and NW–SE directions reflect the influence of regional tectonic forces and align with the broader African tectonic trends that intersect the Nubian Block. This alignment highlights the interconnected nature of geological structures and the importance of understanding their orientations for comprehensive geological analysis.

The Sobel operator, a widely used edge detection algorithm, plays a crucial role in this process. By applying the Sobel operator to the DEM, the algorithm detects edges and transitions in the terrain which correspond to lineaments. When combined with shaded-relief images, this method provides a detailed and accurate representation of the geological features. Furthermore, the variation in vegetation cover also contributes to the visibility of lineaments. Dense vegetation can obscure geological features, while sparse or absent vegetation allows for clearer observation. By considering both the DEM and vegetation cover, the analysis becomes more comprehensive, leading to a better understanding of the geological structures.

When utilizing hill shade radiance for lineament detection, caution is necessary as it may result in the identification of 'false lineaments'. These are features might appear to be lineaments but are not geological structures. Additionally, the appearance of landforms, which are crucial for detecting lineaments can vary significantly depending on the direction of solar illumination. For example, a drumlin viewed from the side may appear linear, but when observed from the front, it can resemble a circular hill.

To elaborate, hill shading is a technique used in digital elevation models (DEMs) to simulate the effect of sunlight on the terrain, creating a three-dimensional appearance. This method enhances the visibility of topographical features making it easier to identify lineaments. However, the direction and angle of the simulated sunlight can greatly influence the appearance of these features. In some cases, certain landforms may cast shadows or highlights that mimic the appearance of lineaments. These false lineaments can lead to incorrect interpretations if not carefully analyzed. For instance, a drumlin, which is an elongated hill formed by glacial action, may appear as a linear feature when illuminated from the side. However, when the illumination direction changes, the same drumlin may appear as a circular hill, leading to potential misidentification.

Therefore, it is essential to consider multiple illumination angles and directions when using hill shading for lineament detection. By doing so, one can differentiate between true geological lineaments and features that merely appear linear due to the lighting conditions. This approach helps in achieving more accurate and reliable geological interpretations.

In abridgment, while hill shading is a valuable tool for enhancing the visibility of topographical features in DEMs, it requires careful application to avoid the misidentification of false lineaments. Understanding the impact of different solar illumination directions on the appearance of landforms is crucial for accurate lineament detection and geological analysis.

6.8 THE FOREMOST INTERROGATION

In light of the aforementioned perspective, we delve into the critical question of how uncertainties impact the modeling of Digital Elevation Models (DEMs) for three-dimensional (3-D) lineament visualization using multispectral satellite data, without the necessity of incorporating digital elevation data. This investigation is exemplified through the application of Sentinel-2 satellite data, utilizing a quantum fuzzy B-spline algorithm.

This chapter examines three key hypotheses:

1. *3-D Reconstruction of lineaments*: The author proposes that lineaments can be effectively reconstructed in three-dimensional visualization. This hypothesis explores the potential of advanced algorithms to render detailed 3-D representations of geological features, enhancing our understanding of their spatial distribution and characteristics.
2. *Quantum Edge Detection Algorithm*: The author also hypothesizes that a quantum edge detection algorithm can serve as an automatic tool to distinguish between lineaments and surrounding geological features in optical remotely sensed satellite data. This approach leverages the principles of quantum mechanics to improve the accuracy and efficiency of edge detection, facilitating the identification of lineaments amidst complex geological backgrounds.
3. *Resolving DEM uncertainties*: The author additionally, contends that the uncertainties inherent in DEM models can be addressed using a quantum fuzzy B-spline algorithm. This innovative method aims to map spatial variations in lineaments in 3-D, providing a more precise and reliable representation of geological structures. By integrating quantum fuzzy logic with B-spline interpolation, this algorithm enhances the robustness of DEM-based analyses, mitigating the impact of data uncertainties.

The Grand Ethiopian Renaissance Dam (GERD) serves as a pertinent case study for this research. The application of these advanced methodologies to Sentinel-2 data over the GERD region demonstrates the feasibility and effectiveness of our proposed approach. The integration of quantum fuzzy B-spline algorithms with multispectral satellite data offers a novel pathway for overcoming the limitations of traditional DEM modeling, paving the way for more accurate and comprehensive geological analyses.

6.9 FUZZY B-SPLINES FOR DIGITAL ELEVATION MODEL RECONSTRUCTION

The uncertainties inherent in ASTER Digital Elevation Models (DEMs) can be effectively addressed using a fuzzy B-spline algorithm. To understand this, we first need to introduce the concept of B-splines within the framework of fuzzy logic. The identification of fuzzy B-splines is based on both triangle-based and edge-based criteria. Consider that the variations in DEM values are constrained by a membership function $\mu_{DEM}(x): X \rightarrow [0,1]$, which defines the degree to which each element of the DEM belongs to the fuzzy set DEM. This relationship can be expressed as a fuzzy set DEM associated with a set $x \in X$ to the fuzzy set *DEM*, such that $0 \leq \mu_{DEM}(x) \leq 1$ has a corresponding membership degree.

By exploiting quantum computing principles, we can represent a fuzzy set DEM on the Bloch Sphere. Here, the North Pole $|0\rangle$ is signifying 0 (false), and the South Pole $|1\rangle$ is expressing 1 (truth). The values of the fuzzy set DEM along the meridian of the Bloch Sphere are interpreted as the interval [0, 1] of fuzzy logic (Figure 6.21). A pure quantum state, represented as (a point on the surface of the sphere) $(1/\sqrt{2})\,(|0\rangle + |1\rangle)$, embodies a fuzzy value of 0.5 upon measurement. This innovative approach allows us to map the uncertainties in DEMs into a quantum framework, providing a robust method for handling and interpreting elevation data with higher precision and reliability.

The simulation of fuzzy quantum computing for a set DEM is accomplished using three fundamental gates: the inverter, conjunction, and disjunction gates. First, the inverter gate, represented as NOT(DEM) or DEM, operates on the principle of the Fuzzy Quantum Not gate. Mathematically, this is expressed as: $(\alpha|0\rangle + \beta|1\rangle) = \beta|0\rangle + \alpha|1\rangle$. In this scenario, the square of the amplitude associated with the $|1\rangle$ represents the probability of obtaining a '1' measurement during DEM reconstruction.

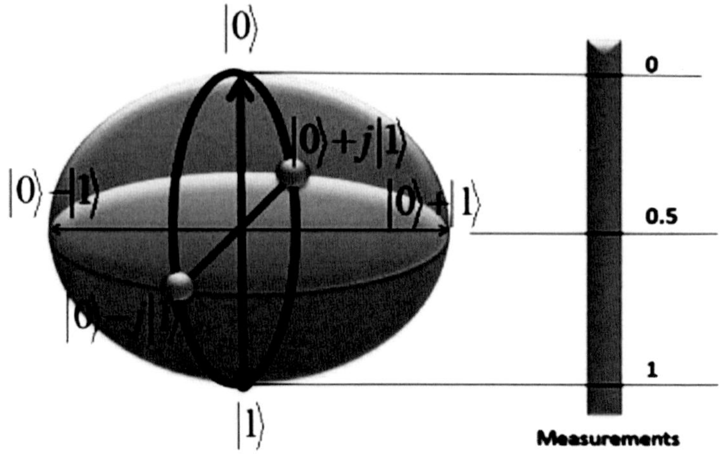

Figure 6.21 Bloch sphere.

Second, the conjunction gate, denoted as AND(*DEM*, *X*) or $\wedge$(*DEM*, *X*) or $DEM \cap X$, is implemented using the quantum AND operation, specifically the Toffoli gate with one qubit initialized to $|0\rangle$. The mathematical formulation of the Fuzzy Quantum Conjunction Operator for DEM reconstruction is obtained by:

$$\begin{aligned} DEM \wedge \left(\alpha_1 |0\rangle + \alpha_2 |1\rangle, \beta_1 |0\rangle + \beta_2 |1\rangle\right) &= DEM\left(\alpha_1 |0\rangle + \alpha_2 |1\rangle, \beta_1 |0\rangle + \beta_2 |1\rangle, 0\right) \\ &= \left(\left(\alpha_1 |0\rangle + \alpha_2 |1\rangle\right) \otimes \left(\beta_1 |0\rangle + \beta_2 |1\rangle\right) \oplus |0\rangle\right). \end{aligned} \quad (6.6)$$

The Kronecker product of three parallel input qubits, as described in Equation 6.6, is formulated as follows:

$$\begin{bmatrix} \alpha_1 \\ \alpha_2 \end{bmatrix} \otimes \begin{bmatrix} \beta_1 \\ \beta_2 \end{bmatrix} \otimes \begin{bmatrix} 1 \\ 0 \end{bmatrix} = \begin{bmatrix} \alpha_1\beta_1 \\ \alpha_1\beta_2 \\ \alpha_2\beta_1 \\ \alpha_2\beta_2 \end{bmatrix} \otimes \begin{bmatrix} 1 \\ 0 \end{bmatrix} = \begin{bmatrix} \alpha_1\beta_1 \\ 0 \\ \alpha_1\beta_2 \\ 0 \\ \alpha_2\beta_1 \\ 0 \\ \alpha_2\beta_2 \\ 0 \end{bmatrix} \quad (6.7)$$

Therefore, Equation 6.7 is utilized as the input to compute the Toffoli Gate as follows:

$$\overset{\text{Toffoli Gate}}{\begin{bmatrix} 1 & 0 & 0 & 0 & 0 & 0 & 0 & 0 \\ 0 & 1 & 0 & 0 & 0 & 0 & 0 & 0 \\ 0 & 0 & 1 & 0 & 0 & 0 & 0 & 0 \\ 0 & 0 & 0 & 1 & 0 & 0 & 0 & 0 \\ 0 & 0 & 0 & 0 & 1 & 0 & 0 & 0 \\ 0 & 0 & 0 & 0 & 0 & 1 & 0 & 0 \\ 0 & 0 & 0 & 0 & 0 & 0 & 0 & 1 \\ 0 & 0 & 0 & 0 & 0 & 0 & 1 & 0 \end{bmatrix}} \bullet \begin{bmatrix} \alpha_1\beta_1 \\ 0 \\ \alpha_1\beta_1 \\ 0 \\ \alpha_1\beta_1 \\ 0 \\ \alpha_1\beta_1 \\ 0 \end{bmatrix} = \begin{bmatrix} \alpha_1\beta_1 \\ 0 \\ \alpha_1\beta_1 \\ 0 \\ \alpha_1\beta_1 \\ 0 \\ 0 \\ \alpha_1\beta_1 \end{bmatrix} \begin{matrix} |000 \\ |001 \\ |010 \\ |011 \\ |100 \\ |101 \\ |110 \\ |111 \end{matrix} \quad (6.8)$$

The illustration of solving equation 6.8 is depicted in this aspect:

$$\alpha_1\beta_1|000\rangle+\alpha_1\beta_2|010\rangle+\alpha_2\beta_1|100\rangle+\alpha_2\beta_2|111\rangle=\left(\alpha_1\beta_1|00\rangle+\alpha_1\beta_2|01\rangle+\alpha_2\beta_1|10\rangle\right)\otimes|0\rangle+\left(\alpha_2\beta_2|11\rangle\right)\otimes|1\rangle \quad (6.9)$$

Equation 6.9 provides the initial simulation of the DEM using the quantum fuzzy conjunction operator, realized through a Toffoli gate (Figure 6.22). This simulation reveals that certain visible features of the *DEM* from ASTER data in GERD, Ethiopia are represented by qubits. These qubits begin to delineate the variations in the *DEM.* This approach highlights the effectiveness of fuzzy quantum computing in representing the complex data of *DEMs*.

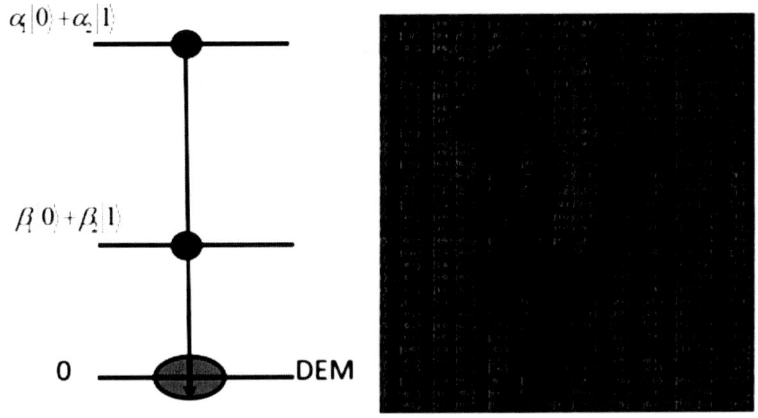

Figure 6.22 Toffoli gate simulation of initial DEM.

Successively, the fuzzy quantum dislocation as a function of OR(*DEM*, *X*) or ∨(*DEM*, *X*) or *DEM U X* is instigated in the form of Toffoli gate to attain:

$$\alpha_2\beta_2|001\rangle+\alpha_2\beta_1|011\rangle+\alpha_1\beta_2|101\rangle+\alpha_1\beta_1|110\rangle=\left(\alpha_2\beta_2|00\rangle+\alpha_2\beta_1|01\rangle+\alpha_1\beta_2|10\rangle\right)\otimes|1\rangle+\left(\alpha_1\beta_1|11\rangle\right)\otimes|0\rangle \quad (6.9)$$

Equation 6.9 enhances the determination of *DEM* features compared to Equation 6.8 (Figure 6.23). In this context, the construction of *DEM* features using the quantum fuzzy disjunction operator realized through a Toffoli gate shows great promise. This approach allows the *DEM* to accurately assemble true *X* values while eliminating false values in the context of *DEM U X.*

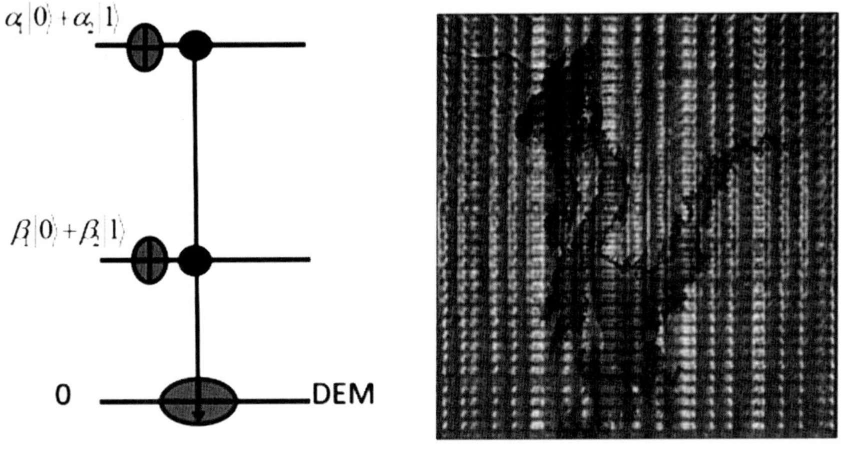

Figure 6.23 Toffoli gate simulation of initial *DEM* as a function of quantum fuzzy disjunction operator.

6.10 FUZZY QUANTUM ENTANGLEMENT FOR DEM RECONSTRUCTION

Let us assume the quantum fuzzy B-spline is *QS* [(α_1, α_2), (β_1, β_2)] that is described as a linear combination of the basis wave functions in two topological parameters $\Psi|\alpha\rangle$ and $\Psi|\beta\rangle$. Then the wave function of non-decreasing sequences of DEM real number is $\Psi|x\rangle$ which $\Psi|x\rangle \subset \Psi|\alpha\rangle$ and $\Psi|x\rangle \subset \Psi|\beta\rangle$ in the circumstance that $\Psi|\alpha\rangle \not\subset \Psi|\beta\rangle$ and $\Psi|\alpha\rangle \cup \Psi|x\rangle \cup \Psi|\beta\rangle$. The quantum polynomial function B-spline $\Psi|P_{i,1}(x)\rangle$ can be given by

$$\Psi|P_{i,1}(x)\rangle = \begin{cases} 1 & \text{If } x_i \le x \le x_{i+1}, \\ 0 & \text{Otherwise;} \end{cases} \tag{6.10}$$

$$\Psi|P_{i,p}(x)\rangle = \frac{x - x_i}{x_{i+p-1} - x_i}\Psi|P_{i,p-1}(x)\rangle + \frac{x_{i+p} - x}{x_{i+p} - x_{i+1}}\Psi|P_{i+1,p-1}(x)\rangle \text{ for } p > 1 \tag{6.11}$$

To exercise more shape controllability over the surface and invariance to perspective transformations, fuzzy B-spline is introduced. Besides having the control point as in the B-spline, fuzzy B-spline also provides a set of weight parameters $w_{i,j}$ that exert more local shape controllability to achieve projective invariance. The fuzzy B-spline surface is composed then the gate complexity is the number of 2-qubit gates; which is exploited. Formally, an algorithm with query complexity Q is gate-efficient if its gate complexity is $O(Q.\text{poly}(\log(qN)))$. Subsequently, the runtime of the quantum algorithm for obtaining precise fuzzy B-spline surface can be demonstrated implementing $O(\kappa^2 \log N\varepsilon^{-1})$, where $\varepsilon > 0$ is the error and κ is condition number. Number [30–33].

Therefore
$$\kappa = \frac{\max_i \Psi(|\alpha\rangle|\beta\rangle)}{\min_i \Psi(|\alpha\rangle|\beta\rangle)}$$

in which, If the condition number is not too much larger than one, the estimated DEM from the quantum B-spline surface is well-conditioned, and it means that its inverse can be computed with great precision.

$$\Psi(|\alpha\rangle|\beta\rangle) = \sum_{i=0}^{M}\sum_{j=0}^{O}\left[\left(\alpha_2\beta_2|00\rangle + \alpha_2\beta_1|01\rangle + \alpha_1\beta_2|10\rangle\right)\otimes|1\rangle + \left(\alpha_1\beta_1|11\rangle\right)\otimes|0\rangle\right]\otimes\left[C_{ij}\right]\otimes \left[\Psi|P_{i,p}(x)\rangle\Psi\langle P_{j,p}(x)|\right]. \tag{6.12}$$

Implementation of equation 6.12 reconstructs a three-dimensional shaded-relief image in higher detail than the Sentinel-2 one. In comparison with the ASTER shaded-relief image, the quantum B-spline reveals other geological features such as faults and drainage streams with them cautiously across GERD, Ethiopia (Figure 6.24).

6.11 QUANTUM-BASED EDGE DETECTION FOR LINEAMENT AUTOMATIC MAPPING

In practice, edge detection algorithms count on the modeling of image gradients by dissimilar categories of filtering masks. The computational complexity of at least $O(2^n)$, therefore, is required for entirely conventional lineament edge detection algorithms, because every pixel has to be handled. A quantum algorithm promises to provide an exponential speed-up equated with standing-edge mining algorithms. Nevertheless, this algorithm involves a COPY operation and a quantum black box for computing the gradients of all the pixels instantaneously [32]. This section introduces novel procedures with a highly competent quantum algorithm that handles the borders

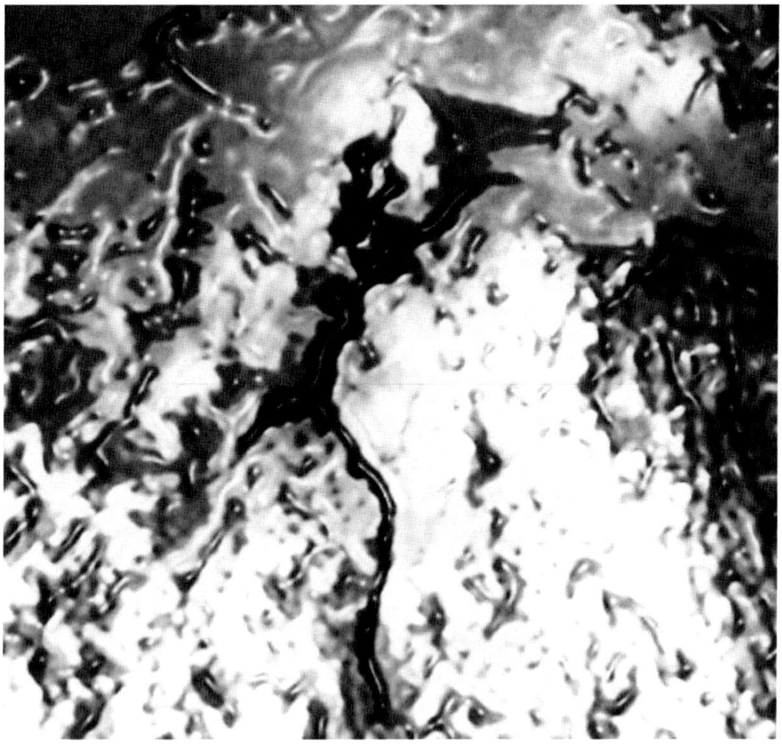

Figure 6.24 Shaded-relief image simulated using quantum fuzzy B-spline.

between several regions in satellite data in $O(1)$ time, autonomous of the image dimensions. Let us assume that the spots of slightly a couple of neighboring pixels in a satellite image column are delivered by the binary sequences $b_1,\ldots, b_{n-1}$ 0 and $b_1,\ldots, b_{n-1}$ 1, with $b_j = 0$ or 1. In this sense, their pixel values are required to be stored as the coefficients $c_{b_1},\ldots, b_{n-1}$ 0 and $c_{b_1},\ldots, b_{n-1}$ 1; respectively, of the matching computational foundation circumstances. In practice, let us implement a Hadamard gate H, which renovates a qubi $\frac{|0\rangle \rightarrow (|0\rangle + |1\rangle)}{\sqrt{2}}$ and $\frac{|1\rangle \rightarrow (|0\rangle - |1\rangle)}{\sqrt{2}}$ is implemented to automatically detect the lineament boundaries. In this understanding, the Hadamard transform on the preceding qubit swaps them to the new-fangled coefficients $c_{b_1},\ldots, b_{n-1}\, 0 \pm c_{b_1},\ldots, b_{n-1} 1$ [33]. Therefore, H the mathematical transformation of $\Psi(|\alpha\rangle|\beta\rangle)$ is given by:

$$\Psi(|\alpha\rangle|\beta\rangle)_{2^n-1} \otimes H : \begin{bmatrix} c_0 \\ c_1 \\ c_2 \\ c_3 \\ \cdots \\ c_{N-2} \\ c_{N-1} \end{bmatrix} \mapsto \frac{1}{\sqrt{2}} \begin{bmatrix} c_0 + c_1 \\ c_0 - c_1 \\ c_2 + c_3 \\ c_2 - c_3 \\ \cdots \\ c_{N-2} + c_{N-1} \\ c_{N-2} - c_{N-1} \end{bmatrix}. \tag{6.13}$$

In equation 6.13, the quantum B-spline image is turned into a unity image matrix as $\Psi(|\alpha\rangle|\beta\rangle)_{2^n-1}$ represents $2^{n-1} \times 2^{n-1}$. In this regard, the output image state $|g\rangle$ is delivered by:

$$|g\rangle = \left(\Psi(|\alpha\rangle|\beta\rangle)_{2^n-1} \otimes H\right)\left|\sum_{k=0}^{N-1} c_k\,|k\rangle\right\rangle \tag{6.14}$$

where $N = 2^n$ pixels, and to acquire the lineament boundaries between the couples of pixels using equation 6.14, the n-qubit amplitude permutation is applied to the input image state $|g\rangle$, yielding a new image state $\left|\left|\sum_{k'=0}^{N'-1} c'_k \left|k'\right\rangle\right\rangle\right\rangle_{new}$. To this end, $\left|\sum_{k=0}^{N-1} c_k \left|k\right\rangle\right\rangle$ odd (even) elements equivalent to the even (odd) elements of the original input one. In other words, $c'_{2k} = c_{2k+1}$ and c'_{2k+2}. Consequently, the quantum amplitude transformation can be proficiently accomplished in time of $O[poly(n)]$. Subsequently, a new image state $\left|\left|\sum_{k'=0}^{N'-1} c'_k \left|k'\right\rangle\right\rangle\right\rangle_{new}$ is transferred into a single qubit using Hadamard rotation for obtaining all lineament boundary values. In this understanding, to yield complete lineament boundary values in a single step, a modification of the quantum edge detection algorithm exploits a secondary qubit for encoding the $\left|\left|\sum_{k'=0}^{N'-1} c'_k \left|k'\right\rangle\right\rangle\right\rangle_{new}$ image (Figure 6.25); which yields an $(n + 1)$-qubit redundant image state as:

$$\left|\left|\sum_{k'=0}^{N'-1} c'_k \left|k'\right\rangle\right\rangle\right\rangle_{new} \otimes \left(\left|0\right\rangle + \left|1\right\rangle\right) = \sqrt{2} \otimes 2^{2-1} \begin{pmatrix} c_0, c_0, c_1, c_1, c_2, c_2, c_3, c_3, \ldots, c_{N-2}, \\ c_{N-2}, c_{N-1}, c_{N-1} \end{pmatrix}^T \tag{6.15}$$

By measuring the last qubit, conditioned on obtaining 1, the n-qubit state can be obtained (Figure 6.25); which is given by :

$$\left|\left|\sum_{k'=0}^{N'-1} c'_k \left|k'\right\rangle\right\rangle\right\rangle_{new} \otimes \left(\left|0\right\rangle + \left|1\right\rangle\right) = \sqrt{2} \otimes 2^{2-1} \begin{pmatrix} c_0 - c_1, c_1 - c_2, c_2 - c_3, \ldots, \\ c_{N-2} - c_{N-1}, c_{N-1} - c_0 \end{pmatrix}^T \tag{6.16}$$

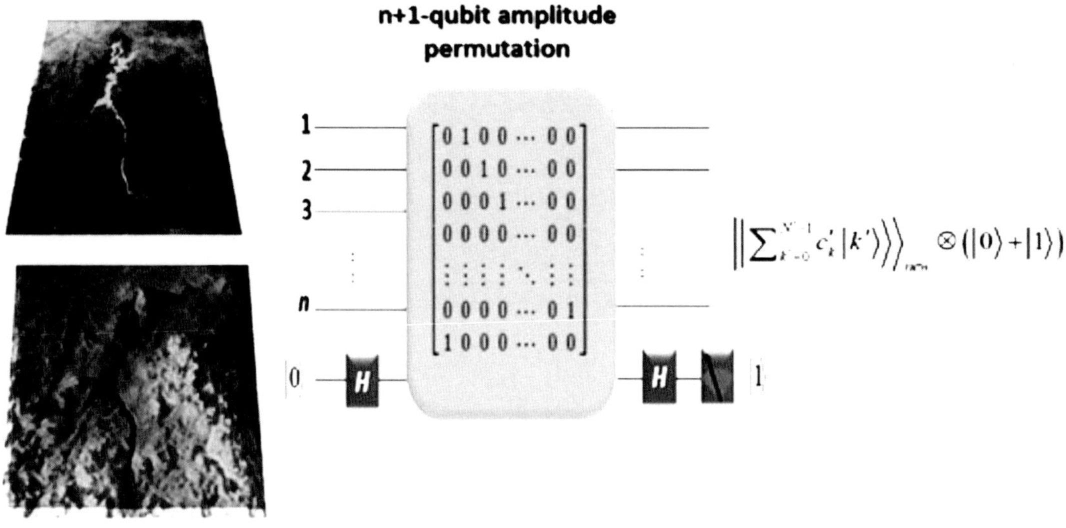

Figure 6.25 Quantum edge detection algorithm based on Hadamard gate.

Equation 6.16 encapsulates the complete boundary information of lineaments in satellite images. Consequently, Figure 6.26 illustrates the 3-D high-density variations of lineaments showing a stronger entanglement with heavy vegetation covers. It is evident that these lineaments are associated with faults, and there is a significant occurrence of lineaments within mountainous regions. These types of lineaments can be referred to as mountain lineaments. Figure 6.26 highlights sharp lineament features, which are linear or curvilinear features on the Earth's surface. These lineaments are often indicative of underlying geological structures such as faults. In the Guba region, these lineaments are crucial for understanding the tectonic framework and potential fault zones. In the Guba region, lineaments are mostly directed in the northeast direction, while the lineaments next to the saddle dam are pointed in the southwest direction (Figure 6.27).

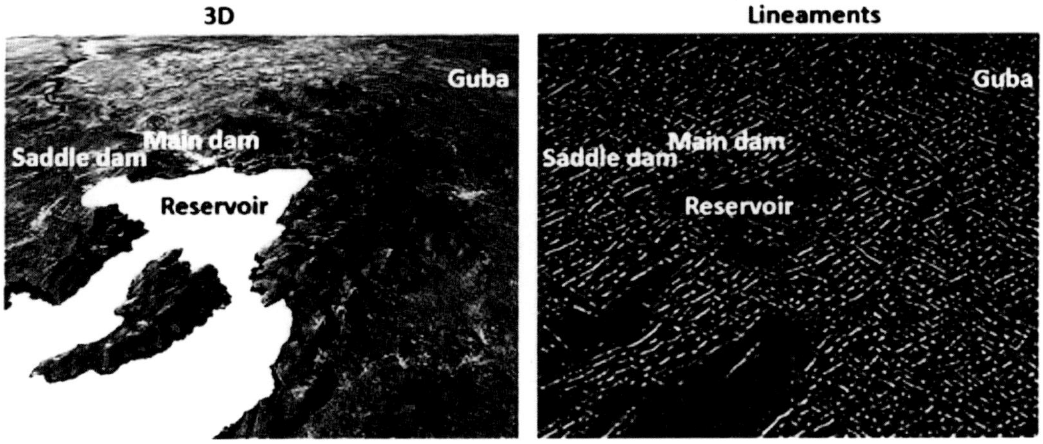

Figure 6.26 Automatic identification of lineament variations using 3-D quantum edge detection algorithm.

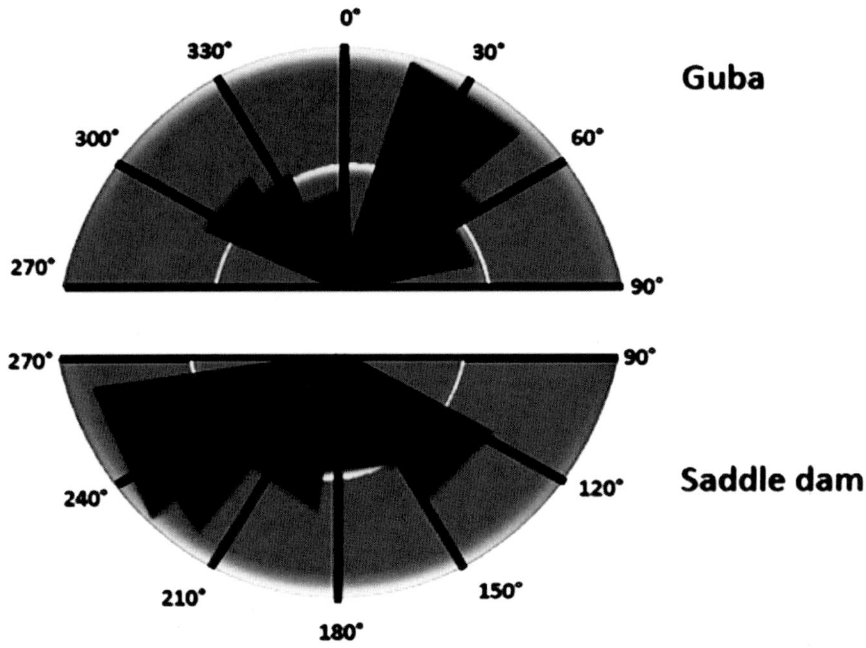

Figure 6.27 Rose diagram of lineament directions.

In the Guba region, the high-density variations of lineaments are more entangled with regions of heavy vegetation, suggesting that natural geological features are more prominent in less developed areas. Therefore, Figures 6.26 and 6.27 also depict mountainous regions, where the occurrence of lineaments is significant. These mountain lineaments are associated with geological processes such as tectonic activity and erosion, which have shaped the terrain over time.

Distinct lineament features and fault lines arise from the interaction between fuzzy quantum B-splines and the quantum edge detection algorithm (Figure 6.28). This quantum entanglement allows for a smoother distribution of lineaments on 3-D quantum fuzzy B-splines, creating a more coherent representation of structural features. In the GERD zone, prominent shear zones divide distinct terranes, playing a key role in deciphering the region's tectonic evolution. Additionally, the area exhibits intricate folding and faulting patterns, including early east-west compression deformations followed by later north-trending folds. Notably, some faults run parallel to the main dam axis, effectively segmenting the reservoir area [35–40].

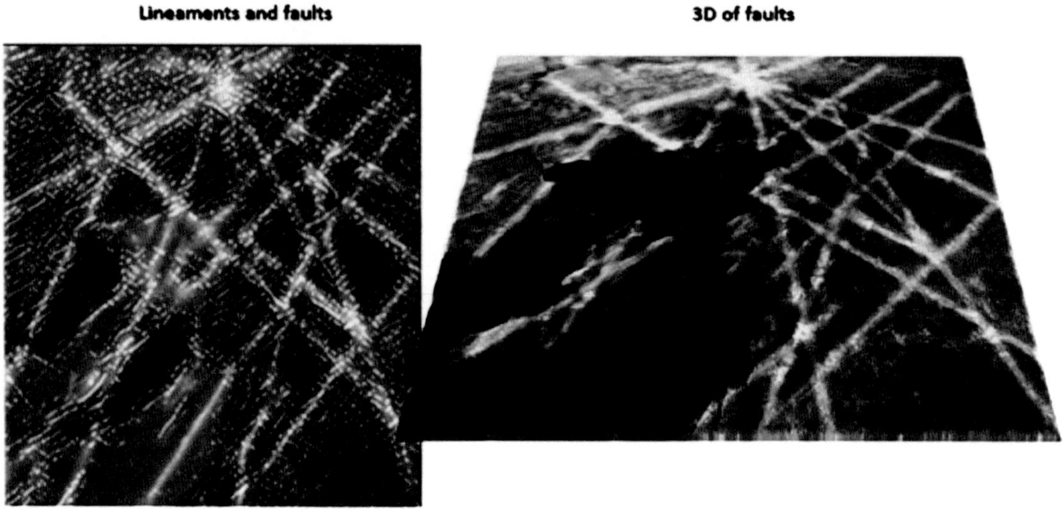

Figure 6.28 Automatic identification of fault variations using 3-D quantum edge detection algorithm.

The geological structure surrounding the Grand Ethiopian Renaissance Dam (GERD) reveals several significant fault lines running parallel to the main dam which divides the reservoir area. These faults, crucial in understanding the region's seismic landscape, are a central feature of Mustafa and Elmahdy's research [35]. They emphasize that the faults along the African Rift are far from dormant; they display persistent seismic and microseismic activity. This continuous movement, even in minor faults, reflects ongoing tectonic dynamics and indicates that the area remains geologically active.

Importantly, some of the region's major earthquakes have occurred along faults located within a 250 km radius of the GERD. This proximity to seismic activity amplifies the need for ongoing monitoring of these faults to safeguard the dam's structural integrity. Faults exhibiting seismic activity, regardless of their magnitude, are generally considered active and, therefore, capable of causing significant geological events.

In seismic risk assessments, it is essential to evaluate both the presence and absence of earthquakes along a fault line. A period without seismic events does not automatically imply inactivity; it may instead indicate a buildup of stress, which could eventually be released as a larger seismic event. Thus, careful caution is needed in assessing the risk of fault lines in the GERD vicinity [37–39].

The geological environment around GERD, marked by parallel faults and persistent seismic activity, requires thorough and continuous monitoring to ensure the dam's stability and safety. Compounding this risk is the combination of faults, lineaments, and high clay concentrations (discussed in Chapter 5), which increases the potential for structural issues. Over time, these conditions could lead to cracks, subsidence, or even a full breach of the dam, which will be examined in subsequent chapters.

In the context of fuzzy quantum B-spline surface measurements, when a qubit returns a value of 1, qubits α and β collapse to the state 1, as indicated by Equation 6.12. These results are then applied in further calculations that incorporate qubits α and β. This interplay between the quantum edge detection algorithm and the fuzzy quantum B-spline model produces a refined and highly accurate mapping of geological features. Quantum entanglement ensures a more uniform distribution of lineaments, enhancing the overall precision and quality of the 3-D quantum fuzzy B-spline models.

6.12 CONCLUSION

In conclusion, this chapter has introduced a groundbreaking approach for the automatic detection of lineaments and faults using multispectral satellite imagery, such as Sentinel-2 data, across the GERD region in Ethiopia. The integration of quantum fuzzy B-spline and quantum edge detection techniques enables the precise delineation of lineament density variations. The quantum fuzzy B-spline method excels in reconstructing a true 3-D representation of the satellite imagery while preserving critical edge boundary details. Furthermore, the application of quantum edge detection, utilizing the Hadamard gate, has proven highly effective in extracting detailed lineament features, even in areas with dense vegetation cover.

It is imperative to emphasize that the geological setting around the GERD, characterized by parallel faults and ongoing seismic activity, demands vigilant monitoring and thorough analysis. This is essential to ensure the safety and stability of the dam and its surrounding areas. The novel methodologies presented in this chapter offer significant advancements in our ability to understand and manage these complex geological environments.

References

[1] Lutgens, F.K., Tarbuck, E.J. and Tasa, D.G. 2017. Essentials of Geology. Pearson.

[2] Ohnaka, M. 2013. The physics of rock failure and earthquakes. Cambridge University Press.

[3] Attoh, K. and Brown, L.D. 2008. The Neoproterozoic trans-saharan/trans-brasiliano shear zones: suggested Tibetan analogs. AGUSM. 2008: S51A-04.

[4] Marshak, S. and Van Der Pluijm, B.A. 2004. Earth Structure: An Introduction to Structural Geology and Tectonics. WW Norton.

[5] Lister, G., Tkalčić, H., Hejrani, B., Koulali, A., Rohling, E., Forster, M., et al. 2018. Lineaments and earthquake ruptures on the East Japan megathrust. Lithosphere. 10(4): 512–522.

[6] Solomon, S. and Ghebreab, W. 2006. Lineament characterization and their tectonic significance using Landsat TM data and field studies in the central highlands of Eritrea. Journal of African Earth Sciences. 46(4): 371–378.

[7] Koch, M. and Mather, P.M. 1997. Lineament mapping for groundwater resource assessment: a comparison of digital Synthetic Aperture Radar (SAR) imagery and stereoscopic Large Format Camera (LFC) photographs in the Red Sea Hills, Sudan. International Journal of Remote Sensing. 18(7): 1465–1482.

[8] Boyer, R. and McQueen, J. 1964. Comparison of mapped rock fractures and airphoto linear features. Photogrammetric Engineering and Remote Sensing. 30(4): 630–635.

[9] Mogaji, K.A., Aboyeji, O.S. and Omosuyi, G.O. 2011. Mapping of lineaments for groundwater targeting in the basement complex region of Ondo State, Nigeria, using remote sensing and geographic information system (GIS) techniques. International Journal of Water Resources and Environmental Engineering. 3(7): 150–160.

[10] Marghany, M. and Hashim, M. 2010. Lineament mapping using multispectral remote sensing satellite data. International Journal of Physical Sciences. 5(10): 1501–1507.

[11] Won-In, K. and Charusiri, P. 2003. Enhancement of thematic mapper satellite images for geological mapping of the Cho Dien area, Northern Vietnam. International Journal of Applied Earth Observation and Geoinformation. 4(3): 183–193.

[12] Robinson, C.A., El-Baz, F., Kusky, T.M., Mainguet, M., Dumay, F., Al Suleimani, Z., et al. 2007. Role of fluvial and structural processes in the formation of the Wahiba Sands, Oman: A remote sensing perspective. Journal of Arid Environments. 69(4): 676–694.

[13] Vassilas, N., Perantonis, S., Charou, E., Tsenoglou, T., Stefouli, M. and Varoufakis, S. 2002. Delineation of lineaments from satellite data based on efficient neural network and pattern recognition techniques. *In*: Proceeding Second Hellenic Conference on AI. 355–365.

[14] Zaineldeen, U.F. 2011. Paleostress reconstructions of Jabal Hafit structures, Southeast of Al Ain City, United Arab Emirates (UAE). Journal of African Earth Sciences. 59(2–3): 323–35.

[15] Majumdar, T.J. and Bhattacharya, B.B. 1988. Application of the Haar transform for extraction of linear and anomalous patterns over part of Cambay Basin, India. International Journal of Remote Sensing. 9(12): 1937–1942.

[16] Mostafa, M.E. and Bishta, A.Z. 2005. Significance of lineament patterns in rock unit classification and designation: a pilot study on the Gharib-Dara area, northern Eastern Desert, Egypt. International Journal of Remote Sensing. 26(7): 1463–1475.

[17] Mostafa, M.E. and Qari, M.Y. 1995. An exact technique of counting lineaments. Engineering Geology. 39(1–2): 5–15.

[18] Koike, K., Nagano, S. and Ohmi, M. 1995. Lineament analysis of satellite images using a Segment Tracing Algorithm (STA). Computers and Geosciences. 21(9): 1091–1104.

[19] Walsh, G.J. and Clark Jr, S.F. 2000. Contrasting methods of fracture trend characterization in crystalline metamorphic and igneous rocks of the Windham quadrangle, New Hampshire. Northeastern Geology and Environmental Sciences. 22(2): 109–120.

[20] Suzen, M.L. and Toprak, V.E. 1998. Filtering of satellite images in geological lineament analyses: an application to a fault zone in Central Turkey. International Journal of Remote Sensing. 19(6): 1101–1114.

[21] Novak, I.D. and Soulakellis, N. 2000. Identifying geomorphic features using LANDSAT-5/TM data processing techniques on Lesvos, Greece. Geomorphology. 34(1–2): 101–109.

[22] Mah, A., Taylor, G.R., Lennox, P. and Balia, L. 1995. Lineament analysis of Landsat Thematic Mapper images, Northern Territory, Australia. Photogrammetric Engineering and Remote Sensing. 61(6): 761–773.

[23] Marghany, M. 2012. Fuzzy B-spline algorithm for 3-D lineament reconstruction. International Journal of Physical Sciences. 7(15): 2294–2301.

[24] Leech, D.P., Treloar, P.J., Lucas, N.S. and Grocott, J. 2003. Landsat TM analysis of fracture patterns: a case study from the Coastal Cordillera of northern Chile. International Journal of Remote Sensing. 24(19): 3709–3726.

[25] Marghany, M. 2012. Three-dimensional lineament visualization using fuzzy B-spline algorithm from multispectral satellite data. Remote Sensing Advanced Techniques and Platforms. Croatia: InTech Open Access Publisher. 13: 213–232.

[26] Oguchi, T., Aoki, T. and Matsuta, N. 2003. Identification of an active fault in the Japanese Alps from DEM-based hill shading. Computers and Geosciences. 29(7): 885–891.

[27] Das, S., Pardeshi, S.D., Kulkarni, P.P. and Doke, A. 2018. Extraction of lineaments from different azimuth angles using geospatial techniques: a case study of Pravara basin, Maharashtra, India. Arabian Journal of Geosciences. 11(8): 160.

[28] Mallast U, Gloaguen R, Geyer S, Rödiger T, Siebert C. 2011. Derivation of groundwater flow-paths based on semi-automatic extraction of lineaments from remote sensing data. Hydrology and Earth System Sciences. 15(8): 2665.

[29] Koike, K., Nagano, S. and Ohmi, M. 1995. Lineament analysis of satellite images using a Segment Tracing Algorithm (STA). Computers and Geosciences. 21(9): 1091–1104.

[30] Raghavan, V., Masumoto, S., Koike, K. and Nagano, S. 1995. Automatic lineament extraction from digital images using a segment tracing and rotation transformation approach. Computers and Geosciences. 21(4): 555–591.

[31] Haeruddin, Irawan J.F. 2020. Identifying of the relationship between lineament density and vegetation index at Tumpangpitu Mining Area, East Java, Indonesia. *In*: AIP Conference Proceedings. 2278(1): 020006. AIP Publishing LLC.

[32] Reiser, R., Lemke, A., Avila, A., Vieira, J., Pilla, M. and Du Bois, A. 2016. Interpretations on quantum fuzzy computing: intuitionistic fuzzy operations× quantum operators. Electronic Notes in Theoretical Computer Science. 324: 135–150.

[33] Yao, X.W., Wang, H., Liao, Z., Chen, M.C., Pan, J., Li, J., et al. 2017. Quantum image processing and its application to edge detection: theory and experiment. Physical Review X. 7(3): 031041.

[34] Marghany, M. 2022. Remote Sensing and Image Processing in Mineralogy. CRC Press.

[35] Mohamed, M.M. and Elmahdy, S.I. 2017. Remote sensing of the Grand Ethiopian Renaissance Dam: a hazard and environmental impacts assessment. Geomatics, Natural Hazards and Risk. 8(2): 1225–1240.

[36] Tadesse, S., Milesi, J.P. and Deschamps, Y. 2003. Geology and mineral potential of Ethiopia: a note on geology and mineral map of Ethiopia. Journal of African Earth Sciences. 36(4): 273–313.

[37] Deyassa, G., Kebede, S., Ayenew, T. and Kidane, T. 2014. Crystalline basement aquifers of Ethiopia: Their genesis, classification and aquifer properties. Journal of African Earth Sciences. 100: 191–202.

[38] Getachew, H. and Tamiru, F. 2023. Delineation of Gold mineralization zones using Induced Polarization and Magnetic Methods at Adare Green Stone Belt, Nejo, Oromia, Ethiopia. Journal of Science, Technology and Arts Research. 12(1): 33–51.

[39] Semaw, F., Zeleke, G. and Balew, A. 2022. Evaluating flood risk management practices and vulnerability mapping in Alawuha watershed (North Wollo Zone, Ethiopia) using GIS and remote sensing. Applied Geomatics. 14(2): 347–367.

[40] Marghany, M. 2021. Advanced Algorithms for Mineral and Hydrocarbon Exploration Using Synthetic Aperture Radar. Elsevier.

Chapter 7

Fundamentals of Microwave Altimetry: Analyzing Satellite Data

7.0 INTRODUCTION

The preceding chapters have introduced groundbreaking methodologies that shed light on the intricate mechanisms governing clay concentration and geological characteristics surrounding the Grand Ethiopian Renaissance Dam (GERD). These elements are critical to ensure the safety and operational efficiency of the dam. One of the most significant factors intertwined with clay concentration and tectonic features, such as faults and lineaments is the water level within the dam's reservoir. Fluctuations in water level are particularly crucial because they interact with high concentrations of clay, which can exert substantial stress on both the dam's structural components and the underlying geological formations.

The challenge, however, lies in accurately measuring these water levels and their associated outflows. Traditional optical data, while useful in certain contexts, fall short in capturing the nuanced changes in water level with the precision required for a structure as pivotal as GERD. This limitation underscores the importance of radar altimetry, a technology capable of providing the detailed measurements needed to monitor water levels effectively in such complex environments.

This chapter, therefore, delves into the principles and mechanisms underlying radar altimetry, with a focus on how this technology can be employed to accurately capture and analyze water levels within dam reservoirs like GERD. By understanding the capabilities and applications of radar altimeters, we can better appreciate their role in safeguarding the structural integrity and operational efficiency of large-scale hydraulic projects.

7.1 RADIO DETECTING AND RANGING

In the effort to understand the secrets hidden in radar altimeter data capturing urban infrastructure mobility and activities, it is crucial to thoroughly explore RADAR's basic characteristics. Active microwave Earth observation, also known as RADAR (Radio Detection and Ranging), has a central role in this fascinating story. The term 'radio' refers to RADAR's origins, where long wavelengths (1 to 10 m) fall within the radio band of the electromagnetic spectrum [1–3].

A complex interplay of advanced technological elements directs the radar system in which a prominent role is played by the antenna, transmitter and receiver. The transmitter acts as the

DOI: 10.1201/9781003327325-7

conductor, producing the energy that propels the radar beam towards its intended target. The transmitted signal, a rapid and rhythmic burst, serves as the pulsating core of radar's communication process. As these signals travel through space at an incredible speed of 300,000 km/s, they interact with targets, which can absorb, reflect, refract, or scatter the energy, resulting in a dynamic and ever-changing exchange [2].

In this complex process, the receiver captures the pulsed return signal and determines signal strength by comparing received and transmitted signals [4–6]. Hence, the term 'active' in microwave imaging takes center stage, representing the system's capability to transmit microwave energy—pulses—across a specific wavelength for a defined duration [1, 3, 6]. This rhythmic transmission, encompassed in the concept of pulsed coherent radar, vividly illustrates the radar system's functionality, as shown in Figure 7.1.

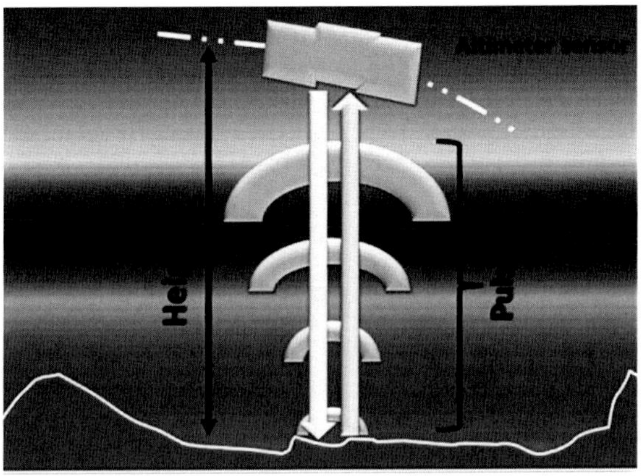

Figure 7.1 Transmitted pulse and its backscatter in a radar system.

The subsequent sections explore the intricate tapestry of radar altimeter data, delving deeper into the realms where microwaves and imagery converge.

7.2 THE ALTIMETER MICROWAVE

An altimeter, often referred to as an altitude meter, is a sophisticated instrument designed to calculate the height of an object relative to a fixed reference point. This process of determining altitude is termed 'altimetry'. The concept bears a close relationship to 'bathymetry', which involves measuring the depth beneath the sea surface, thereby extending the principles of altimetry to the exploration of underwater topography.

The primary function of an altimeter is to measure the vertical distance between its platform—whether a satellite, aircraft or another airborne entity—and the Earth's surface below. This measurement is achieved through the emission of short, concentrated microwave pulses directed towards the ground. The altimeter then meticulously records the time taken for these pulses to travel to the surface and back to the sensor. By calculating the time delay of this round trip, the device determines the precise distance between the altimeter and the target surface.

In essence, altimeters operate as nadir-looking pulse radars (Figure 7.2), meaning they are oriented directly downward, towards the nadir point on the Earth's surface, from their airborne or spaceborne vantage points [3, 5]. This orientation allows them to deliver highly accurate altitude data, which is crucial for a variety of applications, from topographic mapping and climate studies to the management of large-scale hydraulic structures like dams. Through the integration of these measurements, altimetry contributes not only to our understanding of elevation and depth but also to our ability to monitor and respond to changes in the Earth's dynamic environment.

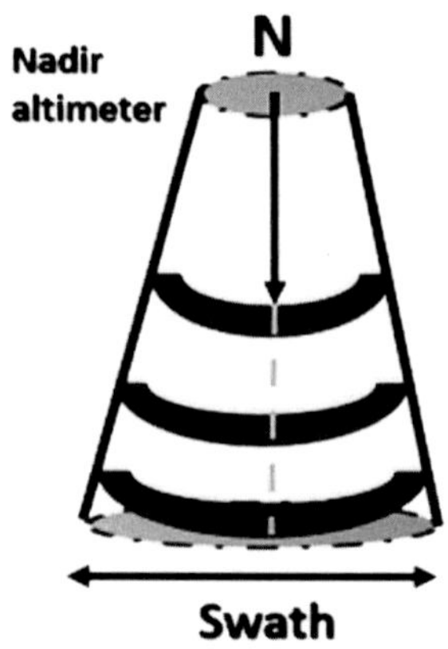

Figure 7.2 Concept of nadir altimeter-looking pulse.

7.3 HOW THE ELECTROMAGNETIC SPECTRUM FUNCTIONS IN RADAR ALTIMETER OPERATION?

The operation of an altimeter, particularly a radar altimeter, relies on the principles of the electromagnetic wave spectrum. The electromagnetic spectrum encompasses all types of electromagnetic radiation, which differ in wavelength and frequency. These range from long-wavelength radio waves to short-wavelength gamma rays, with visible light sitting somewhere in the middle.

Radar altimeters specifically use microwave radiation, which falls within the radio wave portion of the electromagnetic spectrum, typically in the range of 1 to 30 gigahertz (GHz) [1, 6] (Figure 7.3).

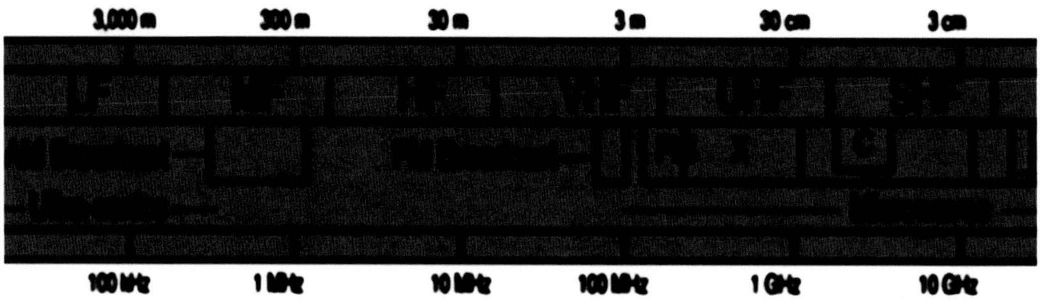

Figure 7.3 Radio wave frequency spectra.

This frequency range; therefore, is chosen for several reasons [2–6]:

Penetration and reflection: Microwaves can penetrate through clouds, rain and other atmospheric conditions that would scatter or absorb higher-frequency waves like visible light. This makes them ideal for consistently measuring distances to the Earth's surface under a variety of weather conditions.

Precision: The wavelength of microwaves is short enough to allow for precise measurements. When a radar altimeter emits a microwave pulse, it can accurately determine the distance to the Earth's surface by measuring the time it takes to travel to the surface and back. The precision of this measurement is directly linked to the wavelength of the microwave; shorter wavelengths can provide more accurate distance measurements.

Resolution: By carefully selecting the frequency of the microwave pulses, radar altimeters can achieve a high spatial resolution. This allows them to distinguish between small features on the Earth's surface, making them useful for detailed topographic mapping, as well as for monitoring water levels and surface deformations.

The process begins when the altimeter's transmitter generates a short burst of microwave energy, which is directed downward toward the Earth's surface. When this energy hits the surface, part of it is reflected toward the altimeter. The altimeter's receiver then detects this reflected signal. By calculating the time difference between the emission of the pulse and the reception of its echo, the altimeter can determine the distance to the surface. This process is repeated thousands of times per second, allowing the altimeter to create a continuous profile of the surface beneath it. The information gathered can be used for a variety of purposes, including determining the height of terrain, the depth of oceans and to monitor the structural integrity of man-made constructions like dams.

Needless to say, the electromagnetic spectrum's microwave portion is critical for the operation of radar altimeters, enabling them to provide precise and reliable measurements of distance to the Earth's surface, regardless of atmospheric conditions.

7.4 PRINCIPLES OF ALTIMETER

An altimeter, much like a Synthetic Aperture Radar (SAR), functions by emitting a radar signal and subsequently recording the backscattered signal from surfaces it encounters. However, the key difference lies in the orientation of this emission. While SAR systems transmit their radar signals at various angles to the object of interest, an altimeter emits and receives its radar waveform perpendicularly, meaning the signal travels directly downward toward the Earth's surface and reflects straight back. In general, the speed of light is about a million times faster than the cruising speed of a typical plane (v), so a radio signal that bounces to the ground and back travels a distance of about twice the plane's altitude ($2h$) (Figure 7.4) [5–7].

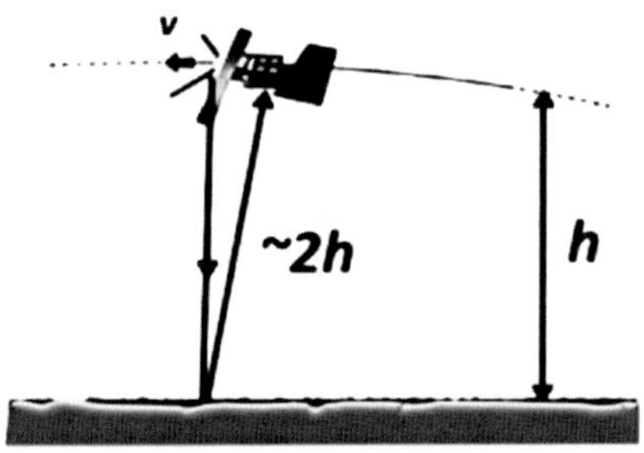

Figure 7.4 Principle of radar altimeter operation.

Mathematically, the mechanism can be described as follows [6, 8]:

Radar signal emission: The altimeter emits a radar pulse at a specific frequency f, corresponding to a wavelength λ. The signal travels from the altimeter to the target surface.

Time of flight: The time t it takes for the radar signal to travel to the surface and return is recorded. This time is related to the distance h between the altimeter and the target surface by the formula:

$$h = \frac{1}{2} c \cdot t \tag{7.1}$$

where c is the speed of light in the medium (approximately 3×10^8 m/s in a vacuum). Therefore, the amplitude A of the backscattered signal is then analyzed. Since the signal is emitted and received perpendicularly, the backscatter amplitude directly correlates with the reflective properties of the surface and the distance from the altimeter. In this sense, the amplitude A can be described by the radar equation, which in its simplified form for a monostatic radar (where the transmitter and receiver are co-located, as in an altimeter) is given by:

$$A \propto \frac{P_t G^2 \lambda^2 \sigma}{(4\pi)^3 h^4} \tag{7.2}$$

Equation 7.2 demonstrates that the amplitude A of the backscattered signal is a function of several factors, primarily: the radar cross-section σ of the surface, which represents its reflective properties; the power of the emitted radar signal P_t; and the antenna gain G. In addition to the distance h between the altimeter and the surface and the wavelength of the radar signal λ, in other words, the distance h between the altimeter and the surface affects the signal strength inversely with the fourth power of h^4 as seen in the radar equation. This implies that even small changes in distance significantly impact the amplitude A [1, 7, 9]. However, the radar cross-section σ depends on the material properties, surface roughness, and angle of incidence. For a perpendicular (nadir) observation, the radar cross-section is primarily a function of the surface's dielectric constant and roughness, which can be modeled as:

$$\sigma = f(\varepsilon_r, R) \tag{7.3}$$

where ε_r is the surface's relative permittivity (dielectric constant), and R represents the surface roughness. Combining these elements, the backscatter amplitude A as a function of distance h and surface properties σ can be expressed as:

$$A(h,\sigma) \propto \frac{P_t G^2 \lambda^2 \sigma}{(4\pi)^3 h^4} \tag{7.4}$$

Equation 7.4 encapsulates how the amplitude of the backscattered signal varies with the distance to the target and its reflective characteristics, providing a basis for interpreting the altimeter's measurements.

Consistent with the above-mentioned perspective, the height h of an object, such as water level, can be estimated by inverting the backscatter signal, taking into account the amplitude and time of flight:

$$h = f(A, t) \tag{7.5}$$

where f represents the function that models the relationship between the backscatter amplitude and time of flight, considering the radar system's calibration and the specific properties of the surface. In this framework, object height is particularly amenable to inversion from the perpendicular backscatter signal. This is because the direct relationship between the signal's amplitude and the surface height is less complex than with synthetic aperture radar (SAR), where varying angles introduce additional variables and require more sophisticated processing. This is due to their perpendicular signal orientation and the specific frequency bands used, which are optimized for the measurement of vertical distances in varying environmental conditions. Consequently, altimeters are preferred for applications requiring high precision in water-level monitoring, inland and coastal waters [6–10].

The altimeter operates across various frequency bands, each chosen for specific applications. The primary signal bands include the E-band, Ka-band, and S-band. With its relatively lower frequency, the S-band is especially well-suited for retrieving advanced water-level parameters due to its ability to penetrate atmospheric disturbances like rain and fog, providing more reliable data.

7.5 RADAR ALTIMETER FREQUENCIES

Radar altimeters, essential tools for measuring altitude, operate through two primary mechanisms: (i) Frequency Modulated Continuous Wave (FMCW) altimeters and (ii) Pulse altimeters. The functionality of each type depends on the nature of the radar signals they employ. FMCW altimeters can be categorized into broad-beamwidth and narrow-beamwidth types, each defined by the width of the antenna beam. Therefore, the signal emitted by an FMCW altimeter is characterized by a continuous wave whose frequency varies linearly with time. The transmitted frequency $f(t)$ can be expressed as:

$$f(t) = f_0 + \Delta f \cdot \frac{t}{T} \tag{7.6}$$

where f_0 is the initial frequency, Δf is the frequency deviation, and T is the modulation period. The received signal, after reflection from the surface, is compared with the transmitted signal to determine the delay time Δt.

Consistent with the above-mentioned perspective, pulse altimeters, also known as short-pulse or pulse-compression altimeters, function based on the transmission of discrete radar pulses. The defining characteristic here is intrapulse modulation (Figure 7.5), which modulates the pulse within its duration to enhance resolution [10–13].

In this context, Figure 7.5 illustrates three different types of intrapulse modulations: (i) Linear Frequency Modulation, (ii) Amplitude Modulation, and (iii) Phase Modulation. For Linear Frequency Modulation, the frequency of the signal increases linearly over time within the pulse duration, creating a chirp effect. The amplitude of the signal is modulated over time, varying within the pulse duration due to amplitude modulation procedures. Similarly, the phase of the signal is modulated, changing according to a specific pattern over time within the pulse during phase modulation. Each type of modulation serves different purposes in enhancing the resolution and accuracy of radar signals [12].

Altimeters, therefore, can also operate in the optical spectrum, such as with laser altimeters, which utilize light waves instead of radio waves. These systems emit short pulses of laser light and measure the time of flight to determine distance. For a laser altimeter, the distance (d) is calculated using the time delay (Δt) of the laser pulse. The formula is:

$$\Delta t = \frac{2 \cdot d}{c} \tag{7.8}$$

Given the much shorter wavelength of light compared to radio waves, laser altimeters can achieve higher resolution as demonstrated in equation 7.8. Suppose the distance d is 10,000 meters. Then:

$$\begin{aligned} \Delta t &= \frac{2 \cdot d}{c} = \frac{2 \times 10{,}000 \text{ meters}}{3 \times 10^8 \text{ meters/second}} \\ &= \frac{20{,}000}{300{,}000{,}000} \text{ seconds} \\ \Delta t &\approx 6.67 \times 10^{-5} \text{ seconds or } 66.7 \text{ microseconds} \end{aligned} \tag{7.9}$$

The value of Δt represents the time it takes for the radar signal to travel to the target surface and back to the altimeter.

A continuous-wave radar system that operates without frequency modulation encounters a fundamental limitation: it cannot determine the distance to a target. This is because such a system lacks a specific timing mark or reference point that would allow it to accurately measure the time interval between the transmission of a signal and the reception of its echo. Without this critical timing reference, the radar system is unable to convert the time difference into a precise range measurement [13–15].

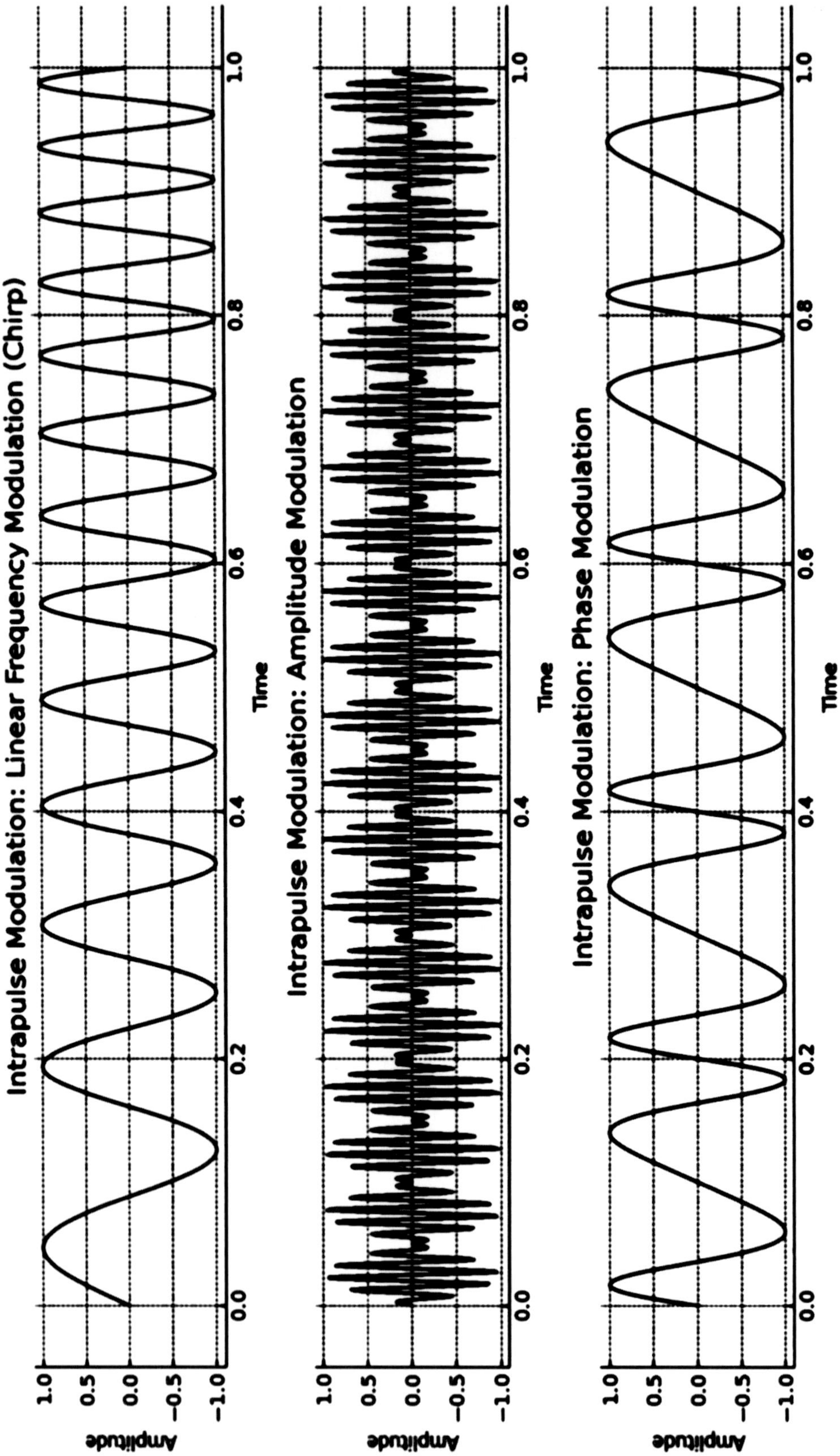

Figure 7.5 Concept of intrapulse modulation.

To overcome this limitation frequency modulation can be applied to the transmitted signal. The radar emits a signal that changes its frequency periodically—either increasing or decreasing over time. When the transmitted signal encounters a target and the echo returns, the frequency of this echo signal will be slightly different from the frequency of the signal currently being transmitted. This difference in frequency corresponds to a time delay, denoted as Δt, which arises due to the travel time of the signal to the target and back [11–13].

In a simple continuous-wave (CW) radar system, the signal is transmitted continuously at a single frequency f_0. Since there is no variation in frequency over time, there is no direct way to measure the time delay Δt, making it impossible to determine the range d. In FMCW radar, the transmitted frequency $f(t)$ varies linearly with time. This can be expressed as:

$$f(t) = f_0 + \alpha t \tag{7.10}$$

where α is the rate of frequency change (modulation rate). The modulation rate α, refers to the rate at which the frequency of a signal changes over time in a frequency-modulated continuous wave (FMCW) radar system. It is a critical parameter that directly influences the resolution and accuracy of distance measurements in such systems. In FMCW radar, the transmitted signal's frequency is not constant; instead, it varies linearly with time. This variation can either be an increase (up-chirp) or a decrease (down-chirp) in frequency. The modulation rate α is the slope of this frequency change and is defined mathematically as the change in frequency per unit time:

$$\alpha = \frac{\Delta f}{T} \tag{7.11}$$

Equation 7.11 says, there is an occurrence of total frequency deviation Δf or the difference between the maximum and minimum frequencies during the modulation period T, which is also known as the sweep time. It can be said that the modulation rate dictates how quickly the frequency of the transmitted signal changes. A higher modulation rate means that the frequency changes more rapidly, while a lower modulation rate implies a slower frequency change.

The question now is: what is the relationship of frequency to range resolution? The modulation rate α influences the radar's ability to resolve targets that are close together. The range resolution Δd of an FMCW radar, which is the minimum distance between two distinguishable targets is inversely proportional to the total frequency deviation Δf and directly proportional to the speed of light c:

$$\Delta d = \frac{c}{2\Delta f} \tag{7.12}$$

Since $\Delta f = \alpha T$, the range resolution can also be expressed as:

$$\Delta d = [2\alpha T]^{-1} c \tag{7.13}$$

Therefore, a higher modulation rate (with a fixed modulation period) leads to better range resolution, allowing the radar to distinguish between targets that are closer together. The modulation rate also affects the calculation of the time delay Δt between the transmitted and received signals. Given the relationship:

$$\Delta t = \frac{\Delta f_r}{\alpha} \tag{7.14}$$

where Δf_r is the frequency difference between the transmitted and received signals, a higher modulation rate results in a smaller time delay for a given frequency difference, indicating a closer target.

In practical terms, the modulation rate must be carefully chosen based on the radar's intended application. In applications requiring high precision, such as automotive radar or detailed terrain mapping, a high modulation rate is desirable to achieve fine-range resolution (Figure 7.6). On the other hand, for applications where detecting targets at greater distances is more important than fine resolution, a lower modulation rate may be sufficient, allowing the radar to maintain accuracy over longer ranges. In this sense, the modulation rate α is a fundamental parameter in FMCW radar systems that controls how rapidly the signal's frequency changes over time [14]. It directly impacts the radar's range resolution and its ability to accurately measure the distance to targets. Therefore, understanding and selecting the appropriate modulation rate is crucial for optimizing radar performance in different scenarios.

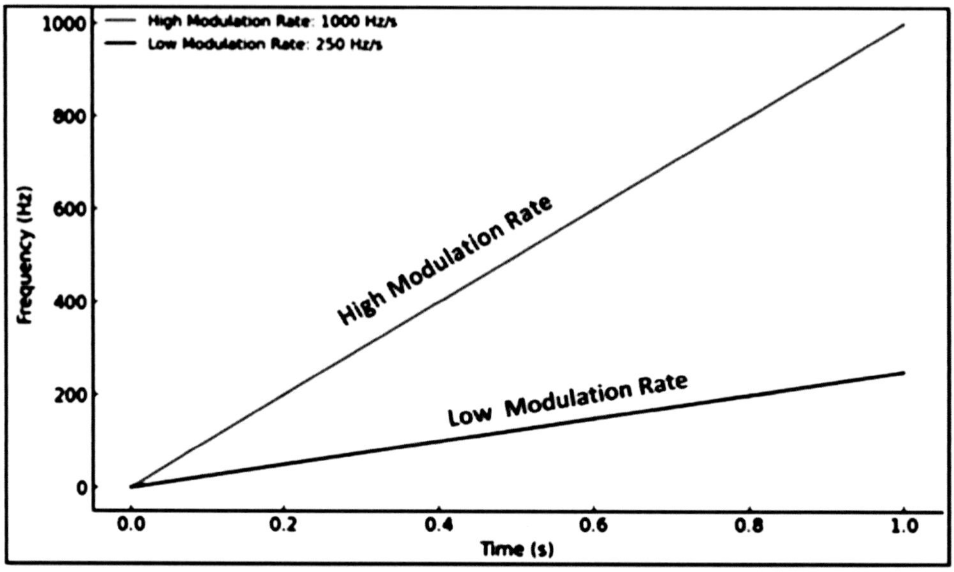

Figure 7.6 Frequency modulation with different modulation rates.

This visual representation of Figure 7.6 also shows how the modulation rate α affects the rate at which the frequency of the signal changes. A higher modulation rate allows for quicker frequency changes, which can improve range resolution in radar systems. Conversely, a lower modulation rate results in a slower frequency change, which may be suitable for applications where fine resolution is less critical.

In addition to frequency differences, FMCW radar can also measure phase differences between the transmitted and received signals [16]. The phase difference $\Delta\phi$ provides another way to determine the range, particularly for small distances, where it is given by:

$$\Delta\phi = 2\pi \cdot \frac{\Delta t}{T} = 2\pi \cdot \frac{d}{\lambda} \tag{7.15}$$

where T is the modulation period, and λ is the wavelength of the signal. By analyzing these differences in frequency and phase, the FMCW radar can accurately determine the distance to a target. In other words, this phase difference is critical in determining the distance to the target in FMCW radar systems [14–16]. The shift is directly related to the time delay caused by the signal traveling to and from the target (Figure 7.7). For instance, the dashed line in Figure 7.7 indicates the phase difference between the transmitted signal and the received signal, which is delayed by a phase shift.

This approach is particularly effective for continuous wave systems where direct time measurements are challenging. The ability to extract range information from frequency and phase differences allows FMCW radar to be highly accurate and effective in various applications, such as automotive radar, altimetry and synthetic aperture radar systems [15–17].

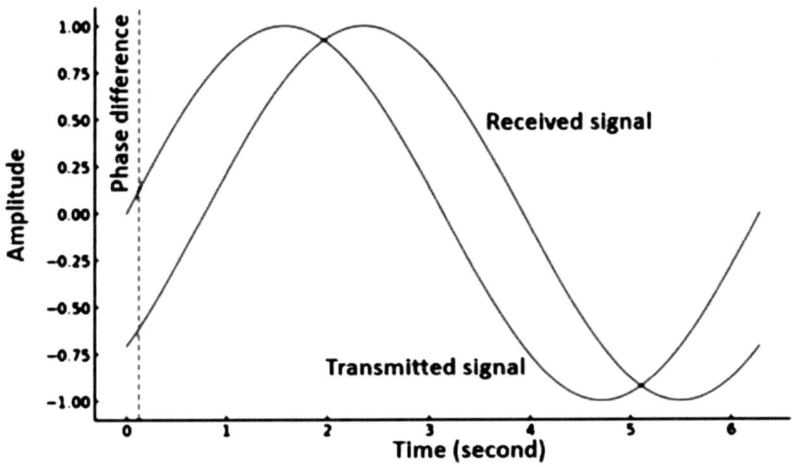

Figure 7.7 The phase difference between the two radar signals.

7.6 RADAR ALTIMETRY PULSES

In radar altimetry, the system directs a pulse of microwave energy straight downwards (towards the nadir) and measures the time delay of the returning echo. The key aspect here is the time delay, which allows the radar altimeter to determine the distance to the Earth's surface directly below the instrument, despite the large footprint that the radar beam covers. What are the impacts of the footprint of a spaceborne altimeter? In a large footprint scenario, a spaceborne microwave altimeter typically has a large footprint, often spanning tens of kilometers. This is due to the broad beam width of the radar antenna at such high altitudes. Therefore, the time delay between the transmission of the pulse and the reception of the echo is a crucial measurement. Since the time delay determines the distance, the first part of the echo will always come from the point on the ground directly below the altimeter (nadir point) [11, 18].

Accordingly, there are two approaches to footprint determination: (i) pulse-limited footprint (Figure 7.8) and (ii) beam-limited footprint (Figure 7.9). In scenarios where the footprint is

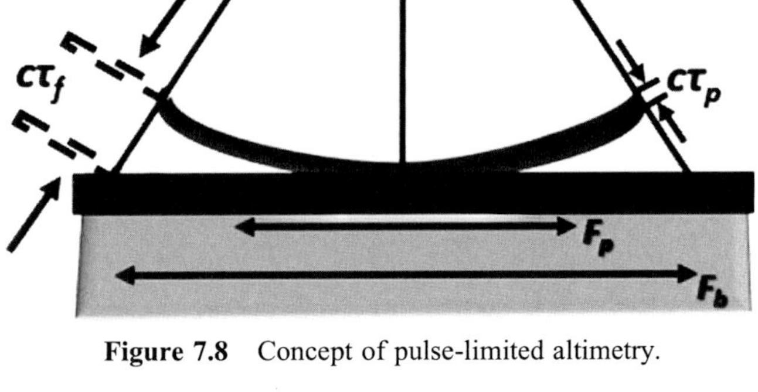

Figure 7.8 Concept of pulse-limited altimetry.

pulse-limited, the size of the area illuminated on the ground is dictated by the length of the radar pulse. Unlike the beam-limited case, the pulse does not simultaneously illuminate the entire region defined by the beamwidth. Instead, it only lights up a fraction of that area, with the extent of illumination being directly proportional to the pulse duration. Consequently, the pulse-limited footprint is derived from the specific portion of the surface that the radar pulse manages to cover at any given moment. Let us assume that τ_p is the length of the pulse and F_p and F_b are the pulse footprint and beam footprint, respectively [15, 18, 20].

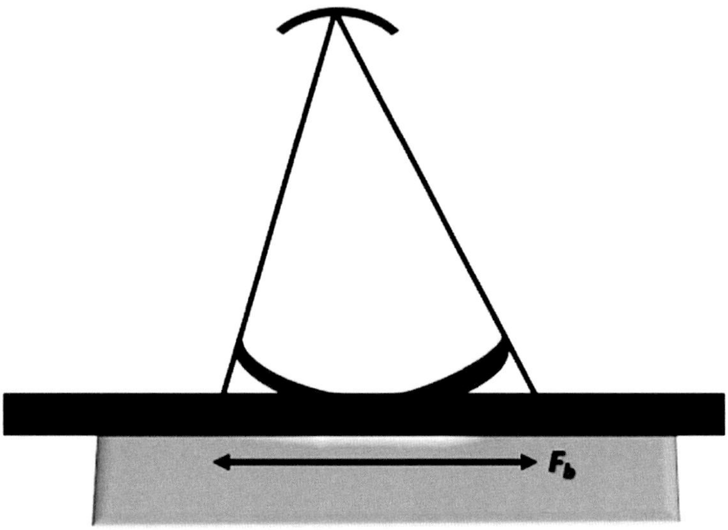

Figure 7.9 Concept of beam-limited altimetry.

Conversely, in the beam-limited footprint, the footprint size is governed by the beamwidth of the radar's antenna. In this case, the entire beamwidth is illuminated at once, and thus, the footprint on the ground is determined by the angular spread of the radar beam and the altitude at which the radar operates. The beam-limited footprint tends to be larger, as it is not only a function of the antenna's beamwidth but is also significantly influenced by the radar's height above the ground. This characteristic makes the beam-limited footprint more sensitive to variations in the radar's design and operational parameters as compared to the pulse-limited footprint [19–21].

When comparing the two approaches, the pulse-limited footprint focuses on a specific fraction of the area at a given time, offering a more confined measurement, while the beam-limited footprint encompasses the entire spread of the beam, resulting in a broader coverage area. The distinction between these methods hinges on whether the radar's footprint is more influenced by the duration of the pulse or by the physical properties of the radar's antenna system and operational height. This difference highlights the importance of understanding the radar's operational context, as it determines which footprint approach is more appropriate for accurate surface measurement.

7.7 DERIVING MATHEMATICAL FORMULAS OF RADAR PULSES

Along with the above-mentioned perspective, the beam surface footprint, F_b, is given approximately by:

$$F_b \simeq H\lambda[\mathrm{D}]^{-1} \quad [\mathrm{m}], \tag{7.16}$$

where D is the antenna diameter. Equation 7.16 emphasizes that the size of the radar's footprint is not solely dependent on the beamwidth, but is also significantly influenced by the altitude of the radar system [H] and radar wavelength λ. As the height increases, the footprint expands, illustrating the proportionality between the flying height and the surface area covered by the radar beam.

The pulse surface footprint F_p is determined by the distance equivalent to the pulse length, denoted as τ_p. To find F_p, let us use the Pythagorean theorem, considering the geometry of the radar beam interacting with the Earth's surface. In this view, the footprint radius R_p is related to the pulse length and altitude as follows:

$$R_p^2 + H^2 = \tau_p^2 \tag{7.17}$$

This can be rearranged to express the pulse length:

$$\tau_p^2 = \sqrt{H^2 + R_p^2} \tag{7.18}$$

Given that R_p is the radius of the footprint (Figure 7.8), we can simplify this equation using the binomial approximation for small values of R_p/H:

$$\tau_p \approx H\left(1 + \frac{R_p^2}{2H^2}\right) \tag{7.19}$$

Since $\frac{R_p^2}{2H^2}$ is typically small, we approximate:

$$\tau_p \approx \left(H + \frac{R_p^2}{2H}\right) \tag{7.20}$$

From this, we can approximate the footprint radius R_p as:

$$R_p^2 \approx \tau_p H \tag{7.21}$$

Taking the square root:

$$R_p \approx \sqrt{\tau_p H} \tag{7.22}$$

Thus, the pulse surface footprint F_p is given by:

$$F_p \approx 2\sqrt{2c\tau_p H} \tag{7.23}$$

Given the constraints on the physical size of satellite altimeter antennas, most systems are pulse-limited, meaning that the pulse length is the primary determinant of the footprint size [11, 16, 18, 20].

An altimeter that operates with a footprint ($F_b > F_p$) is classified as pulse-limited because the size of the footprint is primarily determined by the pulse length rather than the beamwidth of the antenna. In this case, the radar pulse illuminates only a portion of the area defined by the beamwidth at any given time, and thus, the pulse length dictates the footprint size.

On the other hand, if the antenna is sufficiently large, such that the footprint (F_λ) is equal to the pulse-limited footprint (F_p), the altimeter is considered beam-limited. Here, the entire beamwidth is illuminated simultaneously, and the footprint size is determined by the angular spread of the beam and the altitude of the radar. Therefore, owing to the physical limitations on the size of antennas that can be deployed on satellites, most satellite altimeters are pulse-limited. This means that the footprint size is predominantly controlled by the pulse length, as the antennas are not large enough to create a beam-limited footprint [17–19].

7.8 HOW CAN RADAR ALTIMETRY EQUATION BE DERIVED FROM PULSE PRINT F_P?

The radar power equation, also known as the radar range equation, describes the power of the radar signal received by the radar system after it has been transmitted, reflected off a target, and returned. This equation is fundamental in radar theory as it relates various system parameters to the effective range of the radar [10, 18]. Therefore, the equation 7.2 can be modified as follows:

$$P_r = \frac{P_t \cdot G_t \cdot G_r \cdot \lambda^2 \cdot A_r \sigma}{(4\pi)^3 \cdot R^4 \cdot Ls} \tag{7.24}$$

In the context of radar system performance, various parameters interplay to determine the effectiveness of target detection. The received power (P_r) denotes the power of the radar signal that is received back after reflecting off the target, measured in watts (W). Therefore, transmitted power (P_t) represents the power of the radar signal as it is emitted from the transmitter, also measured in watts (W). Consequently, the gain of the transmitting antenna (G_t) quantifies the amplification of the radar signal provided by the transmitting antenna, which enhances the signal's strength in a specific direction. The gain of the receiving antenna (G_r), similarly, measures the amplification of the received signal by the receiving antenna. For monostatic radar systems, where the same antenna is used for both transmission and reception, (G_t) and (G_r) are often equal [19–22].

Subsequently, the radar signal's wavelength λ is measured in meters (m) and is crucial as it influences the radar's resolution and penetration capabilities. Thus, the radar cross-section (σ), which is measured in square meters (m^2), reflects the target's ability to scatter radar waves. This parameter is indicative of the target's reflectivity and size as perceived by the radar. Hence, the Range (R) presents the distance from the radar system to the target, measured in meters (m), and it directly affects the signal attenuation and detection capabilities. Lastly, the system losses (L_s) are the dimensionless factor that accounts for various inefficiencies and losses within the radar system, including atmospheric attenuation, hardware inefficiencies, and other factors that may degrade signal quality [21–22].

Therefore, these parameters collectively determine the radar's ability to detect and accurately measure targets. The balance and interplay between transmitted power, antenna gains, wavelength, radar cross-section, range, and system losses are pivotal in optimizing radar performance and ensuring effective target detection [23].

The main parameter of the area A_r can be estimated using F_p as follows:

$$A_r = \pi\left(\frac{F_p}{2}\right) = 2\pi c\tau_p H. \tag{7.25}$$

Therefore, equation 7.24 can be reformulated as:

$$P_r = \frac{P_t \cdot G_t \cdot G_r \cdot \lambda^2 \cdot 2\pi c\tau_p \sigma}{(4\pi)^3 \cdot H^3 \cdot Ls} \tag{7.26}$$

In our analysis, we have utilized the value of H for height instead of using the range parameter R in equation 7.24. This decision rests on the assumption that the measured range is equivalent to the actual height of the target. Consequently, by equating range with height, we streamline the calculations and simplify the model. This approach presumes that the radar's range measurement directly reflects the target's vertical dimension, thereby aligning the two variables for practical purposes [20–24].

7.9 ALTIMETRY ECHO SHAPE DETERMINATION

The echo captured by an altimeter does not mirror the transmitted pulse exactly but rather appears 'smeared out' over time due to several factors. Primarily, the target surface is rarely perfectly smooth—even in calm ocean conditions. As a result, energy is scattered back to the altimeter from various points within the footprint.

In an ideal scenario, where the surface is a flat, perfectly smooth, specular reflector, the returned signal would indeed be a replica of the transmitted pulse—a true mirror image. In this case, the signal would only be backscattered from directly beneath the sensor at the center of the footprint (Figure 7.10). Although the intensity of the signal would be diminished, its shape would remain unchanged.

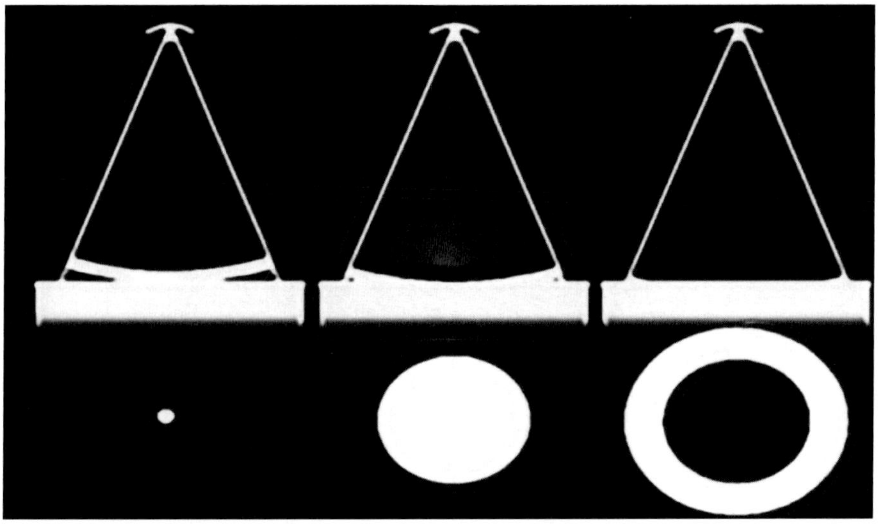

Figure 7.10 Transmitted radar altimeter pulses over a flat surface.

In other words, when a radar altimeter operates over a flat surface, the transmitted pulse spreads out as it travels toward the surface and illuminates progressively larger concentric rings centered on the nadir point (the point directly beneath the radar on the surface). This phenomenon occurs because the radar pulse has a finite beam width, and as the pulse travels farther from the radar, the area it covers increases. Needless to say, for a flat surface, the pulse footprint forms a series of concentric rings as it approaches the surface (Figure 7.10).

However, in practical conditions, signals incident on the surface outside the central region of the footprint are reflected away from the sensor and thus go undetected. This scattering effect contributes to the temporal spreading or 'smearing' of the echo, making it appear less like the original transmitted pulse and more diffused over time.

The shape of the actual echo received by the altimeter consequences from a combination of three primary factors: (a) the shape of the transmitted pulse, (b) the scattering properties across the footprint, (c) and the localized variations in surface height. While the original pulse shape defines the baseline, it is the scattering characteristics and surface height variations that forge the connection between the shape of the echo and the surface properties [22, 24].

As surface roughness increases (Figure 7.11), the length of the echo remains relatively stable but its peak intensity diminishes. This reduction in peak intensity provides insight into the backscatter properties of the surface influenced by the incidence angle across the footprint. The rate at which the echo decays can thus be related to near-surface wind speeds over the ocean or used to characterize the statistical distribution of surface slopes [21, 24].

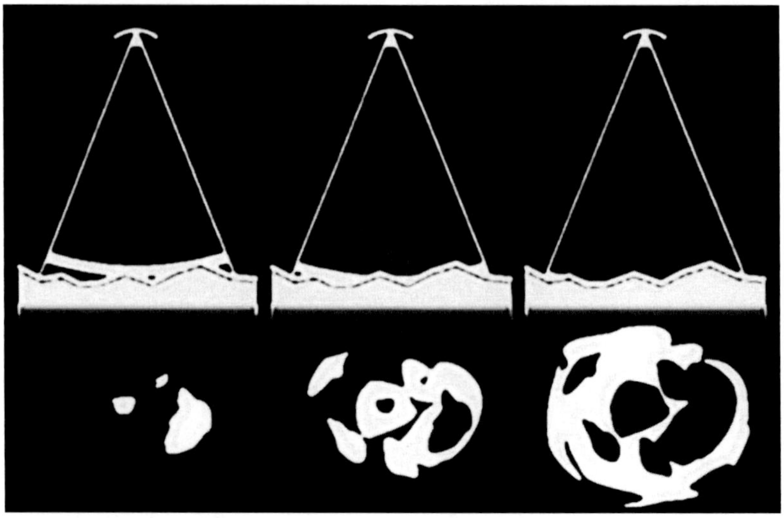

Figure 7.11 Transmitted radar altimeter pulses over a rough surface.

Moreover, the decrease in peak echo power with increased surface roughness also offers a qualitative measure of surface roughness. By examining the peak power, one can infer the degree of roughness, adding another layer of information about the surface characteristics. When surface topography becomes excessively varied and irregular, the echo recorded by the altimeter no longer presents a clear, distinct response. Consequently, it becomes challenging to determine which delay time in the echo corresponds to the altimeter's height, as the leading edge of the echo no longer aligns with the nadir (Figure 7.12).

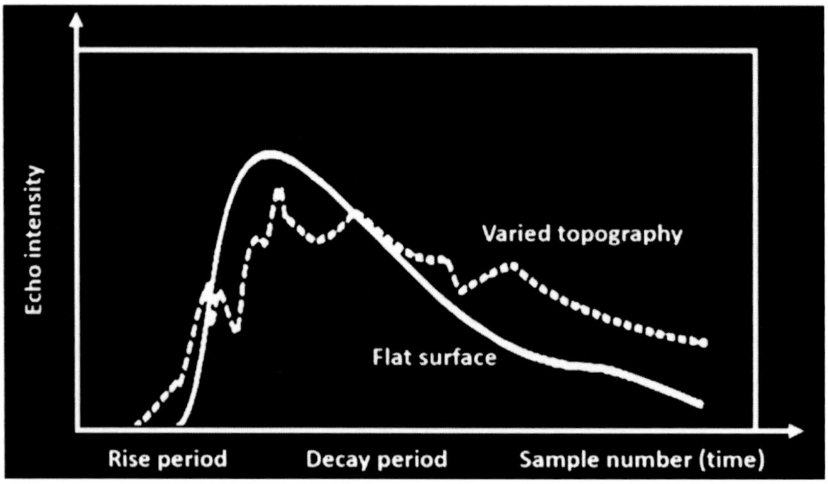

Figure 7.12 Echo intensity variation due to flat surface and varied topography.

The impact of height variations—distinct from wavelength-scale roughness—is observable in the slope of the leading edge of the echo. As the diversity in surface heights increases, the leading edge of the echo exhibits a more gradual rise. This gradual rise time, denoted as Δt, changes proportionally with the root mean square (RMS) height of the surface [23].

In oceanic environments, where variations in surface height can be represented as regular disturbances of an otherwise level surface, this relationship is valuable for estimating significant wave height. By analyzing the rise time of the echo's leading edge, one can infer the average height of waves, providing valuable insights into surface conditions [19, 21, 24].

7.10 DERIVING WAVEFORM ABOVE RIVER AND DAM RESERVOIR

To derive the mathematical equations related to the waveform profile for various surfaces such as oceans, rivers, and dam reservoirs, we can analyze the waveform characteristics based on surface properties.

For an ocean surface, the waveform profile $W(t)$ of the echo is influenced by the surface roughness and scattering properties. The general form of the waveform profile can be modeled as follows:

$$W_{\text{ocean}}(t) = P_t \cdot G_t \cdot G_r \cdot \frac{\sigma_{\text{ocean}}}{(4\pi R^2)} \cdot e^{-\frac{t}{\tau}} \cdot \mathrm{e}^{\left(-\frac{t^2}{2\sigma_t^2}\right)} \tag{7.27}$$

In equation 7.27, τ is the time constant related to system losses, and σ_t is the standard deviation of the pulse width, accounting for pulse broadening. The ocean surface waveform is generally characterized by a broad, smooth profile due to the surface roughness [22].

Therefore, for a river or similar bodies of water, which typically have less roughness compared to the ocean but more variability compared to a reservoir, the waveform profile can be modeled as:

$$W_{\text{river}}(t) = P_t \cdot G_t \cdot G_r \cdot \frac{\sigma_{\text{river}}}{(4\pi R^2)} \cdot e^{-\frac{t}{\tau}} \cdot \mathrm{e}^{\left(-\frac{t^2}{2\sigma_t^2}\right)} \cdot \mathrm{e}^{\left(-\frac{t^2}{2\sigma_r^2}\right)} \tag{7.28}$$

The output of the expression:

$$\mathrm{e}^{\left(-\frac{t^2}{2\sigma_t^2}\right)} \cdot \mathrm{e}^{\left(-\frac{t^2}{2\sigma_r^2}\right)} \tag{7.29}$$

can be simplified using properties of exponential functions. Specifically, the product of two exponentials with the same base can be combined by adding their exponents. Since the base of the exponential functions is the same (e), we add the exponents:

$$\mathrm{e}^{\left(-\frac{t^2}{2\sigma_t^2}\right)} \cdot \mathrm{e}^{\left(-\frac{t^2}{2\sigma_r^2}\right)} = e^{\left(-\frac{t^2}{2\sigma_t^2} - \frac{t^2}{2\sigma_r^2}\right)} \tag{7.30}$$

The terms inside the exponent can be combined into a single fraction:

$$-\frac{t^2}{2\sigma_t^2} - \frac{t^2}{2\sigma_r^2} = -\frac{t^2(\sigma_r^2 + \sigma_t^2)}{2\sigma_t^2\sigma_r^2} \tag{7.31}$$

This result represents the product of two Gaussian functions with different variances, leading to a Gaussian function with a combined variance. Therefore, equation 7.28 can be modified based on equation 7.31 as follows:

$$W_{\text{river}}(t) = P_t \cdot G_t \cdot G_r \cdot \frac{\sigma_{\text{river}}}{(4\pi R^2)} \cdot e^{-\frac{t}{\tau}} \mathrm{e}^{\left(-\frac{t^2(\sigma_r^2+\sigma_t^2)}{2\sigma_t^2\sigma_r^2}\right)} \tag{7.32}$$

Therefore, the final output of multiplying the two exponential expressions in equation 7.32 is:

$$W_{\text{river}}(t) = P_t \cdot G_t \cdot G_r \cdot \frac{\sigma_{\text{river}}}{(4\pi R^2)} \cdot \mathrm{e}^{\left(-\frac{t^2(\sigma_r^2+\sigma_t^2)}{2\sigma_t^2\sigma_r^2} - \frac{t}{\tau}\right)} \tag{7.33}$$

However, radar cross-section for a river σ_{river}, which is less than that of the ocean (equation 7.27) but still includes variability due to surface features such as waves, vegetation, and other topographic elements. In this perspective, rivers generally have a more localized and less diffuse waveform compared to oceans.

On the other hand, for a dam reservoir or similar artificial water bodies, the waveform profile can be approximated as:

$$W_{\text{reservoir}}(t) = P_t \cdot G_t \cdot G_r \cdot \frac{\sigma_{\text{reservoir}}}{(4\pi R^2)} \cdot \delta(t - \Delta t) \cdot e^{\left(-\frac{t^2}{2\sigma_t^2}\right)} \tag{7.34}$$

Equation 7.34 says $\sigma_{\text{reservoir}}$ is the radar cross-section of the reservoir, which is typically higher and more specular due to the smoother surface. Therefore, $\Delta(t - \Delta t)$ represents a Dirac delta function centered at the time delay Δt, indicating that the waveform is sharper and more distinct. This means reservoir surfaces tend to return a more peaked echo due to their relatively smooth and consistent surface compared to natural water bodies. In other words, the Dirac delta function $\delta(t - \Delta t)$ is a mathematical representation of an impulse at $t = \Delta t$. This means that $W_{\text{reservoir}}(t)$ is non-zero only at $t = \Delta t$ and zero elsewhere [22, 25, 28]. Thus:

$$W_{\text{reservoir}}(t) = \begin{cases} P_t \cdot G_t \cdot G_r \cdot \dfrac{\sigma_{\text{reservoir}}}{4\pi R^2} \cdot e^{\left(-\frac{\Delta t^2}{2\sigma_t^2}\right)} & \text{if } t = \Delta t \\ 0 & \text{otherwise} \end{cases} \tag{7.35}$$

Additionally, the Gaussian exponential term $e^{\left(-\frac{t^2}{2\sigma_t^2}\right)}$ describes the pulse broadening due to the radar's pulse width. Since $\delta(t - \Delta t)$ causes $W_{\text{reservoir}}(t)$ to be zero for all t except at $t = \Delta t$, the solution is essentially the amplitude of the radar echo at that specific time. The solution to the equation $W_{\text{reservoir}}(t)$ describes the radar echo as a sharply peaked signal at $t = \Delta t$. For all other times $t \neq \Delta t$, the waveform $W_{\text{reservoir}}(t)$ is zero due to the Dirac delta function. Thus, equation 7.35 provides a detailed description of how the radar signal interacts with a reservoir surface and how the echo is recorded by the radar system.

7.11 WATER HEIGHT ESTIMATION IN RIVER AND DAM RESERVOIRS USING DELAY TIME

For any surface, the relationship between the time delay Δt and the height h of the surface features is given by:

$$\Delta t = \frac{2h}{c} \tag{7.36}$$

Rearranging this equation to solve for the height h:

$$h = \frac{c\Delta t}{2} \tag{7.37}$$

where Δt is the time delay of the echo; h is the height of the surface; and c is the speed of light in the medium (approximately 3×10^8 ms^{-1} in a vacuum). For height variations, the rise time Δt_r of the leading edge of the echo, which is related to the Root Mean Square (RMS) height h_{RMS}, is given by:

$$\Delta t_r \propto \frac{h_{\text{RMS}}}{c} \tag{7.38}$$

here, Δt_r is proportional to the RMS height h_{RMS}. To establish a relationship, we use a proportionality constant k:

$$\Delta t_r = k\frac{h_{\text{RMS}}}{c} \tag{7.39}$$

Rearranging to solve for h_{RMS}:

$$h_{\text{RMS}} = \frac{c\Delta t_r}{k} \tag{7.40}$$

The exponential term $e^{\left(-\frac{t^2}{2\sigma_t^2}\right)}$ in equation 7.35 represents the broadening of the radar pulse due to its width. The standard deviation σ_t of the Gaussian function characterizes the pulse width. The width of the echo can be related to the RMS height if the surface roughness significantly affects the shape of the returned signal. The standard deviation σ_t determines the width of the Gaussian pulse [26–29]. The width of the pulse, often expressed in terms of full width at half maximum (FWHM), can be related to σ_t as follows:

$$\text{FWHM} = 2\sqrt{2\ln(2)} \cdot \sigma_t \tag{7.41}$$

Rearranging the FWHM equation to solve for σ_t:

$$\sigma_t = \frac{\text{FWHM}}{2\sqrt{2\ln(2)}} \tag{7.42}$$

If the rise time Δt_r is related to σ_t by the pulse shape, and for a Gaussian pulse, Δt_r can be approximated by:

$$\Delta t_r = 2\sqrt{2\ln(2)} \cdot \sigma_t \tag{7.43}$$

Substituting σ_t from the FWHM expression into the RMS height formula:

$$h_{\text{RMS}} = c \cdot \left[2\sqrt{2\ln(2)} \cdot \sigma_t\right]\left[k\right]^{-1} \tag{7.44}$$

Equation 7.44 provides a straightforward method to estimate the RMS height from the radar measurements. By measuring the FWHM of the radar pulse, which is affected by the roughness of the surface, you can calculate the RMS height of the surface. The surface roughness causes the radar pulse to spread out, increasing the FWHM. By understanding this relationship and measuring the FWHM, infer the roughness of the surface can be inferred and its RMS height determined. However, this method is practical for remote sensing applications, where direct measurements of surface roughness are challenging. Instead, radar systems can use the characteristics of the returned pulse to estimate surface properties [27–29].

7.12 RANGE AMBIGUITY

Altimeter range ambiguity refers to a situation in radar altimetry where the radar system cannot distinguish between multiple possible distances (ranges) to a target. This ambiguity occurs because the radar pulse repetition frequency (PRF) (Figure 7.13) is such that the radar may receive an echo from a pulse after transmitting a subsequent pulse. As a result, the radar cannot definitively determine which pulse a received echo corresponds to, leading to multiple possible range solutions.

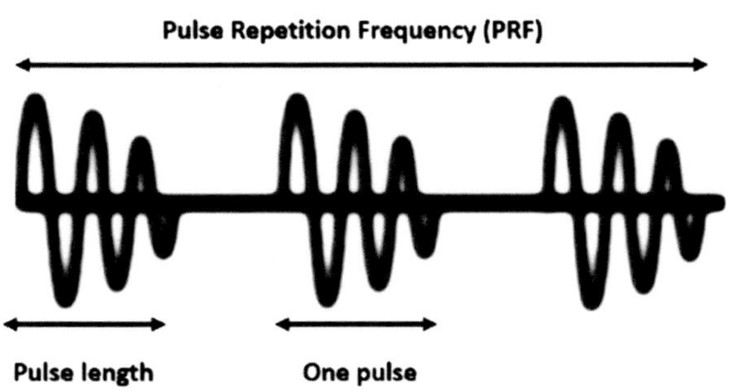

Figure 7.13 Radar pulse repetition frequency (PRF).

The question now is: How does range ambiguity arise? A radar altimeter works by transmitting a pulse of radio waves and measuring the time it takes for the pulse to travel to the target (e.g., the Earth's surface or a water body) and back. The range to the target is calculated from this time delay. In this perspective, PRF is the rate at which successive pulses are transmitted by the radar. If the PRF is too high, the time between successive pulses is short. If the target is far enough away, the echo from one pulse might return after the next pulse has already been transmitted [22–25].

In other words, because the radar does not know which pulse an echo corresponds to, there is ambiguity in determining the true distance to the target. In this sense, the radar might measure a time delay that could correspond to either the current pulse or a previous pulse, leading to different possible calculated ranges. Put differently, suppose the radar transmits a pulse, and after some time, it receives an echo. If the PRF is high, by the time the echo is received, the radar may have already sent out another pulse. In this sense, the radar detects the echo and calculates the range based on the assumption that the echo belongs to the latest pulse. However, the echo might be from an earlier pulse. Consequently, this uncertainty leads to multiple possible ranges: the correct range, plus any number of additional ranges corresponding to echoes from earlier pulses [22, 28].

Due to range ambiguities, the actual range R_{true} can be expressed as multiple ambiguous solutions:

$$R_{\text{true}} = R_{\text{means}} + n \times R_{\text{unamb}} \tag{7.45}$$

where n is an integer (0, 1, 2, ...), representing the ambiguity order. Therefore, the ambiguity function $\Phi(n)$ is used to describe the potential ambiguity in range, considering the multiple possible values of n:

$$\Phi(n) = R_{\text{meas}} + n \times \frac{c}{2 \times \text{PRF}} \tag{7.46}$$

Mathematically, if T_{pulse} is the time interval between consecutive pulses, then PRF is given by:

$$\text{PRF} = \frac{1}{T_{\text{pulse}}} \tag{7.47}$$

The unambiguous range R_{unamb} is the maximum distance that a pulse can travel and return to the radar before the next pulse is transmitted. It is given by:

$$R_{\text{unamb}} = \frac{c}{2 \times \text{PRF}} \tag{7.48}$$

Consequently, the measured range R_{meas} is the distance calculated based on the time delay Δt between the transmission of the pulse and the reception of the echo:

$$R_{\text{meas}} = \frac{c \times \Delta t}{2} \tag{7.49}$$

where Δt is the round-trip time delay. However, range ambiguities can be resolved by using additional information, such as:

Multi-frequency techniques: Using different Pulse Repetition Frequencies (PRFs), we can cross-check and eliminate ambiguous solutions. For two different PRFs, PRF1 and PRF2, the unambiguous ranges $R_{\text{unamb},1}$ and $R_{\text{unamb},2}$ are:

$$R_{\text{unamb},1} = \frac{c}{2 \times \text{PRF}_1}, \quad R_{\text{unamb},2} = \frac{c}{2 \times \text{PRF}_2} \tag{7.50}$$

The measured ranges $R_{\text{meas},1}$ and $R_{\text{meas},2}$ can be used to solve for the true range R_{true} by considering the intersection of the possible range solutions:

$$R_{\text{true}} = R_{\text{meas},1} + n_1 \times R_{\text{unamb},1} = R_{\text{meas},2} + n_2 \times R_{\text{unamb},2} \tag{7.51}$$

where n_1 and n_2 are integers.

Phase information: Utilizing phase coherence of the returned signal. Phase coherence can be used to determine the correct range by analyzing the phase ϕ of the returned signal. The phase difference $\Delta\phi$ between the transmitted and received signals is related to the range by:

$$\Delta\phi = \frac{4\pi R_{\text{true}}}{\lambda} \tag{7.52}$$

where λ is the wavelength of the radar signal. By comparing phase differences across pulses or frequencies, one can constrain R_{true}.

Prior knowledge: Using known terrain or target information to constrain possible ranges. In this sense, if prior knowledge of the terrain or target is available, the possible ranges can be constrained to specific intervals [11, 22, 24, 30]. Let R_{known} be the known approximate range, then the true range R_{true} can be constrained by:

$$R_{\text{true}} \in [R_{\text{known}} - \Delta R, R_{\text{known}} + \Delta R] \tag{7.53}$$

where ΔR is a small uncertainty based on the known information. The combination of these techniques can be represented as follows. For a given measured range R_{meas} with possible ambiguities:

$$R_{\text{true}} = \min_{n_1, n_2}\left(R_{\text{meas},1} + n_1 \times \frac{c}{2\times \text{PRF}_1}, R_{\text{meas},2} + n_2 \times \frac{c}{2\times \text{PRF}_2} \right) \tag{7.54}$$

Subject to:

$$\Delta\phi = \frac{4\pi R_{\text{true}}}{\lambda}, \quad R_{\text{true}} \in [R_{\text{known}} - \Delta R, R_{\text{known}} + \Delta R] \tag{7.55}$$

This expression exploits multi-frequency forms, phase information, and prior knowledge to resolve the range ambiguities and determine the correct R_{true}.

Figure 7.14 illustrates range ambiguities in radar altimetry. The plot shows the true range and the ambiguous ranges that could be measured using two different Pulse Repetition Frequencies (PRF1 and PRF2). Each point on the graph represents a potential measured range due to the ambiguity order n.

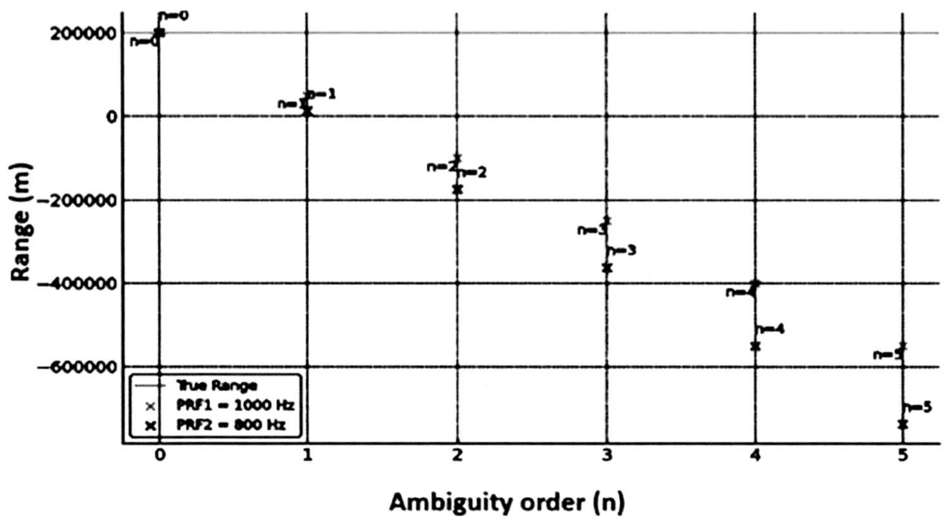

Figure 7.14 Range ambiguities in radar altimetry.

The solid grey line represents the true range, which is the actual distance to the target. The dark and grey dots represent the ambiguous ranges for two different PRFs. These ambiguities arise because the radar cannot distinguish between different potential distances when the PRF is high enough that echoes from previous pulses overlap with subsequent ones. By using different PRFs or combining other techniques, such as phase coherence, the correct range (true range) can be determined despite these ambiguities [28–31].

7.12 CONCLUSION

This chapter has meticulously laid out the mathematical framework underpinning the radar altimeter operating system, specifically designed for imaging variations in water height across diverse aquatic environments, such as oceans, rivers, and dam reservoirs. The theoretical constructs and principles expounded here form the foundation for precise measurement and analysis, essential for understanding dynamic water surfaces.

In the subsequent chapter, we will delve into the practical application of this theoretical work, providing a detailed exploration of advanced methodologies for investigating water height. We will particularly focus on the Grand Ethiopian Renaissance Dam (GERD), where we will employ sophisticated radar interferometry altimetry techniques. These advanced methods will ensure the extraction of highly accurate and consequential data regarding water height variations, thereby enabling informed decision-making and effective management of this critical resource. The upcoming discussion will highlight how the integration of radar interferometry with altimetry enhances the precision and reliability of water height measurements, demonstrating the tangible benefits of the theoretical framework established in this chapter.

References

[1] Ahern, F.J. 1995. Fundamental concepts of imaging radar: Basic level; unpublished manual, Canada Centre for Remote Sensing, Ottawa, Ontario. 87.

[2] Chan, Y.K. and Koo, V.C. 2008. An introduction to synthetic aperture radar (SAR). Progress in Electromagnetics Research B. 2: 27–60.

[3] Bamler, R. 2000. Principles of synthetic aperture radar. Surveys in Geophysics. 21(2): 147–157.

[4] CCRS. 1993. Radar basics—introduction to synthetic aperture radar; unpublished manual, Canada Centre for Remote Sensing, Ottawa, Ontario. 75.

[5] Moreira, A. .1992. Real-time synthetic aperture radar (SAR) processing with a new subaperture approach. IEEE Transactions on Geoscience and Remote Sensing. 30(4): 714–722.

[6] Ramirez, A.B., Rivera, I.J. and Rodriguez, D. 2005. SAR image processing algorithms based on the ambiguity function. *In*: Circuits and Systems, 2005. 48th Midwest Symposium. 1430–1433. IEEE.

[7] Chelton, D.B., Ries, J.C., Haines, B.J., Fu, L.L. and Callahan, P.S. 2001. Satellite altimetry. *In*: International Geophysics. 69: 1–ii. Academic Press.

[8] Shum, C.K., Ries, J.C. and Tapley, B.D. 1995. The accuracy and applications of satellite altimetry. Geophysical Journal International. 121(2): 321–336.

[9] Vignudelli, S., Birol, F., Benveniste, J., Fu, L.L., Picot, N., Raynal, M., et al. 2019. Satellite altimetry measurements of sea level in the coastal zone. Surveys in Geophysics. 40: 1319–1349.

[10] Marsh, J.G. 1983. Satellite altimetry. Reviews of Geophysics. 21(3): 574–580.

[11] Cheney, R.E. 2001. Satellite altimetry. Encyclopedia of Ocean Sciences. 2504.

[12] Srinivasan, M. and Tsontos, V. 2023. Satellite altimetry for ocean and coastal applications: A review. Remote Sensing. 15(16): 3939.

[13] Ndehedehe, C. 2022. Remote Sensing Hydrology. In: Satellite Remote Sensing of Terrestrial Hydrology. 3–17. Cham: Springer International Publishing.

[14] Choi, J.H., Jang, J.H. and Roh, J.E. 2015. Design of an FMCW radar altimeter for wide-range and low measurement error. IEEE Transactions on Instrumentation and Measurement. 64(12): 3517–3525.

[15] Reshma, S., Midhunkrishna, P.R., Joy, S., Sreelal, S. and Vanidevi, M. 2021. Improved frequency estimation technique for FMCW radar altimeters. *In*: 2021 International Conference on Recent Trends on Electronics, Information, Communication and Technology (RTEICT). 185–189. IEEE.

[16] Kuang, L., Pan, Y.J., Shen, X.F. and Huang, J.Y. 2013. A new algorithm for long range FMCW radar altimeter. *In*: 2013 IEEE China Summit and International Conference on Signal and Information Processing. 621–625. IEEE.

[17] Pan, Y., Shen, X., Huang, J., Xu, Z. and He, C. 2013. A novel low probability intercept radar altimeter based on FMCW and frequency hopping hybrid modulation. *In*: 2013 International Conference on Communications, Circuits and Systems (ICCCAS). 2: 269–273. IEEE.

[18] Mantripp, D. 1996. Radar altimetry. 119–171. Institute of Physics Publishing, London, UK.

[19] Smith, W.H. and Scharroo, R. 2014. Waveform aliasing in satellite radar altimetry. IEEE Transactions on Geoscience and Remote Sensing, 53(4): 1671–1682.

[20] Walsh, E.J., Uliana, E.A. and Yaplee, B.S. 1978. Ocean wave heights measured by a high resolution pulse-limited radar altimeter. Boundary-Layer Meteorology. 13: 263–276.

[21] Escudier, P., Couhert, A., Mercier, F., Mallet, A., Thibaut, P., Tran, N., et al. 2017. Satellite radar altimetry: Principle, accuracy, and precision. *In*: Satellite Altimetry Over Oceans and Land Surfaces. 1–70. CRC Press.

[22] Woodhouse, I.H. 2017. Introduction to Microwave Remote Sensing. CRC press.

[23] Nikolova, N.K. 2017. Introduction to Microwave Imaging. Cambridge University Press.

[24] Kurokawa, K. (Ed.). 2014. An Introduction to the Theory of Microwave Circuits. Elsevier.

[25] Liu, K.T., Tseng, K.H., Shum, C.K., Liu, C.Y., Kuo, C.Y., Liu, G., et al. 2016. Assessment of the impact of reservoirs in the upper Mekong River using satellite radar altimetry and remote sensing imageries. Remote Sensing. 8(5): 367.

[26] Cretaux, J.F., Nielsen, K., Frappart, F., Papa, F., Calmant, S. and Benveniste, J. 2017. Hydrological applications of satellite altimetryrivers, lakes, man-made reservoirs, inundated areas. *In*: Satellite Altimetry Over Oceans and land Surfaces. 459–504. CRC Press.

[27] Chander, S., Ganguly, D., Dubey, A.K., Gupta, P.K., Singh, R.P. and Chauhan, P. 2014. Inland water bodies monitoring using satellite altimetry over Indian region. The International Archives of the Photogrammetry, Remote Sensing and Spatial Information Sciences. 40: 1035–1041.

[28] Birkett, C.M. and Beckley, B. 2010. Investigating the performance of the Jason-2/OSTM radar altimeter over lakes and reservoirs. Marine Geodesy. 33(S1): 204–238.

[29] Verma, K., Nair, A.S., Jayaluxmi, I., Karmakar, S. and Calmant, S. 2021. Satellite altimetry for Indian reservoirs. Water Science and Engineering. 14(4): 277–285.

[30] Ha, J.S., Jang, J.H., Lee, S.J., Park, Y.S. and Hong, S.Y. 2021. Angle Ambiguity Analysis of Antenna Baseline for Interferometric Radar Altimeter. The Journal of Korean Institute of Electromagnetic Engineering and Science. 32(7): 626–634.

[31] Lin, F.L. and Steiner, M. 2001. New techniques for radar coherent range ambiguity resolution. *In*: Proceedings of the 2001 IEEE Radar Conference (Cat. No. 01CH37200). 99–104. IEEE.

Chapter 8

Modeling Water Height Level Variations in Lake Millennium Using Advanced Interferometry Altimeter Satellite Data

8.0 INTRODUCTION

Optical data from satellites face limitations in temporal coverage due to cloud presence, especially during the wet season. In contrast, radar altimeter sensors are unaffected by clouds, as radar waves can penetrate through them, providing crucial observations during wet or flooding seasons when reservoir water levels change rapidly.

With the Surface Water and Ocean Topography (SWOT) mission, which aims to enhance altimetry, set to began begin data collection in late 2022, the reliance on Sentinel-derived surface water levels decreased. However, SWOT data have not been globally validated, necessitating a comparison of SWOT-derived water levels to validate the various approaches used in this study.

The motivation for this study is to compare the reservoir surface water extent and water level estimation processes using interferometric radar altimeter techniques. One limitation of Synthetic Aperture Radar (SAR) and optical sensors is their coverage area, which cannot span large dimensions like the Indian Ocean. Satellite microwave altimeters offer a broader coverage scale than SAR data. Additionally, SAR data cannot determine annual water height cycles in reservoirs. Interferometry, a well-recognized technique, typically operates with SAR data, but there is potential for its application with satellite altimeter data.

This chapter addresses the main question: how can satellite altimeters incorporate interferometry techniques? It presents advanced interferometry technology that can be applied to satellite altimeter data. The study focuses on selected areas such as Lake Millennium, El Roseses, and Lake Naser, located in the Blue Nile Basin to investigate annual water height levels.

8.1 MICROWAVE ALTIMETER MEASUREMENTS OF INLAND WATER BODIES

The previous chapter has demonstrated the way radar altimeter in measures object height as a function of delay time. The echoes or reflected waves received by an altimeter are influenced by

DOI: 10.1201/9781003327325-8

the conditions of the ground surface topography, leading to variations in echo power over time. When the leading pulse strikes a flat surface, the echo amplitude increases. However, if the surface is extremely rough, the reflected waves increase gradually. The estimation of object height for example ocean wave height relies on these reflected waves, as the slope of the curve representing echo amplitude over time is proportional to object height or wave height (Figure 8.1) [1–3].

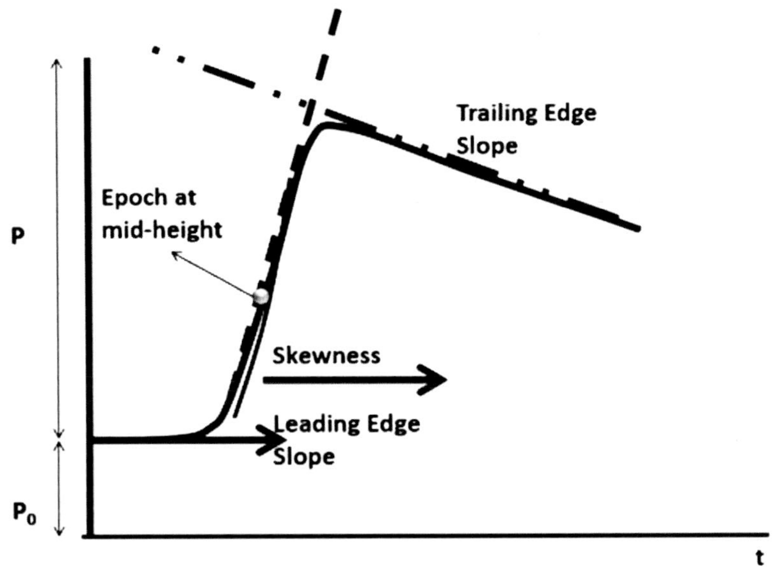

Figure 8.1 The shape of the reflected waves.

In other words, the epoch at mid-height provides the delay time of the reflected pulse, which is a function of range. Therefore, the amplitude of radar signal power reveals the backscatter coefficient and sigma0. Consequently, object height can be estimated from the slope of the leading edge. Subsequently, the rate of leading curvature can be acquired by analyzing the skewness of the waveform. Thus, mispointing of the radar antenna is associated with the trailing edge slope, indicating any deviation from the nadir pointing of the radar altimeter. This indicates that the leading edge slope must be above the thermal noise level [2].

The altimeter irradiates a circular beam over the ocean or land surfaces, with a diameter ranging from 3 to 5 km, depending on sea conditions, wave height or the ruggedness of the land. A rough surface or rugged land results in a wider footprint of approximately 10 km (Figure 8.2). In contrast, a calm sea surface or flat land provides a narrower footprint of about 2 km [1, 3].

Additionally Chapter 7 addressed that range ambiguities occur when the radar system incorrectly identifies the range of the target due to the presence of multiple reflections or echoes (Figure 8.3). This can happen when the radar pulse reflects off multiple surfaces or when the pulse travels a longer path than expected. In this view, what is the impact of range ambiguities on height estimation accuracy? If the radar system receives false echoes (Figure 8.4) from other surfaces (e.g., nearby hills, buildings or even the sea surface in the case of coastal regions), it may incorrectly calculate the round-trip time, leading to errors in height estimation. In urban or complex terrain environments the radar signal can reflect off multiple surfaces before returning to the altimeter. This multi-path interference can cause significant errors in the measured round-trip time, resulting in inaccurate height estimates. In some cases, the radar signal may be attenuated by atmospheric conditions or obstacles, leading to weaker echoes. This can make it difficult to accurately determine the round-trip time and affect height estimation [3].

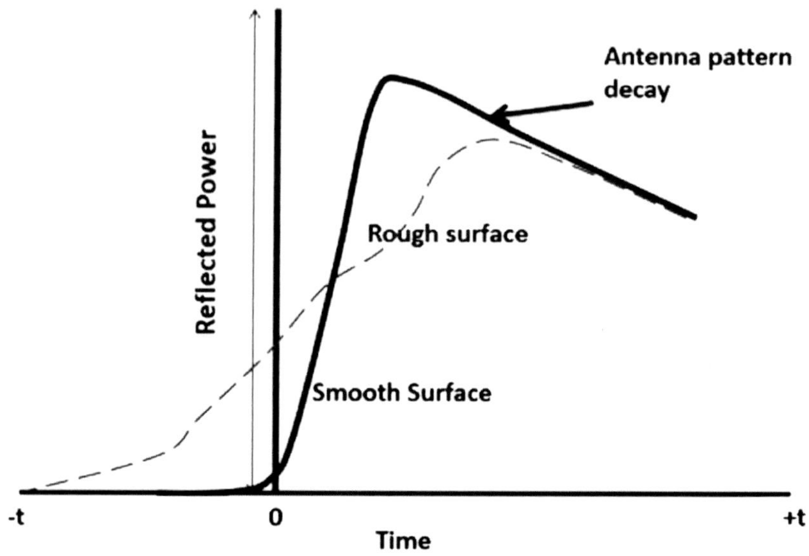

Figure 8.2 Reflected pulses from different surface conditions.

Figure 8.3 Multiple echo reflections.

Additionally, range ambiguities can significantly degrade the signal-to-noise ratio (SNR) in radar systems. This degradation occurs because range ambiguities introduce additional, unwanted signals that interfere with the true target echo. As a result, the radar system finds it more challenging to distinguish the actual target signal from the background noise and other spurious signals.

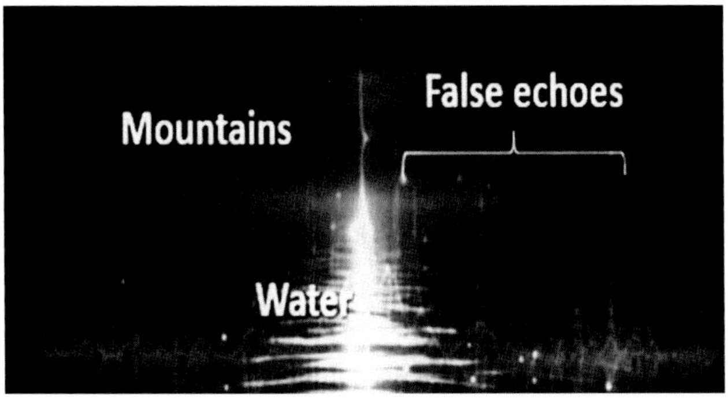

Figure 8.4 False echo reflections.

8.2 TIME DELAY VERSUS WATER HEIGHT AND SNR

Let us assume a linear relationship between delay time (τ) and water height (h) for simplicity. The relationship might be of the form:

$$h = a \cdot \tau + b \tag{8.1}$$

where a and b are constants. Figure 8.5 presents a comprehensive simulation of water height variations along Lake Millennium from 2019 to 2021, employing data acquired from the OSTM/Jason-2 GDR 20Hz and Jason-3 Interim GDR 20Hz multi-radar altimeter sensors. These sophisticated instruments reveal a consistent water height of 10 meters, accompanied by a maximum time delay of 1.4 seconds. The analysis yields a Root Mean Square Error (RMSE) of ±3.5 meters, reflecting the precision of the measurements.

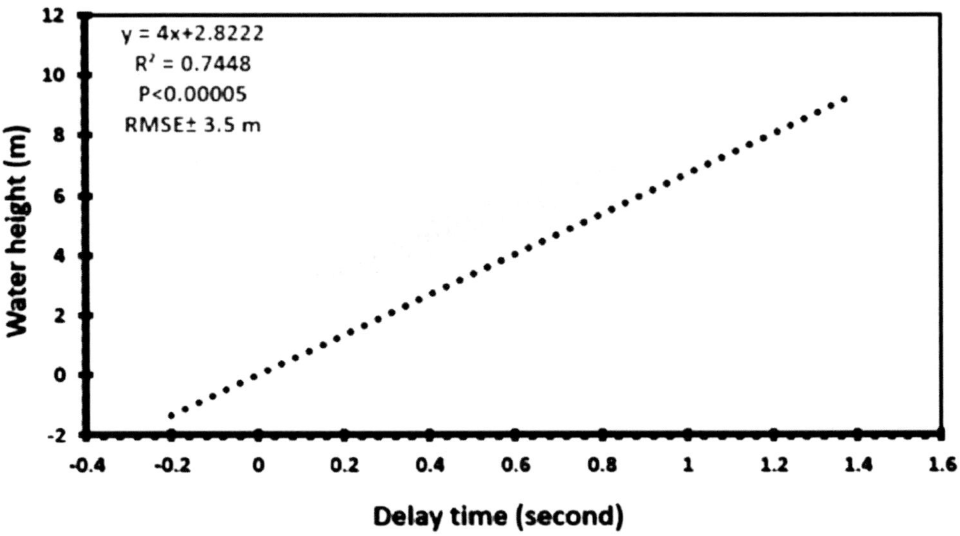

Figure 8.5 Water height simulated from time delay of radar altimeter sensor.

The R-squared value of 0.593 suggests a moderate level of linear correlation between time delay and SNR. This means that about 59.3% of the variance in the SNR can be explained by the time delay. While this is significant, it also indicates that other factors may influence the SNR that are not captured by the time delay alone. This relationship is statistically significant, but the moderate R-squared value and the RMSE of ±23 dB suggest that while time delay is an important factor influencing SNR, it is likely that there are other variables at play that also affect the SNR [1, 4].

Remarkably, this observation coincides with the peak signal-to-noise ratio (SNR) of 40 dB (Figure 8.6), a value significantly influenced by time delay due to range ambiguities. This high SNR underscores the reliability of the radar altimetry data under these conditions. This figure illustrates the relationship between the Signal-to-noise Ratio (SNR) in decibels (dB) and the time delay in seconds. The data points in the graph demonstrate a positive linear correlation between these two variables, which is represented by the dashed trend line [1, 5].

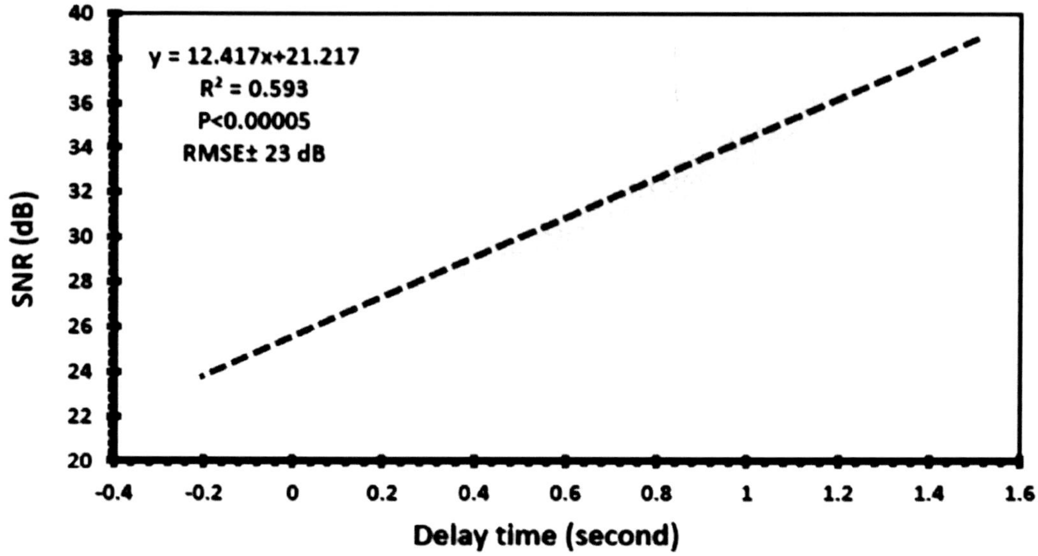

Figure 8.6 The significant relationship between SNR and time delay.

Furthermore, the data exhibits a robust linear relationship between water height and time delay, as evidenced by an r-square value of 0.744, with a statistical significance of ($p < 0.00005$). This strong correlation highlights the efficacy of the radar altimeter sensors in capturing and predicting water height dynamics over the given period, providing a powerful tool for monitoring and managing aquatic resources. The precision and consistency demonstrated by these sensors reinforce their critical role in advancing our understanding of hydrological processes [3, 5].

Consistent with the above perspective, the Signal-to-noise ratio (SNR) significantly impacts the accuracy and reliability of water height measurements in altimeter sensors, especially in simulated environments. When the SNR is high, the signal received by the altimeter sensor is strong, relative to the background noise. This leads to more precise and accurate measurements of the time delay, which directly influences the calculated water height. In high-SNR conditions the altimeter can distinguish between the true signal and noise, leading to minimal errors in water height estimation [4].

Conversely, when the SNR is low, the signal is weak compared to the noise. This condition makes it challenging for the sensor to accurately detect the true signal, resulting in greater uncertainty and potential errors in the measurement of water height. Noise can obscure the signal, leading to incorrect estimations of the time delay, and consequently, the water height.

The RMSE is a measure of the discrepancies between the observed and simulated water heights. A high SNR typically corresponds to a lower RMSE. This means that the measurements of the sensor are closer to the true values. As SNR decreases, the RMSE tends to increase, indicating that the readings of the sensor are less reliable and more prone to error. As seen in Figure 8.6, there is a positive correlation between SNR and time delay. Since time delay is a key factor in determining water height, any changes in SNR can directly influence the estimated water height. Therefore, a high SNR ensures that the time delay is measured with high confidence, resulting in more reliable water height estimates.

The empirical data on water height, corresponding to various signal-to-noise ratio (SNR) values derived from delay times, reveals a precise relationship: for each 1 dB increase in SNR, the water height is observed to rise by approximately 0.40 meters. This relationship is further quantified by an estimated Root Mean Square Error (RMSE) of ±0.23 meters (Figure 8.7), underscoring the model's robust fit and its ability to predict water height with notable accuracy based on SNR [1, 3, 5].

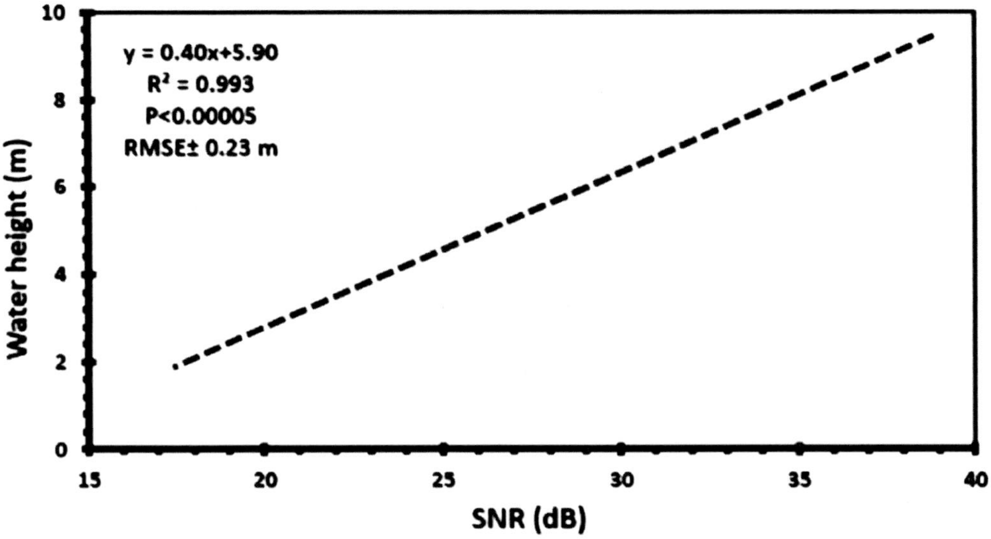

Figure 8.7 Impacts of SNR on water height simulation in radar altimetry data.

Moreover, the analysis yields an R-squared value of approximately 0.993, a figure that signifies an exceptionally strong linear correlation between SNR and water height. This high R-squared value indicates that around 99.3% of the variability in water height can be explained by changes in SNR, affirming the precision and dependability of the linear model employed.

The findings from this analysis provide compelling evidence of a powerful linear relationship between SNR and water height. As SNR increases, so too does the accuracy and reliability of the water height measurements obtained from the altimeter sensor. The close clustering of data points around the regression line, coupled with the low RMSE, further corroborates that SNR is a crucial predictor of water height, with minimal associated error. This high level of predictive accuracy not only highlights the efficacy of the model but also emphasizes the critical role of SNR in enhancing the precision of altimetric measurements in hydrological studies [2, 4].

Needless to say, maintaining a high SNR is crucial for the altimeter's performance, particularly in challenging environments like oceans, rivers, and reservoirs where surface roughness and other factors can affect the signal return.

8.3 PRECISION IN HEIGHT RETRIEVAL

The concept of range ambiguity vividly illustrates one of the fundamental challenges in the field of altimetry: the Earth's surface is inherently uneven. This unevenness introduces two primary issues.

First, the varied topography extends the duration of the return echo. This extension increases the likelihood that the signal might be detected in an incorrect reception window. Essentially, the signal could be misinterpreted due to the extended time it takes to return from different elevations. Second, as the altimeter moves along its flight path, significant changes in the average topography over successive pulses can alter the timing of the echo within the reception window. This shift might cause the echo to overlap with another reception window. If these topographic changes are predictable or occur slowly enough to be identified as a trend, the Pulse Repetition Frequency (PRF) can be adjusted to mitigate this issue. These technical challenges are crucial for accurately detecting the correct echo. However, from an application standpoint, a more pressing concern is the precision with which the true ground height can be estimated [3].

Therefore, there are four major sources of uncertainty in height measurements. First, the quality of the instrumentation is crucial, as the precision and reliability of the altimeter significantly influence the accuracy of the measurements. Second, the location of the platform plays a vital role; the exact position of the satellite or aircraft carrying the altimeter affects the measurement accuracy. Third, the estimate of the speed of travel of the pulse can introduce errors, as variations in the speed at which the radar pulse travels, influenced by atmospheric conditions can affect the results. Fourth and last, variations in surface topography add complexity to the measurement process due to the natural variability of the Earth's surface [5].

A key requirement for accurate altimetry is knowing the precise location of the instrument. Without accurate knowledge of the instantaneous location or flight path, the measurements from an altimeter are merely arbitrary range values that include information on both surface topography and instrument height variations. Inertial navigation and real-time differential GPS make it possible to locate aircraft with high accuracy.

Satellites have the advantage of maintaining fairly steady and predictable flight paths. However, platforms with altimeters usually carry a laser retro-reflector (essentially an optical trihedral corner reflector) so that they can be tracked by ground-based laser range-finding instruments. This allows the satellite position to be known relative to a terrestrial reference system, such as WGS84. Consequently, the range measurements can be translated into a height in a geographical coordinate system [1, 6].

All altimeters require their pulses to travel through the atmosphere, and for satellite systems, additionally through the ionosphere. Both of these media impose a delay on the velocity of the electromagnetic (EM) wave, proportional to their refractive index, as described by Snell's law. The delay imposed on the velocity of the electromagnetic (EM) wave as it travels through the atmosphere and ionosphere can be expressed using the refractive index (n). Consistent with Snell's law, the refractive index (n) is given by:

$$n = c[v]^{-1} \tag{8.2}$$

where c is the speed of light in a vacuum, and v is the speed of light in the medium (atmosphere or ionosphere). The time delay (Δt) introduced by the medium can be calculated as:

$$\Delta t = \frac{d}{v} - \frac{d}{c} \tag{8.3}$$

Equation 8.3 shows that $\frac{d}{v}$ is the time taken for the wave to travel through the medium, while $\frac{d}{c}$ is the time taken for the wave to travel the same distance (d) in a vacuum. Therefore, by substituting equation 8.2 into the delay equation, we obtain:

$$\Delta t = \frac{d}{c}(n-1) \tag{8.4}$$

As said by equation 8.4, the delay (Δt) is proportional to the refractive index (n) of the medium. This equation shows how the refractive index of the atmosphere or the ionosphere affects the delay in the velocity of the electromagnetic wave [3, 6].

The atmosphere's refractive index varies with temperature, pressure and water vapor content. Consequently, the wave velocity assumed in standard calculations is incorrect, leading to a slight overestimation of the range. The difference between the speed of light in a vacuum and the atmosphere is minimal, but this error becomes significant when high accuracy is required. For instance, a dry atmosphere can introduce a range error of a few meters if not corrected using an atmospheric model. A 'wet' atmosphere, containing water vapor, is much more variable both spatially and temporally compared to temperature or pressure. This variability can contribute to an additional error of up to 50 centimeters [1, 6].

To accurately account for the delay in the propagation of radar pulses through the atmosphere, a dry tropospheric correction (DTC) must be applied to the range measurements [7]. This correction considers air density and is calculated using atmospheric pressure through the following standard relationship:

$$\Delta R = R_m\left(\text{dry}\right) + 2.277\left(\frac{P_{\text{atm}}}{h_{\text{tm}}}\right)\left(1 + 0.0026\cos^2\phi\right) \tag{8.5}$$

where (P_{atm}) is the atmospheric pressure at the level of the lake, and ϕ is the latitude of the point (to account for the Earth's curvature). Since it is practically impossible to measure atmospheric pressure everywhere using local instrumentation, the DTC is derived from atmospheric reanalysis data provided by organizations such as the European Center for Medium-Range Weather Forecasts (ECMWF) and the National Center for Environmental Prediction (NCEP). However, this approach can introduce errors over continents if the local altitude is not properly considered in the equation [6–8].

Therefore, the error in the range due to the ionosphere can be mathematically expressed as:

$$\Delta R_{\text{ion}}(f) = \frac{40.3 \times TEC}{f^2}\int_0^a n_e(z)dz \;\; [\text{m}] \tag{8.6}$$

where (TEC) is the total columnar electron content, and (f) is the frequency of the radar pulse. Additionally, a is the altitude of the satellite above the reference ellipsoid. The total columnar electron content (TEC) varies from about (10^{16}) to (10^{18}) electrons per square meter, meaning that the range error ΔR_{ion} is typically in the region of tens of centimeters for most microwave altimeters. This means, the ionospheric influence is highly sensitive to frequency. One effective way to minimize this effect is to use a dual-frequency system. By measuring the small discrepancy in the range between the two frequencies, the ionospheric influence can be estimated. Alternatively, using the highest possible frequency will also help to minimize this effect [6, 8].

Geometric factors, such as surface slope and surface roughness, significantly impact the accuracy of height measurements in altimetry. Variations in topography influence the accuracy because the nearest point, and thus the location of the first echo, may not be directly beneath the altimeter (nadir) [4, 6, 9].

Consider the simplest case of a regular slope with an angle (α), as illustrated in Figure 8.8. The nearest location is at the perpendicular to the surface, denoted as (R_n) in the diagram. This distance (R_n) is what the altimeter measures as the 'height'. The relationship between the measured height (R_n) and the actual height (H) is given by the cosine of the angle (α):

$$\cos\alpha = R_n\,[R]^{-1} \tag{8.7}$$

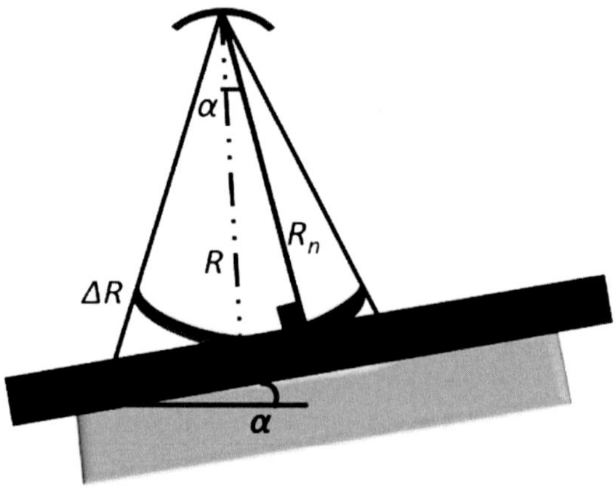

Figure 8.8 Errors induced by surface slope in altimetry measurements.

For small angles (α) (in radians), this can be approximated by expanding the secant term:

$$R = \frac{R_n}{\cos\alpha} = R_n \sec\alpha s \simeq R_n\left(1 + \frac{\alpha^2}{2} + \cdots\right). \tag{8.8}$$

This approximation allows us to express the expected error (ΔR) in the measured height due to the slope:

$$\Delta R = R - R_n \simeq \frac{R_n\alpha^2}{2} \text{ [m]}. \tag{8.9}$$

Thus, equation 8.9 demonstrates that the error in the height measurement increases with the square of the slope angle (α). Therefore, to minimize the impact of surface slope on height measurements, one effective approach is to use a beam-limited altimeter. This type of altimeter reduces the footprint size, thereby limiting the area over which the slope can affect the measurements.

For pulse-limited altimeters, there is some flexibility to account for the slope in the flight-path direction. This can be achieved by combining sequences of measurements to construct a transect of the surface height. To this end, the variations in topography along the flight path can be better understood and accounted for, leading to more accurate height measurements [7–9].

8.4 ESTIMATING SURFACE HEIGHT WITH RADIO ALTIMETER

According to the above-mentioned perspective, when a transmitted pulse reaches the surface, it initially illuminates a small point. As the pulse advances, the illuminated area rapidly expands from a point to a disk, increasing the returned power. The reflecting area is determined by the beamwidth of the antenna. The energy from the center of the main beam travels a shorter path compared to the energy from the edges, forming an annulus. The geometry ensures that the annulus area remains constant as its diameter increases. Consequently, the returned signal strength, which depends on the reflecting area, grows rapidly until the annulus is fully formed, remains constant, and then diminishes as the annulus reaches the edge of the radar beam. The simple mathematical formula to determine the surface height (H) [1, 4, 9] is given by:

$$\text{(Corrected) height} = \text{altitude} - \text{(Corrected) Variety} \tag{8.10}$$

When measuring the surface height of a reservoir, several critical factors must be considered to ensure accuracy. First, the water level can fluctuate due to seasonal variations, precipitation, and water usage. Continuous monitoring of these fluctuations is essential for accurate data. Second, the surface area of the reservoir changes with varying water levels, affecting volume calculations. This variability must be accounted for in the measurements.

Third, environmental factors such as wind speed and direction can create surface waves, introducing noise into the measurements. Filtering out these environmental influences is necessary to obtain precise height data. Additionally, the geoid—the hypothetical sea level influenced by gravity—plays a crucial role. Variations in the geoid, caused by changes in water masses and densities, can distort the measured height. For example, a denser rock zone at the reservoir's bottom can cause noticeable deviations in surface height.

Fourth and last, dynamic topography, influenced by the Earth's rotation and other geophysical processes must be considered. This dynamic topography can introduce variations of about 1 meter. By removing the geoid from the measured surface height, the dynamic topography can be computed, providing a clearer picture of the actual water surface height [10–12].

8.5 NUMERICAL EXAMPLE OF ESTIMATING WATER LEVEL ANOMALIES

In practice, the mean water level is subtracted to yield the variable component (water level anomalies) of the reservoir's surface height. This approach helps in accurately monitoring and managing water resources in reservoirs. Suppose we have the following measurements of the reservoir's surface height over some time (Table 8.1):

Table 8.1 Surface height variations over five days

Time (days)	Measured Surface Height (m)
1	100.5
2	100.7
3	100.4
4	100.6
5	100.8

First, we calculate the mean water level:

$$\text{Mean Water Level} = \frac{100.5 + 100.7 + 100.4 + 100.6 + 100.8}{5} = \frac{503.0}{5} = 100.6 \text{ m} \tag{8.11}$$

Next, we subtract this mean water level from each measured surface height to find the water level anomalies (Table 8.2).

Table 8.2 Scheming of water level anomalies

Time (days)	Measured Surface Height (m)	Mean Water Level (m)	Water Level Anomaly (m)
1	100.5	100.6	100.5 – 100.6 = –0.1
2	100.7	100.6	100.7 – 100.6 = –0.1
3	100.4	100.6	100.4 – 100.6 = –0.2
4	100.6	100.6	100.6 – 100.6 = –0.0
5	100.8	100.6	100.8 – 100.6 = –0.2

These anomalies represent the deviations from the mean water level, providing a vibrant scenario of the variations in the reservoir's surface height over time.

8.6 DELAY-DOPPLER RADAR IMPACT ON ALTIMETER RETURN SIGNALS

As previously discussed, understanding the Doppler signal fundamentally relies on two key aspects: time delay and ionospheric effects. Time delay is crucial for determining the distance from a target. By measuring the time it takes for a radar pulse to travel to the target and return, we can accurately calculate the range. This principle is essential in applications such as altimetry, where precise height measurements are critical. Therefore, the ionosphere can introduce delays and distortions in the radar signal due to its varying electron content. These effects must be accounted for to ensure accurate measurements. Using dual-frequency systems helps mitigate these ionospheric influences by allowing us to estimate and correct for the discrepancies caused by the ionosphere [10–12]. Needless to say, the pulse time fluctuation of the backscatter signal of the ocean or land surfaces is identified as (i) the period before the pulse arrives (ii) the period after the pulse arrives, and before the tail of the pulse has been received by the antenna, and after the tail, the pulse has been received by the antenna [1, 4, 9, 11].

To simplify the analysis of the radar footprint, we consider it as two rectangles, each with a width (W). This approach allows us to treat the illuminated area as a combination of these two rectangles, making the calculations more manageable. Mathematically, the total illuminated area ($A(t)$) at any given time (t) can be expressed as:

$$A(t) = 2 \times W \times L(t) \tag{8.12}$$

where W is the width of each rectangle, and $L(t)$ is the length of each rectangle, which varies with time as the radar pulse travels and reflects off the surface. In this sense, the returned power $P_r(t)$ is proportional to the illuminated area and the transmitted power (P_t). Assuming a constant reflectivity (σ), the returned power can be expressed as:

$$P_r(t) = P_t \times \sigma \times A(t) \tag{8.13}$$

Substituting the expression for ($A(t)$), we obtain

$$P_r(t) = P_t \times \sigma \times 2W \times L(t) \tag{8.14}$$

Equation 8.14 highlights how the radar footprint, modeled as two rectangles, influences the returned power signal. By simplifying the footprint in this manner, we can more effectively analyze the radar's performance and the accuracy of the measurements. The length ($L(t)$) of the rectangles changes with time as the radar pulse travels and reflects off the surface. This time dependency can be modeled based on the radar system's parameters and the surface characteristics [1, 4, 7, 11, 13].

By considering the footprint as two rectangles of width (W), we can simplify the calculation of the power signal as a function of time. This approach allows us to analyze the radar's performance and the accuracy of the measurements more effectively.

$$P(W) = \begin{cases} 0 & t < t_0 \\ 2W_r(t) & t_0 < t < t_0 + t_p \\ 2W_r[L(t) - L(t - t_p)] & t > t_0 + t_p \end{cases} \tag{8.15}$$

An altimeter functions by radiating a pulse towards the Earth's surface. The time interval between the transmission of this pulse and the reception of its backscatter from the Earth's surface is directly proportional to the satellite's altitude. This principle of radar is applied to altimetry to understand the behaviors and characteristics of the pulse as it interacts with surface irregularities.

The magnitude and shape of the echoes or waveforms contain valuable information about the surface characteristics that cause the backscatter. The most accurate results are obtained over the ocean, which is spatially homogeneous and conforms to known statistical models. In contrast, land surfaces which are heterogeneous and contain discontinuities or significant slopes, present greater challenges for precise analysis. Even under optimal conditions, such as over the ocean, the pulse duration must be extremely short—no longer than 70 picoseconds—to achieve an accuracy of a few centimeters. To achieve the necessary technical specifications, an emission power exceeding 200 kW is required, and the altimeter must switch every few nanoseconds. For example, during the radar altimeter's pass over Lake Millennium (Figure 8.9), the broader peak observed indicates that the radar signal has traversed a greater distance (Figure 8.10). As the pulse reaches the target, the signal strength increases rapidly forming an annular shape. This power remains relatively constant until the annulus reaches the edge of the radar beam, after which it begins to diminish. The farther the target, the broader the peak, due to the longer travel time and the larger illuminated area (Figure 8.10).

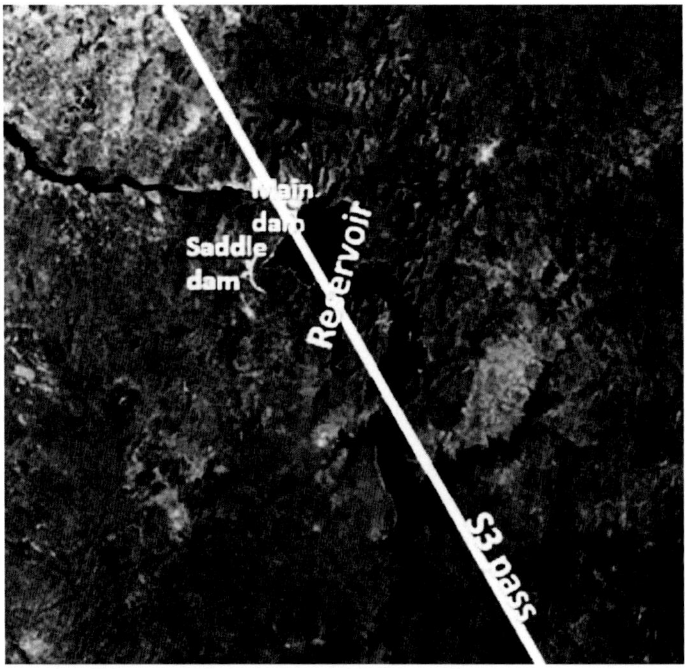

Figure 8.9 Altimeter Sentinel 3 overpass Lake Millennium.

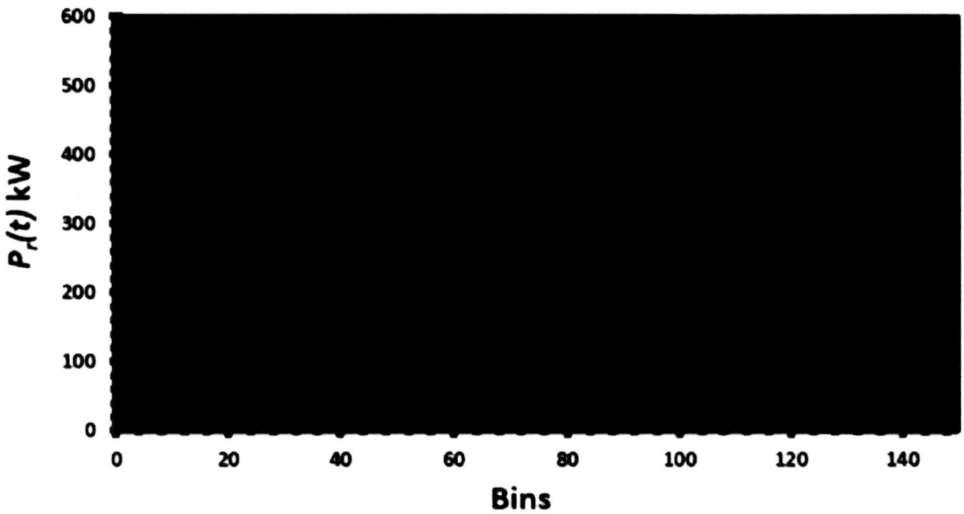

Figure 8.10 Power altimeter emission spectra.

In essence, the requirement for high-emission power and rapid switching is crucial for accurate altimeter readings. The example of Lake Millennium illustrates how the radar signal's travel distance affects the signal's characteristics. The formation of an annulus and the subsequent changes in signal strength highlight the intricate relationship between the radar beam's reach and the target's distance. The broader peak for distant targets underscores the impact of travel time and illuminated area on the radar signal's behavior [8–11].

These technical challenges are addressed by the full deramp technique, which allows the use of only 5 W for emission. The range resolution of the altimeter is approximately half a meter (3.125 ns), but the range measurement performance over the ocean is about an order of magnitude better than this. This enhanced performance is achieved by fitting the shape of the sampled echo waveform to a model function that represents the form of the echo.

Needless to say, this waveform demonstrates how the Doppler effect influences the radar signal based on the distance to the target. By analyzing the shape and magnitude of the echoes, we can infer the distance and velocity of the target, providing valuable information for applications such as altimetry and surface mapping [7, 10, 13].

8.7 ALTIMETER SENSORS

In alignment with the aforementioned perspective, altimeter sensors, especially those designed for measuring water height in lakes and reservoirs epitomize the zenith of technological innovation in remote sensing. These advanced instruments employ radar or laser pulses to precisely gauge the distance between the sensor and the water surface, thereby delivering accurate measurements of water height. Notable altimeter sensors used for measuring water height in lakes and reservoirs include radar altimeters, laser altimeters, and synthetic aperture radar (SAR) altimeters.

The radar altimeter sensors are exemplified by missions such as TOPEX/Poseidon, Jason-1, Jason-2, Jason-3, Sentinel-3, and CryoSat-2. TOPEX/Poseidon (Table 8.3) (Figure 8.11), a collaborative mission by NASA and CNES, was equipped with a radar altimeter that provided precise measurements of sea surface height. Building on this foundation, Jason-1, Jason-2, and Jason-3, as successors to TOPEX/Poseidon, continue to deliver critical data on sea surface height and ocean circulation [6, 13].

More recently, Sentinel-3, part of the European Space Agency's Copernicus program, carries a radar altimeter that measures sea surface height, significant wave height, and wind speed. Consequently, CryoSat-2, equipped with the Synthetic Aperture Interferometric Radar Altimeter (SIRAL), measures ice thickness and sea-ice freeboard and also collects data over large water bodies. Additionally, SARAL/AltiKa is a joint mission by ISRO and CNES, SARAL/AltiKa uses a Ka-band radar altimeter to provide high-resolution measurements of sea surface height and inland water bodies.

In this context, laser altimeters include ICESat, ICESat-2, and the Geoscience Laser Altimeter System (GLAS). Specifically, ICESat and ICESat-2, part of NASA's Ice, Cloud, and land Elevation Satellites utilize laser altimetry to measure ice sheet elevation, sea-ice freeboard and land topography. Additionally, they are capable of measuring water height in lakes and reservoirs. Similarly, GLAS, onboard the ICESat satellite, provides precise elevation data, including measurements of water height. This system enhances our understanding of various geophysical processes by delivering accurate and detailed elevation information [13].

In the realm of synthetic aperture radar (SAR) altimeters, notable examples include Sentinel-1 and TerraSAR-X. Sentinel-1 utilizes SAR technology to produce high-resolution images, which are particularly useful for interferometric applications such as monitoring water levels. In contrast, TerraSAR-X employs SAR to generate detailed images, making it highly effective for measuring water height. Additionally, the Envisat RA-2 (Radar Altimeter 2) has historically provided valuable data on sea surface height, wave height, and wind speed [9, 12, 16].

Collectively, these SAR altimeters significantly enhance our capacity to monitor and analyze various water bodies. Their contributions are pivotal in advancing environmental monitoring and water resource management, thereby supporting a deeper understanding of our planet's hydrological dynamics.

Table 8.3 List of altimeter satellite sensors

Sensors	Band	Radar frequency (GHz)
Geosat	Ku	13.5
ERS-1	Ku	13.8
ERS-2	Ku	13.8
TOPEX	Ku c	13.6 5.3
Poseidon	Ku	13.65
GFO	Ku	13.5
Jason-1	Ku c	13.6 5.3
ENVISAT	Ku s	13.6 3.2
Jason-2	Ku c	13.6 5.3
Jason-3	Ku c	13.6 5.3
SENTINEL-3	Ku c	13.575 5.41
Cryosat-2	Ku	13.575

8.8 PRINCIPLES OF SYNTHETIC APERTURE RADAR ALTIMETER INTERFEROMETRY

The Synthetic Aperture Interferometric Radar Altimeter (SIRAL-2) is an advanced instrument designed to overcome the challenges associated with measuring ice surfaces. Building on existing technology, SIRAL-2 incorporates several significant enhancements. It operates by emitting a radar pulse towards the ground and analyzing the echoes reflected from the Earth's surface. The spacecraft's position is determined using an onboard ranging instrument called DORIS (Doppler Orbitography and Radiopositioning Integrated by Satellite), which allows the signal return time to reveal the surface altitude. Maintaining correct antenna orientation is crucial for this process and is achieved using a trio of star trackers [12].

SIRAL, the primary instrument of the CryoSat-2 mission, was designed and developed for the European Space Agency (ESA) by Thales Alenia Space in France. It inherits technology from the Poseidon-2 altimeter flown on the Jason-1 mission. The primary objective of SIRAL is to observe ice sheet interiors, ice sheet margins, sea ice and other topographical features [11–15].

Since the launch of CryoSat-2 in 2010, new altimeter technologies leveraging Doppler and interferometric capabilities have emerged. These advancements are likely to become standard in future altimeters, as evidenced by the Sentinel-3 altimeter launched in 2016. The concept of the Delay-Doppler Altimeter (DDA) is also known as the SAR altimeter. Delay-Doppler altimeters feature a high pulse repetition frequency (PRF) to ensure pulse-to-pulse coherence, resulting in an along-track resolution of approximately 300 meters, an improved signal-to-noise ratio, and enhanced altimeter ranging performance. For example, the CryoSat-2 SIRAL utilizes the SAR mode over ocean zones [4, 12, 15].

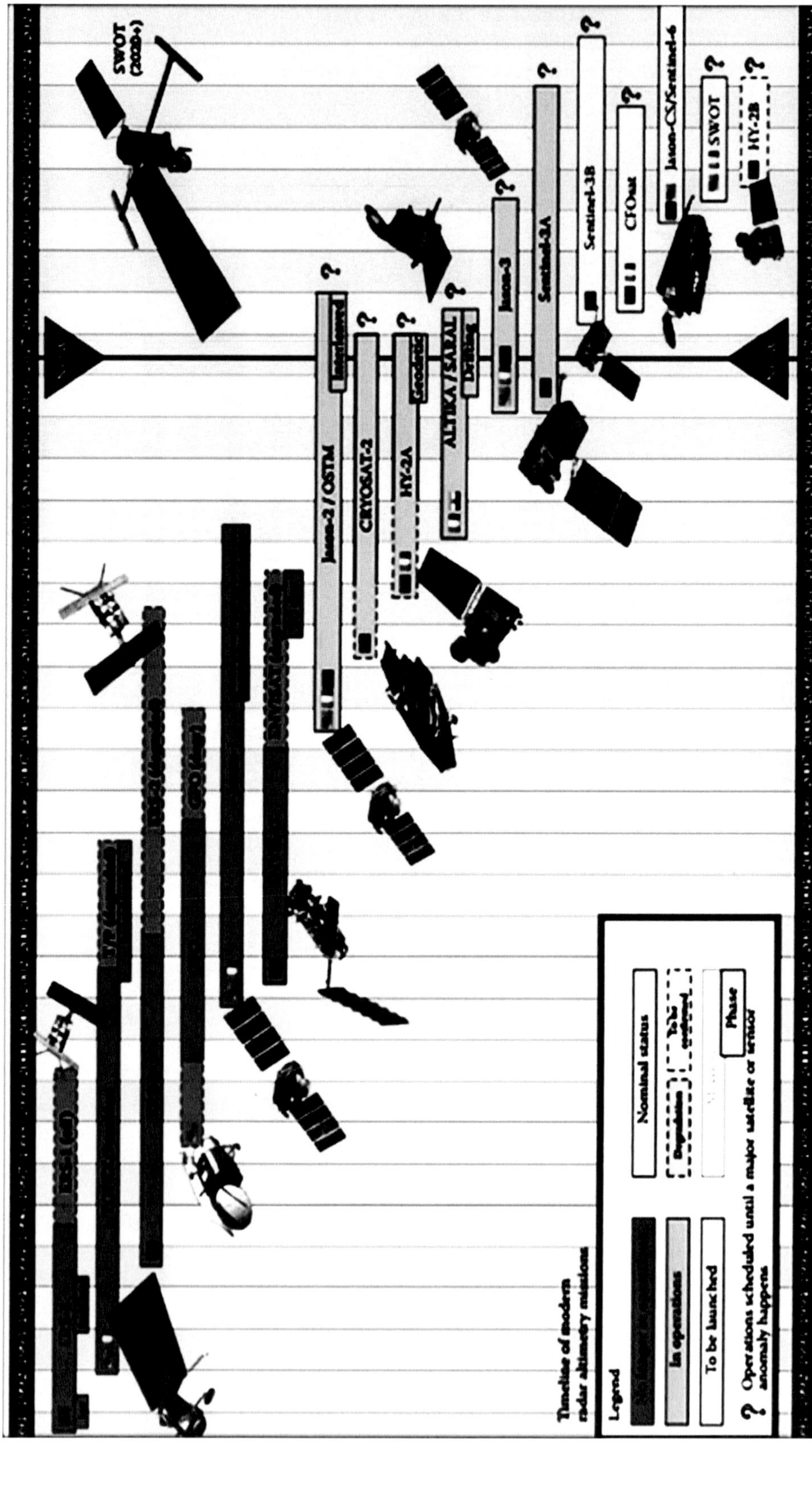

Figure 8.11 Different altimeter satellites.

CryoSat's most advanced mode, the SAR interferometric mode (SARin), is particularly used over ocean surface margins. In this mode, the altimeter performs synthetic aperture processing and uses a second antenna as an interferometer to detect the across-track angle to the earliest radar returns. The SARin mode provides precise surface location measurements on sloping surfaces and can be used to study more contrasted sea surface slopes. Over most rough ocean areas, SIRAL operates in the conventional Low Rate Mode (LRM), which is the standard pulse-limited radar altimeter mode. In this mode, the data rate is significantly lower than in other measurement modes.

In short, while SIRAL is primarily designed for ice and sea-ice measurements, its advanced radar capabilities make it a valuable tool for monitoring lakes and reservoirs. Its high precision and versatility enable effective water resource management and environmental monitoring [16–18].

8.9 ALTIMETER INTERFEROMETRY TECHNIQUE

The concepts of interferometric altimeters were first proposed by Jensen [13] and led to the improvement of the Cryosat mission. A detailed description of the concepts and processing of the Cryosat SARin information is given in Wingham et al. [15]. The primary (left) antenna transmits the radar signal and the two antennas measure the backscattered echo waveform (Figure 8.12). The primary complex waveform is multiplied by the complex conjugate of the second antenna waveform. The phase of the ensuing cross-channel waveform is then defined as the interferometric phase difference, which consequences from the slight range distinction of an off-nadir scatterer for the two antennas. The normalized modulus of the conjugate product offers an estimate of the signal coherence. The stacked SAR echoes for each antenna are computed using the SAR mode processing. The SAR echoes, phase, and coherence are furnished with ESA Level-1B products.

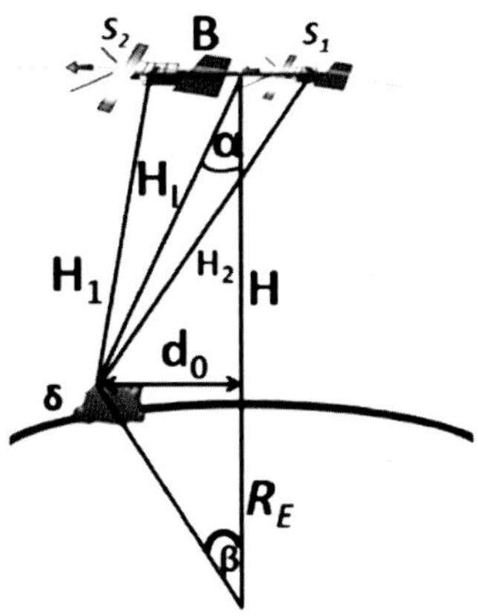

Figure 8.12 Geometry of altimeter interferometry.

In SARin mode, the waveform evaluation window is elevated to 512 bins (240 m) to better sample sloping terrains. In the Baseline-C data, the merchandise was used to assess the performance of the system under typical conditions and establish a reference for comparison with future data sets. The use of zero-padding prior to FFT processing, besides, increases the variety of range bins in 1024 except for altering the range window. Each bin corresponds consequently to 1.565 ns or 0.23 m along with the range.

Let us assume that the two SAR antennas S_1 and S_2 mounted on a satellite platform are separated by a baseline B with a baseline angle α, and observe the complex response at the point δ with a slant range H_1 (Figure 10.12). The mathematical relationships between the interferometric phase difference, $\Delta\psi$, and the off-nadir angle, α, can be given by:

$$\Delta\psi = \frac{2\pi B}{\lambda}\sin(\alpha) \tag{8.16}$$

where λ is the radar wavelength and B is an interferometer baseline (distance between the two antennas). Under the small-angle approximation, the off-nadir angle α is expressed by:

$$\alpha = \frac{\lambda\Delta\psi}{2\pi B} \tag{8.17}$$

Galin et al. [14] are estimated the across-track distance to nadir, d_0 by:

$$d_0 = \frac{ct_i}{2}\bar{\alpha} = H_i\bar{\alpha} \tag{8.18}$$

Equation 8.18 demonstrates that range H_i equals half of the multiplication of pulse time t_i by the speed of light c then multiply by angle scaling $\bar{\alpha}$. Consequently, the freeboard δ due to the earth curvature is defined as:

$$\delta = (H - H_i\cos\alpha_1 + R_E(1-\cos\beta))\cos\beta \tag{8.19}$$

where $\beta = H/R_E\alpha_1$ and $H_i = ct_i/2$. The SARin echoes are similar to the SAR ones, except that the number of range bins in the echoes thermal noise part (TNP) is significantly larger (125×2 vs 50). Altimeters over open waters have a detectable signature in the thermal noise part (TNP), which is accounted as above the sea surface of high resolution (HR) waveforms of pulse-limited altimeters that can be easily detected [12, 15].

The swath over which sea level can be detected which is of the order of 6 km is thus significantly increased to 12 km. The SAR detection algorithm can be applied to the SARin waveforms without modification. However, in the echoes, TNP the signals received by the two antennas are by nature random noise and thus incoherent [16–20].

8.10 INSAR PRECISION PROCEDURES ALTIMETER SCHEME

InSAR altimeter accurate analysis is a function of the accuracies of satellite altitude, slant range, baseline length, baseline angle and phase difference which are denoted as σ_H, σ_B, σ_{H_1}, σ_α, and σ_β respectively. These parameters are autonomous. These independent parameters can be expressed mathematically as:

$$\sigma_{R_E} = \sigma_H\left(\frac{\partial R_E}{\partial H}\right) + \sigma_B\left(\frac{\partial R_E}{\partial B}\right) + \sigma_{H_1}\left(\frac{\partial R_E}{\partial H_1}\right) + \sigma_\alpha\left(\frac{\partial R_E}{\partial \alpha}\right) + \sigma_\beta\left(\frac{\partial R_E}{\partial \beta}\right) \tag{8.20}$$

Equation 8.20 demonstrates that the height approximation accuracy is a function of a decreasing of the look angle α. In this circumstance, α must equal 0.5° to obtain accurate height estimation. On the contrary, decreasing α decreasing causes a decrease in the range resolution. Additionally, a low sea surface leads to quasi-specular scattering, where the surface behaves like a mirror, reflecting radar signals in a specific direction. In this regard, the echo-tracking model has differences in backscattering echoes from the sea surface.

The echo-tracking algorithm is implemented in the slant range enhancement post the phase and geometry computations to deliver a precise slant range estimation. More precisely, Figure 8.13 demonstrates that the location of the sea surface at the point δ_α is the radar look angle and r is a function of α. In other words, the echo waveform is a function of the plane surface impulse response $\delta_{FS}(\tau, r)$, system point target response $S_r(\tau, r)$, and height probability density function of the sea surface scattering elements $Q(\tau)$. The mathematical expression of the sea surface echoes waveform is based on the convolution [12, 16], which is given by:

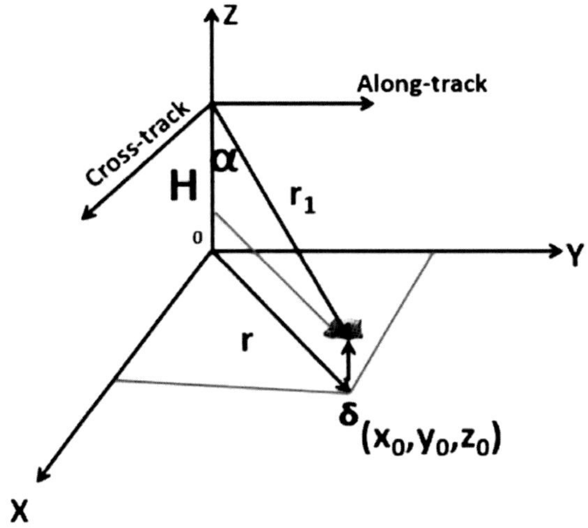

Figure 8.13 Altimeter SAR geometry.

$$\delta_r(\tau, r) = \delta_{FS}(\tau, r)xS_r(\tau, r)xQ(\tau) \tag{8.21}$$

$$\delta_r(\tau,r) = \int_{\text{Water Surface Area}} \frac{\lambda^2 \delta(\tau)\sigma_0(r)}{(4\pi)^3 L_p \|r_1\|^4} \|G(r)\|^2 \, dA \tag{8.22}$$

Equation 8.22 is considered as a presentation of the radar equation. In this context, $\delta(\tau)$ is the transferred waveform in the duration period τ with wavelength λ. At the position r over the changing area dA, the sea surface backscatter is σ_0 and L_p is the double loss of the atmosphere. Further, G is the antenna gain. Consequently, system point target response $S_r(\tau, r)$ is formulated as:

$$S_r(\tau,r) = C^2 \sin c^2\left(\frac{\tau_a}{2\pi}\left(2\pi f_a - 2\pi f_d\left(1 - \frac{k_r t_0}{f_0}\right)\right)\right) \sin c^2\left(\tau_c\left(\frac{k_r}{cH}\|r\|^2 - \tau\right)\right) \tag{8.23}$$

Equation 8.23 indicates a filter length of along-track and cross-track τ_a and τ_c, respectively. Consequently, f_a, f_d and f_0 are azimuth, Doppler and wave frequencies, respectively. Therefore, the linear frequency modulation rate k_r equals the ratio of baseline B and period T, i.e., $k_r = \frac{B}{T}$. Finally, C is a function of SAR along-track and cross-track filter length.

The height probability density function of the sea surface scattering elements $Q(\tau)$ is determined from:

$$Q(\tau) = \frac{1}{\sqrt{2\pi\sigma_s}}\left[1 + \frac{\lambda}{6}\left(\frac{\tau^3}{\sigma_s^3} - 3\frac{\tau}{\sigma_s}\right)\right] e^{\left(-\frac{\tau^2}{2\sigma_s^2}\right)} \tag{8.24}$$

Equation 8.24 is the keystone to estimate the water height H from the backscatter quantity σ_s, which indicates the root-mean-square wave height as follows:

$$\sigma_s = \frac{1}{2C} H_s \tag{8.25}$$

In Figure 8.14, we observe the delay time of the echo waveform model in Synthetic Aperture Radar altimeter observations. Specifically, at the coordinates (10000, 0, 1) with an altitude (H) of 800 km, the simulated water height in Lake Millennium is approximately 10 meters.

This configuration yields a maximum normalized epoch power of 1, with a minimum delay time of less than 0.008 seconds.

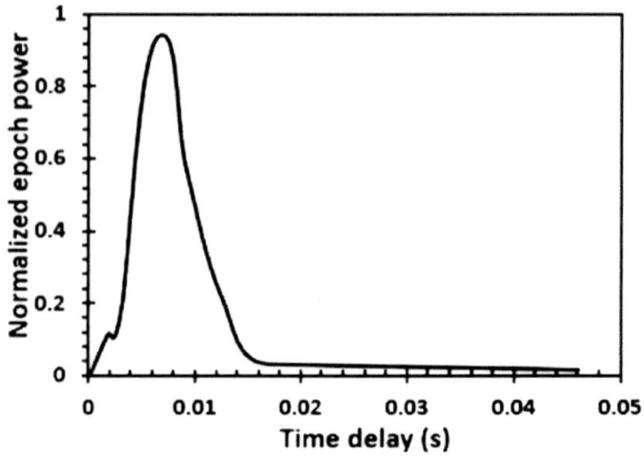

Figure 8.14 The epoch waveform of near-nadir SAR observation with a certain position.

Interestingly, the maximum delay time of 0.04 seconds corresponds to a significantly lower normalized echo power, approaching zero. This inverse relationship highlights the efficiency of the SAR system in capturing accurate data.

Moreover, when compared to the OSTM/Jason-2 GDR 20 Hz and Jason-3 Interim GDR 20 Hz multi-radar altimeter sensors (Figure 8.5), the SIRAL-2 sensor demonstrates a notably shorter delay time. This efficiency translates to more precise water height measurements, with delay times consistently under 0.008 seconds. Thus, the SIRAL-2 sensor's performance underscores its capability to deliver highly accurate water height data with minimal delay, outperforming other contemporary radar altimeter sensors [12, 15, 19].

The time delay of the emitted signals is determined by the mid-power epoch of the echo, which is identified by fitting the backscatter signal to the echo waveform model. This delay is crucial for accurately measuring distances [12, 18, 20]. The slant range, which represents the actual distance between the radar and the target is then calculated by multiplying this time delay by the speed of light.

8.11 DELAY-DOPPLER ALTIMETER

Conventional radar altimeters emit pulses with a long interval period of approximately 500 μs. SIRAL, however, emits a burst of pulses with an interval of only 50 μs between them. The receiving echoes are thus interrelated, and by considering the complete burst of pulses in one campaign, the data processor can discrete the echo into the set of strips across the track by manipulating the slight frequency shifts. The Doppler effect causes a frequency shift in the forward- and aft-looking parts of the beam. The noise reduction is achieved by superimposing and averaging strips on each other.

In interferometric mode (SARIn), however, a second receiving antenna is triggered to determine the receiving angle. SARIn, therefore, qualifies to sense approximately radar backscatter signals receiving from a target not precisely positioned beneath the altimeter sensor. Consequently, the variance in the path-length time of the radar echoes is tiny between radar echoes on the track and radar echoes out of the track. In this circumstance, α must be accurate [12, 21].

The Delay-Doppler Altimeter differs from a conventional radar altimeter in that it uses coherent processing of sets of transmitted pulses. In this view, the full Doppler bandwidth is demoralized to achieve the most efficient use of the backscatter intensity, which is reflected from the Earth's surface (Figure 8.15). On the contrary, the conventional altimeter system determines the distance between the satellite and the mean water height. In this context, the Delay-Doppler Altimeter method differs from conventional altimeter instruments by two techniques. The primary technique involves pulse-to-pulse coherence and full Doppler processing, which permit the quantity of the along-track spot of the range measurement. The latter technique includes the implementation of two antennas and two receiver channels, which allow for the dimension of the across-track angle of the range quantity.

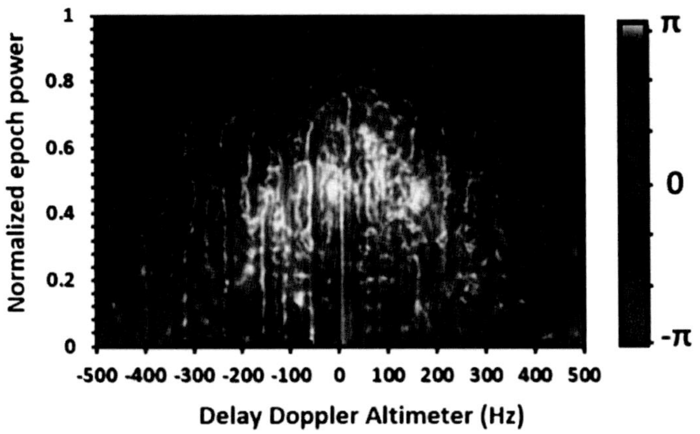

Figure 8.15 Delay-Doppler altimeter interferometry.

The Delay-Doppler Altimeter technology delivers numerous more compensations than a conventional altimeter. The water height precision offered by Delay-Doppler Altimeter technology is approximately twice that of the conventional altimeter. Simulations of the associated signal processing concepts have produced 0.5 cm precision in a smooth water surface, with precision remaining better than 1.0 cm even in water heights as great as 10 m. The Delay-Doppler Altimeter technique reduces the uncertainties caused by water height variations.

For instance, both the Delay-Doppler Altimeter and conventional altimeter experience comparable levels of random noise due to a smooth water surface. Nevertheless, as the water height grows, a conventional altimeter suffers an intense noise level growth. However, the coherent processing of the Delay-Doppler Altimeter experiences only a trivial intensification in random noise with water height. Therefore, the Delay-Doppler Altimeter is appropriate for geodetic uses. The random error owing to water level fluctuations is the dominant error source. Consequently, stream flow and water height retrievals from the Delay-Doppler Altimeter have twice the precision of current sensors [12, 15, 18, 21].

8.12 CRYOSAT-2 SIRAL DATA ACQUISITIONS

In this investigation, CRYOSAT-2 SIRAL data are applied to wave spectra patterns in the Indian Ocean. CryoSat-2 revolves on a non-sun-synchronous polar orbit with an inclination angle of 92° inclination at an altitude of 713 km. Consequently, the repeat cycle for CryoSat is 369 days and 5,344 orbits. This cycle provides full coverage of the earth and the cycle number is provided in the CryoSat product headers. The repeat cycle is made up of approximately 30-day sub-cycles, defined and used primarily for statistic and quality reporting [21–25]. There are two instruments attached to a board of CryoSat-2's mission. The primary two instruments aboard CryoSat-2 are SIRAL-2. Two SIRAL instruments serve as a backup in case the other flops.

The second instrument is Doppler Orbit and Radio Positioning Integration through Satellite, or DORIS, which is used to compute the spacecraft's orbit precisely an array of retroreflectors is also installed aboard the spacecraft, allowing measurements to be taken from the ground to verify the orbital data provided by DORIS [23–25]. In LRM the altimeter is achieved as a conventional pulse narrow altimeter. This mode functions at a pulse repetition frequency (PRF) of 1,970 Hz. This PRF is low enough to ensure that the echoes are decorrelated [12, 25]. Consequently, the echoes obtained may be incoherently extra to decrease speckle noise by $\frac{1}{\sqrt{M}}$ a factor, where M is the average number of pulses in the selected time interval [12, 23].

In SAR mode, the pulses are emitted in bursts. The correlation between echoes is anticipated. Thus, in SAR mode, the PRF within a burst is higher than the LRM one and equal to 17.8 KHz [15]. At this PRF, every burst emits 64 pulses. Every time, the pulses are transmitted, the altimeter delays for the returns and transmits again for the succeeding burst. Hence, there is not merely an 'intra-burst' PRF then similarly a burst repetition frequency (BRF), which, according to SIRAL measurements is 85.7 Hz.

Both SIRAL LRM and SAR modes transmit pulses of equivalent pulse length. The principal distinction between these modes is the PRF and its related effects. In SAR mode, pulse-to-pulse correlation is a consequence of its excessive PRF, while pulse-to-pulse correlation is not nonexistent in LRM, nor preferred [22–25]. In this regard, the decorrelation PRF is obtained by dividing the spacecraft velocity V by the decorrelation distance and correcting for the curvature of the earth R_e [19] as given by:

$$PRF = \frac{V}{0.31\,\lambda H(r^{-1})\left(\frac{R_e + H}{R_e}\right)} \tag{8.26}$$

here λ is the radar altimeter wavelength, H is the altimeter height and r is the radius of the circular uniformly illuminated water lake area. The mathematical relationship between r and water height can be expressed as:

$$r = \sqrt{\frac{(ct + 2H_{\text{water}})H}{1 + \frac{H}{R_e}}} \tag{8.27}$$

being t is a lag time between the leading and trailing edges of the pulse. Consequently, equation 8.27 demonstrates that the increase in PRF is due to the extreme growth of water height H_{water} (Figure 8.16).

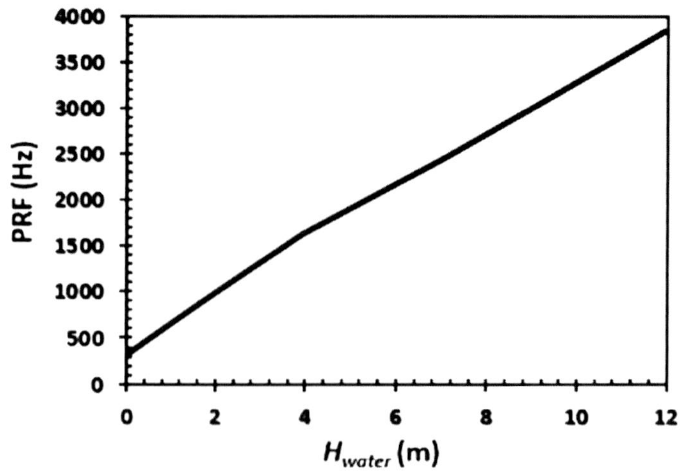

Figure 8.16 PRF as a function of water height H_{water}.

Figure 8.17 shows an example of the water height H_{water} with an area of 300 km × 300 km, with a 12 m water height H_{water}. The map shows how water height varies spatially across Lake Millennium from 2013 to 2024. Areas with different colors highlight the changes in water height, which can be due to various factors such as topography, water flow and other environmental conditions.

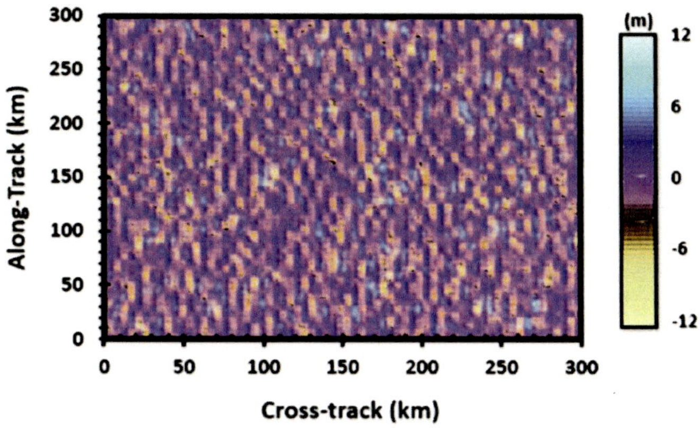

Figure 8.17 Simulation of water height H_{water} using SIRAL LRM data.

Dark blue areas indicate regions where the water height is significantly below the reference level, up to −12 meters. On the contrary, dark red areas indicate regions where the water height is significantly above the reference level, up to 12 m. Therefore, intermediate colors (purple, light blue, green, yellow, and orange) represent varying water heights between these extremes (Figure 8.17).

8.13 CYCLE OF WATER HEIGHTS ACROSS LAKE MILLENNIUM

Waveforms associated with inland water bodies can be significantly influenced by the surrounding land. This influence can cause the waveform to exhibit multiple peaks or become distorted. Such distortions can lead to noisy and potentially erroneous water level estimates. Therefore, it is crucial to consider these factors when deriving water level time series.

One effective approach, as detailed by Nielsen et al. [25], involves assuming that an observation follows a mixed distribution comprising both a Gaussian and a Cauchy distribution. The Cauchy distribution, known for its heavier tail compared to the Gaussian distribution ensures that erroneous observations exert a smaller influence on the estimated mean value. This method enhances the robustness of the water level estimates. The Gaussian distribution, also known as the normal distribution, is characterized by its bell-shaped curve. The probability density function (PDF) for a Gaussian distribution is given by:

$$f(x;\,\mu,\sigma)=\left[\sigma\sqrt{2\pi}\right]^{-1}e^{\left(-\left[2\sigma^{2}\right]^{-1}(x-\mu)^{2}\right)} \quad (8.28)$$

where μ is the mean (location parameter), σ is the standard deviation (scale parameter), and (x) is the variable. The Cauchy distribution, also known as the Lorentz distribution, has a probability density function (PDF) given by:

$$f(x;x_0,\gamma)=\frac{1}{\pi\gamma\left[1+\left(\frac{x-x_0}{\gamma}\right)^{2}\right]} \quad (8.29)$$

where x_0 is the location parameter (the peak of the distribution), γ is the scale parameter describes the width of the distribution, and x is the variable.

Given that water levels fluctuate over time, it is essential to take measurements over short time spans and use an observation model that accurately describes data distribution. By leveraging this information, the final water level time series can be constructed using a state-space model. This model consists of two components: a process model that describes the true height variations and an observation model that accounts for the distribution of the data.

The combination of Gaussian and Cauchy distributions in the observation model, along with the state-space model provides a sophisticated framework for deriving accurate water level time series. This approach effectively mitigates the impact of distorted waveforms and ensures reliable water level estimates, even in the presence of surrounding land influences [12, 18, 23, 25].

The time series analysis reveals that water height variations remained consistent from 2013 to 2020, with the maximum wave height reaching 565 meters (Figure 8.18). However, from 2021 to 2024, there was a rapid increase in water height, rising from 568 meters to 600 meters. From 2013 to 2020, the water height variations were relatively stable (Figure 8.19). This period was characterized by minor fluctuations, with the maximum wave height consistently peaking at 565 meters. Such stability suggests that the environmental conditions and hydrological inputs during these years were relatively constant, allowing for predictable water level patterns.

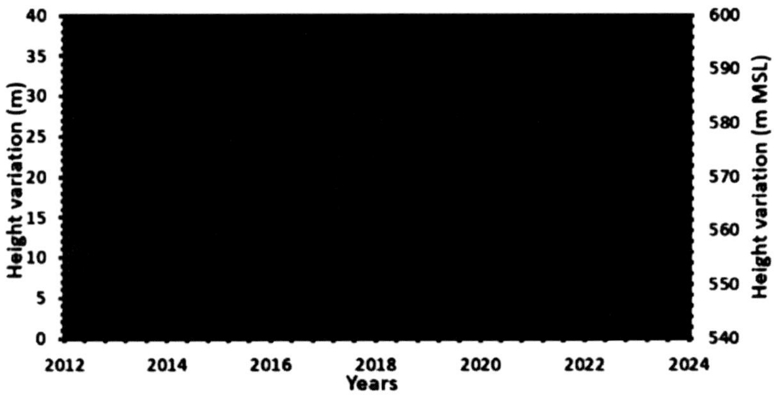

Figure 8.18 Time series of water height variations retrieved using altimeter InSAR technique along Lake Millennium.

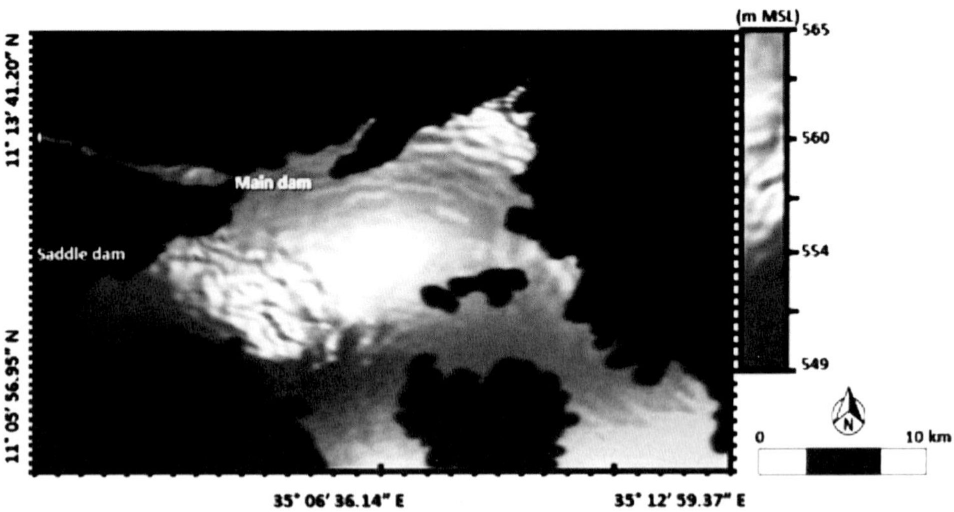

Figure 8.19 Water level stability across Millennium Reservoir from 2013 to 2021.

In stark contrast, the period from 2021 to 2024 witnessed a significant and rapid increase in water height. Starting at 568 meters in 2021, the water level surged to 600 meters by 2024 (Figure 8.20). This dramatic rise indicates a substantial change in the factors influencing the lake's water levels. Possible contributors to this increase could include higher precipitation rates, changes in water management practices or climatic shifts leading to increased runoff and inflow.

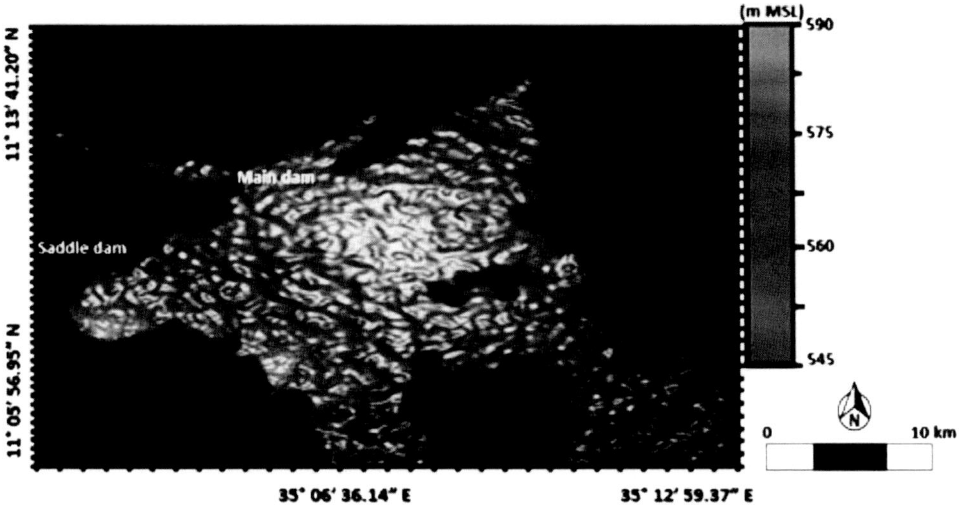

Figure 8.20 Water level surges across Millennium Reservoir from 2021 to 2024.

The filling of the Grand Ethiopian Renaissance Dam (GERD) has been a key aspect of regional geopolitics and hydrology, especially regarding Ethiopia's water resource management and its impact on downstream countries like Sudan and Egypt.

Here is an overview of the key milestones in the filling scenario of GERD until 2024. The phased filling process of the Grand Ethiopian Renaissance Dam (GERD) has been marked by a series of gradual developments, each phase building upon the previous one, while steadily increasing the dam's water storage capacity. The first filling in July 2020 initiated the process by storing approximately 4.9 billion cubic meters (BCM) of water. This initial filling, which coincided with the rainy season, allowed for the dam to begin generating power, though it also sparked regional tensions, particularly between Egypt and Sudan. Consequently, despite the technical success, political disagreements over the lack of an agreement emerged [37].

Following the first phase, the second filling in July 2021 raised the water level to a total of 13.5 BCM. Unlike the first filling, this phase proceeded without a formal accord between the involved nations, exacerbating diplomatic tensions. However, Ethiopia continued filling the dam, leveraging the rainy season's natural water inflow. As a result, while operational progress continued, the political disputes grew sharper.

In comparison, the third filling in August 2022, which raised the total stored volume to 22 BCM, represented a further advancement in the dam's operational capacity. Despite ongoing tensions, this phase marked the point where additional turbines became operational, solidifying Ethiopia's ability to generate hydropower. Thus, despite continued discord with downstream countries, the dam's functionality expanded in parallel with its growing capacity.

The fourth filling, completed in August 2023, further increased the total water volume to approximately 31 BCM. As in previous phases, this filling aligned with the seasonal flooding of the Blue Nile, allowing Ethiopia to stay on course with the dam's expansion. While the fourth phase marked substantial progress toward full capacity, regional disagreements persisted, as no binding agreement had yet been reached. Nevertheless, Ethiopia's progress ensured a continued increase in electricity generation capacity.

Looking ahead, the fifth filling, completed by September 2024, has brought the dam's storage volume closer to its final target capacity of 74 BCM. This phase marks the culmination of years of phased fillings, steadily increasing the dam's water level while maintaining focus on hydropower generation. With the dam nearing full operation, diplomatic discussions are likely to focus on managing water usage in future dry years, as well as mitigating downstream impacts. Thus, each phase of GERD's filling has not only contributed to its operational development but also intensified the regional debate over water resources, making the forthcoming phases critical for both Ethiopia's energy ambitions and regional stability.

The lake of El-Roseries has exhibited remarkable stability over the past decade, with water levels fluctuating between 488 meters and 497 meters (Figure 8.21). This consistent range indicates a relatively balanced hydrological system, where inflows and outflows have maintained an equilibrium, preventing significant deviations in water height. From 2013 to 2020, the water levels in El-Roseries Lake remained within a narrow band of 488 to 497 meters. This stability suggests that the lake's catchment area experienced consistent climatic conditions and water management practices. The minor fluctuations within this range can be attributed to seasonal variations in precipitation and evaporation, which are typical for many natural water bodies.

However, a notable change occurred from 2021 to 2024. During this period, the water level, which was at 498 meters in 2023, experienced a significant decline, dropping to 488 meters by mid-2024. This rapid decrease in water level could be indicative of several underlying factors. Potential causes might include reduced rainfall, increased water extraction for agricultural or industrial use or changes in upstream water management practices that affect the inflow to the lake [37–39].

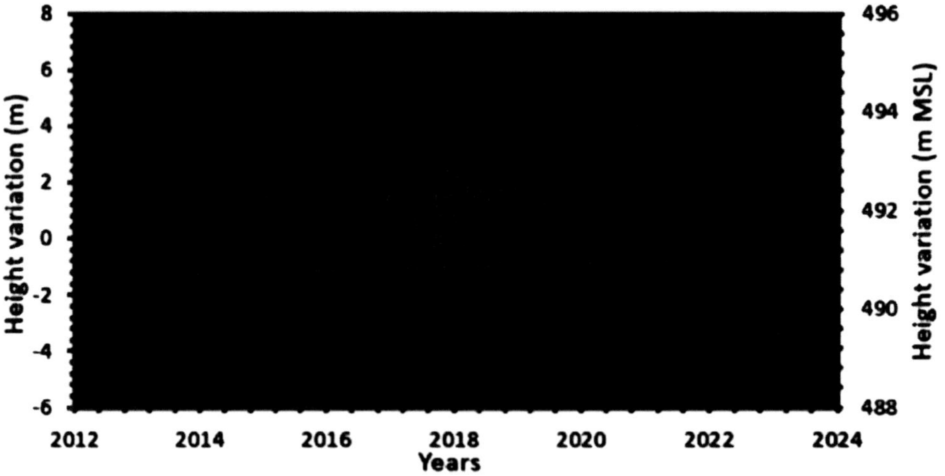

Figure 8.21 Time series of water fluctuation across El-Roseries Lake.

From 1992 to around 2010, the water levels in Lake Nasser show a relatively stable pattern with minor fluctuations. The height variation in meters (left y-axis) oscillates within a narrow range, indicating that the lake maintained a consistent water level during this period. The height variation in meters above sea level (mASL, right y-axis) also reflects this stability, staying between approximately 165 and 175 meters.

From 2021 onwards, the water levels show a significant upward trend. The height variation in meters above sea level rises steadily, reaching close to 180 meters by 2024 (Figure 8.22). This sharp increase suggests a substantial change in the factors influencing the lake's water levels, such as increased precipitation, changes in upstream water flow or enhanced water conservation efforts.

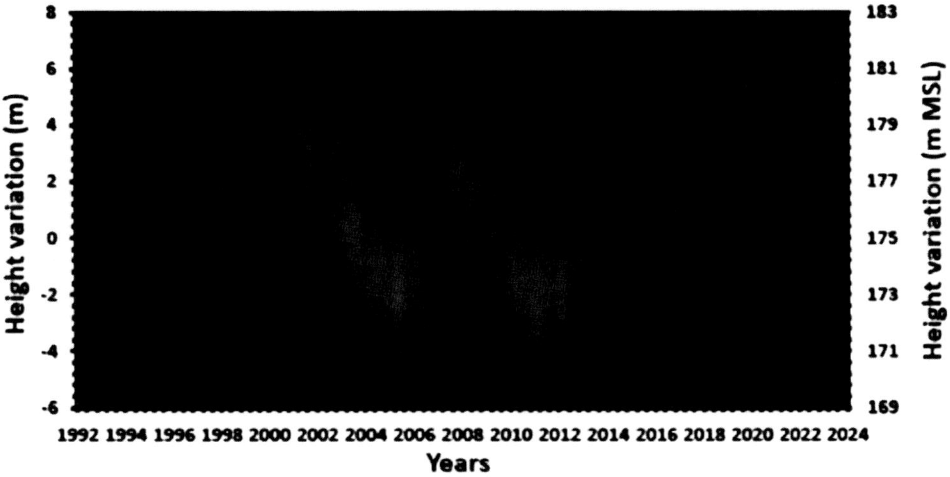

Figure 8.22 Time series of water fluctuation across Nasir Lake.

Consistent with the above perspective, the operations of these reservoirs did not show any abrupt changes, maintaining a consistent pattern of seasonal variations. The mean height variation over the observed period was different for each water body: Lake Roseires had a mean height variation of 4 meters, Lake Nasser had 3 meters, and the Millennium Reservoir had a substantial 9 meters. These variations reflect the differing hydrological characteristics and responses of each lake to environmental factors. The higher mean height variation in the Millennium Reservoir indicates a greater susceptibility to external influences, such as flooding. This result confirmed the work done by Kansara and Lakshmi [39].

8.14 SARIN – AN ACCURATE TOOL

In this comprehensive analysis, we have meticulously evaluated the performance of the new altimetry modes, specifically SAR (Synthetic Aperture Radar) and SARIn (SAR Interferometric), and compared them to the results obtained from conventional altimetry. The SAR and SARIn modes are distinguished by their smaller footprint, approximately 300 meters along-track, which enables a remarkable precision of just a few centimeters. This high level of accuracy is maintained even when measuring small lakes, demonstrating the advanced capabilities of these new modes.

The efficacy of the SARIn becomes evident in its achievement of higher accuracy, as demonstrated by an elevated true positive rate (TPR) in Receiver Operating Characteristic (ROC) curves for retrieved water level height in TOPEX/Poseidon, Jason-1, Jason-3, and Sentinel-6. The ROC area, representing the reliability of the simulated water level height, attains a significant 95% accuracy when employing the SARIn, as illustrated in Figure 8.23.

When we compare these results with other altimeter sensors, the agreement is significantly well pronounced with data obtained from SARIn, a previous satellite mission known for its reliable altimetry measurements. This comparison underscores the enhanced performance of other altimeter sensors and SARIn modes in capturing precise water height data [26–28].

However, it is important to consider the long repeat period of CryoSat-2, which spans 369 days. This extended period means that the satellite's tracks cross a given lake at different locations each time. Consequently, some tracks may be situated near the shoreline, where the potential for land contamination is higher. This variability in track locations can introduce residual geoid errors in the water level estimates, which may explain why, in some cases, the agreement with other altimeter sensors is well pronounced for SARIn [29–33].

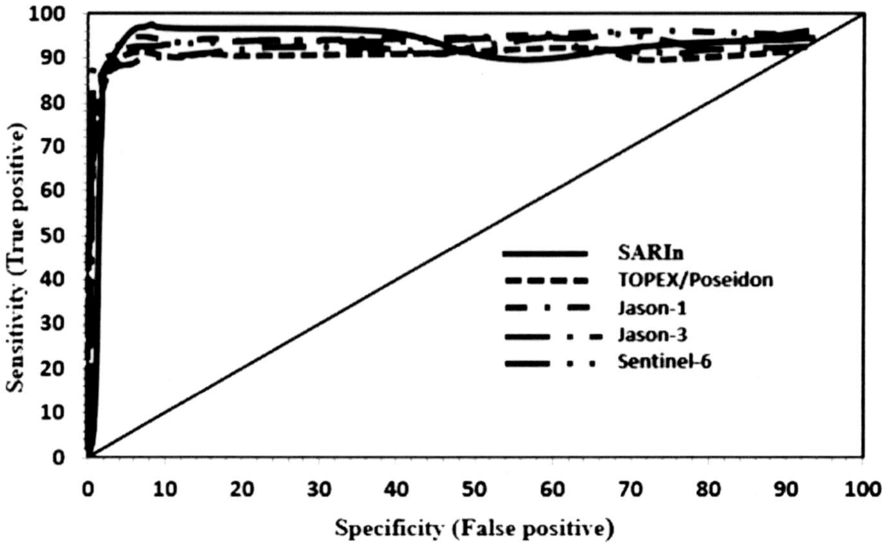

Figure 8.23 Accuracy of simulated water level height from SARIn and other altimeter sensors.

Despite these challenges, CryoSat-2, equipped with SAR and SARIn modes has delivered improved results overall. It is crucial to note that the precision of the mean water levels and the agreement with in situ data presented in this analysis should not be generalized for all scenarios. The quality of the estimated water levels is influenced by several factors, including the surrounding topography, the complexity of the lake shoreline, and the presence of off-nadir water bodies [33–36].

Needless to say, while the new altimetry modes of CryoSat-2 have shown significant advancements, their performance is context-dependent. The surrounding environment and specific conditions of each lake play a critical role in the accuracy of the measurements. This nuanced understanding highlights the importance of considering various factors when evaluating the performance of altimetry technologies.

8.15 CONCLUSION

In this chapter, we have explored the application of interferometric altimeter radar to accurately measure water height levels in Lake Millennium, Lake Roseires, and Lake Nasser. The SARIn technique has proven to be highly effective, as evidenced by its superior accuracy. This is demonstrated by an elevated true positive rate (TPR) in Receiver Operating Characteristic (ROC) curves for water level measurements obtained from TOPEX/Poseidon, Jason-1, Jason-3, and Sentinel-6. Notably, the ROC area, which indicates the reliability of the retrieved water levels, achieved an impressive 95% accuracy when using SARIn.

The operational patterns of these reservoirs remained stable, showing consistent seasonal variations without abrupt changes. The mean height variations observed over the study period differed among the lakes: Lake Roseires exhibited a mean height variation of 4 meters, Lake Nasser 3 meters, and the Millennium Reservoir a significant 9 meters. These differences highlight the unique hydrological characteristics and responses of each lake to environmental factors. The greater mean height variation in the Millennium Reservoir suggests a higher susceptibility to external influences, such as flooding.

In the subsequent chapter, we will delve into the efficiency of dam operations, focusing on the impact of turbulence caused by turbine activity.

References

[1] Chelton, D.B., Ries, J.C., Haines, B.J., Fu, L.L. and Callahan, P.S. 2001. Satellite altimetry. pp. 1-ii. *In*: L.-L. Fu and A. Cazenave (eds). International Geophysics, Vol. 69. Academic Press.

[2] Rosmorduc, V., Benveniste, J., Lauret, O., Maheu, C., Milagro, M. and Picot, N. 2011. Radar altimetry tutorial. ESA, Europe. pp.112–128.

[3] Hayne, G.S. 1980. Radar altimeter mean return waveforms from near-normal-incidence ocean surface scattering. IEEE Transactions on Antennas and Propagation. 28(5): 687–692.

[4] Robinson, I.S. 2004. Measuring the Oceans from Space: the Principles and Methods of Satellite Oceanography. Springer Praxis Books.

[5] Chelton, D.B., Ries, J.C., Haines, B.J., Fu, L.L. and Callahan, P.S. 2001. Satellite altimetry. pp. 1–131. *In*: L.-L. Fu and A. Cazenave (eds). Satellite Altimetry and Earth Sciences. Elsevier, Amsterdam..

[6] Marghany, M. 2018. Advanced Remote Sensing Technology for Tsunami Modelling and Forecasting. CRC Press.

[7] Crétaux, J.F., Calmant, S., Romanovski, V., Shabunin, A., Lyard, F., Bergé-Nguyen, M., et al. 2009. An absolute calibration site for radar altimeters in the continental domain: Lake Issykkul in Central Asia. Journal of Geodesy. 83(8): 723–735.

[8] Crétaux, J.F., Calmant, S., Romanovski, V., Perosanz, F., Tashbaeva, S., Bonnefond, P., et al. 2011. Absolute calibration of Jason radar altimeters from GPS kinematic campaigns over Lake Issykkul. Marine Geodesy. 34(3–4): 291–318.

[9] Crétaux, J.F., Berge-Nguyen, M., Calmant, S., Romanovski, V.V., Meyssignac, B., Perosanz, F., et al. 2013. Calibration of Envisat radar altimeter over Lake Issykkul. Advances in Space Research. 51(8): 1523–1541.

[10] Cretaux, J.F., Nielsen, K., Frappart, F., Papa, F., Calmant, S. and Benveniste, J. 2017. Hydrological applications of satellite altimetryrivers, lakes, man-made reservoirs, inundated areas. *In*: Satellite Altimetry Over Oceans and Land Surfaces. 459–504. CRC Press.

[11] Cretaux, J.F., Berge-Nguyen, M., Calmant, S., Jamangulova, N., Satylkanov, R., Lyard, F., et al. 2018. Absolute calibration or validation of the altimeters on the Sentinel-3A and the Jason-3 over Lake Issykkul (Kyrgyzstan). Remote Sensing. 10(11): 1679.

[12] Marghany, M. 2021. Nonlinear Ocean Dynamics: Synthetic Aperture Radar. Elsevier.

[13] Marghany, M., Hashim, M. and Cracknell, A.P. 2008. Volterra algorithm for modelling sea surface current circulation from satellite altimetry Data. *In*: Computational Science and Its Applications–ICCSA 2008: International Conference, Perugia, Italy, June 30–July 3, 2008, Proceedings, Part II 8 (pp. 119–128). Springer Berlin Heidelberg.

[14] Marghany, M. 2009. Volterra–lax-wendroff algorithm for modelling sea surface flow pattern from jason-1 satellite altimeter data. Transactions on Computational Science VI, pp.1–18.

[15] Wingham, D.J., Francis, C.R., Baker, S., Bouzinac, C., Brockley, D., Cullen, R., et al. 2006. CryoSat: A mission to determine the fluctuations in Earth's land and marine ice fields. Advances in Space Research. 37(4): 841–871.

[16] Vaijayanthi, S. and Vanitha, N. 2015. Aircraft Identification in high-resolution remote sensing images using shape analysis. International Journal of Innovative Research in Computer and Communication Engineering. 3(11): 11203–11209.

[17] Raney, R. Keith. 1998. The delay/Doppler radar altimeter. IEEE Transactions on Geoscience and Remote Sensing. 36(5): 1578–1588.

[18] Wingham, D.J., Francis, C.R., Baker, S., Bouzinac, C., Brockley, D., Cullen, R., et al. 2006. CryoSat: A mission to determine the fluctuations in Earth's land and marine ice fields. Advances in Space Research. 37(4): 841–871.

[19] Raney, R. Keith. 1998. The delay/Doppler radar altimeter. IEEE Transactions on Geoscience and Remote Sensing. 36(5): 1578–1588.

[20] Martin-Puig, C., Giulio, R., Jose, M., David, C., Meric, S., Peter, C., et al. 2008. Theoretical model of SAR altimeter over water surfaces. *In*: Geoscience and Remote Sensing Symposium. IGARSS 2008. IEEE International. 3: III-242. IEEE, 2008.

[21] Jensen, J. Robert. 1999. Angle measurement with a phase monopulse radar altimeter. IEEE Transactions on Antennas and Propagation. 47(4): 715–724.

[22] Galin, N., Duncan, J.W., Robert, C., Marco, F., Walter, H.F.S., and Saleh, A. 2013. Calibration of the CryoSat-2 interferometer and measurement of across-track ocean slope. IEEE Transactions on Geoscience and Remote Sensing. 51(1): 57–72.

[23] Tournadre, J., Fanny, G.-A. and Benoit, L. 2012. Antarctic icebergs distributions, 2002–2010. Journal of Geophysical Research: Oceans. 117: C5.

[24] Tournadre, J., Bouhier, N., Boy, F. and Dinardo, S. 2018. Detection of ice berg using Delay Doppler and interferometric Cryosat-2 altimeter data. Remote Sensing of Environment. 212: 134–147.

[25] Nielsen, K., Zakharova, E., Tarpanelli, A., Andersen, O.B. and Benveniste, J. 2022. River levels from multi mission altimetry, a statistical approach. Remote Sensing of Environment. 270: 112876.

[26] Martin-Puig, Cristina, Jose Marquez, Giulio Ruffini, R. Keith Raney, and Jérôme Benveniste. 2008. SAR altimetry applications over water. arXiv preprint arXiv:0802.0804.

[27] CryoSat Mission and Data Description, Doc No. CS-RP-ESA-SY-0059, 2007.

[28] Pandey, P.C., Gairola, R.M. and Gohil, B.S. 1986. Wind-wave relationship from SEASAT radar altimeter data. Boundary-layer meteorology. 37(3): 263–269.

[29] Fujisaki, K., Tanaka, K. and Tateiba, M. 1999. A simulation of satellite altimeter return pulses from three-dimensional ocean waves. *In*: IEEE International Geoscience and Remote Sensing Symposium. IGARSS'99 (Cat. No. 99CH36293). 2: 995–997. IEEE.

[30] Fukuda, K., Fujisaki, K. and Tateiba, M. 2007. A computer simulation of relative backscattering coefficients of sea surfaces in satellite altimetry. Electronics and Communications in Japan (Part II: Electronics). 90(12): 121–128.

[31] Portilla, J., Ocampo-Torres, F.J. and Monbaliu, J. 2009. Spectral partitioning and identification of wind sea and swell. Journal of atmospheric and oceanic technology. 26(1): 107–122.

[32] Michailides, C., and Angelides, D.C. 2018. Optimization of a flexible floating structure for wave energy production and protection effectiveness. Engineering Structures. 85: 249–263.

[33] Steedman, R. 1993. Collection of wave data and the role of waves on nearshore circulation. Report prepared for the Water authority of Western Australia, Perth.

[34] Sanderson, P.G. and Eliot, I. 1999. Compartmentalisation of beachface sediments along the southwestern coast of Australia. Marine Geology. 162(1): 145–164.

[35] Louis, J.P., and Radok, J.R.M. 1975. Propagation of tidal waves in the Joseph Bonaparte Gulf. Journal of Geophysical Research. 80(12): 1689–1690.

[36] Hemer, M.A., Simmonds, I. and Keay, K. 2008. A classification of wave generation characteristics during large wave events on the Southern Australian margin. Continental Shelf Research. 28(4–5): 634–652.

[37] Wheeler, K.G., Jeuland, M., Hall, J.W., Zagona, E. and Whittington, D. 2020. Understanding and managing new risks on the Nile with the Grand Ethiopian Renaissance Dam. Nature Communications. 11(1): 1–9.

[38] Salhi, A. and Benabdelouahab, S. 2023. The downstream impact of the first and second filling of the Grand Ethiopian Renaissance Dam. Sustainable Water Resources Management. 9(1): 14.

[39] Kansara, P. and Lakshmi, V. 2022. Water levels in the major reservoirs of the Nile River Basin—A comparison of Sentinel with satellite altimetry data. Remote Sensing. 14(18): 4667.

Chapter 9

Unveiling Dam Turbulence and Efficiency Using Marghany's Quantum-Optimized Multi-Objective Algorithm in SAR Imaging

9.0 INTRODUCTION

In the previous chapter, we demonstrated the use of altimeter interferometry to monitor the time series of water levels in Lake Millennium, El El-Roseries, and Lake Naser. This chapter addresses the question: Can water level data be used to understand the operation and efficacy of the Grand Ethiopian Renaissance Dam (GERD)?

The altimeter water level measurements in the previous chapter will be utilized to analyze the water inflow mechanisms from GERD. Furthermore, the turbulence in water flow will be used as a key index for monitoring the operation of GERD's turbines, given that the primary purpose of this mega project is to generate hydroelectric energy. To achieve this, high-resolution radar imaging techniques will be implemented to monitor turbulence flow, which is a crucial index for assessing dam operations and efficiencies.

9.1 TURBULENCE ENHANCES DAM OPERATION MONITORING

Turbulence, the chaotic and irregular movement of water (Figure 9.1), plays a crucial role in the operation and efficiency of dams. Proper management of turbulent flow is essential to ensure efficient water passage and to minimize damage to the dam infrastructure. First, flow management is crucial [1]. Turbulence affects the flow of water through dam structures, including spillways and turbines (Figure 9.2.). Effective management of this turbulent flow is necessary to maintain smooth water passage and prevent potential damage. Second, during high-flow events, energy dissipation becomes vital. Turbulence aids in dissipating the kinetic energy of water which is essential to prevent erosion and structural damage to the dam and its spillways. In other words, during high-flow events, such as heavy rainfall or snowmelt, the kinetic energy of water increases dramatically. Turbulence helps dissipate this energy, which is vital to prevent erosion and structural damage to the dam and its spillways. By breaking down the energy into smaller, less destructive forces, turbulence acts as a natural buffer, protecting the dam's structural integrity [2].

DOI: 10.1201/9781003327325-9

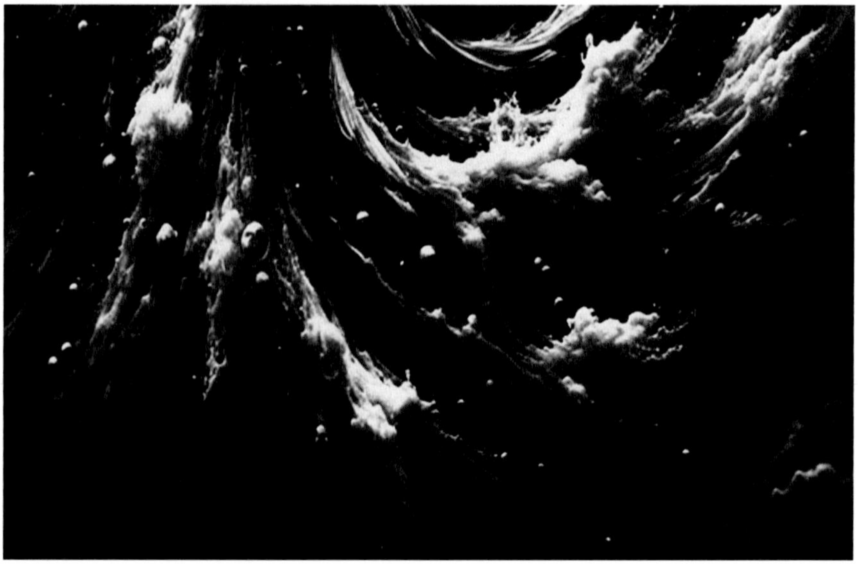

Figure 9.1 Irregular and chaotic turbulence movement.

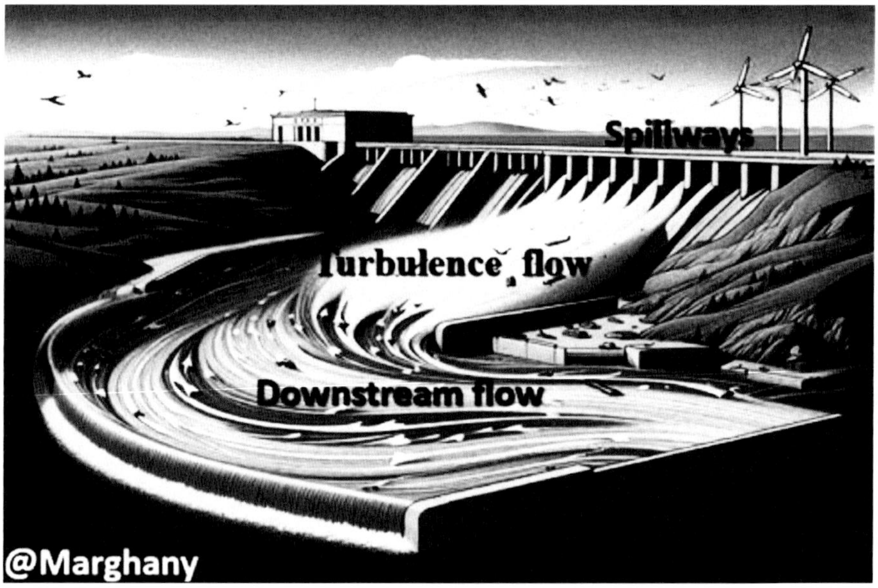

Figure 9.2 Dam turbulence flow.

Moreover, sediment transport is significantly influenced by turbulent flow. Managing sediment transport effectively is crucial to maintaining reservoir capacity and preventing blockages in the dam's outlets. In the context of hydropower dams, turbulence directly impacts the efficiency of turbines. Turbines are designed to convert the kinetic energy of flowing water into mechanical energy, which is then transformed into electrical energy. However, excessive turbulence can reduce the efficiency of this process by causing irregular flow patterns around the turbine blades. By understanding and controlling turbulence, operators can optimize flow conditions, leading to more efficient energy generation. This optimization not only enhances power output but also extends the lifespan of the turbines by reducing wear and tear [1, 3].

Lastly, structural integrity is affected by turbulence-induced forces. Engineers must account for these forces in their designs to ensure the long-term stability and safety of the dam. By monitoring and managing turbulence, dam operators can enhance the overall efficiency and safety of dam operations, ensuring that these critical infrastructures continue to function effectively.

Needless to say, turbulence is a key factor in dam operations and efficiency. From flow management and energy dissipation to sediment transport and hydropower efficiency, understanding and controlling turbulence is essential for maintaining the functionality and safety of these critical infrastructures [1–4]. By exploiting advanced monitoring techniques, dam operators can ensure their facilities continue to operate effectively, providing essential services such as water supply, flood control, and energy generation. Therefore, the question now is: how is turbulence generated within and around the dam?

9.2 TURBULENCE GENERATION MECHANISMS WITHIN AND AROUND THE DAM

Turbulence in dams is generated through several mechanisms, primarily due to the interaction between water flow and the dam's structural features.

9.2.1 Flow Over Spillways

When water flows over a spillway, it transitions from a relatively smooth, laminar flow to a more chaotic, and turbulent flow (Figure 9.3). This transition occurs because the water accelerates as it moves over the spillway crest and then plunges into the downstream pool, creating a turbulent hydraulic jump.

Figure 9.3 Transition phases from laminar flow to turbulent flow.

The transition from laminar to turbulent flow can be described using the Reynolds number Re (Figure 9.4), which is a dimensionless quantity used to predict flow patterns in different fluid flow situations. The Reynolds number is given by:

$$Re = \frac{\rho v L}{\mu} \tag{9.1}$$

here ρ is the fluid density (kg/m³); v is the flow velocity (ms^{-1}); L is the characteristic length (m); and μ is the dynamic viscosity of the fluid (Pa·s). For flow over a spillway, if (Re) exceeds a critical value (typically around 2,000 for open channel flow), the flow transitions from laminar to turbulent. Therefore, as water flows over the spillway crest, it accelerates due to gravity. The velocity of the water at the crest can be approximated using the Bernoulli equation:

$$0.5\, v^2 + gh = c \tag{9.2}$$

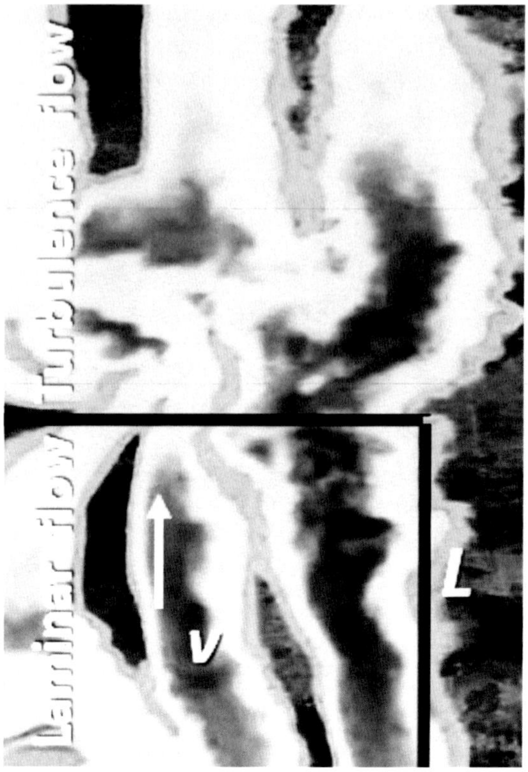

Figure 9.4 Simplification of Reynolds number equation.

where v is the flow velocity (ms^{-1}); g is the acceleration due to gravity (9.81 ms^{-2}); c is constant; and h is the height of the water above the spillway crest (m) (Figure 9.5). Assuming negligible initial velocity and solving for (v):

$$v = \sqrt{2gh} \tag{9.3}$$

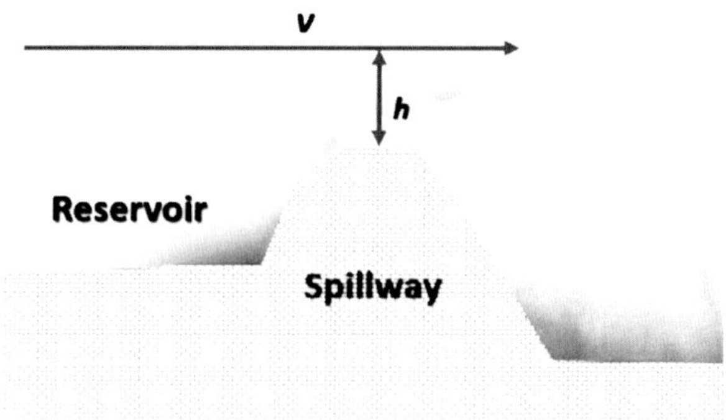

Figure 9.5 Water flows over the spillway.

When the accelerated water plunges into the downstream pool, it creates a hydraulic jump, a phenomenon where the flow transitions from supercritical (high velocity, low depth) to subcritical (low velocity, high depth) [5, 7]. The Froude number ((Fr)) is used to characterize this flow:

$$Fr = \frac{v}{\sqrt{gh}} \tag{9.4}$$

For a hydraulic jump to occur, the Froude number must be greater than 1 (supercritical flow). The energy loss ((ΔE)) in the hydraulic jump can be calculated using the specific energy equation:

$$\Delta E = \frac{(y_2 - y_1)^3}{4 y_1 y_2} \tag{9.5}$$

where y_1 is the initial depth of the flow (m), and y_2 is the sequent depth after the jump (m).

9.2.2 Obstructions and Structural Features

The presence of structural features such as gates, piers, and baffles within a dam can disrupt the smooth flow of water, leading to turbulence [7]. These obstructions cause the water to change direction and speed abruptly, generating vortices and eddies. The abrupt changes in velocity and direction create shear forces within the flow. Shear forces arise due to the velocity gradient between different layers of water. The shear stress τ can be expressed as:

$$\tau = \mu \frac{du}{dy} \tag{9.6}$$

Equation 9.6 shows that $\frac{du}{dy}$ is the velocity gradient perpendicular to the flow direction (s^{-1}). High shear forces contribute to the breakdown of laminar flow and the onset of turbulence. Consequently, vortices are swirling motions of fluid that occur when the flow is disturbed by obstacles [4, 7]. The strength of a vortex can be quantified by its circulation Γ:

$$\Gamma = \oint_C \mathbf{v} \cdot d\mathbf{l} \tag{9.7}$$

In equation 9.7, $\mathbf{v}$ is the velocity vector (ms^{-1}), and $d\mathbf{l}$ is the differential length element along the closed contour (C) (m). This means the presence of gates, piers, and baffles creates regions of high circulation, leading to the formation of vortices and eddies [5–7].

9.2.3 High Flow Velocities

During flood events or large water releases, the flow velocity v increases substantially. This increase directly affects the Reynolds number, as seen in the formula 9.1. For instance, consider a scenario where the flow velocity doubles. The Reynolds number will also double, pushing the flow regime towards turbulence. This indicates, turbulent flow is marked by chaotic and irregular fluctuations in velocity and pressure. These fluctuations can be described using statistical measures such as the root mean square (RMS) of velocity fluctuations (u') (Figure 9.6):

$$u' = \sqrt{\frac{1}{N}\sum_{i=1}^{N}(u_i - \bar{u})^2} \tag{9.8}$$

where u_i is the instantaneous velocity at point (i); $\bar{u}$ is the mean velocity; and (N) is the number of measurements. Therefore, the pressure fluctuations p' can similarly be quantified as:

$$p' = \sqrt{\frac{1}{N}\sum_{i=1}^{N}(p - \bar{p})^2} \tag{9.9}$$

Equation 9.9 presents p_i is the instantaneous pressure at point (i), and $\bar{p}$ is the mean pressure. Therefore, turbulent flow enhances the mixing and energy dissipation within the fluid. The rate of energy dissipation per unit mass ε in turbulent flow can be expressed as:

$$\varepsilon = \nu \left(\frac{\partial u_i}{\partial x_j} \right)^2 \tag{9.10}$$

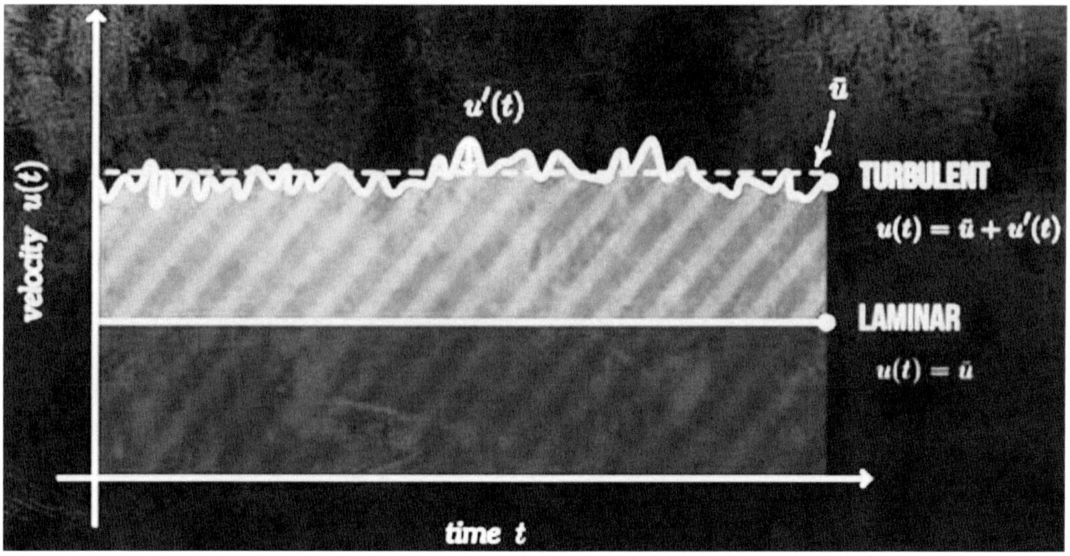

Figure 9.6 Turbulence fluctuation velocity.

The term $\frac{\partial u_i}{\partial x_j}$ captures the rate of change of the velocity field in different spatial directions. In turbulent flows, these gradients are typically large due to the chaotic and irregular nature of the flow. The rate of energy dissipation is proportional to the kinematic viscosity and the square of the velocity gradients. This means that higher velocity gradients (which are common in turbulent flows) lead to greater energy dissipation [1, 4, 7].

9.2.4 Shear Force

Shear forces are a primary mechanism for generating turbulence within fluid flow. These forces arise when there is a velocity gradient, such as between the fast-moving water near the center of a channel and the slower-moving water near the banks or bed. The intensity of the turbulence increases with the strength of the shear forces [1, 3, 7, 9]. In this perspective, the turbulent kinetic energy ((*k*)) is a measure of the intensity of turbulence and is defined as:

$$k = \frac{1}{2}(\overline{u}'^2 + \overline{v}'^2 + \overline{w}'^2) \tag{9.11}$$

here (u'), (v'), and (w') are the velocity fluctuations in the (x), (y), and (z) directions, respectively (m/s). In this view, the overbar denotes a time average. Therefore, the production of turbulent kinetic energy due to shear ((P)) can be expressed as:

$$P = -\overline{u}'v'\frac{\partial \overline{u}}{\partial y} \tag{9.12}$$

Equation 9.12 provides a quantitative measure of the production of turbulent kinetic energy due to shear forces in a fluid flow. It highlights the role of Reynolds stress and mean velocity gradients in generating turbulence. Understanding this relationship is crucial for predicting and managing turbulent flows in various engineering applications, including dam operations [3, 8, 10].

9.2.5 Flow Direction Changes and Turbulence Generation

When water encounters abrupt changes in direction, such as bends or obstacles, it experiences rapid variations in velocity and pressure. These variations disrupt the smooth, laminar flow and induce turbulence. The key factor here is the velocity gradient, which can be quantified using the Navier-Stokes equations. The Navier-Stokes equations govern the motion of fluids (liquids and gases) and are essential for studying phenomena like turbulence, vortices, and flow patterns. These equations combine Newton's second law with fluid dynamics, relating changes in velocity, pressure, viscosity, and external forces acting on a fluid.

For incompressible flow, where the fluid density (ρ) is constant, the general Navier-Stokes equation simplifies to:

$$\rho\left(\frac{\partial t}{\partial u}+u\cdot\nabla u\right)=-\nabla p+\mu\nabla 2u+f \tag{9.13}$$

The Navier-Stokes equation consists of several important terms that describe fluid motion. First, ρ represents the fluid density, which is the mass per unit volume. The term $\partial u/\partial t$ denotes the rate of change of velocity concerning time, also known as unsteady acceleration. Next, $u \cdot \nabla u$ refers to convective acceleration, illustrating how the velocity at a point changes due to the movement of the surrounding fluid. The $-\nabla p$ term represents the pressure gradient, which is the force that drives fluid motion. Additionally, $\mu\nabla^2 u$ is the viscous term, which accounts for internal friction within the fluid and is proportional to the dynamic viscosity μ. Finally, f symbolizes any external forces acting on the fluid, such as gravity [10–13].

9.2.6 Interaction with the Riverbed

The interaction between flowing water and the riverbed (or the bottom of a reservoir) plays a significant role in generating turbulence. When water flows over a rough or uneven surface, the irregularities in the bed disrupt the smooth flow of water, creating chaotic and turbulent motion. This interaction can be explained using the boundary layer theory and the Navier-Stokes equations for fluid flow. In this view, as water flows over the riverbed, a thin layer of fluid near the surface moves more slowly due to friction (no-slip condition). This forms the boundary layer, where the velocity of the fluid increases from zero at the riverbed to the free-stream velocity U at some height y above the bed [13]. Therefore, the velocity profile $u(y)$ near the bed can be described by:

$$u(y)=U\left(\frac{y}{\delta}\right)^n \tag{9.14}$$

where $u(y)$ is the velocity at height y from the bed, U is the free-stream velocity away from the boundary layer, and δ represents the boundary layer thickness. The parameter n depends on the flow conditions; for instance, in laminar flow, n = 1/7 [14–16].

Understanding how turbulence is generated in dams is crucial for effective dam design and operation. By managing and mitigating the effects of turbulence, engineers can enhance the efficiency and safety of dam operations, ensuring that these critical infrastructures continue to function effectively.

9.3 TURBINES CONNECTED TO A DAM PRODUCE TURBULENT FLOW

A turbine connected to a dam can generate turbulent flow through a combination of factors including high-speed water flow, blade interaction, flow constriction and expansion, rotational effects, and pressure fluctuations. As water flows through the turbine (Figure 9.7), its velocity increases due to the pressure drop from the dam. The high velocity can lead to turbulence. The velocity of water v at different sections of the turbine can be calculated using the Bernoulli equation:

$$P_1+\frac{1}{2}\rho v_1^2+\rho g h_1=P_2+\frac{1}{2}\rho v_2^2+\rho g h_2 \tag{9.15}$$

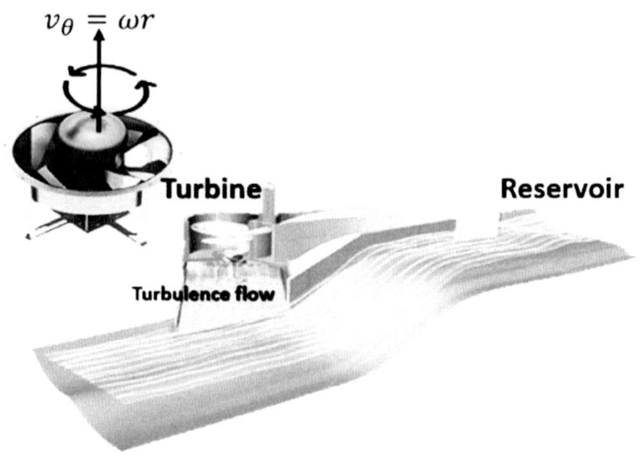

Figure 9.7 The flow of turbulence is created by a turbine.

where P_1 and P_2 represent the pressures at points 1 and 2, respectively; ρ denotes the density of water; v_1 and v_2 are the velocities at these respective points; g signifies the acceleration due to gravity; and h_1 and h_2 indicate the heights at points 1 and 2; respectively. Therefore, the interaction between the water flow and the turbine blades can cause turbulence. The flow over the blades can be described using the blade element theory. For a turbine blade, the angle of attack α affects the local flow velocity and pressure distribution. The lift L and drag D forces on a blade can be expressed as:

$$L = \frac{1}{2}\rho v^2 c C_L D = 0.5\rho v^2 c C_D \tag{9.16}$$

where c is the chord length of the blade, C_L and C_D are the lift and drag coefficients, respectively. Consequently, changes in the cross-sectional area cause velocity variations that can induce turbulence. For an incompressible fluid, the continuity equation relates the velocity v to the cross-sectional area A:

$$A_1 v_1 = A_2 v_2 \tag{9.17}$$

where A_1 and A_2 are the areas at different points, v_1 and v_2 are the velocities at these points. This indicates that the rapid changes in through turbine cause the flow to accelerate or decelerate, potentially leading to turbulence. The rotational effect of the turbine can introduce swirling motions and vortices. The rotational speed ω of the turbine affects the tangential velocity component v_θ:

$$v_\theta = \omega r \tag{9.18}$$

where ω is the angular velocity, r is the radial distance from the axis of rotation. The interaction of the tangential velocity with the radial flow can generate vortices and contribute to turbulence. Therefore, the pressure fluctuations in the turbine can cause unstable flow conditions. The pressure drop ΔP across a turbine can be related to the turbulent kinetic energy k using the turbulence intensity I:

$$I = \frac{k}{\frac{1}{2}\rho v^2} \tag{9.19}$$

where k is related to the turbulence intensity and the pressure drop can be significant in regions of high turbulence. Pointless to add, a turbine connected to a dam generates turbulent flow through the interplay of high-speed water flow, blade interactions, flow constriction and expansion, rotational effects and pressure fluctuations. Each factor contributes to the turbulence observed in the flow, which can be mathematically analyzed using fluid dynamics and turbine mechanics equations [5, 10, 12, 16].

9.4 TURBINES ENHANCE THE OPERATION AND EFFICIENCY OF DAMS

The functionality of a turbine is integral to the effective operation and efficiency of a dam. Firstly, turbines are crucial for converting the kinetic and potential energy of flowing water into mechanical energy, essential for generating electricity in hydroelectric dams. The efficiency of this energy conversion directly impacts the dam's overall effectiveness; a well-functioning turbine maximizes energy extraction and enhances electricity generation capacity.

In addition to energy conversion, turbines play a significant role in regulating water flow through the dam. By managing the flow rate and pressure, turbines ensure consistent and manageable water flow, which is vital for maintaining operational stability and adapting to varying water levels and demands. Proper flow management also helps prevent potential overflows and mechanical failures.

The efficiency of the turbine affects the dam's operational performance. High-efficiency turbines convert a greater percentage of available energy into useful mechanical energy, reducing energy losses and lowering operational costs. This efficiency is influenced by factors such as turbine design, blade shape and operating conditions.

Moreover, turbines contribute to the structural safety of the dam. By controlling water flow, they help manage the hydraulic forces acting on the dam structure, thereby preventing excessive pressure and potential damage. This contributes to the dam's longevity and safety.

Turbine operation also impacts the environmental footprint of the dam. Efficient turbines ensure that water flow is managed to minimize ecological disruption, such as maintaining appropriate flow rates downstream to support aquatic habitats and local ecosystems. Finally, turbines enable the dam to adapt to changing conditions, such as fluctuations in water levels and energy demands. Modern turbines can dynamically adjust their operation based on real-time conditions, enhancing the dam's operational flexibility and responsiveness to varying needs [5, 12].

It can be said that the functionality of turbines is essential for maximizing a dam's energy conversion, flow regulation, operational efficiency, structural safety, environmental impact and adaptability. Ensuring that turbines are well-maintained and efficiently designed is crucial for optimizing the performance and benefits of a dam.

9.5 ADVANCED REMOTE SENSING TECHNOLOGY HELPS MONITOR AND TRACK TURBULENCE DUE TO TURBINE FUNCTIONALITY

The question at hand is how advanced remote sensing technology can aid in monitoring and tracking turbulence caused by turbine functionality. There are significant challenges in studying turbulence flow in the context of the GERD due to the considerable distances involved and the potential impact on neighboring countries downstream, particularly regarding water flow shortages such as those experienced in Egypt.

Advanced remote sensing technology offers innovative solutions for monitoring and tracking turbulence caused by turbine functionality in dams. This is particularly crucial for large-scale projects like the Grand Ethiopian Renaissance Dam (GERD), where studying turbulence flow presents significant challenges due to vast distances and geopolitical tensions among neighboring countries like Egypt, which are affected by water flow shortages.

Firstly, satellites equipped with high-resolution imaging and radar sensors can monitor large areas and provide valuable data on water flow patterns and turbulence. For instance, Synthetic Aperture Radar (SAR) is particularly useful for detecting surface roughness and wave patterns, which are indicative of turbulence [17–19]. Additionally, LIDAR (Light Detection and Ranging) technology uses laser pulses to measure distances and create high-resolution maps of water surfaces. It can detect changes in water elevation and flow velocity, which are critical to understanding turbulence. For example, LIDAR can be deployed on drones or aircraft to scan the water surface near the dam, providing real-time data on flow velocities and turbulence intensity [19].

In addition, Unmanned Aerial Vehicles (UAVs), or drones, equipped with cameras and sensors, can fly over the dam and surrounding areas to collect high-resolution data on water flow and surface conditions. They offer flexibility and can cover large areas quickly. For instance, drones can be used to conduct regular surveys of the dam and its surroundings, capturing detailed images and data on water flow and turbulence [20].

Moreover, Doppler radar systems measure the velocity of moving water particles, making them ideal for tracking flow rates and turbulence. These systems can be installed on the dam structure or nearby locations to continuously monitor water flow. For instance, Doppler radar can provide continuous data on the speed and direction of water flow, helping to identify areas of high turbulence caused by turbine operations. Furthermore, Acoustic Doppler Current Profilers (ADCPs) use sound waves to measure water current velocities at various depths. They are particularly effective for monitoring subsurface turbulence and flow patterns. For example, ADCPs can be deployed in the reservoir and downstream areas to measure vertical and horizontal flow velocities, providing a comprehensive picture of turbulence [21].

However, the use of remote sensing technology in regions with geopolitical tensions, such as the GERD, requires careful coordination and cooperation among affected countries to ensure data sharing and transparency. Ensuring high accuracy and resolution of remote sensing data is crucial for effective monitoring, and advanced calibration and validation techniques are necessary to achieve reliable results. Moreover, remote sensing data should be integrated with existing monitoring and management systems to provide a comprehensive view of dam operations and turbulence impacts.

Pointless to state, advanced remote sensing technologies, including satellite imaging, LIDAR, Doppler radar, ADCPs, and UAVs, offer powerful tools for monitoring and tracking turbulence due to turbine functionality in dams. These technologies can provide critical data for managing water flow, ensuring structural safety, and minimizing environmental impacts, even in challenging and geopolitically sensitive regions like the GERD.

9.6 MICROWAVE DATA OFFERS ADVANTAGES OVER OPTICAL DATA IN ASSESSING DAM TURBULENCE FLOWS AS AN INDEX OF DAM OPERATION AND EFFICIENCY

Microwave data offers distinct advantages over optical data when analyzing dam turbulence flows, particularly as indices of dam operation and efficiency. These advantages stem from the inherent properties of microwave radiation, which make it more suitable for remote sensing in complex and dynamic environmental conditions often associated with dam operations. First, microwaves can penetrate clouds, fog, and rain, conditions that frequently hinder optical data collection. Since dam turbulence often coincides with variable weather patterns, the ability of microwaves to provide consistent data under all weather conditions ensures a continuous and reliable monitoring system. In contrast, optical data are highly susceptible to interference from atmospheric conditions, making them less reliable in regions with frequent cloud cover or precipitation [21].

Second, microwave sensors, particularly those operating in the radar spectrum, are adept at capturing surface roughness and wave patterns in water. These features are closely linked to turbulence caused by dam operations such as water discharge, spillover or turbine flow. By analyzing the variations in surface roughness and wave dynamics, microwave data can effectively model and quantify turbulence. Optical sensors, on the other hand, are more limited in their ability to detect these subtle variations, as they primarily rely on reflectance and may miss critical surface deformation patterns indicative of turbulent flow. Additionally, microwave data are less affected by ambient lighting conditions. Dams often operate at all times of the day, including during nighttime or low-light periods, which makes optical sensing less effective. Microwaves, being an active sensing technology, do not depend on external light sources and can therefore operate continuously, ensuring comprehensive data collection regardless of the time of day [19, 22].

Moreover, the interaction of microwave signals with water surfaces provides information on the dielectric properties of the water, which can be correlated with parameters such as salinity, temperature, and water content. These parameters are essential for understanding the efficiency of dam operations, especially in terms of water management and energy production. Optical data, although useful for visual observation, are not as effective in capturing such physical properties at a granular level. Consequently, microwave data provide a more robust, consistent, and detailed analysis of dam turbulence flows, making them a superior choice for assessing the operational efficiency of dams. This capability ensures that operators can monitor the health of the dam infrastructure, optimize water discharge protocols, and maximize energy generation efficiency under various environmental conditions [22–24].

9.7 THE SYNTHETIC APERTURE RADAR (SAR)

Synthetic Aperture Radar (SAR) is an advanced radar technology used for creating high-resolution images of the Earth's surface, crucial in diverse applications such as environmental monitoring, mapping and dam operation analysis. The term 'synthetic aperture' refers to SAR's ability to simulate a much larger antenna by combining signals from multiple positions along the radar sensor's flight path, effectively overcoming the limitations imposed by the physical size of conventional radar antennas (Figure 9.8). SAR operates by using the relative motion between the radar and the target to simulate a larger antenna and achieve high spatial resolution [25]. The mathematical basis behind SAR imaging involves principles of wave propagation, aperture synthesis and signal processing. Let us break this down with some relevant formulas and a numerical example to demonstrate how SAR achieves a high resolution in comparison to a conventional radar system.

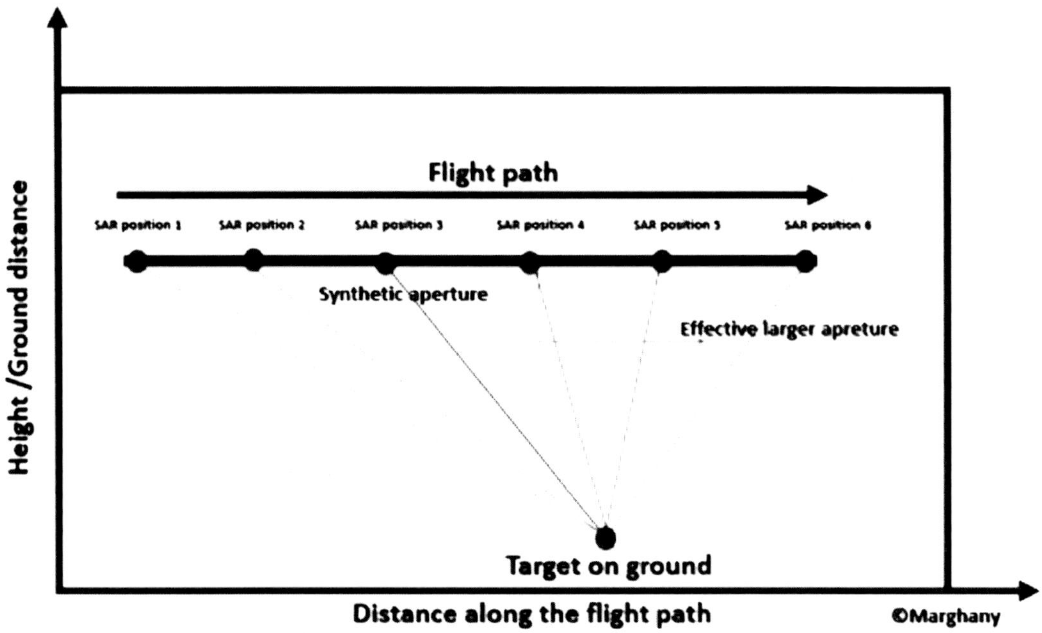

Figure 9.8 SAR concept.

Therefore, the key concepts in synthetic aperture radar (SAR) technology, specifically focus on range resolution and ground range resolution, are essential for understanding how SAR images are generated and how accurately they can represent surface features (Figure 9.9). The slant range is the direct line-of-sight distance from the satellite to the Earth's surface. It is depicted as a diagonal line from the satellite to a point on the ground. The ground range is the horizontal distance on the Earth's surface, extending from the point directly below the satellite to the point where the radar signal hits the ground. Thus, range resolution refers to the ability of the radar to distinguish between two objects that are close together in the slant range direction. In this sense, ground range resolution demonstrates the ability to distinguish between two objects that are close together in the ground range direction.

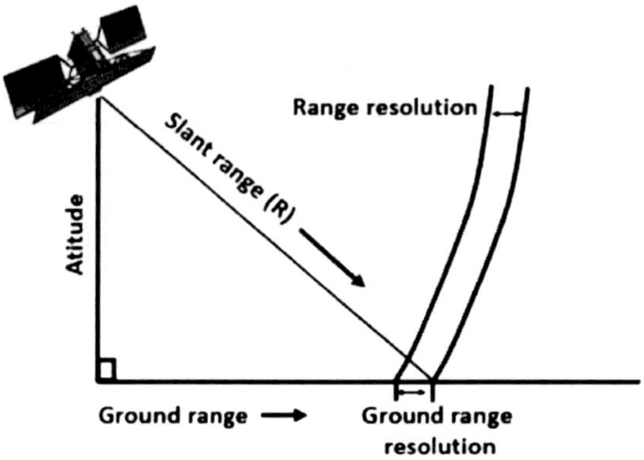

Figure 9.9 The key concepts in synthetic aperture radar (SAR) technology.

Consistent with the above perspective, let us assume that the spatial resolution Δr of a conventional radar system is determined by the size of the antenna and the radar wavelength. The azimuth resolution (along the flight direction) Δa of radar with a real aperture is given by:

$$\Delta a = \frac{\lambda R}{L} \tag{9.20}$$

where λ is the radar wavelength, R is the range to the target (distance from the radar to the surface), and L is the physical length of the antenna. For a large radar antenna, a smaller Δa (better resolution) can be achieved. However, large antennas are impractical to deploy, especially in space. SAR, therefore, synthesizes a larger aperture by combining the echoes received over a longer distance. The azimuth resolution of SAR, which is independent of range, is approximately:

$$\Delta a_{\text{SAR}} = \frac{L}{2} \tag{9.21}$$

Equation 9.21 shows that the azimuth resolution of SAR depends only on the length of the synthetic aperture, not on the distance to the target, and SAR can achieve the resolution of a large antenna even with a much smaller physical antenna [25–27]. The range resolution Δr (across the radar beam) is given by the pulse bandwidth B of the radar:

$$\Delta r = \frac{c}{2B} \tag{9.22}$$

here c is the speed of light (approximately $3 \times 10^8 \text{ms}^{-1}$), and B is the radar bandwidth.

A larger bandwidth B leads to finer range resolution. Let us illustrate the resolution of SAR versus traditional radar with a concrete numerical example. Let us assume that the wavelength of radar $\lambda = 3$ which corresponds to a frequency of about 10 GHz; typical for X-band radar); ranges to target $R = 10\text{km}$; the physical antenna length $L = 2$ m; and lastly the bandwidth $B = 100\text{MHz}$.

Using the formula for azimuth resolution of traditional radar:

$$\Delta a_{\text{SAR}} = \frac{L}{2} = \frac{2}{2} = 1\,\text{m} \tag{9.23}$$

This shows that SAR achieves a much finer azimuth resolution of 1 meter, even with a small 2-meter physical antenna. In other words, SAR mimics a large aperture through synthetic processing, resulting in significantly improved resolution. Therefore, for the range resolution, using the bandwidth formula:

$$\Delta r = \frac{c}{2B} = \frac{3\times 10^8}{2\times 10^8} = 1.5\ \text{m} \tag{9.24}$$

This means that the range resolution (across the beam) would be 1.5 meters with a 100 MHz bandwidth radar. This fine spatial resolution achieved by SAR allows it to detect small-scale surface roughness and turbulence patterns caused by dam operations, such as water discharge from turbines. In the context of dam turbulence flow, SAR can provide imagery at a resolution of 1 meter or less, allowing for precise monitoring of flow structures, turbulence, and other hydrodynamic phenomena that are critical for assessing dam efficiency and operational performance.

To unify these concepts into a novel high-resolution radar formula, we can combine both range and azimuth resolutions into a single expression that describes the overall spatial resolution Δ of the radar system [26–28]. The total spatial resolution Δ for a SAR system can be thought of as a combination of the range and azimuth resolutions:

$$\Delta_{\text{total}} = \sqrt{\Delta r^2 + \Delta a_{\text{SAR}}^2} \tag{9.25}$$

Substituting the expressions for Δr and Δa_{SAR}:

$$\Delta_{\text{total}} = \sqrt{\left(\frac{c}{2B}\right)^2 + \left(\frac{\lambda}{2}\right)^2} \tag{9.26}$$

This formula combines the bandwidth-determined range resolution and the wavelength-determined azimuth resolution to give an overall resolution of the SAR system. The fine azimuth resolution achieved by SAR allows for much sharper images than traditional radar, and this is essential for high-precision monitoring applications such as dam flow analysis and other environmental sensing tasks [25, 28].

According to the above-mentioned perspective, SAR's high resolution makes it possible to detect fine-scale variations in water surface roughness, which can be related to flow velocity changes, energy dissipation and possible areas of erosion. This high precision is indispensable for optimizing dam operation, predicting potential structural damage and ensuring environmental sustainability.

9.8 QUANTIZED MARGHANY SPECULATION FOR HIGH-RESOLUTION SAR DATA

To derive a quantized version of the high-resolution radar formula using concepts from **quantum mechanics**, we must first recognize the classical nature of the original formula 9.26 and then introduce quantum principles, such as the quantization of energy, momentum and wave-particle duality to reformulate the system in a quantum context. In quantum mechanics, we can quantize variables such as energy, momentum, and wavelength. From the point of view of photon energy, each radar pulse can be thought of as a collection of photons with energy E related to the frequency f by Planck's relation:

$$E = hf = \frac{hc}{\lambda} \tag{9.27}$$

where h is Planck's constant ($h = 6.626 \times 10^{-34}$ Js), and λ is the wavelength. Therefore, the momentum p of each photon is given by:

$$p = h\lambda^{-1} \tag{9.28}$$

Let us denote the uncertainty in position (spatial resolution) as Δx. According to the uncertainty principle, this uncertainty is related to the uncertainty in momentum (Δp) as follows:

$$\Delta x \cdot \Delta p \geq \frac{\ddot{y}}{2} \tag{9.29}$$

where Δx is the uncertainty in position, Δp is the uncertainty in momentum, and $\hbar = h\,[2\pi]^{-1}$ is the reduced Planck constant. In this view, the bandwidth B in quantum terms can be thought of as quantized energy levels or transitions between states. For the radar pulse, the bandwidth can be related to the number of quantized energy levels available [29]. Thus, the azimuth resolution Δa_{SAR}, which classically depends on the wavelength λ, can now be written in terms of the photon momentum p:

$$\Delta a_{\text{SAR}} = \frac{h}{2p} \tag{9.30}$$

This introduces a quantum mechanical dependence on momentum rather than classical wavelength. In quantum terms, consequently, the range resolution Δr, originally dependent on the bandwidth B, can be linked to the energy of photons. Using $E = hf$, and considering f is related to bandwidth by $B = \Delta f$, the range resolution can be expressed in terms of the energy:

$$\Delta r = \frac{hc}{2E} \tag{9.31}$$

The Heisenberg Uncertainty Principle gives a lower bound on the spatial resolution. Since $\Delta x \cdot \Delta p \geq \frac{\ddot{y}}{2}$, we can substitute $\Delta p = \lambda h$, yielding a minimum spatial resolution Δx:

$$\Delta x \geq \frac{\ddot{y}}{\Delta p} = \frac{\lambda}{2\pi} \tag{9.32}$$

This constraint limits how finely the radar can resolve objects, introducing quantum limits to the resolution [30]. With these quantum relations, the quantized version of the high-resolution radar formula can now be written as a combination of the momentum-based azimuth resolution and the energy-based range resolution:

$$\Delta_{\text{quantum}} = \sqrt{\left(\frac{hc}{2E}\right)^2 + \left(\frac{h}{2p}\right)^2} \tag{9.33}$$

This formula expresses the total radar resolution in terms of fundamental quantum mechanical quantities—energy and momentum—rather than classical parameters like wavelength and bandwidth. It incorporates quantum effects like the uncertainty principle, making it more fundamental for applications where quantum limits on resolution may be relevant, such as ultra-high-resolution imaging systems [28, 31].

In spotlight SAR, the radar beam is steered to continuously focus on a specific area of interest, leading to extended integration time T, which in turn improves the azimuth resolution. The azimuth resolution equation 9.30 in classical spotlight SAR can be expressed as:

$$\Delta a_{\text{SAR}} = \frac{\lambda}{2\theta} \tag{9.34}$$

here θ is the angular beamwidth of the antenna. By increasing the integration time T, we are effectively extending the synthetic aperture length L_{syn}, which improves the resolution. The synthetic aperture length L_{syn} in spotlight SAR is related to the platform velocity v and the integration time T by:

$$L_{syn} = vT \tag{9.35}$$

The angular beamwidth θ is inversely proportional to L_{syn}, so:

$$\theta \approx \frac{\lambda}{D_{syn}} = \frac{\lambda}{vT} \tag{9.36}$$

Substituting this into the azimuth resolution equation gives:

$$\Delta a_{spotlight} = \frac{\lambda}{2} \times \frac{vT}{\lambda} = \frac{vT}{2} \tag{9.37}$$

This shows that the azimuth resolution improves as the integration time T increases, leading to finer details in the image [30, 32]. Using these quantum principles, we can now derive a quantized azimuth resolution for spotlight SAR. The classical azimuth resolution, $\Delta a_{spotlight}$, depends on the platform velocity and integration time, and the quantum uncertainty introduces a lower bound on this resolution. Incorporating both the classical resolution and the uncertainty principle gives:

$$\Delta a_{quantum} = \frac{vT}{2} \cdot \sqrt{1 + \left(\frac{\ddot{\mathbf{y}}}{pL_z}\right)^2} \tag{9.38}$$

This equation shows that the azimuth resolution $\Delta a_{spotlight}$ is influenced by both the classical parameters (velocity, integration time) and quantum constraints (photon momentum, angular uncertainty). The quantum term introduces a fundamental limit to how finely SAR can resolve targets, which becomes significant for very high-resolution imaging or very small targets [33].

9.9 RADAR IMAGING MECHANISMS FOR DAM-GENERATING TURBULENCE

Exploring the complexities of range rate measurement leads us to consider the intricate concept of Doppler frequency. This phenomenon arises from subtle variations in frequency due to differences in the spacing between transmitted and received signals. An illuminating scenario arises when analyzing the linear velocity v of an aircraft over discrete time intervals, represented by the variables t_1 and t_2. During this interval dt, the aircraft dynamically transitions from point A to point B, as depicted in Figure 9.10. This dynamic motion is captured by the variables dR and dt, symbolizing the change in range and time, respectively. Within this nuanced context, the range rate, which signifies the rate of change in range, is precisely expressed by the following mathematical equation [29]:

$$\dot{R} = \frac{dR}{dt} \tag{9.39}$$

In mathematical terms, the transmitted pulse can be represented as:

$$v_T(t) = \text{rect}\left[\frac{t}{\tau_p}\right]\cos(2\pi f_c t) \tag{9.40}$$

where

$$\text{rect}\left[\frac{t}{\tau_p}\right] = \begin{cases} 1 & 0 \le t < \tau_p \\ 0 & \text{elsewhere} \end{cases}. \tag{9.41}$$

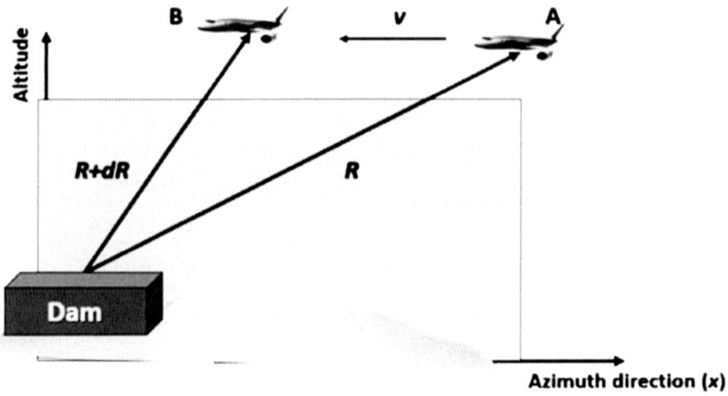

Figure 9.10 Range rate measurement by Doppler.

In the complex realm of radar dynamics, the carrier frequency denoted by (f_c) emerges as a central player, intricately linked with the pulse width (τ_p = 1 μs) and the temporal intricacies represented by (t), signifying the duration of the transmitted pulse. Conversely, the received pulse, a multifaceted entity, reveals its essence through the nuanced perspectives of delay time or the dynamic parameter of range rate denoted by (v_R). This intricate interplay of frequencies and temporal dimensions unveils an ironic tapestry of complexities, where each component assumes a distinct role in shaping a narrative that transcends traditional radar debate.

$$v_R(t) = \xi v_T(t - \tau_R) \tag{9.42}$$

Within the tangle of radar intricacies, the amplitude scaling factor, elegantly symbolized by (ξ), assumes a central role, converging with the elusive notion of range delay represented by the symbiotic (τ_R). This symbiotic relationship between mathematical elements weaves a narrative of precision, where the amplitude factor modulates the symphony of signals, and the temporal nuances of range delay unfurl, forming a mathematical tableau that reverberates with the rhythm of radar intricacies. In this context, the range delay is delineated by the equation:

$$\tau_R(t) = \frac{2R(t)}{c} \tag{9.43}$$

Expressed mathematically, the Doppler frequency as a function of range delay is formulated as:

$$v_R(t) = \xi \operatorname{rect}\left[\frac{t - \tau_R}{\tau_p}\right] \cos\left(2\pi\left(f_c + f_d\right)t + \phi_R\right) \tag{9.44}$$

The complex function $\phi_R = -2\pi f_c(2R/c)$ within this equation symbolizes the phase shift resulting from the range delay, while the enigmatic Doppler frequency shift $f_d = -f_c(2\dot{R}/c)$ elegantly narrates its story as a harmonic function of the radar wavelength denoted by λ. This equation unveils the subtle interplay between the movements of satellites and objects, influenced by the Doppler frequency shift. The comprehension of radar signal behavior in various scenarios, including the monitoring of moving objects or the examination of satellite data, hinges upon these intricate interactions [30–32]. By incorporating both the phase shift and Doppler frequency shift, this equation provides a robust foundation for analyzing the intricate dynamics inherent in radar systems which is given by:

$$f_d = -2\dot{R}/\lambda. \tag{9.44.1}$$

Equation 9.44 elucidates that the frequency of the reflected signal, denoted by $f_c + f_d$, transcends mere numerical significance to embody a celestial symbol f_c. The Doppler frequency, symbolized by f_d encapsulates the dynamic interplay of these frequencies, depicting how the transmitted and received signals coalesce in motion. Upon grasping this concept, Equation 9.44.1 becomes instrumental, facilitating the calculation of the range rate.

In theory, calculating the Doppler frequency, as Equation 9.44 proposes, presents more complexity than anticipated. This challenge stems from the differing magnitudes of f_d and f_c. Despite the formidable nature of the Doppler frequency computation, it remains achievable. To overcome this hurdle, the transmit signal must possess considerable duration, typically extending to milliseconds (ms) or longer, rather than microseconds (μs). Alternatively, the Doppler frequency can be determined through signal processing involving multiple signals.

In turbulent flow, particularly under dam structures, the vorticity, shear forces, and velocity gradients produce a highly complex pattern of surface disturbances. These disturbances scatter the radar waves back to the SAR sensor, creating a unique Doppler shift for each turbulent eddy.

The Doppler shift f_d is related to the relative motion of the radar platform and the turbulent structures in the flow. This shift is crucial for obtaining cross-range (azimuth) resolution and is expressed as:

$$f_d = \frac{2v_r \sin(\theta)}{\lambda} \tag{9.45}$$

where θ is the angle between the radar beam and the velocity vector of the moving eddies. The scattered radar signal from the turbulent flow is inherently random, but by focusing on the region over an extended integration time, SAR resolves these patterns. In this context, quantizing the spatial resolution takes on a profound meaning. The turbulent flow, which would otherwise produce a continuous, blurry image due to overlapping eddies, is now discretized into identifiable features, each associated with a specific quantization level. Therefore, each spatial interval Δx represents the smallest resolvable eddy or vortex in the dam's turbulent outflow.

The cross-range resolution in SAR, specifically in the spotlight mode, is determined by the Doppler bandwidth, which is proportional to the size of the synthetic aperture. With quantized spatial resolution, the cross-range resolution $\Delta x_{\text{quantum}}$ becomes:

$$\Delta x_{\text{quantum}} = \frac{L}{N} = \frac{\lambda R}{2vT} \tag{9.46}$$

here R is the range (distance from the radar to the target), λ is the wavelength of the radar, v_r is the platform velocity, and T is the integration time. This formula quantizes the spatial resolution by relating the radar parameters to the number of quantization levels N. Importantly, as N increases, the spatial interval $\Delta x_{\text{quantum}}$ decreases, leading to improved resolution. This is especially valuable when imaging the turbulent flow under a dam, where small-scale eddies and flow irregularities need to be resolved to assess operational efficiency.

According to the above perspective, dam-induced turbulence creates a spectrum of surface waves, vortices, and eddies of various sizes, all of which contribute to surface roughness. The velocity fluctuations in the water generate turbulent eddies, whose scales can be described by the Kolmogorov turbulence theory. The energy dissipation rate ε in turbulent flows influences the size of the eddies:

$$l_{\text{eddy}} \sim \left(\frac{v^3}{\varepsilon}\right)^{1/4} \tag{9.47}$$

where l_{eddy} is the characteristic eddy size, v is the kinematic viscosity, and ε is the turbulence energy dissipation rate. In radar imaging, these turbulent structures generate varying Doppler shifts and backscattered intensities depending on the size and velocity of the eddies. Since Bragg scattering preferentially interacts with surface roughness features that match the radar wavelength, certain eddies and rough patches will scatter more strongly than others, enhancing the radar cross section (RCS).

From the perspective of quantum mechanics, a deeper layer of interaction might exist between the radar backscatter and the water surface, specifically in the form of quantum entanglement. This notion suggests that the radar photons interacting with the turbulent water surface could become entangled with the surface states of the water itself. To understand this mechanism, we can draw an analogy to quantum entanglement between particles, where the measurement of one particle's state instantaneously affects the state of another.

In this case, the surface roughness of the water, represented by quantum states ψ_{water}, may become entangled with the incoming radar wave's quantum states ψ_{radar}, particularly during the scattering event (Figure 9.11). The backscatter from the rough surface and the radar wave would then be correlated in a quantum-entangled manner. The total wave function Ψ for the system can be described as:

$$\Psi = \psi_{\text{water}} \otimes \psi_{\text{radar}}. \tag{9.48}$$

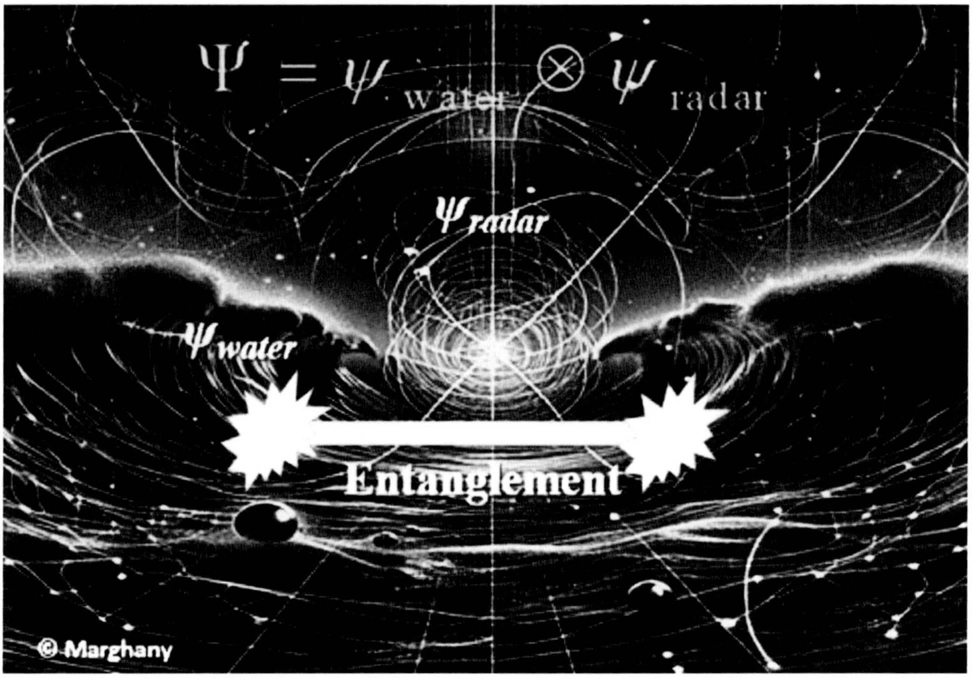

Figure 9.11 Quantum entanglement between radar wave and water surface roughness.

After the interaction, due to entanglement, the final state of the system may remain correlated in such a way that the measurement of the backscattered radar photons would provide insight into the quantum state of the surface roughness. This implies that radar backscatter not only conveys macroscopic information about roughness and turbulence but may also reflect quantum-correlated properties of the turbulent water surface.

In highly turbulent water flows, **quantum coherence** may play a role in modulating the radar backscatter. When quantum coherence is present between the radar wave and the water's surface state, the interference effects of multiple coherent backscattering events could lead to enhanced or diminished RCS (Figure 9.12).

The coherence length L_c, which characterizes the extent of quantum coherence in the radar photons, can modulate the intensity of the backscatter. When L_c is comparable to the size of the turbulent eddies, coherent backscatter is more likely, resulting in strong interference effects. If the coherence length is much shorter than the characteristic eddy size, the backscatter will exhibit classical behavior with little quantum interference [30, 33]. We can formulate a quantized model

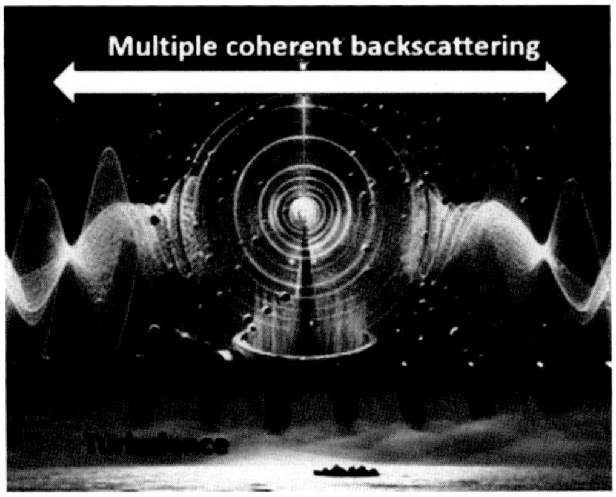

Figure 9.12 Quantum coherence in radar backscatter from dam's turbulence flow.

for radar backscatter from a turbulent water surface, taking into account quantum entanglement and Bragg scattering. The total radar cross-section $\sigma_{\text{quantized}}$ in this model can be written as:

$$\sigma_{\text{quantized}} = \left| \sum_{n=1}^{N} \psi_{\text{water}}^{n} \cdot \psi_{\text{radar}}^{n} \right|^2. \tag{9.49}$$

In equation 9.49, ψ_{water}^{n} and ψ_{radar}^{n} represent the quantum state of the n-th surface roughness feature, and the radar wave's quantum state for the n-th backscatter event; respectively. Therefore, N is the number of quantized surface features that contribute to the total backscatter. Thus, the total backscatter is the result of the sum of contributions from each quantized surface roughness element, with each contribution modulated by quantum entanglement and coherence effects.

The correlation between dam turbulence, RCS backscatter, and quantum mechanics reveals a sophisticated relationship that combines classical fluid dynamics with quantum phenomena. Through Bragg scattering, the radar wave interacts with surface roughness generated by turbulent flow, producing backscatter modulated by the eddy structure. The introduction of quantum entanglement between the radar photons and the water surface enhances our understanding of how radar backscatter encodes information about the turbulent surface, potentially offering new insights into the quantum properties of turbulent systems. This framework opens the door to quantum radar techniques that could achieve unprecedented precision in imaging complex fluid surfaces like those beneath dams.

9.10 MARGHANY QUANTIZED MULTI-OBJECTIVE ALGORITHM FOR DETECTING DAM TURBULENCE IN SAR IMAGES

Developing a quantum multi-objective algorithm for the automatic detection of dam turbulence flows in Synthetic Aperture Radar (SAR) images requires a fusion of quantum computing principles with radar signal processing and fluid dynamics modeling. SAR images provide high-resolution radar data, useful for detecting surface roughness variations caused by turbulence. Turbulence in dam environments involves complex, multi-scale phenomena that can benefit from quantum-enhanced optimization due to the computational intensity of detecting subtle features across vast datasets.

In classical computation, multi-objective optimization typically involves balancing trade-offs between these objectives using algorithms such as genetic algorithms (GA) or particle swarm optimization (PSO). A quantum-enhanced approach leverages quantum parallelism to process multiple solutions simultaneously, allowing for more efficient exploration of the solution space.

The quantum algorithm for this task can be derived using a hybrid approach, combining Quantum Approximate Optimization Algorithm (QAOA) with Quantum Pareto Front Generation. The steps below describe the process mathematically.

Let $X \in \mathbb{R}^{m \times n}$ represent the SAR image, where $m \times n$ denotes the pixel dimensions. The intensity values of each pixel, corresponding to backscattered radar signals, form the amplitude inputs to a quantum state. Let us encode the SAR image into a quantum state $|\psi_{\text{SAR}}\rangle$, representing the pixel values as probability amplitudes:

$$|\psi_{\text{SAR}}\rangle = \sum_{i=1}^{m}\sum_{j=1}^{n}\alpha_{ij}\,|i,j\rangle \otimes |f_d\rangle \tag{9.50}$$

here α_{ij} are the normalized pixel values, i.e., $\alpha_{ij} = \frac{X_{ij}}{\sqrt{\sum_{ij} X_{ij}^2}}$, and $|i, j\rangle$ represents the quantum basis state corresponding to the pixel location. In this scenario, we are tasked with analyzing how a radar pulse, represented as a quantum state $|\psi_{\text{in}}\rangle$, interacts with a turbulent water surface, creating a backscattered quantum state $|\psi_{\text{out}}\rangle$. The objective is to formulate the quantum mechanical framework for detecting and measuring Doppler shifts that indicate turbulence using Quantum Measurement Theory [33]. Upon interacting with the turbulent water surface, the radar signal undergoes scattering and transforms into a new state, $|\psi_{\text{out}}\rangle$, which is a superposition of quantum states representing different Doppler shifts. This scattering process can be written as:

$$|\psi_{\text{out}}\rangle = \int \psi(k)\,|k\rangle\, dk \tag{9.51}$$

where k is the wavenumber, related to the Doppler frequency shift f_d by the relation $k = 2\pi f\,[c^{-1}]$ with c being the speed of light. Therefore, $\psi(k)$ is the probability amplitude corresponding to each wavenumber k. Additionally, $|k\rangle$ is the quantum state corresponding to a specific Doppler shift. This superposition represents how different Doppler shifts contribute to the overall backscattered signal due to variations in surface velocities induced by the turbulent water. In this sense, the detection process involves measuring the outgoing quantum state $|\psi_{\text{out}}\rangle$, specifically focusing on the distribution of Doppler shifts, which are directly linked to the radial velocity of the turbulent water surface. This measurement process is described using quantum measurement theory [32].

Here, the author invents a novel operator named "Marghany quantum observable $\widehat{\mathfrak{M}}_d$", which serves as a key tool in the quantum mechanical analysis of the SAR signal. It allows for the detection and measurement of Doppler shifts in the backscattered radar signal, which is crucial for characterizing turbulent water surfaces, such as those near a dam. The operator $\widehat{\mathfrak{M}}_d$ measures the Doppler frequency shift $f_d(k)$ associated with each wavenumber k in the backscattered signal [33]. This shift occurs due to the motion of the turbulent water surface, which modifies the radar signal's frequency via the Doppler effect. The operator is defined as:

$$\widehat{\mathfrak{M}}_d = \int f_d(k)\,|k\rangle\langle k|\, dk \tag{9.52}$$

where $|k\rangle$ represents the quantum state corresponding to a specific wavenumber (and thus a specific Doppler shift). This operator projects the quantum state $|\psi_{\text{out}}\rangle$ (the scattered radar signal) onto the space of wavenumbers, allowing us to quantify the Doppler shifts induced by the water's turbulent motion. One of the primary uses of $\widehat{\mathfrak{M}}_d$ is to calculate the average Doppler shift in the backscattered signal [34–36]. The expectation value of $\widehat{\mathfrak{M}}_d$ in the outgoing quantum state $|\psi_{\text{out}}\rangle$ gives the average Doppler shift:

$$|f_d\rangle = |\psi_{\text{out}}\rangle \widehat{\mathfrak{M}}_d\, |\psi_{\text{out}}\rangle. \tag{9.53}$$

This expectation value corresponds to the mean radial velocity of the turbulent water surface in the region being observed by the radar. Since the Doppler shift is directly related to the velocity of the scattering surface, this value provides crucial information about the flow dynamics. Beyond the average Doppler shift, the variance and higher-order moments of the Doppler shifts can be extracted using the powers of the operator $\widehat{\mathfrak{M}}_d$. These statistical measures give further insight into the turbulence intensity and variability of the surface velocities:

$$\sigma_{f_d}^2 = \langle \psi_{\text{out}} | \widehat{\mathfrak{M}}_d^2 | \psi_{\text{out}} \rangle - \langle f_d \rangle^2. \tag{9.54}$$

Here $\sigma_{f_d}^2$ is the variance of Doppler shifts. This quantity indicates the spread of velocities on the water surface. A higher variance suggests stronger turbulence, with more significant variations in surface velocities. Thus, higher-order moments can be obtained using $\widehat{\mathfrak{M}}_d^3$ for skewness and $\widehat{\mathfrak{M}}_d^4$ for kurtosis:

$$\langle f_d \rangle^3 = \langle \psi_{\text{out}} | \widehat{\mathfrak{M}}_d^3 | \psi_{\text{out}} \rangle. \tag{9.55}$$

$$\langle f_d \rangle^4 = \langle \psi_{\text{out}} | \widehat{\mathfrak{M}}_d^4 | \psi_{\text{out}} \rangle. \tag{9.56}$$

Quantum multi-objective algorithm alternates between applying the unitary Marghany operators $e^{-i\widehat{\mathfrak{M}}_{d_1} t}, e^{-i\widehat{\mathfrak{M}}_{d_2} t}$, etc., and a cost Hamiltonian representing the objective function [30,34,36]. The parameterized quantum state after $\mathfrak{p}$ rounds is:

$$\left|\psi_{\mathfrak{p}}\left(\vec{\gamma},\vec{\beta}\right)\right\rangle = U\left(\widehat{\mathfrak{M}}_{d_1},\gamma_{\mathfrak{p}}\right)U\left(\widehat{\mathfrak{M}}_{d_2},\beta_{\mathfrak{p}}\right)....U\left(\widehat{\mathfrak{M}}_{d_1},\gamma_1\right)U\left(\widehat{\mathfrak{M}}_{d_2},\beta_1\right)|\psi_{SAR}\rangle \tag{9.57}$$

where $\vec{\gamma}$ and $\vec{\beta}$ are variational parameters to be optimized, and $U\left(\widehat{\mathfrak{M}}_{d_i},\theta\right) = e^{-i\widehat{\mathfrak{M}}_{d_i},\theta}$ is the unitary evolution of $\widehat{\mathfrak{M}}_d$ under each objective function.

The multi-objective optimization leads to a Pareto front of solutions, where no single solution dominates across all objectives. Quantum computing enables the efficient generation of these Pareto-optimal points using quantum superposition and entanglement [34–37]. The final Pareto front $\mathcal{P}$ is obtained by measuring the quantum state and evaluating the expectation values of each Hamiltonian:

$$\mathrm{P} = \left\{\left\langle\psi_{\mathfrak{p}}\middle|\widehat{\mathfrak{M}}_{d_1}\middle|\psi_{\mathfrak{p}}\right\rangle,\left\langle\psi_{\mathfrak{p}}\middle|\widehat{\mathfrak{M}}_{d_2}\middle|\psi_{\mathfrak{p}}\right\rangle,\left\langle\psi_{\mathfrak{p}}\middle|\widehat{\mathfrak{M}}_{d_3}\middle|\psi_{\mathfrak{p}}\right\rangle,\left\langle\psi_{\mathfrak{p}}\middle|\widehat{\mathfrak{M}}_{d_4}\middle|\psi_{\mathfrak{p}}\right\rangle\right\}. \tag{9.58}$$

In multi-objective optimization, the Pareto front represents the set of optimal solutions where no objective can be improved without degrading another. In the quantum context, the goal is to find a quantum state $|\psi_{\text{out}}\rangle$ that optimally balances the different objectives, i.e., the mean Doppler shift, variance, skewness and kurtosis. In other words, the Pareto front is the collection of quantum states where any improvement in one objective (e.g., minimizing variance to reduce noise) results in a trade-off in another (e.g., reducing the accuracy of the mean Doppler shift estimation). These states represent the best possible trade-offs between competing objectives, and no single state can be considered universally 'better' than the others [34, 36, 38].

In practice, finding the Pareto-optimal quantum state means solving a multi-objective quantum optimization problem, where the goal is to maximize or minimize certain objectives (e.g., maximizing accuracy in measuring the mean Doppler shift while minimizing the variance). This leads to a set of quantum states $|\psi_{\text{out}}\rangle$ that lie on the Pareto front, where:

(a) No further reduction in the uncertainty of one measurement (e.g., reducing variance) is possible without increasing uncertainty in another (e.g., reducing the mean shift accuracy).
(b) These Pareto-optimal states represent the best trade-offs in capturing different aspects of the turbulent water flow.

In mathematical terms, the problem can be described as:

$$\text{Minimize/Maximize}\left\{\left\langle\psi_{\mathrm{p}}\middle|\hat{\mathfrak{M}}_{d_1}\middle|\psi_{\mathrm{p}}\right\rangle,\left\langle\psi_{\mathrm{p}}\middle|\hat{\mathfrak{M}}_{d_2}\middle|\psi_{\mathrm{p}}\right\rangle,\left\langle\psi_{\mathrm{p}}\middle|\hat{\mathfrak{M}}_{d_3}\middle|\psi_{\mathrm{p}}\right\rangle,\left\langle\psi_{\mathrm{p}}\middle|\hat{\mathfrak{M}}_{d_4}\middle|\psi_{\mathrm{p}}\right\rangle\right\}. \tag{9.59}$$

Subject to the quantum state $|\psi_{\text{out}}\rangle$ being a valid state in the Hilbert space (i.e., normalized and physically realizable) [34–38].

9.11 EVALUATING DAM EFFICIENCY THROUGH MARGHANY QUANTIZED MULTI-OBJECTIVE ALGORITHM

To derive a normalization equation for the wave function that can estimate the dam efficiency using a Pareto-based multi-objective quantum algorithm, let us link the wave function normalization with an efficiency metric. This metric will allow us to quantify the efficiency of the dam's turbulent flow control based on the quantum measurements of Doppler shifts and other turbulence characteristics. We will utilize the wave function's probability amplitude and its expectation values for various objectives related to turbulence and flow control.

In this perspective, a quantum wave function $\psi(k)$, where k is the wavenumber related to Doppler shifts, must satisfy the normalization condition:

$$\int |\psi(k)|^2\, dk = 1. \tag{9.60}$$

This condition ensures that the total probability (the integral over all wavenumbers k is 1, as is typical in quantum mechanics. The function $|\psi(k)|^2$ represents the probability density of detecting a particular wavenumber (or Doppler shift), which is related to the surface velocity of the water in a dam's outflow region.

Let us assume that the efficiency of the dam is tied to the quantum observables associated with the turbulent flow characteristics, such as the mean Doppler shift, variance, and higher-order moments (skewness and kurtosis). Each of these objectives provides information about the flow dynamics and can be combined into a single efficiency parameter η, where:

(a) High efficiency (close to 1) means that the turbulent flow is well-managed, with optimal energy dissipation and controlled water movement.

(b) Low efficiency (around 0.5) means that the turbulence is not well-managed, leading to suboptimal energy dissipation or chaotic flow.

(c) Poor efficiency (less than 0.5) or no efficiency (close to 0) indicates that the dam is failing to control turbulence, leading to energy losses or operational issues.

According to the above-mentioned perspective, $\langle f_d \rangle$ represents the mean Doppler shift, $\langle \sigma_{f_d}^2 \rangle$ the variance, and $M_n(f_d)$ the higher-order moments (e.g., skewness and kurtosis) of the Doppler shifts, which can describe the water flow's velocity profile. These quantum observables are derived from the wave function ψ(k) and are used to define the objectives in a multi-objective optimization problem [35, 37, 38].

For the quantum Pareto optimization of dam efficiency, we want to combine these observables into a single efficiency metric. The efficiency η can be expressed as a function of these quantum observables:

$$\eta = \eta\left(\langle f_d \rangle, \langle \sigma_{fd}^2 \rangle, M_n(f_d)\right). \tag{9.61}$$

To develop a normalization condition for dam efficiency based on the quantum wave function, we propose that the overall efficiency η should be tied to the degree of normalization of the wave function $\psi(k)$, where certain flow characteristics are optimized. Let us define the efficiency metric η as a function of the normalized expectation values of the Doppler shift operator and its higher-order moments. The wave function is normalized so that the efficiency ranges from 0 to 1. Therefore, the dam efficiency η can be expressed as a ratio of the expectation value of the Doppler shift to the maximum possible value:

$$\eta = \frac{\langle f_d \rangle}{f_d^{\max}} \tag{9.62}$$

where $f_d^{\max}$ is the maximum Doppler shift that could be observed under optimal conditions. Since the Doppler shift $\langle f_d \rangle$ is related to the surface velocity, the ratio $\frac{\langle f_d \rangle}{f_d^{\max}}$ provides a normalized measure of the dam's ability to manage water flow. However, efficiency also depends on the variance of the Doppler shifts, as large variances indicate more chaotic, less efficient flows. We can adjust the efficiency equation to penalize high variance:

$$\eta = \frac{\langle f_d \rangle}{f_d^{\max}} \times \left(1 - \frac{\sigma_{f_d}^2}{\sigma_{f_d}^{\max}}\right) \tag{9.63}$$

where $\sigma_{f_D}^{\max}$ is the maximum allowable variance before the flow is considered too chaotic [34, 39]. This expression ensures that flows with high mean velocities but excessive turbulence (high variance) are considered less efficient. Therefore, equation 9.63 can be further generalized by including higher-order moments like skewness (third moment) and kurtosis (fourth moment), which can capture additional features of the flow's structure:

$$\eta = \frac{\langle f_d \rangle}{f_d^{\max}} \times \left(1 - \frac{\sigma_{f_d}^2}{\sigma_{f_d}^{\max}}\right) \times \left(1 - \frac{M_3(f_d)}{M_3^{\max}}\right) \times \left(1 - \frac{M_4(f_d)}{M_3^{\max}}\right). \tag{9.64}$$

The normalization equation for dam efficiency based on the quantum wave function and multi-objective optimization provides a metric for assessing the efficiency of turbulent flow management. This equation incorporates the mean Doppler shift, variance, and higher-order moments to quantify how well the dam controls the flow of water. The efficiency η, ranging from 0 to 1, offers a direct, normalized measure of the dam's performance in controlling turbulence, with values close to 1 indicating high efficiency and values near 0 signaling poor or no efficiency [17, 34, 37, 40].

9.12 TESTED HIGH-RESOLUTION SAR IMAGE OF GERD

The SAR images under examination capture the area encompassing the main GERD construction, delineated by geographical coordinates ranging from latitude 11°13'36.2"N to 11°11'59.9"N latitude and 35°05'9.6"E to 35°06'1.0"E longitude. These datasets were obtained using ICEYE Spotlight mode technology, operating at X-band with VV polarization and boasting a resolution of 3 m and scene size of 30 × 50 km. These images were acquired on May 9, 2022 (Figure 9.13), and July 10, 2023 (Figure 9.14) respectively. The ICEYE's Spotlight mode offers a high-resolution imaging capability. The standard Spotlight mode provides a ground resolution of 1 meter. Additionally, ICEYE has an extended version of the Spotlight mode that can achieve a resolution of up to 50 centimeters. Therefore, ICEYE's Spotlight mode typically operates with an incidence angle range of 25.98° to 26.50°. This range allows for high-resolution imaging, making it suitable for detailed analysis of surface features, such as those found in dam structures or other critical infrastructure.

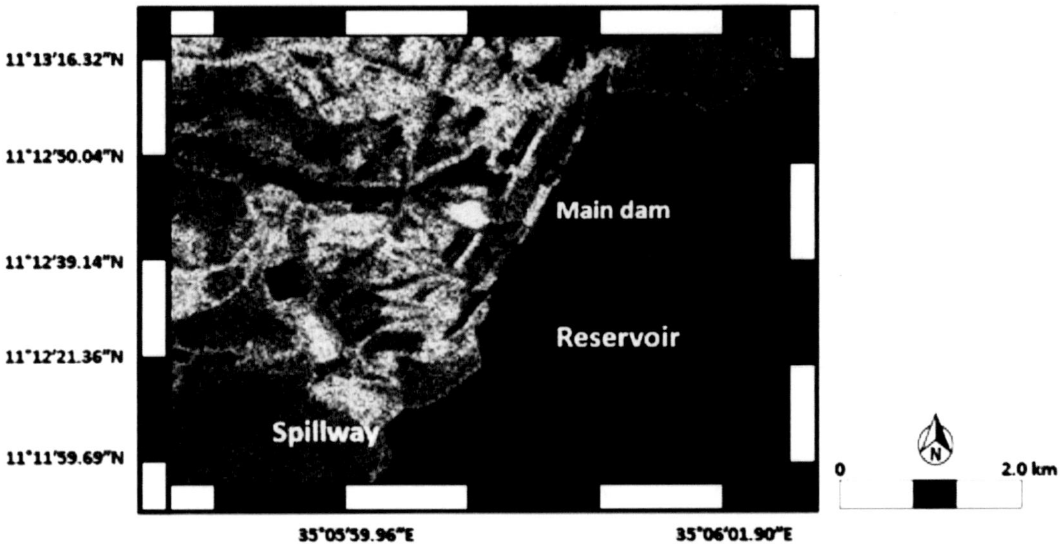

Figure 9.13 ICEYE Spotlight mode was acquired in May 2022.

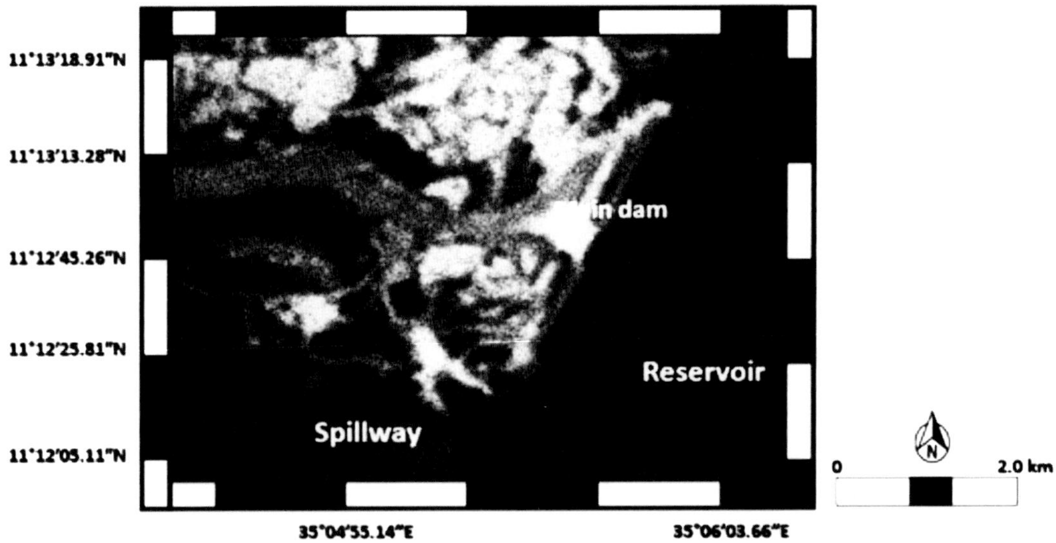

Figure 9.14 ICEYE Spotlight mode was acquired in July 2023.

Figure 9.15 demonstrates the generation of a quantum multi-objective algorithm that involves a process where the system alternates between applying unitary Marghany operators $\widehat{\mathfrak{M}}_d$ and a cost Hamiltonian that represents the objective function. The algorithm starts by preparing an initial quantum state, often a superposition of all possible states. This process leverages the principles of quantum mechanics, such as superposition and entanglement to solve complex optimization problems more efficiently than classical algorithms.

Figure 9.15 Generation of quantum unitary Marghany operators in ICEYE data.

The final Pareto front $\mathcal{P}$ demonstrates the exact determination of turbulence flow across the GERD dam from ICEYE data by measuring the quantum state and evaluating the expectation values of each $\hat{\mathfrak{M}}_d$. The central gate of the main dam is associated with the sharpest Pareto front. On the other hand, the spillway is in alliance with the mildest Pareto front as an indicator of less or no turbulent water flows (Figure 9.16).

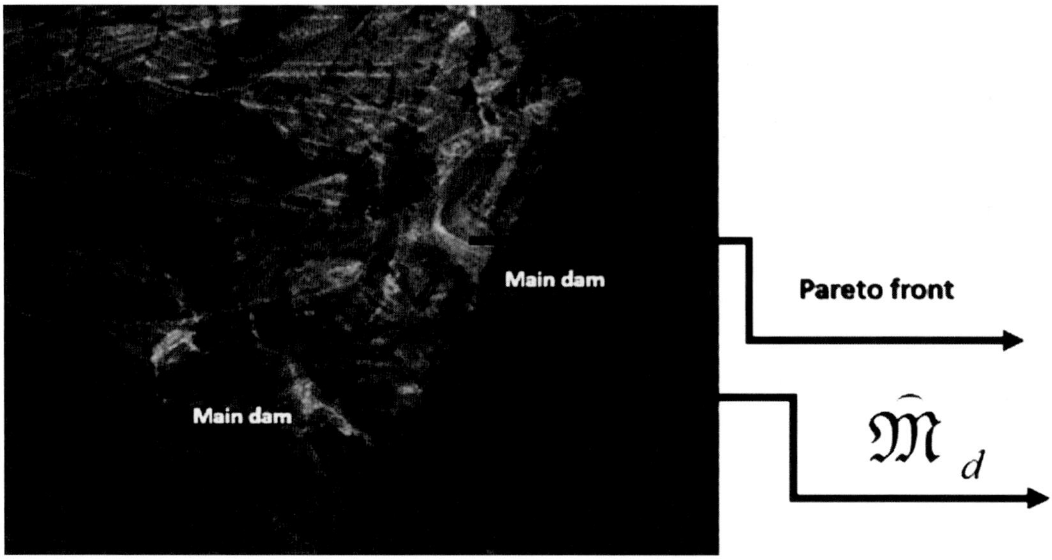

Figure 9.16 Pareto front curves engendered by unitary Marghany operators in SAR data.

Figure 9.17 demonstrates the automated detection of a confined zone of turbulence flow, which spans an area of 0.10 km^2. This turbulence is localized specifically along the central wall of the main dam. Intriguingly, this dynamic flow pattern does not extend to other critical regions of the dam structure, notably the left spillway, as well as the remaining surfaces along the opposite side of the dam, as illustrated in Figure 9.18.

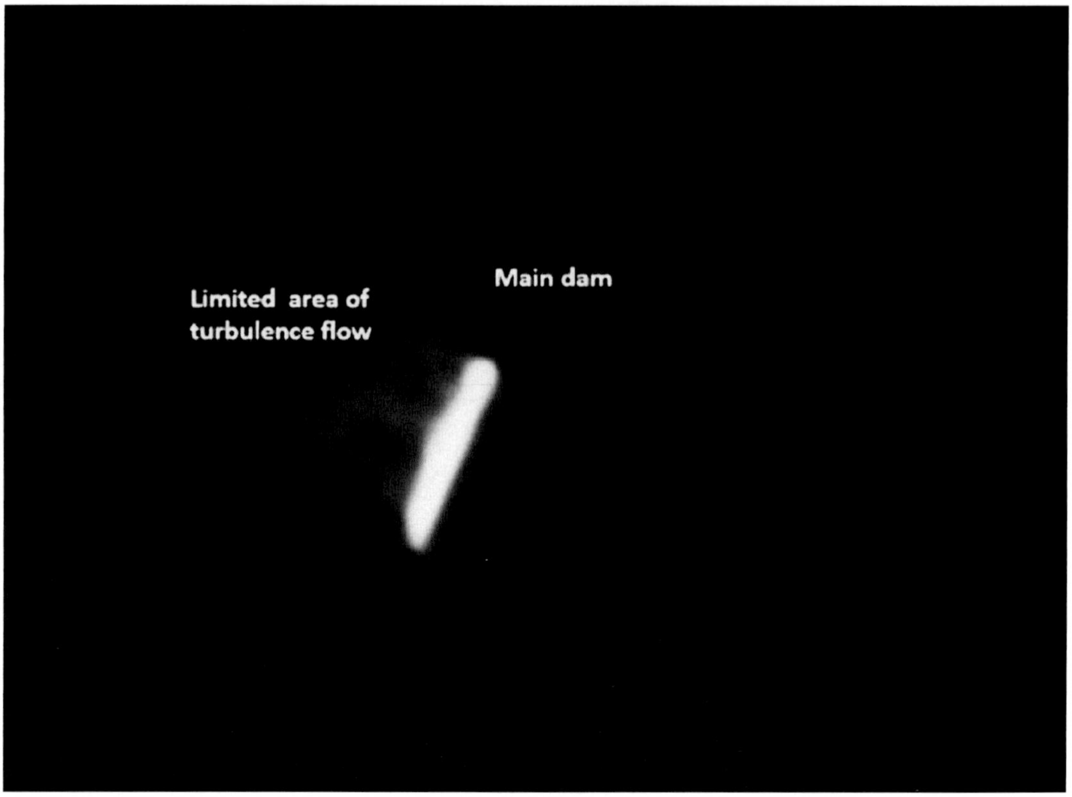

Figure 9.17 Automatic detection of turbulence flows across GERD using Marghany quantized multi-objective Algorithm.

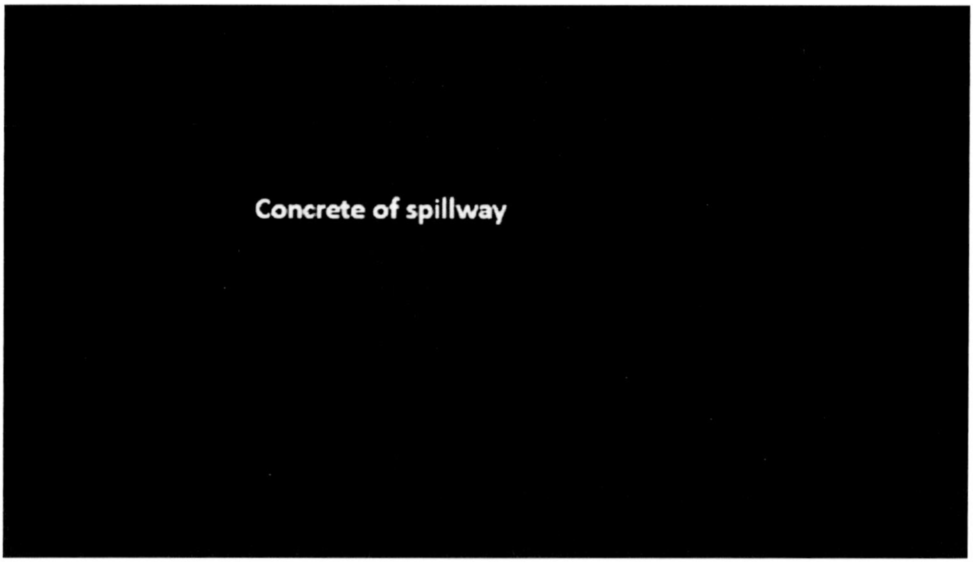

Figure 9.18 Absence of turbulence along spillway in ICEYE data.

The detection of such a concentrated turbulent flow is particularly significant in the context of hydraulic engineering and dam safety. Turbulence, while a natural byproduct of high-velocity water discharge, can have varying implications depending on where it manifests. The fact that the turbulence is restricted to a specific section—namely, the central wall of the dam—warrants closer scrutiny of the structural and hydrodynamic characteristics unique to that area.

Several factors might contribute to this observed localization. One plausible explanation is the geometry of the dam wall, which may cause uneven flow distribution, with certain regions—such as the central wall—experiencing higher fluid velocities and greater shear forces. Additionally, variations in surface roughness, or minor deviations in wall alignment, could amplify the formation of eddies and vortices, leading to the observed turbulence. This stands in contrast to the left spillway, which shows no signs of such flow disturbances, possibly due to more optimal flow conditions or differing structural alignments that prevent turbulence from forming.

From a fluid mechanics perspective, the absence of turbulence along the other side of the dam and in areas such as the left spillway suggests that flow dynamics in those regions is more streamlined. This could be the result of more efficient water management through the spillway design or differences in boundary layer development across different parts of the dam. Further investigation, potentially through computational fluid dynamics (CFD) simulations or in-situ measurements, would be beneficial to fully understand the mechanisms driving these localized turbulence patterns.

Additionally, a probability density function (PDF) of the Doppler shift shows a peak around 0 Hz, which indicates that the most probable Doppler shift is close to zero (Figure 9.19). This suggests that the majority of the observed frequency shifts are small. In this sense, the relatively narrow spread of the distribution implies that there is limited variation in the Doppler shifts. This is characteristic of limited or poor turbulence flow. In highly turbulent conditions, one would expect a broader distribution due to more significant variations in velocity [28, 33, 36, 39].

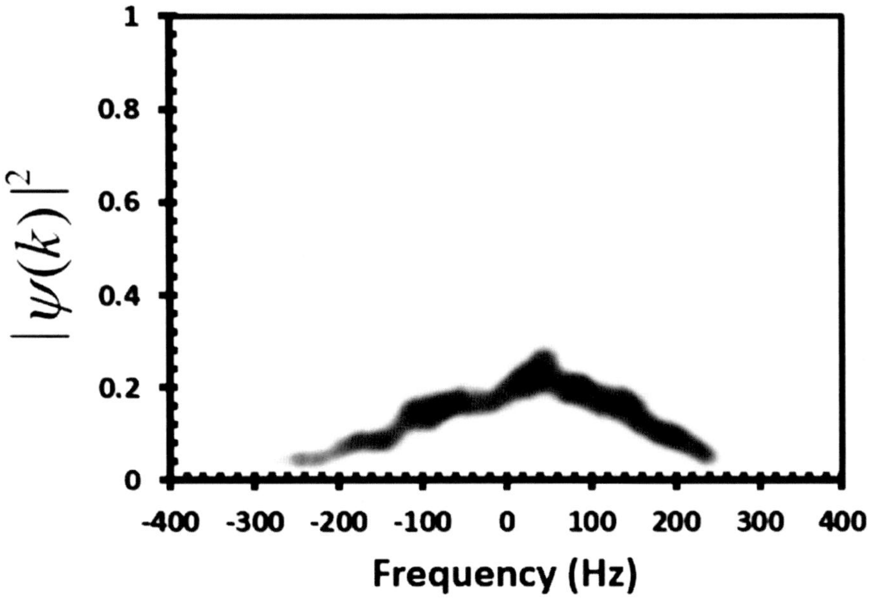

Figure 9.19 Probability density function (PDF) of the Doppler shift of turbulence flow across GERD.

Therefore, the dam's efficiency registers a notably low value of 0.43, which correlates directly with a turbulence probability density function of 0.3 (Figure 9.20). This finding aligns with the observations presented in Figure 9.19, reinforcing the conclusion that turbulence plays a critical role in modulating dam performance. Specifically, the presence of low turbulence flow can significantly reduce the dam's overall efficiency by impairing energy transfer mechanisms, promoting increased sediment deposition, causing uneven water flow distribution, and weakening hydrodynamic forces that drive turbine operations.

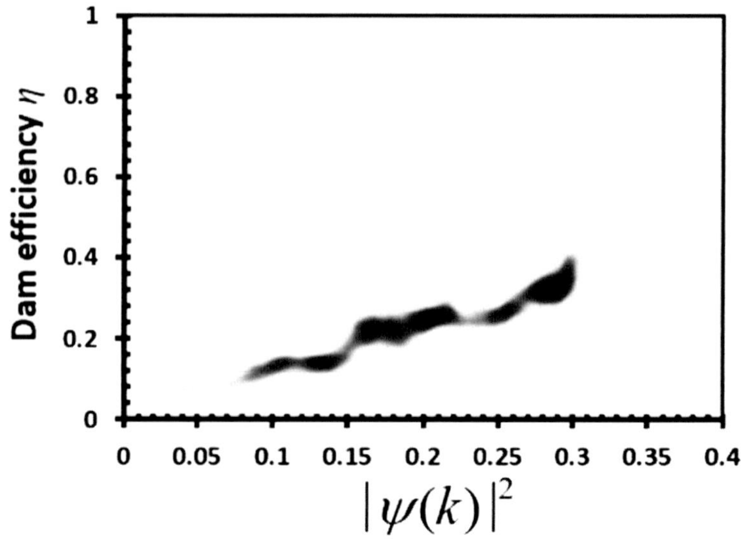

Figure 9.20 Status of GERD's efficiency as a function of water turbulence flow in ICEYE data.

The operational limitations of the dam are further underscored by the fact that only two of the six turbines are currently functional. These two turbines, designated as the 'major turbines', each possess a capacity of 400 MW, making them the primary contributors to energy generation under the existing flow conditions. The operation of these turbines appears to be a key factor behind the limited turbulence flow that has been automatically detected through ICEYE satellite imagery from 2022 to 2023.

The restricted turbulence detected by these images suggests a flow regime that is insufficient to activate all turbines effectively. In optimal conditions, a well-distributed and higher-energy flow would engage more turbines, improving overall efficiency. However, the current low-turbulence state results in suboptimal energy production, as only the most powerful turbines can function under these circumstances. This, in turn, leads to a feedback loop where reduced hydrodynamic forces diminish the dam's capacity to transfer kinetic energy to the turbines effectively, thus lowering the dam's efficiency rating. Additionally, the uneven flow conditions foster localized sediment buildup, which can further obstruct water flow and turbine operation over time [13, 35, 39].

Needless to say, the interplay between turbulence, energy transfer, and turbine functionality is evident from both the statistical correlation between the dam's efficiency and the turbulence probability, and the satellite observations of flow patterns. This nuanced understanding highlights the critical importance of managing turbulence to optimize dam performance and maximize energy output.

In the quest to address the challenges associated with turbulent flow intensity variations, optimizing efficiency, reducing turbulence, and evaluating unitary behavior across different objective functions in SAR data, Marghany Quantized Multi-Objective Algorithm (MQMA) emerges as a sophisticated solution. By applying a multi-objective optimization framework, MQMA effectively maintains a high degree of diversity within the solution space, ensuring a robust exploration of potential solutions. This diversity preservation is crucial for enhancing the accuracy of results, particularly in complex, multi-dimensional optimization problems [33–37].

A key strength of the MQMA lies in its ability to distribute population diversity along the Pareto front—a curve that represents the optimal trade-off between competing objectives. Rather than converging prematurely to suboptimal solutions, MQMA ensures that the solutions found are not only diverse but also positioned more accurately along the front, which is essential for achieving efficient optimization outcomes. This allows for the identification of solutions that strike an ideal balance between multiple objectives, leading to a more precise optimization of the Pareto front.

Furthermore, MQMA significantly enhances the core characteristics of the non-dominated set, which refers to the group of solutions that are not outperformed by any others in all objectives. The stability of this population, even in the presence of conflicting objectives such as SAR texture analyses, is indicative of the algorithm's ability to manage the complexities inherent in multi-objective optimization problems [37–40].

What distinguishes MQMA is its speculative approach, leveraging key principles from quantum computing, such as uncertainty, coherent superposition, and constructive interference. These quantum principles allow the algorithm to explore a vast solution space more efficiently, accurately identifying optimal solutions for automatic detection tasks, such as those involving turbulent flow magnitudes, non-turbulence flow, dam efficiency as a function of Doppler shift, and of the unitary of $\widehat{\mathfrak{M}}_d$ under each objective function within SAR. This capability is crucial for scenarios where accurate delineation of dam efficiency as a function of turbulence flow patterns is required, as seen in the SAR data evaluations presented in Figure 9.21.

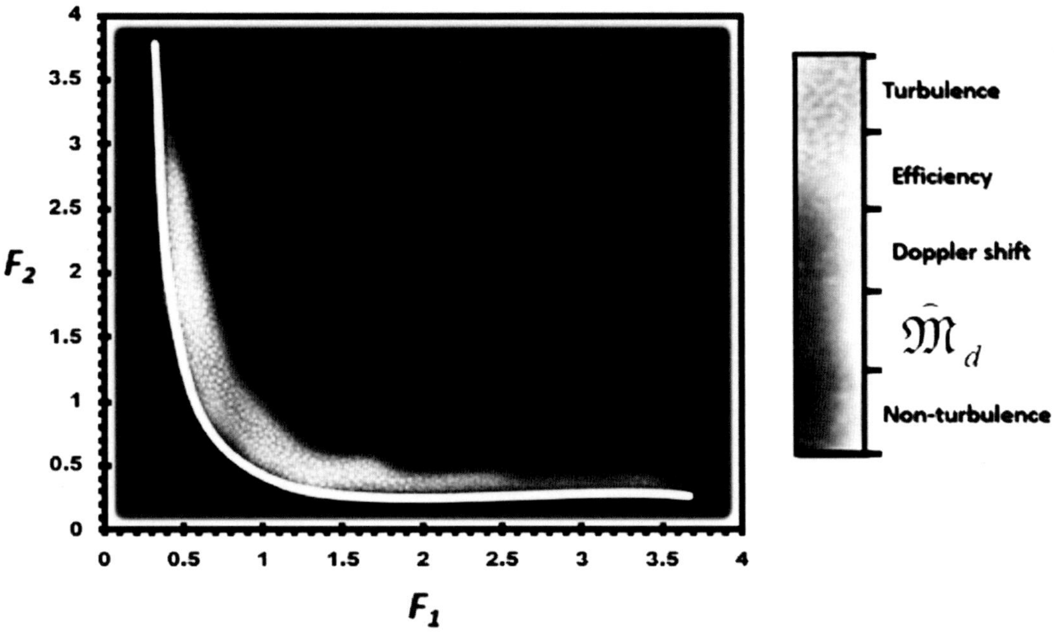

Figure 9.21 Accuracy of MQMA using Pareto front evaluation curve.

The algorithm's precision is further underscored by the numerical performance metrics observed in the Pareto front, with a Pareto front value of 3.7 in both dimensions—front one (F1) and front two (F2) axes. This consistency across dimensions highlights MQMA's superior capacity to achieve solutions that are not only mathematically optimal but also physically meaningful in real-world scenarios, ensuring that the solutions are as close as possible to the optimal trade-off between competing objectives.

Marghany Quantized Multi-Objective Algorithm represents a breakthrough in multi-objective optimization, particularly in the context of SAR data analysis, by leveraging quantum principles to enhance diversity, stability, and accuracy. Its ability to fine-tune solutions along the Pareto front ensures more efficient optimization, making it a powerful tool for solving complex, real-world problems where multiple competing objectives must be addressed simultaneously.

9.13 CONCLUSION

This chapter introduced a novel algorithm named after the author 'Marghany Quantized Multi-objective Algorithm (MQMA)' to serve as a keystone index for determining dam efficiency. The efficiency of the dam is directly linked to the turbulence generated by its turbines. The chapter highlights a specific case of limited turbulence flow of the order probability density function of 0.3, confined to an area of 0.10 km^2, which is used as an indicator of the low operational efficiency of the Grand Ethiopian Renaissance Dam (GERD). This reduced efficiency arises from the fact that only two of the six turbines are currently in operation, a factor that significantly limits the dam's overall energy output.

Furthermore, the chapter underscores the non-operational status of the GERD spillway, which remains unused. Ethiopia continues to withhold water in the reservoir, choosing not to release it downstream, as discussed in the previous chapter. This ongoing water storage is a strategic decision that further constrains the dam's operational efficiency, especially in the context of regional water management and hydropower production. The absence of active water release through the spillway underscores the delicate balance between energy generation, water storage, and international obligations regarding water-sharing with downstream countries.

By combining these elements, the chapter successfully demonstrates how turbulence flow—both in terms of its extent and intensity—can be a critical metric for assessing the performance and efficiency of large dam systems like GERD. The novel algorithm presented offers a comprehensive approach to quantifying this relationship, allowing for more accurate and dynamic assessments of dam efficiency, particularly when operational constraints, such as the number of functional turbines and water management strategies, are factored into the equation.

The upcoming chapter will delve into determining the water flow discharge at the Grand Ethiopian Renaissance Dam (GERD) using the advanced method of along-track interferometry. This technique will provide a detailed analysis of the water movement, offering deeper insights into the factors contributing to GERD's low operational efficiency.

By leveraging along-track interferometry, which enables the precise measurement of water velocity and flow dynamics over time, the chapter will aim to elucidate the underlying reasons for the dam's current performance limitations. This approach will allow for a more refined understanding of how flow patterns—both within the reservoir and downstream—are influenced by turbine operations, water storage decisions, and structural constraints.

Consequently, this analysis will shed light on how the restricted use of turbines and the non-operational spillway affect the dam's ability to generate energy efficiently, reinforcing the conclusions drawn in the previous chapters. Through this detailed examination, the chapter will provide a comprehensive perspective on GERD's hydrodynamic behavior and offer critical insights into optimizing its performance in the future.

References

[1] M. Jánosi, I., Jan, D., Szabó, K.G. and Tél, T. 2004. Turbulent drag reduction in dam-break flows. Experiments in Fluids. 37: 219–229.

[2] Lobovský, L., Botia-Vera, E., Castellana, F., Mas-Soler, J. and Souto-Iglesias, A. 2014. Experimental investigation of dynamic pressure loads during dam break. Journal of Fluids and Structures. 48: 407–434.

[3] Aureli, F., Maranzoni, A., Petaccia, G. and Soares-Frazão, S. 2023. Review of experimental investigations of dam-break flows over fixed bottom. Water. 15(6): 1229.

[4] Issakhov, A., Abylkassymova, A. and Issakhov, A. 2023. Numerical study of the dam-break flood over natural rivers with macroscopic rocks on movable beds. Computers and Geotechnics. 164: 105793.

[5] Tabbara, M., Chatila, J. and Awwad, R. 2005. Computational simulation of flow over stepped spillways. Computers and Structures. 83(27): 2215–2224.

[6] Cassidy, J.J. 1965. Irrotational flow over spillways of finite height. Journal of the Engineering Mechanics Division. 91(6): 155–173.

[7] Gaden, D.L. 2007. An Investigation of River Kinetic Turbines: Performance Enhancements, Turbine Modelling Techniques, and an Assessment of Turbulence Models (Master of Science in Mechanical Engineering). University of Manitoba, Canada.

[8] Lee, K., Kim, Y., Ko, J. and Jeong, S. 2019. A study on the debris flow-induced impact force on check dam with-and without-entrainment. Computers and Geotechnics. 113: 103104.

[9] dos Santos, I.F.S., Camacho, R.G.R. and Tiago Filho, G.L. 2021. Study of the wake characteristics and turbines configuration of a hydrokinetic farm in an Amazonian river using experimental data and CFD tools. Journal of Cleaner Production. 299: 126881.

[10] Nago, V.G., dos Santos, I.F.S., Gbedjinou, M.J., Mensah, J.H.R., Tiago Filho, G.L., Camacho, R.G.R., et al. 2022. A literature review on wake dissipation length of hydrokinetic turbines as a guide for turbine array configuration. Ocean Engineering. 259: 111863.

[11] Bailly, C. and Comte-Bellot, G. 2015. Turbulence. Springer.

[12] Lumley, J. and Blossey, P. 1998. Control of turbulence. Annual Review of Fluid Mechanics. 30(1): 311–327.

[13] Davidson, P. 2015. Turbulence: An Introduction for Scientists and Engineers. Oxford University Press.

[14] Yu, M., Shen, K., Wu, S. and Wei, H. 2013. An experimental study of interaction between bank collapse and river bed evolution. Advances in Water Science. 24(5): 675–682.

[15] Rahuel, J.L., Holly, F.M., Chollet, J.P., Belleudy, P.J. and Yang, G. 1989. Modeling of riverbed evolution for bedload sediment mixtures. Journal of Hydraulic Engineering. 115(11): 1521–1542.

[16] Jin, G., Chen, Y., Tang, H., Zhang, P., Li, L. and Barry, D.A. 2019. Interplay of hyporheic exchange and fine particle deposition in a riverbed. Advances in Water Resources. 128: 145–157.

[17] Dudhani, S., Sinha, A.K. and Inamdar, S.S. 2006. Assessment of small hydropower potential using remote sensing data for sustainable development in India. Energy Policy. 34(17): 3195–3205.

[18] Gao, H., Birkett, C. and Lettenmaier, D.P. 2012. Global monitoring of large reservoir storage from satellite remote sensing. Water Resources Research. 48(9).

[19] Villarreal-Arango, A.F., Hughes, K.S. and Morales-Vélez, A.C. 2019. Multi-decadal Earth dam deformation monitoring using airborne LiDAR and structure from motion at Lago Guajataca, Puerto Rico. *In*: Eighth International Conference on Case Histories in Geotechnical Engineering. 36–44. Reston, VA: American Society of Civil Engineers.

[20] Strelnikova, D., Paulus, G., Käfer, S., Anders, K.H., Mayr, P., Mader, H., et al. 2020. Drone-based optical measurements of heterogeneous surface velocity fields around fish passages at hydropower dams. Remote Sensing. 12(3): 384.

[21] Liu, K.T., Tseng, K.H., Shum, C.K., Liu, C.Y., Kuo, C.Y., Liu, G., et al. 2016. Assessment of the impact of reservoirs in the upper Mekong River using satellite radar altimetry and remote sensing imageries. Remote Sensing. 8(5): 367.

[22] Lee, Y.K., Hong, S.H. and Kim, S.W. 2021. Monitoring of water level change in a dam from high-resolution SAR data. Remote Sensing. 13(18): 3641.

[23] Bandini, F., Frías, M.C., Liu, J., Simkus, K., Karagkiolidou, S. and Bauer-Gottwein, P. 2022. Challenges with regard to Unmanned Aerial Systems (UASs) measurement of river surface velocity using Doppler radar. Remote Sensing. 14(5): 1277.

[24] Fulton, J. and Ostrowski, J. 2008. Measuring real-time streamflow using emerging technologies: Radar, hydroacoustics, and the probability concept. Journal of Hydrology. 357(1–2): 1–10.

[25] Chan, Y.K. and Koo, V. 2008. An introduction to synthetic aperture radar (SAR). Progress in Electromagnetics Research B. 2: 27–60.

[26] Franceschetti, G. and Lanari, R. 2018. Synthetic Aperture Radar Processing. CRC Press.

[27] Moreira, A., Prats-Iraola, P., Younis, M., Krieger, G., Hajnsek, I. and Papathanassiou, K.P. 2013. A tutorial on synthetic aperture radar. IEEE Geoscience and Remote Sensing Magazine. 1(1): 6–43.

[28] Fitch, J.P. 2012. Synthetic Aperture Radar. Springer Science & Business Media.

[29] Marghany, M. 2022. Remote Sensing and Image Processing in Mineralogy. CRC Press.

[30] Marghany, M. 2021. Advanced Algorithms for Mineral and Hydrocarbon Exploration Using Synthetic Aperture Radar. Elsevier.

[31] Chandrasekar, N. 2012. Quantum mechanics of photons. Advanced Studies in Theoretical Physics. 6(8): 391–397.

[32] Marghany, M. 2024. Remote Sensing Image Processing Algorithms for Detecting Air Turbulence Patterns. CRC Press.

[33] Marghany, M. 2024. Synthetic Aperture Radar Image Processing Algorithms for Nonlinear Oceanic Turbulence and Front Modeling. Elsevier.

[34] Marghany, M. 2024. Quantum multi-objective algorithm and radar satellite data for automatic detection of nonlinear turbulent flow in the Red Sea. *In*: Najeeb M.A. Rasul and Ian C.F. Stewart (eds). Oceanographic and Marine Environmental Studies around the Arabian Peninsula. Taylor & Francis Online.

[35] Marghany, M. 2015. Flock 1 data multi-objective evolutionary algorithm for turbulent flow detection. *In*: CD of 36th Asian Conference on Remote Sensing: Fostering Resilient Growth in Asia, ACRS. 1–6.

[36] Marghany, M. and Hakami, M. 2019. Automatic detection of coral reef induced turbulent boundary flow in the red sea from flock-1 satellite data. Oceanographic and Biological Aspects of the Red Sea. 105–122.

[37] Sadeghi Hesar, A., Kamel, S.R. and Houshmand, M. 2021. A quantum multi-objective optimization algorithm based on harmony search method. Soft Computing. 25(14): 9427–9439.

[38] Kim, Y., Kim, J.H. and Han, K.H. 2006. Quantum-inspired multiobjective evolutionary algorithm for multiobjective 0/1 knapsack problems. *In*: 2006 IEEE International Conference on Evolutionary Computation. 2601–2606. IEEE.

[39] Tao, F., Laili, Y.J., Zhang, L., Zhang, Z.H. and Nee, A.C. 2014. QMAEA: a quantum multi-agent evolutionary algorithm for multi-objective combinatorial optimization. Simulation. 90(2): 182–204.

[40] He, W. and Shi, Y. 2019. Multiobjective construction optimization model based on quantum genetic algorithm. Advances in Civil Engineering. 2019(1): 5153082.

Chapter 10

Quantum Along-Track Interferometry: Modeling Downstream Flows and Water Discharge Using TanDEM-X Data

10.1 DAM DISCHARGE

The main fundamental question now is: What is a dam discharge? In the context of hydrology and dam operations, dam discharge refers to the volume of water released from a dam over a specific period (Figure 10.1). This is typically measured in cubic meters per second (m^3/s) or cubic feet per second (cfs). Dam discharge is crucial in managing water resources, flood control, and hydroelectric

Figure 10.1 Concept of dam discharge.

DOI: 10.1201/9781003327325-10

power generation. Therefore, what are the key aspects of dam discharge? The discharge rate is controlled to manage water levels in the reservoir and downstream rivers. This helps to prevent floods, ensure a steady water supply, and maintain ecological balance (Figure 10.2).

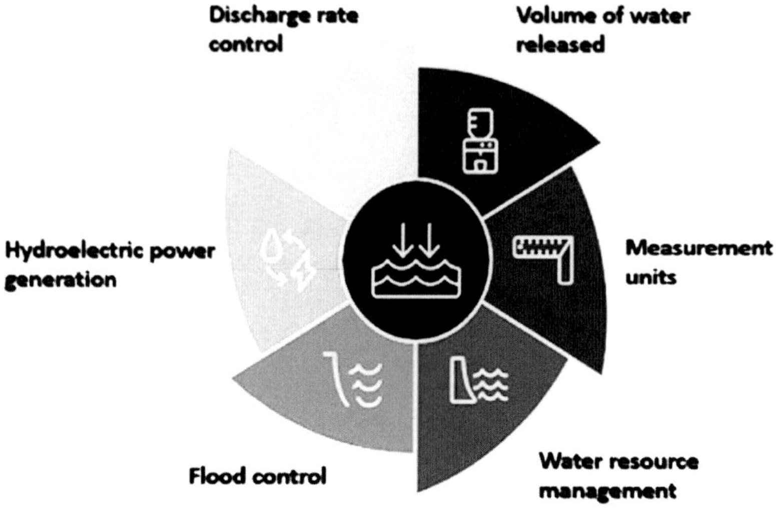

Figure 10.2 Key aspects of dam discharge.

The regulation of dam discharge (Figure 10.3) is a critical aspect of hydrological management, necessitating precise control to maintain optimal water levels in both the reservoir and downstream river systems (Figure 10.4). This regulation is indispensable for (a) flood prevention, (b) ensuring a reliable water supply, (c) and preserving ecological balance. In hydroelectric dams, the discharge rate is meticulously adjusted to maximize electricity production, often synchronized with peak demand periods to enhance energy efficiency [1–3].

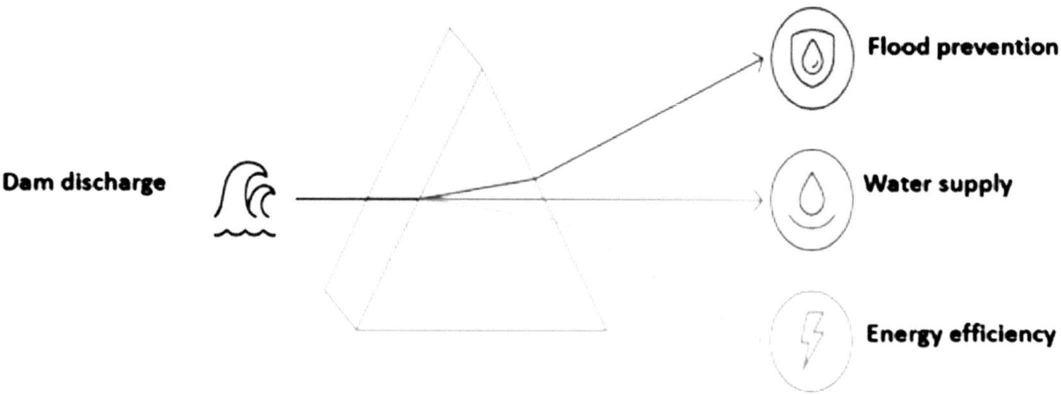

Figure 10.3 Maintaining dam discharge.

Controlled discharge is also vital for sustaining downstream ecosystems. Ensuring adequate water flow supports aquatic life and mitigates riverbed erosion, thereby maintaining the ecological integrity of the river system (Figure 10.5). Furthermore, discharge management is employed to address sediment accumulation within the reservoir. Periodic water releases help to flush out sediments, thereby preserving the reservoir's storage capacity and operational efficiency. In this regard, the discharge rate is inherently variable, influenced by weather conditions, water demand, and maintenance needs. For instance, during heavy rainfall, the discharge rate may be increased to prevent reservoir overflow, ensuring the safety and functionality of the dam infrastructure.

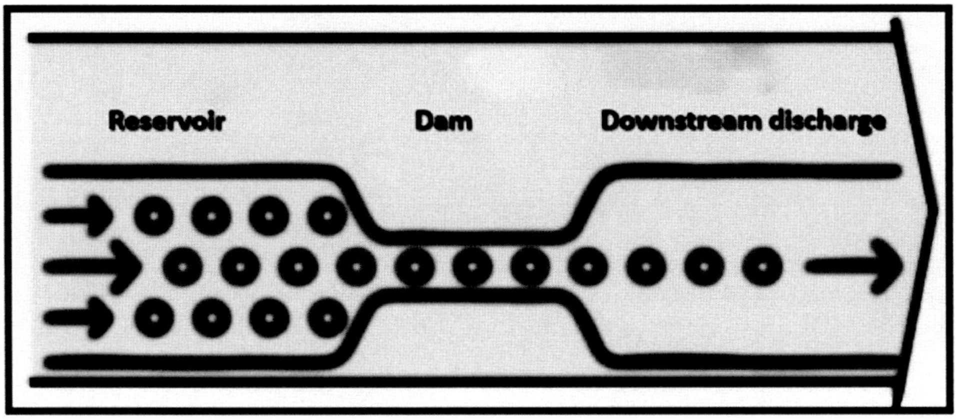

Figure 10.4 Downstream discharge.

A comprehensive understanding and effective management of dam discharge are essential for balancing the multifaceted needs of water supply, energy production and environmental conservation. This holistic approach ensures the sustainable operation of dam systems, benefiting both human and ecological communities [1, 3].

10.2 THE FASCINATING PROCESS BEHIND RELEASING WATER FROM A DAM

The release of water from a dam is a meticulously engineered process, integral to the dam's functionality and safety. Central to this process is the controlled spillway, a sophisticated system equipped with mechanical structures or gates designed to regulate the flow rate of water. The controlled spillway is a marvel of modern engineering, embodying the principles of precision and adaptability. These gates, which can be adjusted to varying degrees allow for the meticulous regulation of water flow. This capability is crucial, as it enables the dam to utilize nearly its full height for water storage throughout the year. By maintaining optimal water levels, the dam can efficiently manage water resources, ensuring a consistent supply for various needs, from agricultural irrigation to urban consumption [1, 4].

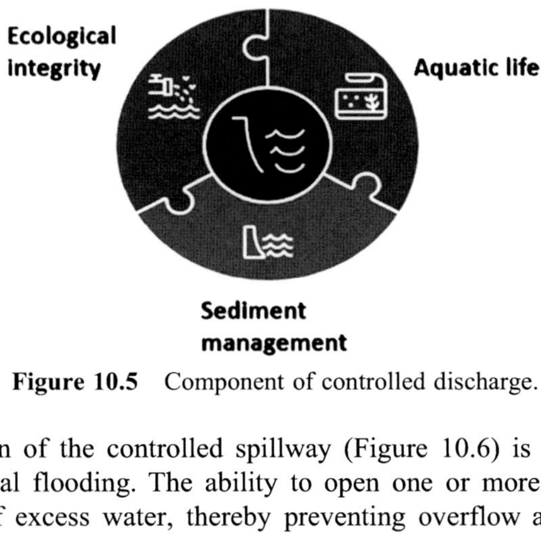

Figure 10.5 Component of controlled discharge.

Moreover, the design of the controlled spillway (Figure 10.6) is pivotal during periods of heavy rainfall or potential flooding. The ability to open one or more gates as required allows for the timely release of excess water, thereby preventing overflow and mitigating the risk of

Figure 10.6 Water discharge through a spillway.

downstream flooding. This controlled release not only safeguards the structural integrity of the dam but also protects the surrounding communities and ecosystems.

In essence, the controlled spillway exemplifies the intersection of engineering ingenuity and environmental stewardship. It reflects a deep understanding of hydrological dynamics and a commitment to sustainable water management. By harnessing the power of precise flow regulation, we can balance the demands of human consumption with the imperatives of ecological preservation, ensuring that our water resources are managed wisely and sustainably for future generations [5].

10.3 THE DIFFERENCE BETWEEN STREAMFLOW AND DISCHARGE IN THE CONTEXT OF HYDROLOGICAL STUDIES

In hydrological studies, streamflow and discharge are often used interchangeably, yet they describe different aspects of water movement in rivers and streams. Understanding the distinction between the two is essential for accurately analyzing water systems and managing resources. Both terms are crucial but apply to different scopes and levels of precision.

Streamflow refers to the overall movement of water within a natural channel, such as a river, stream, or creek. This concept encompasses both the rate of water flow and the broader patterns of movement within the stream system. Streamflow varies depending on factors such as seasonality, precipitation, and land use changes, capturing the entire dynamic behavior of water as it moves through the landscape.

Typically expressed in volume per unit of time (e.g., cubic meters per second or cubic feet per second), streamflow is influenced by various factors, including precipitation, surface runoff, groundwater inputs and evapotranspiration. It also depends on the geological and topographical features of the region. Streamflow, thus, represents a critical part of the hydrological cycle, reflecting how water moves from land surfaces into rivers and streams. In contrast, as stated in section 10.1, discharge refers to the specific volume of water passing through a defined cross-sectional area of a stream or river per unit of time. Discharge offers a more precise measure, quantifying the flow at a particular location. This makes it essential for applications that require exact measurements, such as hydraulic engineering, flood control, and dam operation.

Mathematically, discharge (Q) is determined using the equation (Figure 10.7):

$$Q = A \times v \tag{10.1}$$

where A represents the cross-sectional area of the stream, and v is the average velocity of the water. Like streamflow, discharge is measured in volume per unit of time, but it focuses on the water flowing through a specific point in the river. This specificity is crucial for monitoring, managing, and controlling water resources, particularly when precise data on water volume is necessary [1, 3, 6].

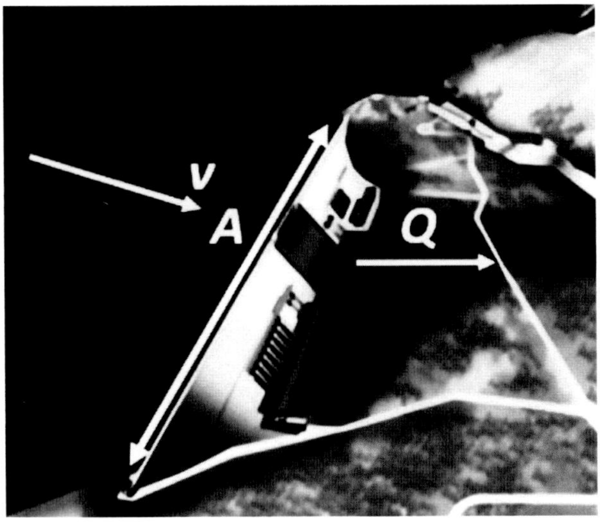

Figure 10.7 Simplification of discharge across the dam.

The primary difference between the two concepts lies in their scope and measurement focus (Table 10.1). Streamflow provides a broader view of how water moves through the entire river or stream system, accounting for variations in precipitation, groundwater and runoff. It is concerned with the overall dynamics of the flow, offering insight into long-term trends and patterns in water movement.

Discharge, however, narrows the focus to a precise quantifiable measurement of water flowing through a specific cross-section of the river at a given moment. It emphasizes accuracy and localization, making it indispensable for practical applications like flood prediction, dam regulation and water resource management. While streamflow describes general water movement, discharge offers a technical measure necessary for specific tasks.

Table 10.1 Comparison between streamflow and discharge

Aspect	Streamflow	Discharge
Definition	The flow of water within a natural or artificial channel.	The volume of water passing a given point in the channel over a specified period.
Measurement	Qualitative, reflecting the continuous movement of water.	Quantitative, typically expressed in cubic meters per second (m^3/s).
Scope	Broader, encompassing various factors influencing water movement.	Specific, providing a precise measure of water volume.
Influencing Factors	Precipitation, topography, land use, and other environmental factors.	The cross-sectional area of the channel and the velocity of the water.
Applications	Understanding the dynamics of water movement within channels.	Essential for water resource management, flood forecasting, and ecological studies.
Nature	Dynamic and variable, reflecting changes in water flow.	A snapshot measure, offering precise data at a given moment.

In practical usage, streamflow is more frequently employed in hydrological studies focused on broader processes, such as watershed modeling and climate analysis. It helps scientists understand how water flows through ecosystems and how changes in the environment affect water distribution. Discharge, on the other hand, is vital for engineering and management tasks, where precise flow rates are needed. For instance, operating reservoirs, managing irrigation, or ensuring environmental flow all rely on accurate discharge measurements to maintain balance and avoid risks like flooding or drought [3–6].

10.4 TYPES OF DAM DISCHARGES

Dams employ a diverse array of discharge systems, each tailored to serve specific purposes, ranging from flood prevention and energy production to environmental stewardship and emergency management. These mechanisms work in tandem to ensure the dam's efficiency and safety, protecting both the infrastructure and the surrounding environment from the risks posed by uncontrolled water flow. The strategic use of these discharge types allows for precise water management, balancing the needs of human activity with the preservation of natural ecosystems. Below is a detailed exploration of these discharge methods, organized logically to demonstrate their roles in water management.

10.4.1 Spillway Discharge

First and foremost, spillways serve as essential safety features that protect the dam during times of excess water (Figure 10.8). When reservoir levels exceed their safe holding capacity, spillways act as relief valves, diverting the surplus water and preventing overtopping. This, in turn, safeguards the structural integrity of the dam. There are two main types of spillways: controlled spillways, which are equipped with gates or valves that allow operators to modulate water release, and uncontrolled spillways, where water flows freely over a fixed crest without manual regulation once a critical level is reached. The primary function of spillways is to manage overflow during periods of high inflow, such as intense rainfall or snowmelt, thereby averting flooding and preventing damage to the dam structure [1, 5].

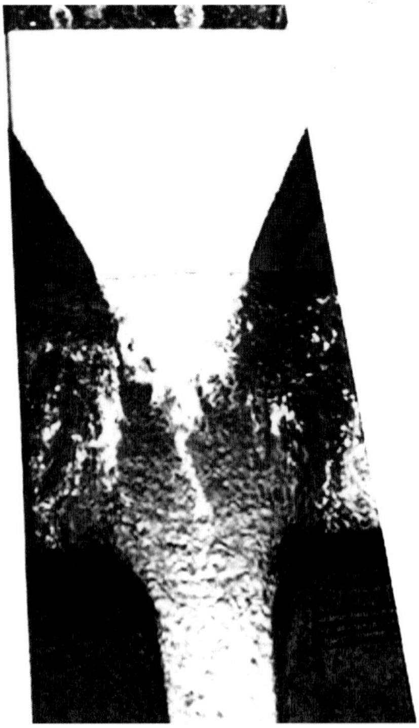

Figure 10.8 Down view of spillway discharge.

10.4.2 Sluice Gate Discharge

Another vital discharge mechanism is the sluice gate, located near the base of the dam (Figure 10.9). These adjustable openings are designed to release water from lower levels of the reservoir in a controlled manner. The sluice gate discharge is crucial for managing water levels and reducing sediment build-up. By releasing sediment-laden water from the bottom of the reservoir, sluice gates help to maintain the reservoir's capacity and prolong the operational lifespan of the dam. Consequently, this form of discharge not only regulates water flow but also addresses sediment management, which is critical to long-term dam maintenance [3, 7].

Figure 10.9 Sluice gate discharge.

10.4.3 Outlet Works Discharge

Just as important are the outlet works that consist of pipes, tunnels, or conduits that allow for controlled water discharge, typically from the lower sections of the dam (Figure 10.10). This system ensures consistent low-flow releases, which are particularly vital during dry periods when water levels may fall. The outlet works support downstream water needs by maintaining flow for ecological purposes, agricultural irrigation, municipal use and hydropower generation, while preventing excessive reservoir depletion. Thus, they play a dual role in sustaining both the dam's operations and the downstream ecosystems.

10.4.4 Powerhouse Discharge (Hydropower Release)

In hydropower dams, powerhouse discharge involves the release of water through turbines, converting its potential energy into electricity (Figure 10.11). This process exemplifies the dam's role in energy production, where water released through the turbines is returned downstream. While the primary objective of this type of discharge is electricity generation, it also helps maintain downstream flows, ensuring that water remains available for other uses such as agriculture, municipal needs and environmental preservation. Consequently, powerhouse discharge integrates energy generation with sustainable water management [1, 5, 7].

Figure 10.10 Outlet work discharge.

10.4.5 Bottom Outlet or Drainage Discharge

For more specialized needs, bottom outlets or drainage discharge are located at the lowest point of the dam and allow for the complete or near-complete draining of the reservoir (Figure 10.12). These outlets are particularly important for emergency drainage or during maintenance periods. In emergencies, bottom outlets enable a rapid reduction in water levels, while during routine maintenance, they allow for dam inspections and repairs by significantly lowering the reservoir water. This function is critical for maintaining the dam's integrity and operational efficiency [4–6].

Figure 10.11 Dam powerhouse discharge.

Figure 10.12 Dam bottom outlet discharge.

10.4.6 Flood Discharge

When faced with extreme weather events, such as storms or floods, dams may require rapid discharge of large volumes of water to prevent overtopping. This is typically achieved via flood discharge systems, often using spillways or auxiliary channels to release water swiftly. By releasing substantial quantities of water, flood discharge systems reduce the risk of dam failure during periods of extreme inflows, thereby protecting both the dam and downstream communities. This form of discharge is essential for safeguarding the structure during high-risk events [2–5].

10.4.7 Environmental Flow Release

In some cases, dams are obligated to release water to preserve the health of the downstream ecosystem. Environmental flow releases are designed to mimic the natural flow patterns of the river, supporting aquatic life, vegetation, and water quality. This type of discharge is critical for maintaining environmental sustainability, especially in areas where the natural flow regime has been altered by the dam. By ensuring ecological balance, environmental flow releases contribute to the long-term health of riverine ecosystems [1, 4, 7].

10.4.8 Emergency Discharge

Emergency discharge refers to the rapid release of water in critical situations, such as imminent dam failure or severe flooding. Both spillways and bottom outlets may be utilized to relieve pressure on the dam quickly. The primary goal of emergency discharge is to avert structural failure by reducing water levels and preventing catastrophic damage. In these situations, the dam's ability to discharge water rapidly is crucial to ensuring public safety and preserving the infrastructure.

Needless to say, dams employ a range of discharge systems, each designed to fulfill specific operational and safety needs. From spillway discharges that protect against overflow to sluice gates and outlet works that regulate water levels and maintain downstream flows, each mechanism plays a critical role. Moreover, powerhouse discharges link energy generation with water management, while flood, environmental, and emergency discharges ensure the dam's ability to respond to extreme conditions and ecological needs. Together, these systems form an integrated approach to water management, balancing human utility, structural safety, and environmental sustainability.

10.5 DO DAMS INFLUENCE THE DISCHARGE OF WATER: DO THEY INCREASE OR DECREASE IT?

Dams, by their very design, typically reduce downstream discharge, particularly during periods of normal or low precipitation. To begin with, when dams collect and store water within their reservoirs, a significant portion of the river's natural flow is withheld. This impoundment, especially during the initial phase—which can stretch over months or even years—drastically diminishes the volume of water available to downstream communities. The result is often profound, as people, agriculture, and ecosystems dependent on a steady flow face the stark reality of water scarcity [3–6].

Moreover, when dams are not merely storage systems but also mechanisms for water abstraction—whether for irrigation, industrial, or domestic purposes—the strain on downstream flow is exacerbated. By diverting significant quantities of water for human use, these activities further compromise the river's natural discharge, leading to a marked decrease in the water volume available downstream.

However, dams are not solely instruments of reduction. In contrast, they possess the capability to regulate discharge, serving as guardians against the extremes of flooding. By capturing excess water during periods of heavy rainfall and releasing it in controlled measures, dams prevent the catastrophic inundation of downstream regions. Yet, this flood control comes at a cost: during dry spells, when the need for conservation prevails, discharge is often curtailed to preserve reservoir levels. As a result, downstream communities may experience critically low water flow during these already vulnerable periods [2–4].

Furthermore, while the controlled release of water can smoothen the natural fluctuations of the river, reducing the extremes of flood and drought, this stabilization can have unintended consequences. Indeed, the regular seasonal high flows, which many ecosystems rely on for replenishment and sustenance are reduced. Floodplains, wetlands, and aquatic habitats that depend on these periodic surges suffer, as the natural rhythm of the river is altered [1, 3, 5, 7].

Finally, the broader environmental ramifications of reduced downstream discharge are profound. Not only does the diminished flow limit the availability of water for drinking, irrigation, and industry, but it also disrupts the natural sediment transport—a key process that sustains riverine ecosystems and geomorphology. Over time, this disruption can degrade habitats, alter landscapes, and weaken the ecological health of entire regions [5–7].

Thus, while dams undoubtedly confer benefits—offering flood control, water storage, and resource allocation—they often do so at the expense of the natural flow and health of downstream systems. In particular, when large-scale abstraction or reservoir filling is involved, the reduction in downstream discharge becomes a significant and lasting challenge.

10.6 MATHEMATICAL EXPLANATION OF WATER DISCHARGE FROM A DAM

The process of water discharge from a dam is a critical balancing act to ensure the safety of the structure and manage the downstream flow. Mathematically, this process can be expressed through fundamental equations that account for physical properties such as water head (height difference), discharge area, and the characteristics of the dam's outlets, such as spillways, gates, or turbines. In this perspective, the flow rate, or discharge Q, from a dam can be expressed using principles derived from Bernoulli's equation and the continuity equation. The general formula for water discharge through an outlet, such as a spillway [1, 3, 7], is:

$$Q = C_d A \sqrt{2gh} \tag{10.2}$$

The discharge rate, denoted as (Q), represents the volume of water flowing per unit of time. This rate is influenced by several factors, each playing a crucial role in determining the overall discharge. First, the discharge coefficient, (C_d), accounts for frictional losses and inefficiencies within the system. This coefficient is essential as it adjusts the theoretical discharge to reflect real-world conditions, ensuring accuracy in calculations. Additionally, the cross-sectional area of the outlet, (A), directly impacts the discharge rate. A larger cross-sectional area allows more water to flow through the outlet, thereby increasing the discharge rate.

Moreover, the acceleration due to gravity, (g), which is approximately (9.81 ms^{-2}), influences the velocity of the water as it exits the outlet. This gravitational force drives the water flow, contributing to the overall discharge rate. Finally, the water head, (h), or the height of water above the outlet, plays a significant role. A higher water head results in greater pressure at the outlet, which in turn increases the discharge rate.

By considering these factors—discharge coefficient, cross-sectional area, acceleration due to gravity, and water head—one can comprehensively understand and calculate the discharge rate of water from a system. According to the above-mentioned perspective, equation 10.2 proves that discharge depends on both the physical dimensions of the outlet and the height of the water above the gate. The higher the water head, the greater the potential energy that is converted into kinetic energy as water flows through the opening.

This means Dams are equipped with mechanisms to regulate the flow of water, ensuring downstream flooding is avoided, while also maintaining appropriate water levels behind the dam. By adjusting the cross-sectional area A, for example, by opening or closing gates or controlling turbines, operators can influence the discharge rate Q. If the discharge rate is not properly managed, it can lead to serious consequences, both for the dam structure and the surrounding environment. The inability to discharge water effectively results in rising water levels behind the dam, which in turn increases the pressure on the dam itself [1, 5, 8].

The critical question now is: what happens if a dam does not discharge water?

10.6.1 Water Level Rise and Pressure Increase

When the dam fails to release water, the water level behind the dam increases, leading to a corresponding increase in hydrostatic pressure exerted on the dam. The pressure P exerted by the water at a given depth h is:

$$P = \rho g h \tag{10.3}$$

In equation, 10.3, the pressure, denoted as (P), is a fundamental parameter in fluid dynamics. It is influenced by several key factors, each contributing to the overall pressure exerted by the fluid. Firstly, the density of water, (ρ), which is approximately (1,000 kgm^{-3}), plays a crucial role. The density determines the mass of water per unit volume, directly affecting the pressure. Additionally, the acceleration due to gravity, g influences the force exerted by the water. This gravitational force acts on the water's mass, contributing to the pressure. Moreover, the water height, (h), or the height of the water column, is a significant factor. A greater water height results in higher pressure at the base of the column, as the weight of the water above exerts a downward force [1, 8].

By considering these factors—pressure, density of water, acceleration due to gravity, and water height—one can comprehensively understand and calculate the pressure exerted by a fluid in a given system. In this understanding, as the water level rises, the pressure increases linearly with height, exerting greater force on the dam. If left unchecked, this can compromise the dam's structural integrity.

10.6.2 Structural Overloading of the Dam

The total horizontal force F exerted by the water on the dam is given by:

$$F = \frac{1}{2}\rho g h^2 L \tag{10.4}$$

where F is the total force, and L is the length of the dam along the river. This equation illustrates that the force increases quadratically with the height h. A small rise in water height leads to a significant increase in the force exerted on the dam. If the force exceeds the dam's design capacity, the structure may crack, deform, or fail entirely [1, 5, 9].

10.6.3 Potential Dam Break

If the dam cannot discharge water, the risk of overtopping or structural failure becomes imminent. In the worst-case scenario, a dam break can occur, resulting in catastrophic downstream flooding. The outflow following a dam break can be described by the Manning equation, which models open-channel flow:

$$Q_{\text{break}} = A\left(\frac{1}{n}\right)R^{2/3}S^{1/2} \tag{10.5}$$

The discharge after a dam break, denoted as (Q), is a critical parameter in understanding the dynamics of water flow following such an event. This discharge is influenced by several key factors, each contributing to the overall flow rate. Firstly, Manning's roughness coefficient, (n), plays a significant role. This coefficient accounts for the frictional resistance exerted by the channel's surface, affecting the velocity and discharge of the water. A higher roughness coefficient indicates greater resistance, thereby reducing the flow rate.

Additionally, the hydraulic radius, (R), is a crucial factor. It is defined as the cross-sectional area of flow, (A), divided by the wetted perimeter. The hydraulic radius influences the efficiency of the flow, with a larger radius typically resulting in a higher discharge rate. Moreover, the slope of the water surface, (S), is another important parameter. This slope represents the gradient of the water surface and directly impacts the gravitational force driving the flow. A steeper slope increases the velocity and discharge of the water. Finally, the cross-sectional area of flow, (A), is fundamental in determining the discharge rate. A larger cross-sectional area allows more water to pass through, thereby increasing the discharge after a dam break [7–10].

By counting these factors—discharge after a dam break, Manning's roughness coefficient, hydraulic radius, slope of the water surface, and cross-sectional area of flow—we can comprehensively understand and calculate the flow dynamics following a dam break.

According to the above-mentioned perspective, if a dam fails to discharge water properly, upstream flooding becomes an immediate consequence. As water accumulates behind the dam, areas upstream are at risk of flooding, which can severely damage infrastructure, inundate farmlands and displace local communities. Moreover, as water pressure increases, the dam itself faces heightened risk due to excessive structural load. The growing pressure can deform or even crack the dam, weakening its integrity. In extreme cases, if the water level overtops the dam, the downstream side may suffer from erosion, potentially leading to a structural breach.

Additionally, the dam's critical components, such as spillways and floodgates, can be overwhelmed if they fail to handle the excess water. When these mechanisms are overloaded, they may not function properly, resulting in uncontrolled flooding downstream. Furthermore, insufficient water discharge allows sedimentation to accumulate in the reservoir. As sediment builds up, the storage capacity of the dam is reduced, which can impair the efficiency of hydropower generation and increase the likelihood of blockages. Finally, downstream drought is a severe outcome of insufficient discharge. When water is not adequately released, downstream ecosystems, agricultural lands, and water supplies suffer. This can lead to drought conditions, a loss of biodiversity, and significant socio-economic challenges for communities that depend on the river's flow for their livelihoods [4, 5, 9].

10.7 CHALLENGES IN MODELING GERD DISCHARGE

Modeling the discharge of the Grand Ethiopian Renaissance Dam (GERD) presents several challenges, particularly when it comes to obtaining in situ field data. In this sense, there is often a lack of continuous and reliable hydrological data. In this perspective, for instance, the measured rainfall data are not freely available from the National Meteorological Agency of Ethiopia. This unavailability can lead to uncertainties in model predictions.

Therefore, the limited availability of advanced monitoring technologies and infrastructure can restrict the quality and quantity of data collected. Consequently, the Nile River flows through multiple countries, making data sharing and collaboration challenging due to political and legal constraints. Additionally, political instability and security issues in the region can hinder field data collection efforts. To mitigate these challenges, researchers often rely on remote sensing and satellite data, which can provide valuable insights despite the limitations of in situ data [10–15].

However, the remote sensing techniques currently employed to monitor the Grand Ethiopian Renaissance Dam (GERD) reservoir are primarily limited to medium-resolution data sources. These include vegetation indices, digital elevation models (DEMs), and data from Sentinel-1; Sentinel-2; and Landsat satellites.

In this perspective, using vegetation indices to monitor water discharge from the GERD presents several challenges. Vegetation indices, such as the Normalized Difference Vegetation Index (NDVI), primarily measure vegetation health and density. They do not directly measure water discharge, making it difficult to accurately infer water flow from these indices. Furthermore, the temporal variability of vegetation indices can be influenced by seasonal changes, agricultural activities and natural vegetation cycles. This variability can introduce noise and reduce the accuracy of water discharge estimates. Additionally, vegetation indices derived from low or medium-resolution satellite data may not capture fine-scale variations in vegetation that could be indicative of changes in water discharge. Subsequently, DEMs are crucial for understanding the topography and elevation changes in the reservoir area. They help in modeling water flow and storage capacity. However, the resolution of available DEMs may not capture fine-scale changes accurately [10, 13, 15].

Monitoring the Grand Ethiopian Renaissance Dam (GERD); therefore, using optical remote sensing and Sentinel-1A presents several challenges. Firstly, optical sensors rely on clear skies to capture images. Consequently, cloud cover, which is common in the region, can obstruct the view, leading to gaps in data collection. Additionally, optical sensors can only capture images during daylight hours, which limits the temporal resolution of the data. Furthermore, water bodies can reflect sunlight, causing glare and reducing the accuracy of measurements [12, 14, 16].

On the other hand, Sentinel-1A, which uses Synthetic Aperture Radar (SAR), also faces its own set of challenges. For instance, SAR data can suffer from geometric distortions such as foreshortening, layover, and shadowing, especially in areas with complex topography like the GERD region [16]. Moreover, changes in the surface properties between successive SAR acquisitions can lead to temporal decorrelation, affecting the accuracy of measurements. Additionally, dense vegetation can interfere with the radar signal, complicating the interpretation of SAR data [11, 17].

The vital question now is: what is an extremely accurate remote sensing technique for modeling millimeter-scale changes in river surface flow, as opposed to the challenges posed by both optical and SAR data?

10.8 PRINCIPLES OF INTERFEROMETRIC SYNTHETIC APERTURE RADAR (INSAR)

For modeling millimeter-scale changes in river surface flow, Interferometric Synthetic Aperture Radar (InSAR) is a highly accurate remote sensing technique. InSAR uses the phase differences between multiple SAR images to detect minute changes in the Earth's surface, including water flow. This method can measure deformations with millimeter precision over time, making it suitable for detailed hydrological studies. InSAR overcomes some of the limitations of optical and traditional SAR data by providing high-resolution measurements regardless of weather conditions or daylight. This makes it particularly effective for continuous monitoring of river surface flow changes.

Interferometric Synthetic Aperture Radar (InSAR) is a radar technology based on the interference of at least two coherent radar pulses, as illustrated in Figure 10.13. The principle underlying this interference can be mathematically captured through the superposition principle, which governs the behavior of electromagnetic waves. The general expression for the electric field resulting from two interfering waves, denoted as S_1 and S_2, is given by:

$$S_{12} = S_1 + S_2 + 2\sqrt{S_1 S_2}\cos(\phi) \tag{10.6}$$

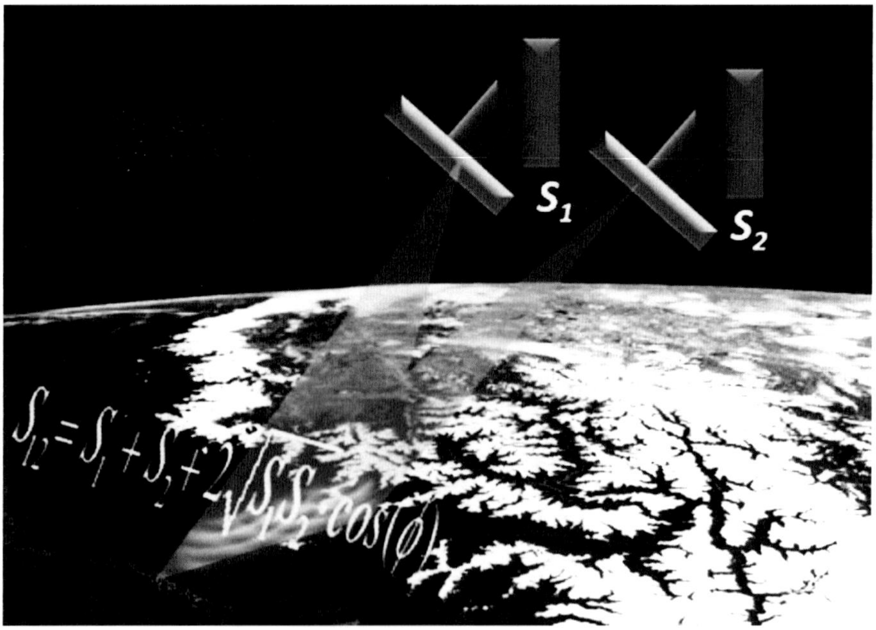

Figure 10.13 Interference of dual SAR pulses.

Here, the term involving the cosine of the phase difference, φ, determines the resulting interference pattern. This interference leads to the backscattering of the phase difference between the pulses, which, in turn, enables the detection of deformations on the millimeter scale over time intervals that can span from days to years. To understand the phase mechanism in more detail, consider a real variable, such as time (t). The phase of a periodic function F, analogous to an angle, represents the number of periods completed by the variable, and is denoted by:

$$\phi(t) = 2\pi \frac{(t - t_0)}{f} \tag{10.7}$$

According to Equation (10.7), the phase difference depends on the rate of change of time in inverse proportion to the frequency of the radar pulse, represented by f. When the frequencies of the two pulses differ slightly, the phase deviation, $\delta(t)$, grows linearly with time. This linear phase progression leads to periodic fluctuations, a phenomenon known as beating. Beating manifests as an interference pattern between two signals of slightly different frequencies, perceived as a periodic variation in intensity, with the rate of fluctuation determined by the frequency difference between the two signals.

In scenarios where the frequencies are not identical, the temporal phase deviation grows linearly with time, and the periodic reinforcement and cancellation of the waves produce the 'beating' pattern. Essentially, beating is the interaction of two signals with marginally different frequencies, generating periodic changes in signal strength, where the rate of these fluctuations corresponds to the difference in the two frequencies.

The concept of radar interferometry can be seen as a natural extension of Young's classical interferometry. In the Young double-slit experiment, light waves passing through two slits create an interference pattern of bright and dark bands. This pattern arises from constructive and destructive interference, which occur when waves overlap in or out of phase, confirming the wave nature of light. Both in Young's experiment and radar interferometry, interference occurs when waves from two sources combine. In Young's case, light waves create visible interference fringes; in radar interferometry, radar waves create phase differences that indicate surface details. In this sense, just as Young's experiment shows constructive and destructive interference based on phase alignment, but radar interferometry uses phase differences between radar waves to infer details about the terrain. Changes in these patterns over time reveal movements such as land deformation, tectonic shifts, or subsidence. Young's experiment indirectly measures wavelength and other properties of light. In radar interferometry, phase shifts between returning radar signals allow precise distance measurements and three-dimensional mapping. To put it differently, Young's scattering refers more broadly to the scattering of waves (light or radar) from different points on a surface, resulting in complex interference patterns that encode information about the surface. In radar interferometry, scattered radar waves from surface features contribute to the overall interference pattern, offering highly detailed surface measurements [33].

In the radar context, the two Synthetic Aperture Radar (SAR) antennas act as coherent point sources. Due to the relatively long wavelengths of radar signals, the signal can be digitized and processed coherently, allowing for the direct measurement of phase information [18–21]. When imaging a surface, the phase fronts from these two sources interfere, and the resulting phase differences encode topographic information. The surface topography effectively slices through the interference pattern, revealing detailed features of the landscape.

Interferometric radar systems can be fundamentally divided into two categories, distinguished primarily by the geometric configuration of their baseline vector. The baseline, defined as the vector between the positions of the antennas in the interferometric pair, serves as the critical parameter that determines the system's classification. Thus, radar interferometry is categorized into two types: (i) cross-track interferometers (see Figure 10.14) and (ii) along-track interferometers (see Figure 10.15).

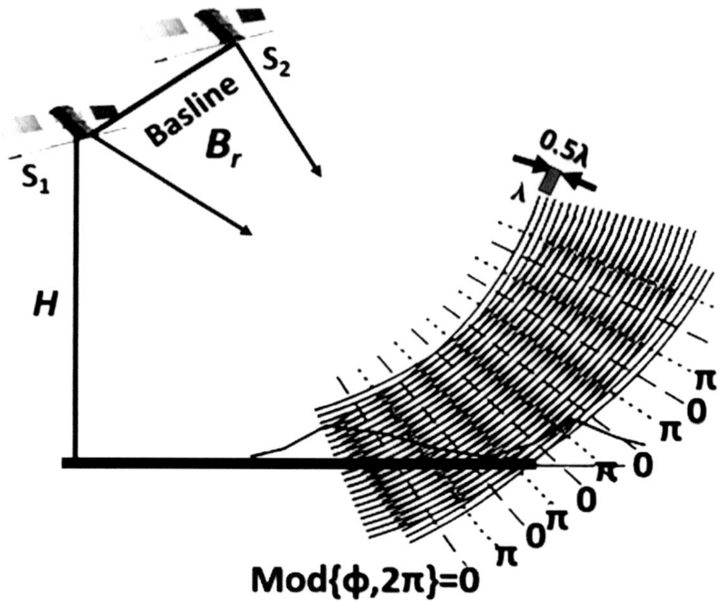

Figure 10.14 Interferometric radar signal analysis for topographic.

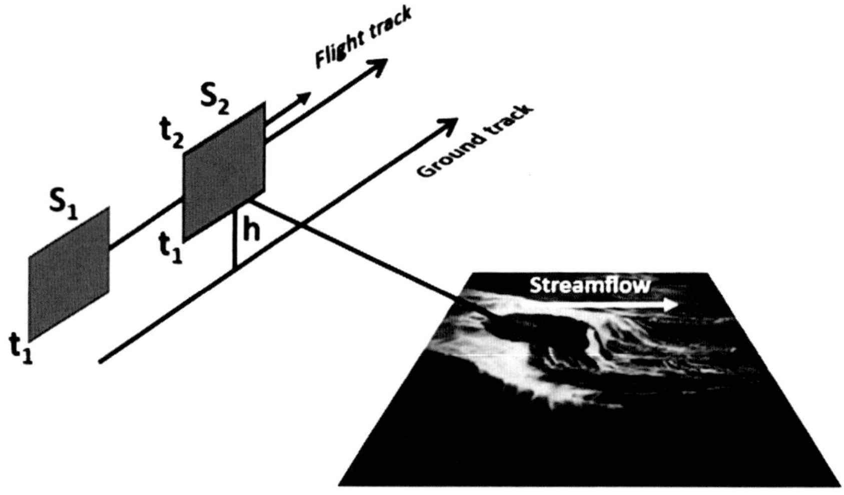

Figure 10.15 Along-track interferometry.

Cross-track interferometers are widely employed in applications requiring precise measurements of topography and surface deformation. In this configuration, the antennas are spaced apart along the cross-track direction—perpendicular to the radar's line of sight—thereby enabling the system to capture the height and surface displacement of a target area with remarkable accuracy.

By comparison, along-track interferometers, which position their antennas in the direction parallel to the radar's path, offer distinct advantages for different applications, such as velocity estimation of moving objects. Consequently, while both types rely on the fundamental principle of interferometry, the geometric arrangement of their baselines imbues them with unique capabilities, making each suited to specific scientific and operational objectives [19, 21].

In contrast, along-track interferometers are primarily designed for measuring radial velocity, with their antennas arranged along the track direction, parallel to the radar's trajectory. Both cross-track and along-track radar systems utilize dual-antenna single-pass interferometry, but the distinction in their operational focus lies in their geometric configurations. Cross-track interferometry employs single-antenna repeat-pass systems, which are adept at capturing topographic details and detecting surface deformations over time. Along-track interferometry, on the other hand, leverages its unique along-track separation geometry to excel in measuring radial velocity [18, 22].

10.9 DERIVING THE MATHEMATICAL EQUATIONS OF ALONG-TRACK INTERFEROMETRY FOR WATER DISCHARGE FROM A DAM

Along-track interferometry (ATI) is a radar remote sensing technique that involves two antennas separated along the platform's velocity direction (usually airborne or spaceborne) to measure the Doppler shift caused by the motion of surface scatterers, such as water, thus allowing the retrieval of surface velocity. In the context of monitoring water discharge from a dam, the ATI method can be particularly useful in capturing surface water velocity over a river or reservoir.

To derive the mathematical equations to calculate water discharge, we need to follow a series of steps that link the radar measurement, surface velocity, and discharge through physical and geometric relationships. Consider two radar antennas positioned along the platform's velocity vector. The signal backscattered from the water surface is received by both antennas at slightly different times [20]. The phase difference between these signals, known as the interferometric phase, can be related to the velocity of the water surface. Therefore, the fundamental measurement in ATI is the Doppler shift, which can be written as:

$$f_D = \frac{2v_r}{\lambda} \tag{10.8}$$

where f_D is the Doppler frequency shift; v_r is the radial velocity component of the surface scatterer (water); and λ is the radar wavelength. For ATI, the radial velocity v_r is primarily due to the movement of the water on the surface, and the relationship between the interferometric phase ϕ and the velocity is:

$$\phi = \frac{4\pi v_r \Delta t}{\lambda} \tag{10.9}$$

In equation 10.9, is the interferometric phase, and Δt is the time delay between the two antenna signals, which is the along-track baseline divided by the platform velocity. Thus, solving for the radial velocity v_r:

$$v_r = \frac{\varphi_{ATI} \cdot \lambda v_s}{4\pi \cdot B_{ATI} \cdot \sin(\theta)} \tag{10.10}$$

Since the radial velocity v_r is only the component of the total water velocity v in the direction of the radar line-of-sight (LOS). Therefore, β_{ATI} is the baseline between two SAR images (Figure 10.16). In this view, we need to transform this into the actual surface velocity. The surface velocity v_s is related to the radial velocity by:

$$v_s = \frac{v_r}{\cos\theta} \tag{10.11}$$

where θ is the incidence angle (the angle between the radar LOS and the normal to the surface). Hence:

$$v_s = \frac{4\pi \cdot v_r \cdot \beta_{ATI} \cdot \sin\theta}{\phi_{ATI}\lambda} \tag{10.12}$$

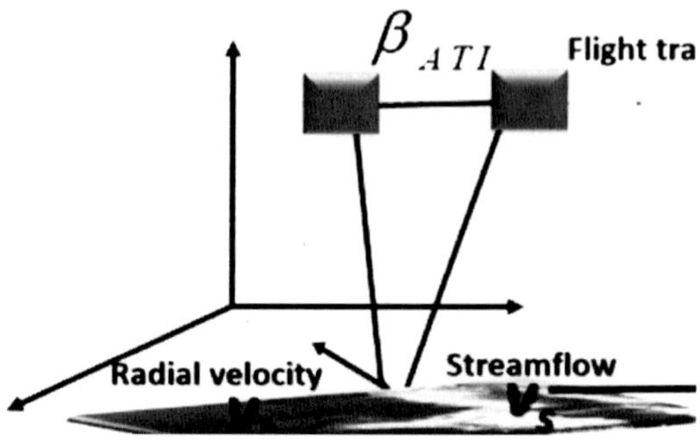

Figure 10.16 Geometry of ATI for radial velocity imaging.

According to equation 1.10, water discharge Q is typically defined as the volume of water flowing through a cross-sectional area of the river or channel per unit of time. To connect the interferometric measurement with discharge, we approximate average velocity $\overline{v}$ from v, acknowledging that v_S represents the surface velocity and that water velocity varies with depth. This can be modeled using a depth-velocity relationship, often assuming a logarithmic or power-law distribution for flow in open channels. A common approximation is:

$$\overline{v} = \alpha v_s \tag{10.13}$$

where α is a correction factor that accounts for the velocity profile in the water column (typically α ranges from 0.85 to 0.95, depending on the flow characteristics). Substituting v_S into the discharge equation, the water discharge Q_{ATI} from along-track-interferometry is then given by:

$$Q_{ATI} = A\alpha v_s \frac{4\pi \cdot v_r \cdot \beta_{ATI} \cdot \sin\theta}{\phi_{ATI}\lambda} \tag{10.14}$$

Here, A can be measured or estimated based on the river width and depth at the location of interest. The other parameters—interferometric phase ϕ, radar wavelength λ, time separation Δt, and incidence angle θ are determined by the radar system and its geometry. Cross-sectional area A, in practice, can vary due to changes in the water level or riverbed morphology. This needs to be measured independently, either through ground-based surveys or remote sensing data. Factors such as radar noise, surface roughness, wind effects and temporal decorrelation can introduce uncertainties into the interferometric phase and thus in the velocity measurements.

Equation 10.14 provides a direct connection between the along-track interferometric phase measurements and the water discharge from the dam, incorporating both radar geometry and surface water flow dynamics. In this sense, the derived equation demonstrates how along-track interferometry can be employed to estimate water discharge from radar phase measurements, by combining interferometric surface velocity retrievals with cross-sectional area estimates. By applying corrections for flow depth profiles and carefully managing uncertainties, ATI offers a robust approach for monitoring river discharge, particularly in remote or inaccessible regions.

The passage outlines the challenges faced when using ATInSAR (Alon-Track Interferometric Synthetic Aperture Radar) data for tracking fast-moving objects or dynamic surfaces, such as ocean waves or free-flowing vehicles. Specifically, Equation 10.14 highlights that when dealing with high-velocity targets, the low sensitivity of ATInSAR interferograms becomes problematic. This occurs due to spatial averaging (often referred to as multilooking), which blurs the interferometric phase, reducing the system's ability to detect rapid or detailed changes [27–30].

When applied to fast-moving objects or fluctuating surfaces, the interferogram phase undergoes significant averaging across spatial regions, limiting the precision with which high-velocity phenomena are captured. Consequently, this spatial averaging diminishes the ability to accurately detect fine-scale movement or fast-changing phases in the radar data, an issue well-documented in previous research [22–24].

In response to this limitation, the chapter suggests exploring alternative methods to address these challenges, notably proposing the use of quantum ATI. This technique incorporates principles of quantum computing into interferometric applications and may offer a more refined approach to overcoming the sensitivity issues encountered in traditional InSAR processing. By leveraging quantum neural network techniques, this method can likely improve phase resolution and the overall ability to track high-velocity targets more effectively, thus addressing the critical shortcomings of InSAR in dynamic environments [23].

This chapter is focused on mitigating the limitations of spatial averaging in InSAR and potentially presenting quantum neural network techniques interferometry as a promising solution for enhancing accuracy in such scenarios.

10.10 QUANTUM OF ALONG-TRACK INTERFEROMETRY

Let us assume the dual antenna and a delay phase ϕ_{delay} is acquainted with one of the lengths of the interferometer. Then the quantity ϕ_{delay} is computed by evaluating the strength of the microwave signal of the dual output beams. The value ϕ_{delay} can be computed with a statistical error proportional to $\frac{1}{\sqrt{N}}$, as N is a non-entangled microwave signal. Precisely, the direct simulation of the mean traveling time, which is determined for N a non-entangled signal. In this instance, the miscalculation is expressed as:

$$\partial R \approx O\left[\Delta\omega\sqrt{N}\right]^{-1} \tag{10.15}$$

Equation 10.15 confirms that the inverse of the bandwidth $\Delta\omega$ effects the inaccuracy that arose on account of the oscillation of the range ∂R. Nonetheless, entangled signals can be utilized to overwhelm the conventional quantum constraint and achieve the Heisenberg limit. To handle the quantum mechanism of ATI, let us deliberate a dual-mode, path-entangled, signal-number state, a sort of Schrödinger cat state, which is thoroughly recognized as the N00N state and mathematically is given by:

$$|\psi_{NOON}\rangle = \left[\sqrt{2}\right]^{-1}\left(|NO\rangle + |ON\rangle\right) \tag{10.16}$$

Equation 10.16 designates the number of backscattered signals generated by ATI, which are either completely in the first complex SAR image S_1 or entirely in the second complex SAR image S_2. On the other hand, it is entirely removed into the entangled-signal source, for generating superpositions of these dual states in the form:

$$|NOON\rangle = |S_1\rangle + |S_2\rangle = |N\rangle_{S_1}|O\rangle_{S_2} + |O\rangle_{S_1}|N\rangle_{S_2} \tag{10.17}$$

Consequently, each part of the entangled state tolerates diagonally a different dimension of the phase changing, which is accurately articulated as:

$$|\psi_{NOON}\rangle = \left(\sqrt{2}\right)^{-1}\left[|N\rangle_{S_1}|O\rangle_{S_2} + |O\rangle_{S_1}|N\rangle_{S_2}\right] \tag{10.18}$$

To commence, the quantity states exist in non-conventional states. Their accomplishment in the phase-shifter is extremely unrelated. When a monochromatic beam of measure states approves across a phase shifter, the phase shift is straight proportional to N, the number of photons. Therefore, there is no n-dependence in the coherent state, where *induce n* is the steady signal intensity at ATI. In manifestations of a unitary development of the state, the growth of any interferometry state transient through a phase shifter ϕ is standardized by:

$$\widehat{U}(\phi) \equiv e^{(i\phi\hat{n})} \tag{10.19}$$

where $\hat{n}$ is the average interferometry fringes are generated by a phase shift in the quantum along-track interferometry technique. The phase shift operator can be exposed to have the consequent dual disparate influences on coherent in opposition to quantize the phase shift states as:

$$\widehat{U}_\phi |O\rangle = e^{i\phi} |O\rangle, \tag{10.20}$$

$$\widehat{U}_\phi |N\rangle = e^{i\phi} |N\rangle, \tag{10.21}$$

Equations 10.20 and 10.21 expose that the phase shift for the coherent state is enlightened of quantity, nonetheless, there is an N dependence on the exponential for the quantum phase shift state. The quantum phase shift state cultivated in phase N-times is further rapidly than the coherent state [32]. In this instance, the N00N state develops into:

$$\left|\psi_{NOON}^{\phi}\right\rangle = \frac{1}{\sqrt{2N!}}\left(|N\rangle_{S_1}|O\rangle_{S_2} + e^{iN\phi}|O\rangle_{S_1}|N\rangle_{S_2}\right) \tag{10.22}$$

Consequently, equation 10.22 can be expressed in terms of annihilation operators:

$$\left|\psi_{NOON}^{(\phi)}\right\rangle = \frac{1}{\sqrt{2N!}}\left((\hat{a}_{S_1}^{\dagger})^N + e^{iN\phi}(\hat{a}_{S_2}^{\dagger})^N |0\rangle_{S_1}|0\rangle_{S_2}\right) \tag{10.23}$$

where the bosonic creation operator $a_\alpha^\dagger$ is given by

$$a_\alpha^\dagger |n_1,\ldots,n_{\alpha-1},n_\alpha,n_{\alpha+1},\ldots\rangle = \sqrt{n_{\alpha+1}}\,|n_1,\ldots,n_{\alpha-1},n_\alpha,n_{\alpha+1},\ldots\rangle, \tag{10.23.1}$$

Let us suppose the ATI dam discharge $\Psi|Q_{ATI}\rangle$ is delivered by:

$$Q_{ATI} = \frac{1}{N!}\left((\hat{a}_{S_1}^{\dagger})^N |0\rangle\langle 0|(\hat{a}_{S_2}^{\dagger})^N + (\hat{a}_{S_2}^{\dagger})^N |0\rangle\langle 0|(\hat{a}_{S_2}^{\dagger})^N\right) \tag{10.24}$$

The ATI observation measurements are vital to approximate the phase shift ground on annihilation operators. In this concern, the fluctuation in dam discharge flow ΔQ_{ATI} can create to inaccuracy in the phase shift as obeys:

$$\partial\phi = \frac{\Delta Q_{ATI}}{|-N\sin N\varphi|} \tag{10.25}$$

Equation 10.25 exposes that the Heisenberg constraint as the ATI phase is deliberated as enormously entangled states. In other words, equation 15.25 can be conveyed as [25]:

$$\lim_{\alpha_1\to 1}\lim_{\alpha 2\to 1}\partial\phi = \frac{\Delta Q_{ATI}}{|-N\sin N\varphi|} \tag{10.26}$$

10.11 QUANTUM HOPFIELD ALGORITHM FOR ATI PHASE UNWRAPPING

To create quantum neural networks, let us deliberate on the task of expanding multi-qubit quantum systems. The accurate correlation between neurons and qubits [26], unlocking access to quantum properties of entanglement and coherence of SAR data is instigated. Let us instead encode the neural network into the energy of a quantum state. This is accomplished by accommodating a memory regulation between the activation patterns of the neural network and pure states $|\phi\rangle$ of L-level of a quantum system. The determined state features of the current pattern can be expressed as $\phi \to |\phi\rangle|\phi|_2$ with $|\phi|_2 = \sqrt{\sum_{i=1}^{L}\phi_i^2}$ and $|\phi\rangle := |\phi|_2^{-1}\sum_{i=1}^{L}\phi_i\,|i\rangle$ which is encoded concerning the standard basis such that $\langle\phi|\phi\rangle = 1$. The L-level quantum system can be realized by a register of $N = [\log_2 L]$ qubits with the intention of the qubit directly above signifying such a network scale logarithmically with the number of neurons [27, 30, 33, 37].

The quantum Hebbian learning algorithm (qHob) is exploited for weighting matrix M of the Hopfield network. It relies on dual vital perceptions: (i) a direct alliance of the weighting matrix with a mixed-state ρ, and (ii) one can efficiently perform quantum algorithms that harness the information contained in W. Consequently, ρ registers of N qubits in relative to:

$$\rho := \omega + \frac{\Pi_L}{L} = W^{-1}\sum_{m=1}^{W}\left|\phi^{(m)}\right\rangle\left\langle\phi^{(m)}\right| \tag{10.27}$$

where Π_d is the L-dimensional identity matrix and W is a training set of activation patterns of phase ATI ϕ, with $m \in \{1, 2, \ldots, W\}$. In this view, the Hopfield neural network can be taught training set using the Hebbian learning rule, which sets the weighting matrix elements ω_{ij} along with the number of occasions in the training set that the neurons i and j mix. In this sense, Hebbian learning is employed to set the weighting matrix M from the L-length training phase unwrapping data $\{\phi^{(m)}\}_{m=1}^{W}$. In this regard, the qHop algorithm continues to compute $|v\rangle = \frac{1}{\Delta\phi_{ATI}}|\phi\rangle$ where the pure state $|\phi\rangle$ is primarily arranged, which comprises user-identified neuron thresholds of the ATI data inverse matrix ϕ_{ATI}^{-1} of feature identifications and a partial memory pattern. In other words, the inverse matrix ϕ_{ATI}^{-1} containing information on the training data and regularization γ. To this end, the output of the pure state $|v\rangle$ contains information on the reconstructed state of the phase $|\phi_{ATI}\rangle$ and Lagrange multiplier vector $\vec{L}_v$. The feature state matrix of ATI phase data can be expressed as a quantum state by [30, 33]:

$$\phi_{ATI}\,|v|_2\,|v\rangle = |w|_2\,|w\rangle \tag{10.28}$$

where

$$|v\rangle := \frac{1}{|v|_2}\left(|\phi_{ATI}|_2\,|0\rangle \otimes \left||\phi_{ATI}|\right\rangle + \left|\vec{L}_v\right|_2 |1\rangle \otimes \left|\vec{L}_v\right\rangle\right) \tag{10.28.1}$$

$$|w\rangle := \frac{1}{|w|_2}\left(|\Xi|_2\,|0\rangle \otimes |\Xi\rangle + \left|\varphi_{ATI}{}^{(inc)}\right|_2 |1\rangle \otimes \left|\varphi_{ATI}{}^{(inc)}\right\rangle\right) \tag{10.28.2}$$

where $\varphi_{ATI}{}^{(inc)}$ is the normalized quantum state corresponding to the incomplete activation pattern and $\left|\varphi_{ATI}{}^{(inc)}\right|_2 = l$. The objective of equations 10.28.1 and 10.28.2 is to optimize the energy function E in equation 10.28. Besides, Ξ presents each element of qHop which should be set so that its magnitude is of order at most 1 qubit. In other words, $\Xi =: \{\Xi_i\}_{i=1}^{d} \in \mathbb{R}^d$ is a user-specified neuronal threshold vector that determines the switching threshold for each neuron. Let us identify the set of M unitary operators $\{U_k\}_{k=1}^{M}$ acting on an $N + 1$ register of qubits consistent with

$$U_k := |0\rangle\langle 0| \otimes \Pi + |1\rangle\langle 1| \otimes e^{-i\left|\phi_{ATI}{}^{(k)}\right\rangle\left\langle\phi_{ATI}{}^{(k)}\right|\Delta t} \tag{10.29}$$

Equation 10.29 reveals that the unitary implements various phase pattern detections in ATI data for a small difference in time Δt under the circumstance of $\left|\phi_{ATI}{}^{(k)}\right\rangle\left\langle\phi_{ATI}{}^{(k)}\right|$. Conversely, these unitary can be simulated using the following mathematical equation:

$$U_s := e^{-i|1\rangle\langle 1|\otimes s\Delta t}$$
$$|0\rangle\langle 0| \otimes \Pi + |1\rangle\langle 1| \otimes e^{-is\Delta t} \tag{10.30}$$

here, $|1\rangle\langle 1|\otimes s\Delta t$ is 1-sparse and efficient simulated. Conversely, the trace tr_2 of U_s is over the second subsystem containing the state feature $|\phi_{ATI}\rangle$ in the ATI data, which is expressed by [30–33]:

$$tr_2\left\{U_s\left(|q\rangle\langle q|\otimes|\phi_{ATI}{}^{(k)}\rangle\langle\phi_{ATI}{}^{(k)}|\otimes\sigma\right)U_s^{\dagger}\right\}$$
$$=U_k\left(|q\rangle\langle q|\otimes\sigma\right)U_k^{\dagger}+\mathcal{O}\left(\Delta t^2\right). \tag{10.31}$$

In equation 10.31, the subsystem of ancilla qubit $|q\rangle\langle q|$ and σ efficiently experiences time evolution $\mathcal{O}(\Delta t^2)$ with U_k. The unitary $e^{i\phi_{ATI}t}$ is simulated to fix the error ε for arbitrary t. Indeed ε can be estimated by:

$$\varepsilon:=\left\|\left(e^{-i|\phi_{ATI}{}^{(1)}\rangle\langle\phi_{ATI}{}^{(1)}|\frac{t}{nM}}..e^{-i|\phi_{ATI}{}^{(M)}\rangle\langle\phi_{ATI}{}^{(M)}|\frac{t}{nM}}\right)^n-e^{-i\phi_{ATI}t}\right\|\in\mathcal{O}\left(\frac{t^2}{n}\right). \tag{10.32}$$

Equation 10.32 indicates that the repetition of $n\in\mathcal{O}\left(\frac{t^2}{\varepsilon}\right)$ is required alongside with M-sparse Hamiltonian simulations. The advantages of this approach are that the copies of the training states $|\phi_{ATI}{}^{(m)}\rangle$ can be used as quantum software states and, in addition, it does not require superpositions of the training states. Conversely, the ATI feature matrix ϕ_{ATI} can be identified as [33]:

$$\arg(S_1\ S_2^*):=\begin{pmatrix}0 & P\\ P & 0\end{pmatrix}+\begin{pmatrix}-\left(\gamma+L_v^{-1}\right)\Pi_{L_v} & 0\\ 0 & 0\end{pmatrix}+\begin{pmatrix}\rho & 0\\ 0 & 0\end{pmatrix} \tag{10.33}$$

The simulation time is split into n small time steps $t = n\Delta t$. In this circumstance, the error ε can be extended into Taylor expansion as [30, 33]:

$$\varepsilon_{\Delta t}:=\left\|e^{i\phi_{ATI}\Delta t}-U_B\left(\Delta t\right)U_C\left(\Delta t\right)U_D\left(\Delta t\right)\right\|\in\mathcal{O}\left(\Delta t^2\right) \tag{10.34}$$

where $U_B(\Delta t)$, $U_C(\Delta t)$, and $U_D(\Delta t)$ are the operators of $\begin{pmatrix}0 & P\\ P & 0\end{pmatrix}$, $\begin{pmatrix}-\left(\gamma+L_v^{-1}\right)\Pi_{L_v} & 0\\ 0 & 0\end{pmatrix}$, and $\begin{pmatrix}\rho & 0\\ 0 & 0\end{pmatrix}$, respectively. In this understanding, at most $\mathcal{O}(\Delta t^2)$, the errors $e^{i\begin{pmatrix}0 & P\\ P & 0\end{pmatrix}\Delta t}$, $e^{i\begin{pmatrix}-\left(\gamma+L_v^{-1}\right)\Pi_{L_v} & 0\\ 0 & 0\end{pmatrix}\Delta t}$, and $e^{i\begin{pmatrix}\rho & 0\\ 0 & 0\end{pmatrix}\Delta t}$ are simulated respectively, which allows the simulation of the unitary $e^{i\phi_{ATI}t}$ to an error of $\varepsilon\in\mathcal{O}(n\Delta t^2)$. Then the matrix of spectral energy of the qHop can be mathematically expressed as [35–37]:

$$\phi_E=\sum_{j:|\lambda_j(\Delta\phi)|>\lambda}\lambda_j(\Delta\phi)|E_j(\Delta\phi)\rangle\langle E_j(\Delta\phi)|+\sum_{j:|\lambda_j(\Delta\phi)|<\lambda}\lambda_j(\Delta\phi)|E_j(\Delta\phi)\rangle\langle E_j(\Delta\phi)| \tag{10.35}$$

Equation 10.35 demonstrates how to simulate the gradient energy of the various feature pattern variations in the ATI as a function of the size of the eigenvalues λ_j in comparison to a fixed user-defined number $\lambda > 0$. In this circumstance, qHop maintains the polylogarithmic efficiency in run time whenever λ is such that $\lambda^{-1}\in O(poly(\log d))$. Therefore, the primary matrix inversion algorithm returns (up to normalization) as [25–33]:

$$A_E^{-1}|E\rangle=\sum_{j:|\lambda_j(\Delta\phi)|\geq\lambda}\langle E_j\left(A_{SAR}\right)|E\rangle\left(\lambda_j(\Delta\phi)\right)^{-1}|E_j(\Delta\phi)\rangle\left(\lambda_j(\Delta\phi)\right) \tag{10.36}$$

The input state of the energy gradient $|E\rangle$ is first achieved and it contains the threshold data and incomplete activation pattern) and considers it in the eigenbasis of the matrix of ATI image energies A_E. Therefore, obtaining information from each energy pattern involves $O(K)$ operations of qHop. The qHop algorithm is then initialized along with sparse Hamiltonian simulation [28–30] to perform quantum phase estimation, allowing $\sum_j \langle E_j(\Delta\phi)|E\rangle |\tilde{\lambda}_j(\Delta\phi)\rangle \otimes |E_j(\Delta\phi)\rangle$ to be obtained with $\tilde{\lambda}_j(\Delta\phi)$ an approximation of the eigenvalue $(\lambda_j(\Delta\phi))$ to precision ε [30, 33].

10.12 QUANTUM ATI HOPFIELD ALGORITHM APPLICATION TO TANDEM-X SATELLITE

TerraSAR-X, a German Earth observation satellite, was launched in June 2007, marking a significant advance in satellite imaging technology. Equipped with a sophisticated Synthetic Aperture Radar (SAR) operating in the X-band frequency (f = 9.65 GHz and λ = 3.1 cm), TerraSAR-X provides high-resolution imagery with impressive precision. In 2010, the German Aerospace Center (DLR) supplied a total of 18 images, all acquired with consistent HH horizontal polarization, allowing for cohesive and repetitive observation cycles. The satellite's SAR system offers incidence angles from 27.3° to 53.3°, enhancing revisit frequency within the satellite's 11-day orbital period [28–30].

To achieve varied spatial resolutions, two acquisition modes were employed: StripMap (SM) and SpotLight (SL). These modes feature pixel spacings of approximately 3 meters and 1.5 meters, respectively, balancing broader spatial coverage with finer detail. The StripMap mode allows for continuous image capture along the satellite's ground track, while SpotLight mode provides highly detailed imagery by focusing on smaller areas. These combined modes enable TerraSAR-X to deliver both expansive and high-resolution data, making it invaluable for applications requiring precision, such as urban mapping, environmental monitoring, and geological analysis [30–32].

This advanced imaging capability reflects the high standard set by German space research and emphasizes the transformative potential of radar-based Earth observation in capturing detailed insights about our planet's surface and monitoring it with unprecedented accuracy [28].

Tenth of TanDEM satellite data acquisitions have been obtained between the latitudes of 11° 12′ 04.88″ N and 11° 13′ 21.61″ N, and the longitudes 35°03′ 31.65″E, and 35°06′ 02.42″E (Figure 10.17). Between August 2023 and May 2024, a series of targeted satellite images, known as spot mode images, were acquired as part of a comprehensive monitoring effort over the Grand Ethiopian Renaissance Dam (GERD). The precision and frequency of these images reflect the critical importance of assessing both the dam's construction progress and the broader hydrological implications of its operation. The satellite captures began with two key acquisitions on August 1 and August 7, 2023, during a period when the third phase of the dam's filling was either completed or nearing completion.

Subsequent images were captured in 2024, with eight significant acquisitions on April 3, April 10, April 14, April 21, May 5, May 6, May 17, and May 24. These captures coincide with the latter stages of the fourth filling phase, a period marked by substantial hydrological shifts due to increased water levels in the GERD reservoir. The satellite used for this data collection, TanDEM-X, is equipped with radar sensors capable of high-resolution imaging in all weather conditions, making it a vital tool for understanding the evolving water storage and release patterns from the dam.

Figure 10.17 visually represents this timeline of data acquisition, providing crucial insights into the dynamic interactions between the dam's infrastructure and the surrounding environment. The continuous monitoring through such high-resolution images enables detailed analysis of changes in surface water, sedimentation patterns, and the structural integrity of the dam over time, while also aiding in the assessment of potential downstream impacts on the Nile River basin.

Figure 10.17 TanDEM-X satellite spot mode data were acquired along GERD from August 2023 to May 2024.

The coherence image reveals a fascinating pattern, where the highest coherence values—exceeding 1.0—are observed along the main dam infrastructure and regions affected by overtopping caused by turbulent flow. This anomaly highlights the significant role of turbulence in shaping these regions, which was detailed in the previous chapter, particularly concerning the operation of just two out of the sixteen turbines at the Grand Ethiopian Renaissance Dam (GERD). The presence of such elevated coherence values indicates stable, unmoving structures, which reflect radar signals strongly and consistently, particularly in areas of intense flow dynamics.

In stark contrast, areas dominated by vegetation show much lower coherence values, typically around 0.2, indicating minimal structural movement or activity. Vegetation, by its very nature, is subject to subtle but constant changes—such as wind movement, seasonal growth, and hydrological variations—which result in weak radar returns and low coherence. This distinction between man-made infrastructure and natural elements like vegetation underscores the coherence of imaging's ability to differentiate between highly stable, engineered surfaces and dynamic, fluctuating natural environments [8, 10].

The coherence data not only offers insight into structural stability and performance but also provides a visual and quantitative assessment of the surrounding environment. Figure 10.18 exemplifies these differences, with the sharp contrast between the dam's robust, high-coherence regions and the softer, low-coherence areas dominated by vegetation [30, 33]. This stark divergence is crucial for ongoing assessments of both the dam's operational effects and the environmental impact of such a large-scale infrastructure project.

The interferogram phase, as calculated using qHop Along-Track Interferometry (ATI) phase measurements, reveals a distinctive pattern of overtopping turbulence flow, ranging from –180° to +180° (Figure 10.19). This phase range is indicative of complex flow dynamics, particularly in the vicinity of the main Grand Ethiopian Renaissance Dam (GERD), where turbulence is most intense. The presence of such flow characteristics aligns with the observed high coherence values, nearing 1, which is typical of regions dominated by stable, engineered structures like the main dam.

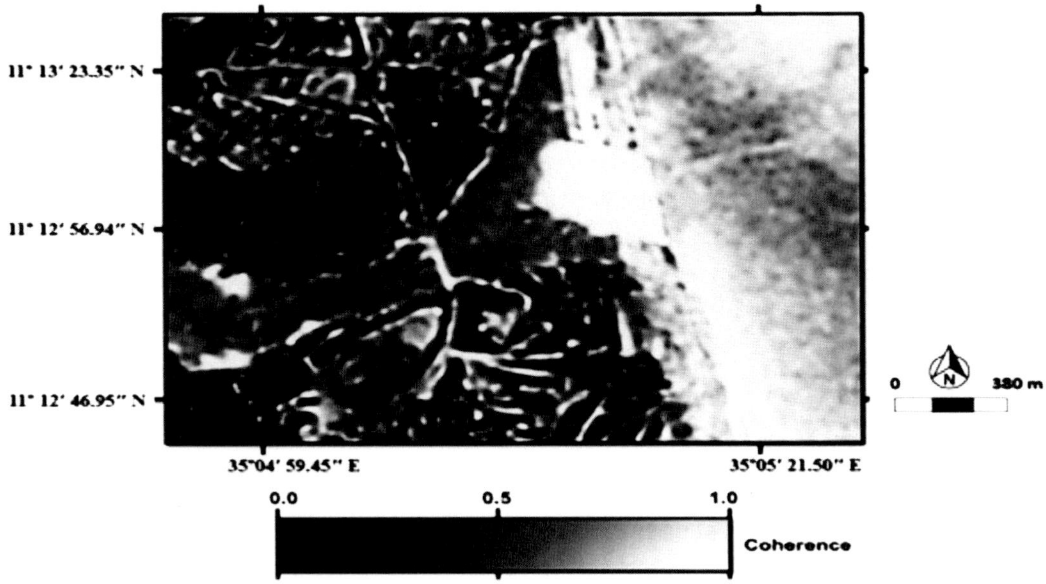

Figure 10.18 Coherence of TanDEM-X data from 2023 to 2024.

The turbulent overtopping observed in this phase range is directly linked to the operational aspects of the dam, particularly concerning water flowing over its crest or through designated outlets. This turbulent behavior not only affects the structural aspects but also highlights the significant role of hydrodynamic forces at play. The coherent signal detected by the interferogram suggests a strong reflection of radar waves, confirming the presence of highly stable surfaces interacting with dynamic water flow. The qHop ATI method, which effectively captures both spatial and phase variations, allows for the precise measurement of such phenomena, providing insights into how the dam's infrastructure interacts with the fluid dynamics of the Blue Nile.

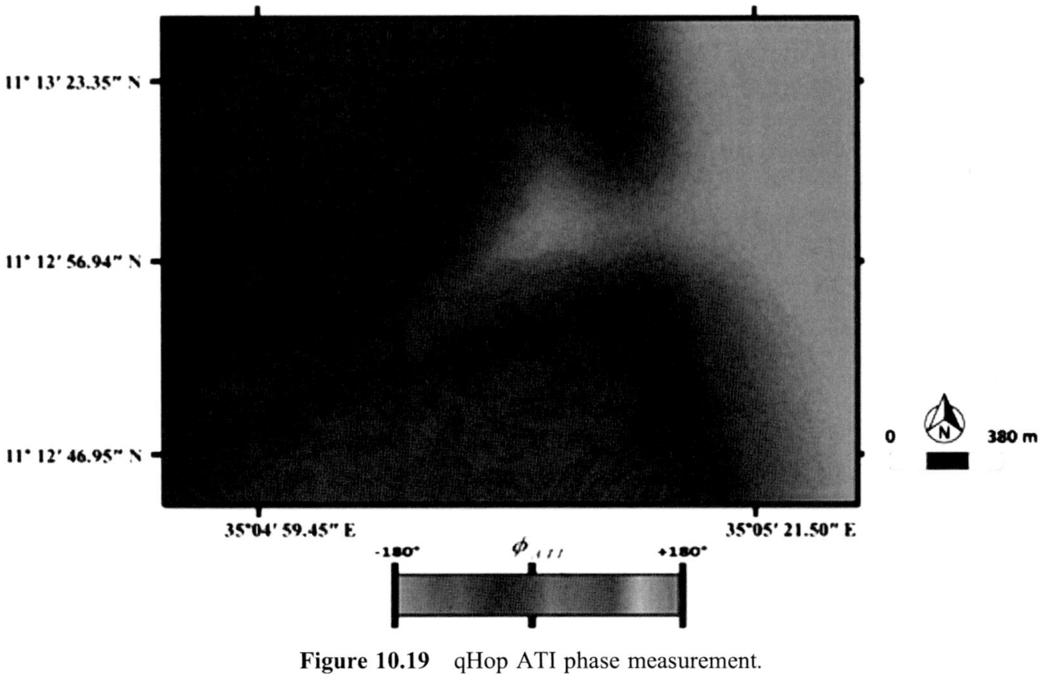

Figure 10.19 qHop ATI phase measurement.

Consequently, the Doppler frequency shift increases to approximately 30 Hz across the overtopping region of the dam, primarily due to the intensified turbulence caused by the rapid water flow (Figure 10.20). This heightened turbulence, a direct result of the operation of only two turbines, induces significant variations in water velocity as it flows over the dam's crest, generating a pronounced frequency shift. The central overtopping region, where the flow is most dynamic, experiences this elevated Doppler shift, which is a clear indicator of the high kinetic energy associated with turbulent water flow.

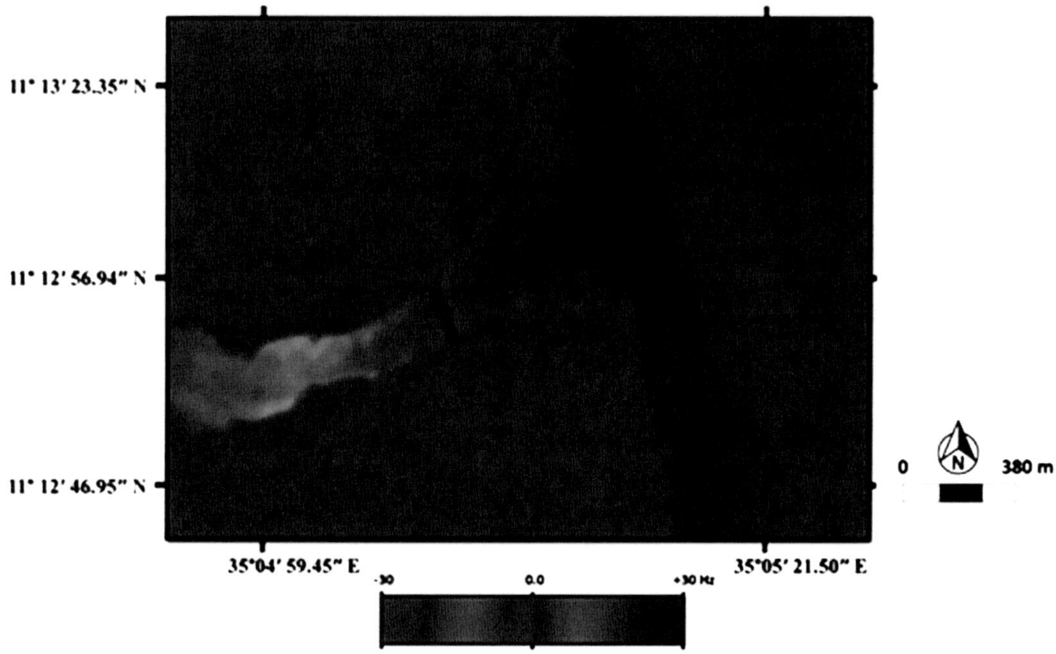

Figure 10.20 Doppler frequency shift of TanDEM-X data.

Conversely, as the distance from the central dam increases, the Doppler frequency shift decreases gradually, reaching as low as 10 Hz. This reduction signifies slower water flow in areas farther from the dam's core, where the turbulence is less pronounced. The simultaneous variation in Doppler frequency—from the overtopping region's high-frequency shifts to the much lower shifts downstream—provides critical insights into the spatial distribution of water velocities and flow dynamics around the dam. The qHop algorithm, which measures water flow velocities, provides critical insights into the hydrodynamic behavior near the Grand Ethiopian Renaissance Dam (GERD). It demonstrates that water velocities in the overtopping region can reach up to 15 ms^{-1}, while water that flows farther from this region slows down significantly, ranging between 0.2 ms^{-1} and 1.1 ms^{-1}. This reduction in velocity away from the overtopping region is notable, especially when considering that the water velocity within the GERD reservoir itself is approximately 0.6 ms^{-1} (Figure 10.21). Furthermore, downstream, the flow decreases even further, to about 0.2 ms^{-1}, indicating a substantial weakening of water flow.

The correlation between turbine operation and the frequency shift is significant. Since only two out of sixteen turbines are operational, the overtopping flow is intensified, and this concentration of water discharge leads to the observed Doppler frequency shift pattern. Understanding this distribution is crucial for monitoring the dam's hydrodynamic behavior and its impacts on downstream ecosystems and water management efforts.

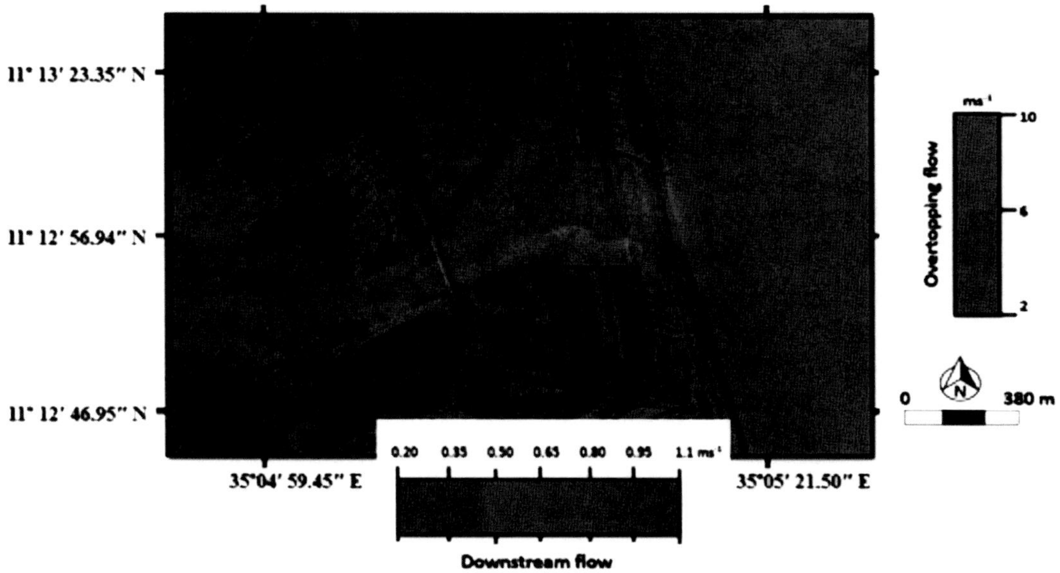

Figure 10.21 Downstream water flows through GERD using qHOP along-track interferometry of TanDEM-X data from August 2023 to May 2024.

The key question, therefore, is whether this weakened downstream flow can affect the overall water volume discharge. The short answer is yes—slower water velocities downstream can indeed impact water volume discharge, particularly in terms of how efficiently the water is transported and distributed. In regions where water flow is significantly slower, such as areas downstream of the Grand Ethiopian Renaissance Dam (GERD), the discharge volume over a given time interval can drop to a minimum rate of around 18 billion cubic meters (BCM) (Figure 10.22). This reduction contrasts sharply with areas of higher flow rates, such as the dam's overtopping regions or during peak turbine operation. The slow water discharge can be influenced by various factors, including terrain, hydraulic resistance and operational management of the dam.

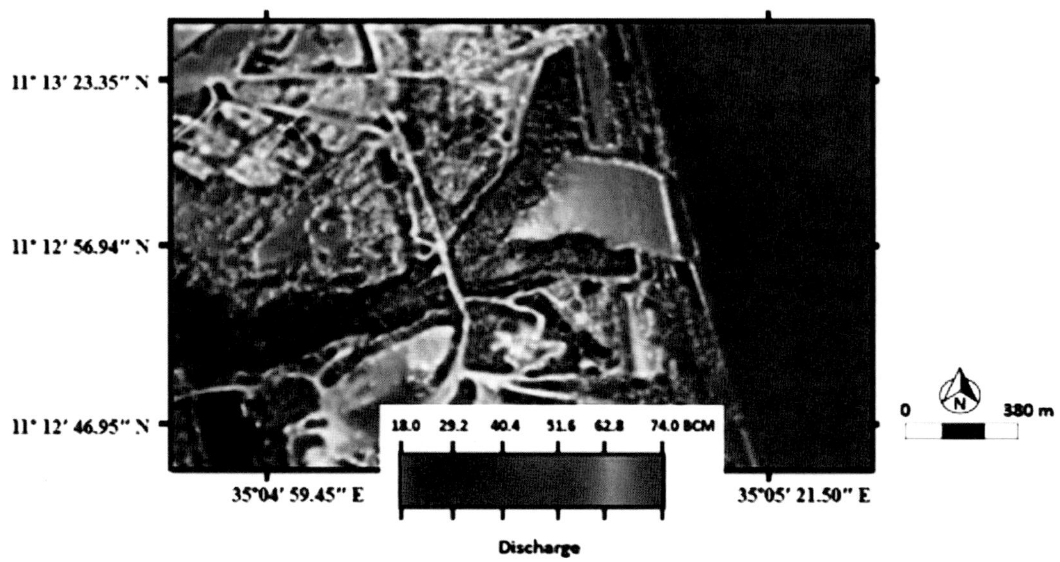

Figure 10.22 Volumetric water discharge through GERD was analyzed using qHOP along-track interferometry from TanDEM-X data (August 2023 to May 2024).

On the other hand, the GERD reservoir itself has a maximum storage capacity of 74 BCM, a figure that underscores the scale of water management at the dam. This vast storage capability is designed to regulate the flow of the Blue Nile, providing both hydropower generation and strategic water management. The discrepancy between the minimum discharge rate of 18 BCM and the maximum storage volume of 74 BCM highlights the dam's role in modulating water release, particularly in terms of controlling the balance between power generation needs, downstream water requirements, and ecological impacts.

This dynamic allows the GERD to act as a buffer, accumulating vast amounts of water during periods of high inflow while releasing it at controlled rates. However, managing this balance is complex, as maintaining consistent downstream flows is crucial to avoid ecological disruptions and geopolitical tensions, especially with Egypt and Sudan, two countries that depend on the Nile's steady discharge for agriculture and water supply. Thus, the storage and discharge mechanisms at GERD are pivotal not only for Ethiopia's energy security but also for the broader hydrological stability of the Nile Basin.

This diminished discharge can have several potential implications [13–15]:

Hydropower efficiency: The slower downstream water velocity can reduce the efficiency of energy generation, especially if the flow is insufficient to drive turbines at optimal capacity. GERD relies on controlled water releases for power production, and weak water flow could limit the energy output, especially during periods when more turbines are brought online.

Sediment transport: Slow water flow downstream may lead to increased sediment deposition. The weak flow can cause sediment to accumulate, rather than being carried downstream, which could gradually reduce the dam's storage capacity and lead to other ecological consequences.

Ecological and agricultural impacts: Reduced water flow downstream may also affect agricultural activities and ecosystems reliant on the consistent flow of water. Weak flows can result in lower water availability for irrigation and disrupt the aquatic habitats that depend on steady current regimes.

Nile River water agreements: From a geopolitical perspective, the decreased water discharge downstream could further complicate negotiations with countries downstream of GERD, such as Sudan and Egypt, where water security is closely tied to the consistent and sufficient flow of the Nile.

10.13 APPROVALS FOR PRECISION OF QHOP ALONG-TRACK INTERFEROMETRY TECHNIQUE

The Along-Track Interferometry (ATI), enhanced by the qHop algorithm, represents a highly precise and advanced method for monitoring streamflow and volumetric water discharge at the Grand Ethiopian Renaissance Dam (GERD) using TanDEM-X satellite data. This technique's ability to generate detailed two-dimensional models, when integrated with sophisticated monitoring tools provides vital insights into the dam's operational efficiency. By capturing the dynamic fluctuations in streamflow and volumetric discharge, qHop allows for a more nuanced understanding of how GERD manages water resources and its implications for the downstream environment.

Building on this framework, the qHop algorithm serves as a powerful instrument to simulate and quantify accumulative streamflow and volumetric discharge, which are key indices for evaluating the overall efficiency of GERD. For instance, the observed minimal discharge of 18 BCM, alongside a downstream flow velocity of just 0.2 ms^{-1}, suggests reduced operational efficiency of the dam. This raises a critical question: can the qHop algorithm effectively discern the impact of GERD's low efficiency on streamflow velocity and volumetric discharge across the Blue Nile basin?

The answer lies in the algorithm's remarkable precision. As demonstrated by the Receiver Operating Characteristic (ROC) curve, qHop achieves an impressive 98% accuracy in correlating the lowest downstream flow of 0.2 ms^{-1} with the corresponding volumetric discharge of 18 BCM, as shown in Figure 10.23. This level of precision is further enhanced through the integration of auxiliary GRACE satellite data, which provides additional validation and reinforces the reliability of qHop's modeling capabilities. This high level of accuracy underscores the algorithm's ability to provide critical insights into the hydrodynamic behavior of the dam under suboptimal operating conditions. Notably, the slow downstream velocity, as determined by the phase unwrapping process in qHop [30, 33] serves as a key indicator of the reduced volumetric discharge, particularly when only two turbines are operational, as discussed in Chapter 9. This finding agrees with the work of Abera et al. [13] and Ali et al. [15].

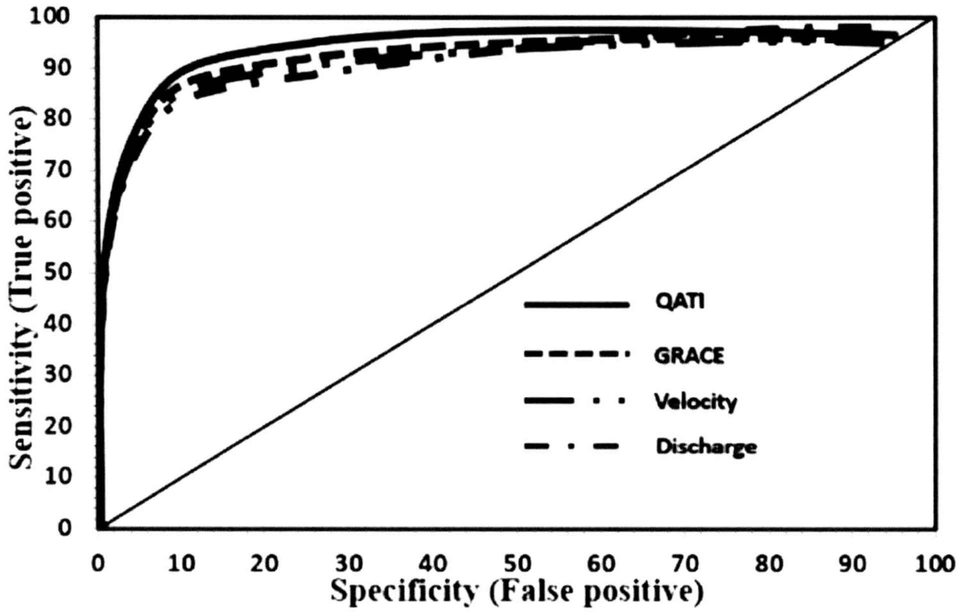

Figure 10.23 ROC curve for confirming the cause of low efficiency of GERD using qHOP algorithm.

The integration of quantum computing principles, particularly the qHop algorithm, into the phase unwrapping process has revolutionized interferometric analysis. By leveraging quantum Hebbian learning to adjust the weight matrix of the Hopfield network, qHop significantly enhances the accuracy of phase interferogram data, allowing for precise measurement of water flow velocity downstream of GERD at rates as low as 0.2 m/s. This integration, known as Quantum Along-Track Interferometric (QATI) radar, provides a robust methodology for assessing the dam's performance and its broader hydrological impacts. This result proves the previous studies that were done by Lanzagorta [25], Behrman [26], Shukla and Vedula [27] and Marghany [30].

10.14 CONCLUSION

This chapter introduces a novel approach to Along-Track Interferometric Synthetic Aperture Radar (SAR) processing applying principles from quantum mechanics to achieve unprecedented accuracy in phase measurement. The QATI technique, through the use of the quantum Hebbian learning algorithm [35–37], reveals that GERD is not yet fully operational, as evidenced by the lowest recorded volumetric water discharge. This groundbreaking integration of quantum mechanics with remote sensing technologies thus offers a powerful new tool for monitoring large-scale water management infrastructures, such as GERD, with unparalleled precision and insight.

In conclusion, the integration of the qHop algorithm with Along-Track Interferometry (ATI) using TanDEM-X satellite data has significantly advanced our ability to monitor and analyze streamflow dynamics and volumetric water discharge at the Grand Ethiopian Renaissance Dam (GERD). The precision offered by this quantum-enhanced approach not only provides critical insights into the dam's operational efficiency but also reveals the broader hydrodynamic implications of its performance on the Blue Nile basin.

Through the novel application of quantum Hebbian learning within the interferometric framework, qHop has demonstrated exceptional accuracy in correlating low downstream velocities with minimal volumetric discharge, offering a robust tool for evaluating the effects of limited turbine operation on water flow. This method, particularly when integrated into the Quantum Along-Track Interferometric (QATI) radar processing technique, represents a transformative step in the remote sensing of large-scale water infrastructures.

As GERD continues to evolve, the use of such advanced algorithms will be indispensable in managing water resources efficiently, ensuring sustainable power generation, and maintaining downstream ecosystems. Ultimately, this research underscores the profound potential of combining quantum computing principles with modern geospatial technologies to tackle the complex challenges of water management in the 21st century.

Therefore, the central question now arises: can the maximum volumetric water storage capacity of 74 BCM in GERD cause potential dam deformation? This critical issue will be thoroughly explored in the next chapter, utilizing the Interferometric Synthetic Aperture Radar (InSAR) technique to detect any displacement or ground movement that may occur due to the immense stress exerted by the 74 BCM of stored water on the dam's structure. InSAR's ability to measure millimeter-scale deformations will provide precise insights into whether the dam is susceptible to structural shifts or subsidence under such pressure, offering a comprehensive evaluation of the dam's long-term stability.

References

[1] Herzog, M.A. 1999. Practical Dam Analysis. London: Thomas Telford.

[2] Jansen, R.B. 2012. Advanced Dam Engineering for Design, Construction, and Rehabilitation. Springer Science & Business Media.

[3] Zhang, L., Peng, M., Chang, D. and Xu, Y. 2016. Dam Failure Mechanisms and Risk Assessment. John Wiley & Sons.

[4] McManamay, R.A., Samu, N., Kao, S.C., Bevelhimer, M.S. and Hetrick, S.C. 2015. A multi-scale spatial approach to address environmental effects of small hydropower development. Environmental Management. 55: 217–243.

[5] Gemechu, E. and Kumar, A. 2022. A review of how life cycle assessment has been used to assess the environmental impacts of hydropower energy. Renewable and Sustainable Energy Reviews. 167: 112684.

[6] Wahlstrom, E. 2012. Dams, Dam Foundations, and Reservoir Sites (Vol. 6). Elsevier.

[7] Cassidy, J.J. 1965. Irrotational flow over spillways of finite height. Journal of the Engineering Mechanics Division. 91(6): 155–173.

[8] Scheumann, W. and Hensengerth, O. 2014. Dams and norms: Current practices and the state of the debate. pp. 1–12. *In*: Scheumann, W. and Hensengerth, O. (eds). Evolution of Dam Policies: Evidence from the Big Hydropower States. Springer Berlin, Heidelberg.

[9] Jackson, D.C. 2017. Dams. Routledge.

[10] Adamo, N., Al-Ansari, N., Sissakian, V., Laue, J. and Knutsson, S. 2020. Dam safety: general considerations. Journal of Earth Sciences and Geotechnical Engineering. 10(6): 1–21.

[11] Solomon, T. and Lukas, P. 2024. Land use land cover analysis of the Great Ethiopian Renaissance Dam (GERD) catchment using remote sensing and GIS techniques. Geology, Ecology, and Landscapes. 8(3): 368–380.

[12] Salama, A., ElGabry, M., El-Qady, G. and Moussa, H.H. 2022. Evaluation of Grand Ethiopian Renaissance Dam Lake Using Remote Sensing Data and GIS. Water. 14(19): 3033.

[13] Abera, W., Haregeweyn, N., Dile, Y., Fenta, A.A., Berihun, M.L., Demissie, B., et al. 2021. Scientific Misconduct and Partisan Research on the Stability of the Grand Ethiopian Renaissance Dam: A Critical Review of a Contribution to Environmental Remote Sensing in Egypt (Springer, 2020). Nile and Grand Ethiopian Renaissance Dam: Past, Present and Future. 273–293.

[14] Dandrawy, M.E. and Omran, E.S.E. 2020. Integrated watershed management of Grand Ethiopian Renaissance Dam via watershed modeling system and remote sensing. Environmental remote sensing in Egypt. 533–574.

[15] Ali, A.M., Melsen, L.A. and Teuling, A.J. 2023. Inferring reservoir filling strategies under limited data availability using hydrological modelling and Earth observation: the case of the Grand Ethiopian Renaissance Dam (GERD). Hydrology and Earth System Sciences Discussions. 1–39.

[16] Chen, X., Sun, Q. and Hu, J. 2018. Generation of complete SAR geometric distortion maps based on DEM and neighbor gradient algorithm. Applied Sciences. 8(11): 2206.

[17] Samuele, D.P., Filippo, S., Orusa, T. and Enrico, B.M. 2021. Mapping SAR geometric distortions and their stability along time: A new tool in Google Earth Engine based on Sentinel-1 image time series. International Journal of Remote Sensing. 42(23): 9135–9154.

[18] Li, S., Xu, W. and Li, Z. 2022. Review of the SBAS InSAR time-series algorithms, applications, and challenges. Geodesy and Geodynamics. 13(2): 114–126.

[19] Lyzenga, D. and Wackerman, C. 1997. Detection and classification of ocean eddies using ERS-1 and aircraft SAR images. European Space Agency (Special Publication) ESA SP. 414: 1267–1271.

[20] Macchiarulo, V., Milillo, P., Blenkinsopp, C., Reale, C. and Giardina, G. 2023. Multi-temporal InSAR for transport infrastructure monitoring: Recent trends and challenges. Proceedings of the Institution of Civil Engineers—Bridge Engineering. 176(2): 92–117.

[21] Marghany, M. 2013. DInSAR technique for three-dimensional coastal spit simulation from RADARSAT-1 fine mode data. Acta Geophysica. 61(2): 478–493.

[22] Woodhouse, I.H. 2017. Introduction to Microwave Remote Sensing. CRC Press.

[23] Romeiser, R., Johannessen, J., Chapron, B., Collard, F., Kudryavtsev, V., Runge, H., et al. 2010. Direct surface current field imaging from space by along-track. *In*: SAR and conventional SAR. Oceanography from space: Revisited. 73–91.

[24] Suchandt, S. and Runge, H. 2015. Ocean surface observations using the TanDEM-X satellite formation. IEEE Journal of Selected Topics in Applied Earth Observations and Remote Sensing. 8(11): 5096–5105.

[25] Lanzagorta, M. 2011. Quantum radar. Synthesis Lectures on Quantum Computing. 3(1): 1–39.

[26] Behrman, E.C., Nash, L.R., Steck, J.E., Chandrashekar, V.G. and Skinner, S.R. 2000. Simulations of quantum neural networks. Information Sciences. 128(3–4): 257–269.

[27] Shukla, A. and Vedula, P. 2019. Trajectory optimization using quantum computing. Journal of Global Optimization. 1–27.

[28] Mittermayer, J. and Runge, H. 2003. Conceptual studies for exploiting the TerraSAR-X dual receive antenna. *In*: IGARSS 2003: 2003 IEEE International Geoscience and Remote Sensing Symposium. 3: 2140–2142. IEEE.

[29] Marghany, M. and Samnat, N. 2019. Along-Track Interferometry TanDEM-X Satellite Data for Modelling Sea Surface Level Change and Sea Surface Current Velocity. *In*: 10th International Conference on Robotics, Vision, Signal Processing and Power Applications: Enabling Research and Innovation Towards Sustainability. 489–496. Springer Singapore.

[30] Marghany, M. 2021. Nonlinear Ocean Dynamics: Synthetic Aperture Radar. Elsevier.

[31] Marghany, M. 2019. Along Track Interferometry TANDEM-X Satellite Data for Modelling Sea Surface Pattern Variations. Modern Approaches in Oceanography and Petrochemical Sciences. 2(4). MAOPS. MS. ID. 144.

[32] Marghany, M. 2022. Remote Sensing and Image Processing in Mineralogy. CRC Press.

[33] Marghany, M. 2021. Advanced Algorithms for Mineral and Hydrocarbon Exploration Using Synthetic Aperture Radar. Elsevier.

[34] Rebentrost, P., Bromley, T.R., Weedbrook, C. and Lloyd, S. 2018. Quantum Hopfield neural network. Physical Review A. 98(4): 042308.

[35] Marghany, M. 2019. Automatic Detection Algorithms of Oil Spill in Radar Images. CRC Press.

[36] Fang, D. 2021. Time-dependent Hamiltonian Simulation of Highly Oscillatory Dynamics. CoRR.

[37] Lin, H., Wang, C., Yu, F., Sun, J., Du, S., Deng, Z., et al. 2023. A review of chaotic systems based on memristive Hopfield neural networks. Mathematics. 11(6): 1369.

Chapter 11

Quantum Convolutional Neural Network for Monitoring Deformation of the Grand Ethiopian Renaissance Dam Using Persistent Scatterer Interferometry Technique

11.0 INTRODUCTION

In previous chapters, we discussed the significant accumulation of sediment deposition in the GERD reservoir, along with the high water level of 590 meters (mMSL), corresponding to an approximate storage volume of 74 BCM, and a low downstream discharge of 18 BCM. The critical question that now arises is: can these physical factors—high sediment deposition, large water storage, and low discharge—contribute to dam or land deformation?

While we previously explored the fundamentals of Interferometric Synthetic Aperture Radar (InSAR) to measure displacement, we delved deeper into Along-Track Interferometry to gain a comprehensive understanding of downstream volumetric discharges. In this chapter, we aim to address the potential impact of GERD's high-volume water storage and reduced operational efficiency on both land and dam structure deformation.

To achieve this, we employ Persistent Scatterer Interferometry (PSI), a highly effective technique renowned for its ability to monitor minute land and structural deformations with exceptional precision. PSI holds great promise in revealing any subtle shifts or stresses that may arise as a result of GERD's current operational conditions.

11.1 PERSISTENT SCATTERER INTERFEROMETRY (PSI)

Persistent Scatterer Interferometry (PSI) is an advanced technique within the realm of Interferometric Synthetic Aperture Radar (InSAR) that offers an exceptionally precise method to detect and monitor ground deformation over time. What distinguishes PSI from other methods is its focus on 'persistent scatterers'—stable, highly reflective objects such as buildings, bridges or exposed rocks that consistently reflect radar signals to the satellite. This allows PSI to track

DOI: 10.1201/9781003327325-11

ground or structural deformations with millimeter-level accuracy, making it a crucial tool for long-term monitoring.

The core strength of PSI lies in its ability to conduct temporal analysis. By analyzing radar data collected from multiple satellite passes over several months or even years, PSI can detect subtle shifts that occur gradually. Unlike traditional InSAR techniques, which may be affected by surface changes such as vegetation growth or moisture variations, PSI is specifically designed to isolate these persistent scatterers, thereby offering a high degree of precision. This stability enables PSI to accurately measure ground or infrastructure deformation, even in complex environments where other methods might struggle [1].

Further, PSI offers extensive area coverage, making it suitable to monitor deformation on both regional and local scales. This wide applicability, coupled with its millimeter-scale precision, renders PSI an ideal method for observing critical infrastructures like dams and bridges. In urban settings, PSI can be used to monitor subsidence caused by groundwater extraction, underground construction or natural geologic processes. Since it is less influenced by environmental factors such as vegetation or weather conditions, PSI ensures consistent, reliable results across a variety of landscapes and urban areas [2].

Thus, the advantages of PSI extend beyond just its accuracy. The technique is particularly valuable in applications such as dam monitoring—where it can detect small but significant shifts in structures like the Grand Ethiopian Renaissance Dam (GERD)—as well as in urban environments experiencing subsidence. Additionally, PSI is used in landslide and geohazard detection, providing early warnings in regions at risk of ground movement. It also plays a vital role in industries like mining and oil extraction, where subsidence or uplift due to resource extraction needs careful monitoring.

From the aforementioned perspective, the strength of Persistent Scatterer Interferometry lies in its ability to consistently track minute ground and structural movements over time, making it an invaluable tool for geotechnical assessments and infrastructure management. Its precision, wide area coverage, and resilience to environmental changes make it essential for applications where high accuracy is paramount, ensuring that potential hazards are identified and mitigated before significant damage can occur.

11.2 HOW DOES PSI OPERATE?

Persistent Scatterer Interferometry (PSI) works by using radar signals from satellites to monitor and measure ground or structural deformation over time with incredible accuracy. The technique relies on a series of radar images acquired over multiple satellite passes to detect slight movements of the Earth's surface or structures. PSI begins with the acquisition of radar images from satellites, which operate using Synthetic Aperture Radar (SAR). Unlike optical imagery, radar is not affected by cloud cover or darkness, allowing continuous monitoring in all weather conditions. The satellite sends radar signals to the Earth's surface, and these signals are reflected to the satellite from objects on the ground [2, 3].

By focusing on these stable objects, PSI reduces noise and errors associated with natural surface changes, such as vegetation growth or soil moisture, which can affect radar signal reflection in conventional InSAR techniques. Persistent scatterers serve as reference points, enabling precise measurements of any small movements or deformations. Once the radar images are acquired, the next step is interferometric processing. This involves comparing the phase of radar signals between different satellite passes. Radar signals carry two main pieces of information: amplitude (the strength of the return signal) and phase (the relative position of the wave). This phase is particularly important because it allows PSI to detect small changes in the distance between the satellite and the ground.

By calculating the phase difference between successive radar images, PSI can detect even minute displacements in the position of persistent scatterers. These phase differences, known as interferograms, reveal any ground or structural movement that has occurred between satellite passes often down to millimeter-level precision. Therefore, PSI uses multiple radar images acquired over a long period—typically months or years—to track deformation trends. Time series analysis is performed on the phase data from the persistent scatterers, allowing the detection of slow-moving, long-term ground deformations [1, 5]. This analysis helps to distinguish between temporary changes (such as seasonal variations) and consistent, ongoing displacement.

11.3 INTERFEROMETRIC PROCESSING

Interferometric Synthetic Aperture Radar (InSAR) is a powerful remote sensing technique that capitalizes on the phase differences between at least two Synthetic Aperture Radar (SAR) images. These images are typically captured from different satellite orbit positions or at different times. In essence, InSAR measures the phase difference between radar signals that have traveled to the ground and back, using this information to detect minute changes in the Earth's surface such as displacement or deformation. In repeat-pass interferometry, therefore, the SAR system flies over the same area along nearly parallel tracks, viewing the terrain from slightly different angles at different times. The satellite positions during the first and second SAR acquisitions are denoted as S_1 and S_2, respectively. The distance between these positions is known as the baseline (B), while the perpendicular baseline ($B_\perp$) refers to the component of the baseline perpendicular to the radar look direction from SAR 2. Additionally, the slant ranges to the ground target from S_1 and S_2 are represented as R_1 *and* R_2, respectively. In this perspective, the keystone of SAR interferometry lies in the phase information from each SAR image, which is directly related to the range difference (Figure 11.1) [1–3]:

$$\Delta R = R_2 - R_1 \tag{11.1}$$

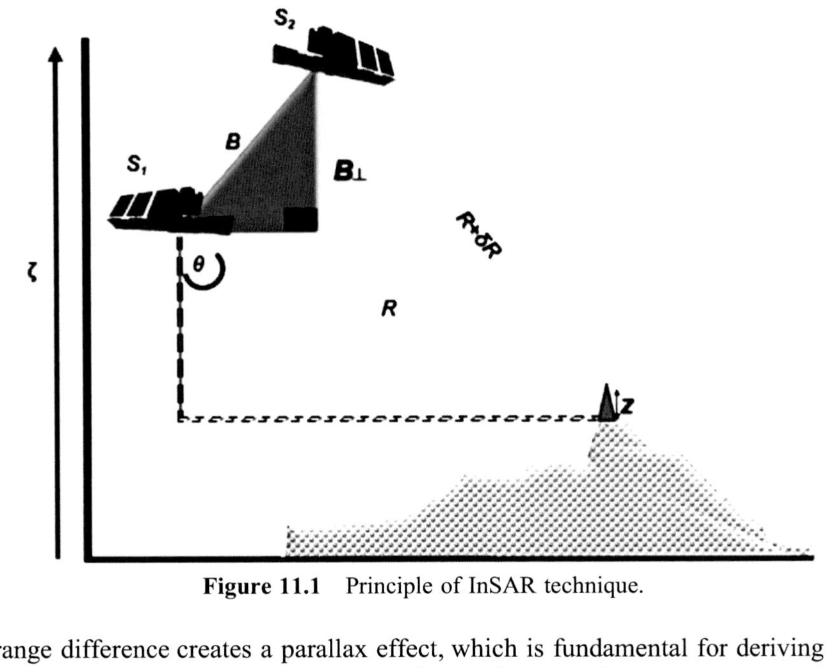

Figure 11.1 Principle of InSAR technique.

This range difference creates a parallax effect, which is fundamental for deriving topography or detecting surface deformation. Consequently, mathematically, the two co-registered SAR images can be expressed in terms of amplitude and phase components. Let the first image be represented as:

$$f(R,x) = |f(R,x)| \exp\{j\phi_1(R,x)\} \tag{11.2}$$

where $|f(R, x)|$ denotes the amplitude, and $\phi_1(R, x)$ is the phase. Similarly, the second image is represented as:

$$g(R,x) = |g(R,x)| \exp\{j\phi_2(R,x)\} \tag{11.3}$$

here amplitude is presented by $|g(R, x)|$ while $\phi_2(R, x)$ is phase. To extract the relevant phase information, the interferogram is formed by multiplying the first SAR image with the complex conjugate of the second image, as follows:

$$I(R,x) = f(R,x)g^*(R,x) \tag{11.4}$$

Therefore, the resulting interferogram contains both amplitude and phase components, where the phase difference:

$$\phi(R,x) = \phi_1(R,x) - \phi_2(R,x) \tag{11.5}$$

Equation 11.5 represents the interferometric phase. The phase response ϕ of SAR image from a point scatterer is directly proportional to the distance, or range from the radar to the scatterer. In this relationship, k serves as the proportionality constant. Additionally, a possible phase shift, ϕ_{scat}, can be introduced due to the inherent properties of the scatterer itself [2, 4]. Mathematically, this can be expressed as follows:

$$\phi_1 = -2kR_1 + \phi_{scat,1} \tag{11.6}$$

$$\phi_2 = -2kR_2 + \phi_{scat,2} \tag{11.7}$$

where ϕ_1 and ϕ_2 are the phases corresponding to two SAR acquisitions taken at times 1 and 2, respectively. In this sense, R_1 and R_2 represent the range from the radar to the scatterer during these two acquisitions. Additionally, ϕ_{scat} and ϕ_{scat} are the phase shifts caused by the scatterer for the two observations [2]. Assuming there is no ground displacement between the two observations, meaning the position of the scatterers remains unchanged, the interferometric phase can be described by the following equation:

$$\phi = \phi_{topo} + \Delta\phi_{scat} \tag{11.8}$$

where ϕ_{topo} is the topographic term, accounting for the elevation differences of the terrain, and $\Delta\phi_{scat}$ represents the phase shift difference due to variations in the scatterer itself. In this view, the topographic term ϕ_{topo} describes how elevation contributes to the interferometric phase and is given by the following equation:

$$\phi_{topo} \approx \frac{4\pi}{\lambda} \cdot \frac{B_\perp}{R\sin\theta} \cdot z \tag{11.9}$$

where λ is the radar wavelength, $B_\perp$ is the perpendicular component of the baseline (the distance between the two radar positions during the acquisitions); R is the slant range to the target; θ is the incidence angle of the radar signal; and z represents the elevation of the point scatterer. In this sense, the topographic term ϕ_{topo} is crucial for measuring terrain elevation or detecting changes in surface deformation using interferometric synthetic aperture radar (InSAR). It is derived from the geometric configuration of the radar system, the perpendicular baseline, and the incidence angle. By analyzing this term, one can infer elevation variations across the observed terrain [1–5].

11.4 SIMULATION OF INSAR COHERENCE

To assess the similarity between a pair of SAR images and to mitigate the effects of random noise fluctuations, InSAR coherence is commonly employed. Coherence serves as a quantitative measure, reflecting the degree of similarity between two SAR images. It is defined as the magnitude of the estimated sample complex cross-correlation coefficient between the two images and is expressed as a value in the range [0, 1]. A coherence value closer to 1 indicates high similarity between the images, while values closer to 0 suggest significant decorrelation. The sample coherence, denoted as $\gamma_N e^{j\phi'}$ is given by the following equation [5–7]:

$$\gamma_N e^{j\phi'} = \frac{\sum_{i=1}^{N} f_i g_i^*}{\sqrt{\sum_{i=1}^{N} |f_i|^2 \sum_{i=1}^{N} |g_i|^2}} \tag{11.10}$$

where γ_N is the sample coherence derived from N measurements; ϕ' is the filtered interferometric phase; f_i and g_i represent the complex pixel values in the two co-registered SAR images; g_i^* denotes the complex conjugate of the pixel values in the second SAR image [5, 7].

The formula calculates the cross-correlation between the two SAR images over a set of pixels, encoding how similar the images are based on the alignment of their phases and magnitudes. The numerator, $\sum_{i=1}^{N} f_i g_i^*$ represents the complex cross-correlation between the images, while the denominator normalizes the correlation by considering the magnitudes of the pixel values in each image [2, 6].

Coherence in Synthetic Aperture Radar (SAR) interferometry; therefore, is influenced by several factors that determine the degree of similarity between two SAR images. These factors can be broadly categorized into three main types: (i) signal-to-noise ratio (SNR) (ii) baseline decorrelation and (iii) temporal Decorrelation.

Firstly, signal-to-noise ratio (SNR) plays a significant role in affecting coherence. It refers to the relative strength of the backscatter signal compared to the noise present in the radar receiver. A higher SNR generally results in better coherence between the SAR image pair, as the signal is clearer and less affected by noise. In such cases, the scene appears more stable, and the images maintain higher fidelity, leading to improved coherence values [1, 5, 7].

Secondly, baseline decorrelation pertains to the physical separation between satellite tracks during successive image acquisitions. When the baseline—the distance between the two satellite positions—is too large, the radar views the scene from significantly different angles. This variation in viewing geometry introduces decorrelation, causing a reduction in coherence. Therefore, maintaining an optimal baseline is crucial to preserving image similarity and coherence. Lastly, temporal decorrelation is often the most dominant factor in repeat-pass SAR interferometry. This type of decorrelation is caused by changes in the land surface between acquisition times. For example, human activities such as the construction of buildings or roads changes in vegetation due to growth or seasonal variation and environmental disturbances like ploughing or wind erosion in desert areas can significantly alter surface properties. As a result, the coherence between the two SAR images decreases over time, reflecting the dynamic nature of the terrain [1, 3, 5, 7, 11, 14].

Needless to say, coherence in SAR interferometry is primarily affected by SNR, baseline separation and temporal surface changes. While high SNR and optimal baselines contribute to maintaining coherence, temporal decorrelation due to surface alterations remains the most challenging factor to manage, particularly in dynamic environments [8–12].

Between January 2022 and August 2024, and 50 Sentinel-A datasets were analyzed using both VV and VH polarimetry bands. The coherence patterns observed differed significantly between these two bands. In the VV band, coherence captures more detailed information about topographical features such as mountains, urban structures, and buildings. These regions exhibit high coherence values, often approaching 1, reflecting the stability and consistency of these surfaces. Conversely, areas dominated by vegetation show lower coherence, typically less than 0.5, while water bodies register the lowest coherence values, close to 0. This demonstrates that the VV band is highly sensitive to stable, rigid structures, while softer, dynamic surfaces like vegetation and water lead to decorrelation (Figure 11.2).

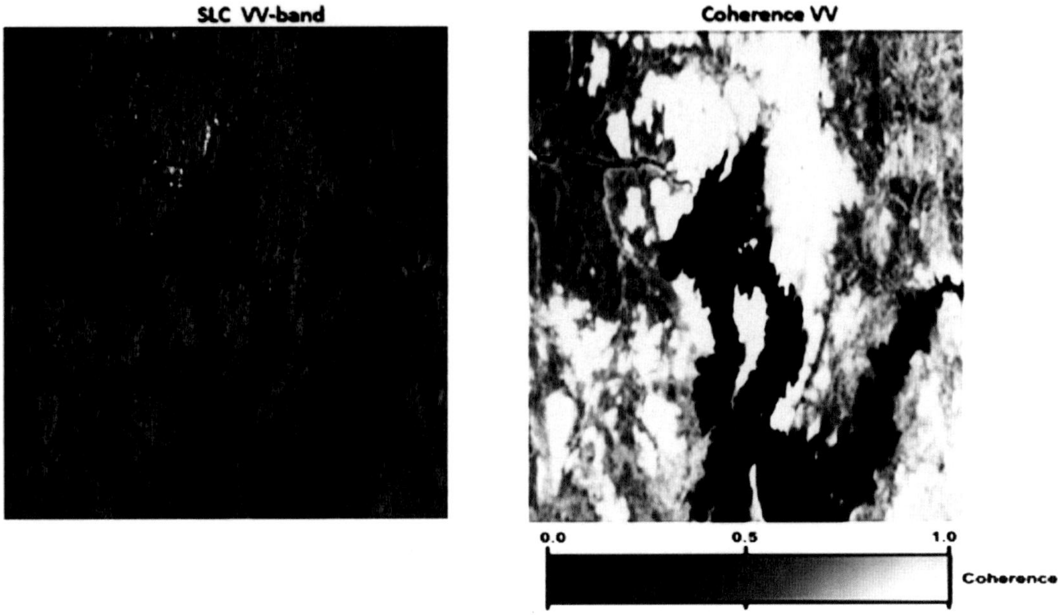

Figure 11.2 Coherence of Sentinel-1A VV-band.

In contrast, the VH band exhibits different behavior. Vegetation covers display even lower coherence values, around 0.2, which is significantly less than those recorded in the VV band. Furthermore, urban areas and buildings also show reduced coherence compared to the VV band, with values closer to 0.7. This suggests that while the VH band is useful for detecting certain surface features, it is less effective than the VV band in capturing structural details of mountainous or urban zones (Figure 11.3). Therefore, in this chapter, we implanted VV polarization for retrieving displacement of the main dam and saddle dam.

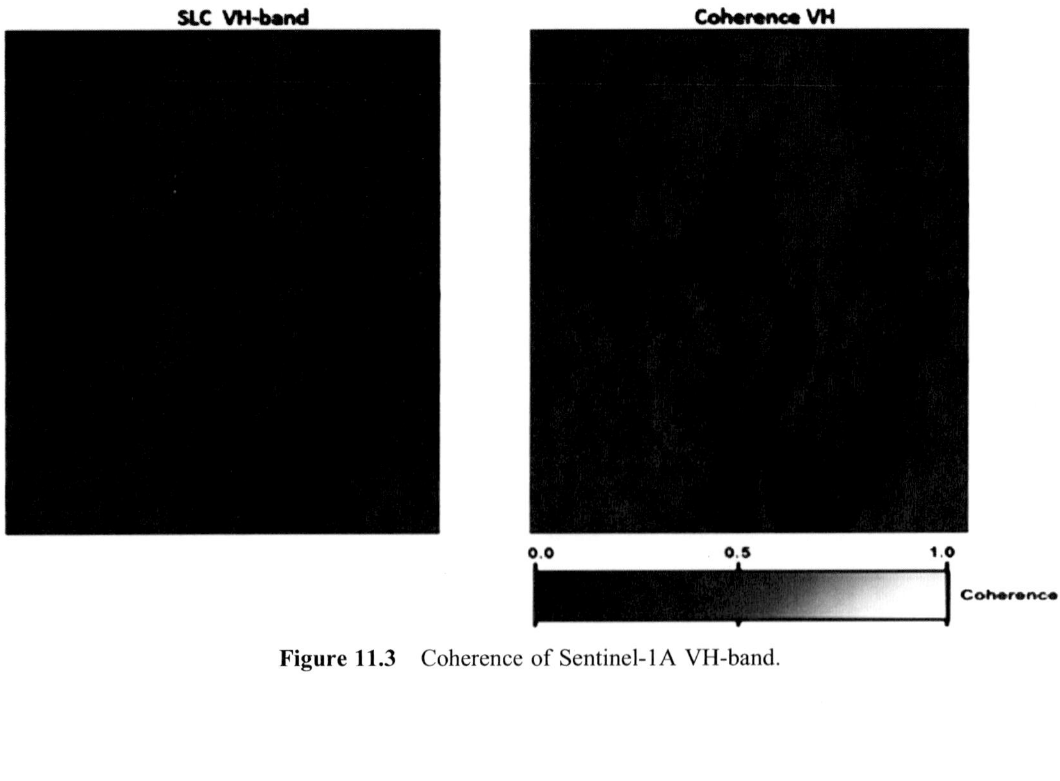

Figure 11.3 Coherence of Sentinel-1A VH-band.

In particular, the coherence behavior between the VV and VH bands highlights how different polarizations can reveal varying aspects of surface characteristics, with the VV band being more suited for stable, man-made structures and the VH band being less sensitive to such features [2, 11, 13, 15].

11.5 QUANTIZED COHERENCE SAR SIMULATION

To quantize the coherence equation using a Quantum Convolutional Neural Network (QCNN), the approach involves translating the classical coherence equation into a quantum algorithm that utilizes the QCNN framework. This framework is useful for efficiently processing quantum data, and, in principle, it can also approximate classical functions or statistics in quantum states, like coherence, with quantum computation. To transform equation 11.10 into the quantum domain, let us first represent the SAR images f_i_and g_i as quantum states $|f_i\rangle|$ and $|g_i\rangle$. These states are encoded into qubits, such that the pixel values are now quantum amplitudes. For example, a pixel in SAR image f_i_ could be encoded as:

$$|f_i\rangle = \sum_j a_j |j\rangle \tag{11.11}$$

where a_j are the amplitudes, and j indexes the possible states for each pixel. Similarly, for g_i, as well as $|g_i\rangle$.

The Quantum Convolutional Neural Network (QCNN) architecture is designed to process quantum data by leveraging hierarchical convolution and pooling layers. These layers play a critical role in reducing the dimensionality of the input quantum state, while crucially preserving its essential features. This approach is notable for its efficiency, as it is characterized by *O(log(N))* parameters, which makes it exponentially more efficient than traditional quantum circuits. Consequently, the QCNN structure allows for a powerful means of capturing correlations between qubits, which is analogous to the pixel-wise correlations between synthetic aperture radar (SAR) images that are fundamental in classical coherence calculations [16].

The input to the QCNN consists of quantum states, denoted as $|f_i\rangle|$ and $|g_i\rangle$, which correspond to the pixel values of SAR images. These input states are represented as N-qubit states and fed into the QCNN for processing. The convolutional layers in the network perform localized operations on these quantum states, analogous to how pixel values in classical coherence equations are summed to compute coherence. As the input quantum states pass through these convolutional layers, they are systematically analyzed for their local correlations, which reflect the interactions between corresponding pixel values in SAR image pairs. The pooling layers then follow, effectively reducing the dimensionality of the system while retaining the critical coherence features, thereby ensuring the computational efficiency of the network [17].

The QCNN further enhances its effectiveness through the application of unitary operations U_i at each layer. These unitary operations correspond to quasilocal transformations on the qubits and they are designed to mimic the pairwise cross-correlation process between the pixel values of quantum SAR states $|f_i\rangle|$ and $|g_i\rangle$. By implementing these unitary operations, the QCNN efficiently processes the input states, enabling a precise representation of coherence across the quantum states. Thus, the hierarchical structure of the QCNN not only allows for efficient dimensionality reduction but also ensures that essential coherence information is preserved and accurately reflected in the output [16–19].

Given a set of SAR data, the QCNN can be trained using Mean Squared Error (MSE) to classify input quantum states based on their coherence. The training process involves initializing the unitary matrices U_i and optimizing them through a learning process, such as gradient descent. For the coherence equation, the QCNN is trained to map SAR input states to coherence values, which can be binary-classified or output as continuous values. The loss function for this QCNN model; therefore, would be:

$$L_{MSE} = \frac{1}{2M}\sum_{\alpha=1}^{M}\left(\gamma_{\alpha}e^{j\phi'} - f(U_i, V_j, F|\psi_{\alpha}\rangle)\right)^2 \quad (11.12)$$

where $\gamma_{\alpha}e^{j\phi'}$ are the expected coherence values for each SAR pair. Therefore, the term $f(U_i, V_j)$ and $F|\psi_{\alpha}\rangle$ is the QCNN output for the input state $|\psi_{\alpha}\rangle$. Additionally, U_i,V_j are the optimized unitaries in the QCNN layers. The QCNN, after sufficient training, classifies SAR coherence values by processing quantum states through its layered architecture. The hierarchical pooling in the QCNN ensures that it efficiently captures long-range correlations, similar to how coherence in SAR interferometry represents the similarity between pixel pairs.

11.6 RELATION TO MULTISCALE ENTANGLEMENT RENORMALIZATION ANSATZ (MERA)

The QCNN structure is deeply connected to the MERA framework, which is used to represent many-body quantum wavefunctions efficiently. A MERA applies unitary and isometric layers, similar to the QCNN, but runs in the reverse direction. This structure allows the QCNN to capture the multiscale correlations in the quantum states, analogous to how coherence measures similarity between large areas of SAR images. By applying the MERA-like architecture, the QCNN can detect critical regions of high or low coherence, capturing both large-scale and fine-grained changes in the SAR data.

The QCNN's pooling layers can be viewed as a form of quantum error correction, where local quantum errors (incoherent areas in the SAR images) are corrected as the quantum states pass through successive layers. This is similar to how QEC optimizes encoding and decoding to correct quantum errors without collapsing the quantum state. By mimicking the renormalization-group flow, the QCNN classifies quantum phases that correspond to coherent or incoherent regions in SAR images, making it particularly powerful for SAR coherence estimation. Therefore, using QCNN, the coherence equation in quantum terms is approximated as:

$$\gamma_N = f\left(U_i, V_j, F|\psi\rangle\right) \quad (11.13)$$

here $f(U_i, V_j, F|\psi\rangle)$ represents the QCNN output based on the trained unitaries and the input quantum state $|\psi\rangle$, corresponding to SAR pixel values. The hierarchical convolution ensures efficient extraction of coherence features, and the pooling layers reduce dimensionality, preserving key coherence information [16, 20].

The final output of the QCNN provides the coherence classification (Figure 11.4). Quantum measurement of the output state can be used to extract the coherence value, which is then used to calculate the degree of similarity between SAR images. In this view, The QCNN coherence image shows interference patterns characteristic of coherent quantum states. These patterns arise when waves overlap and either reinforce or cancel each other out. Consequently, QCNN delivers variations in shading or intensity in the image to represent different coherence values. Darker and lighter areas correspond to different coherence shifts experienced by the quantum states [16, 19].

The coherence observed in the main dam construction is notably characterized by its highest possible value of 1.0, particularly when compared to the VH polarization band (Figure 11.4). However, the introduction of a QCNN enhances the visualization of VH coherence features. What is especially striking is that the dam receiver, previously showing little detail in the VH band, now reveals subtler aspects of the scene. Specifically, the presence of suspended sediments appears to modulate what was originally a near-zero coherence over the water surface, increasing it to approximately 0.4 (Figure 11.5). This phenomenon is likely due to the quantum interference effects inherent in the QCNN processing, which captures and amplifies complex coherence patterns within the data [17–20].

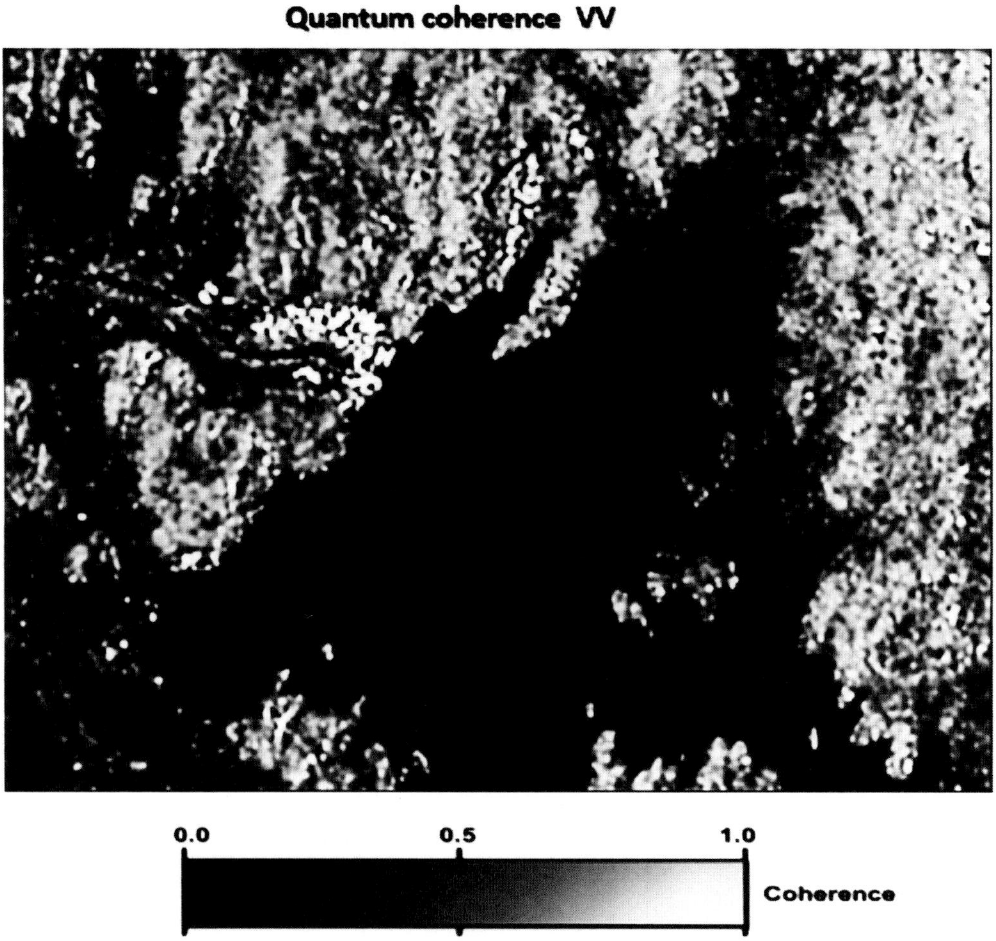

Figure 11.4 Quantum coherence VV-band image generated by QCNN.

The improvement brought about by the QCNN can be understood through the fundamental principles of quantum mechanics, where superposition—the ability of particles to exist in multiple states simultaneously—plays a central role. In classical SAR imagery, coherence measures the similarity between two signals or observations. Analogously, in the quantum domain, coherence between qubits refers to the degree to which they are in a superposed state. The QCNN capitalizes on the inherent quantum mechanical property of interference—both constructive and destructive—to discern hidden features in the data, making it possible to visualize phenomena like suspended sediments affecting water surfaces in ways classical methods might overlook [17–19].

This enhancement is rooted in the QCNN's ability to perform hierarchical quantum operations. By applying unitary transformations across multiple layers, the network processes local regions of quantum states in a manner analogous to classical convolution operations, where the QCNN analyzes qubit correlations in a similar way to classical methods that analyze pixel intensities in SAR imagery (Figure 11.6). The pooling layers further reduce the dimensionality of the quantum state, enabling the model to retain essential coherence information while discarding irrelevant noise (Figure 11.7). Thus, the QCNN effectively reveals intricate details in the coherence of VH bands that would otherwise remain hidden.

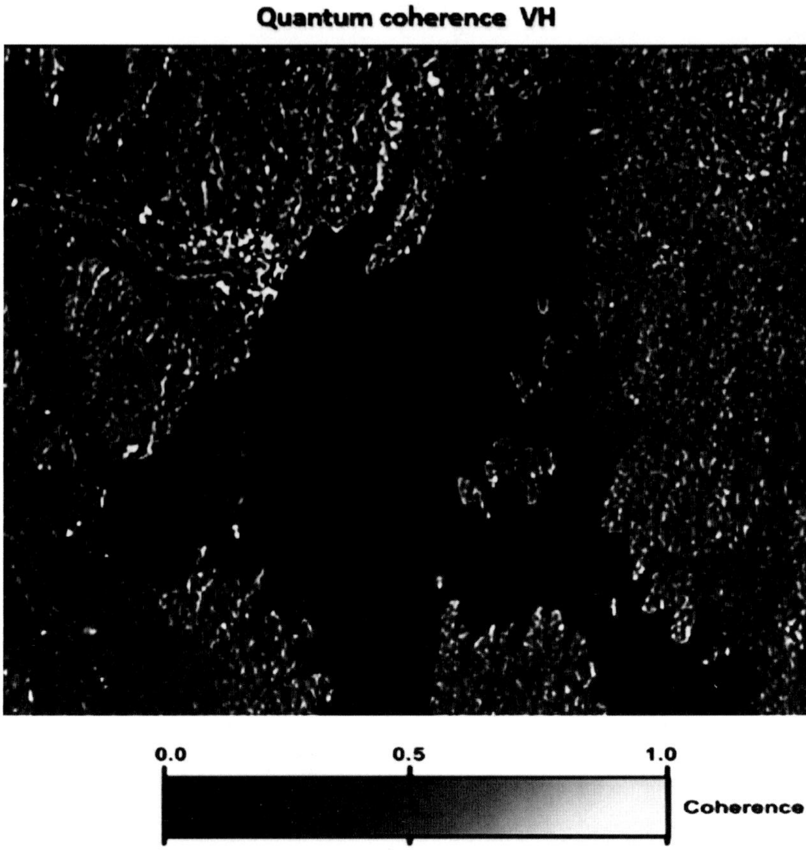

Figure 11.5 Quantum coherence VH-band image generated by QCNN.

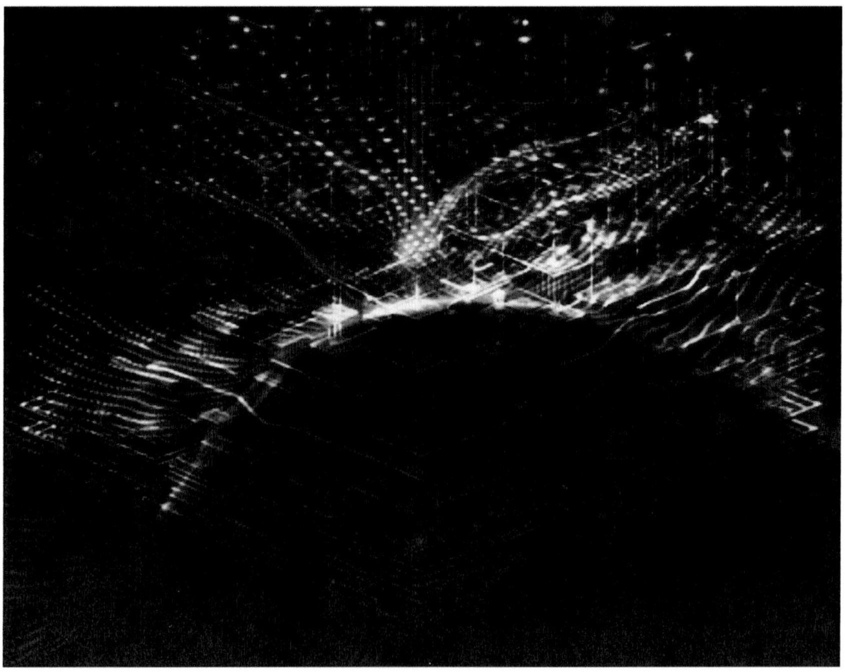

Figure 11.6 Mechanism of unitary transformations across multiple layers encoded in qubit.

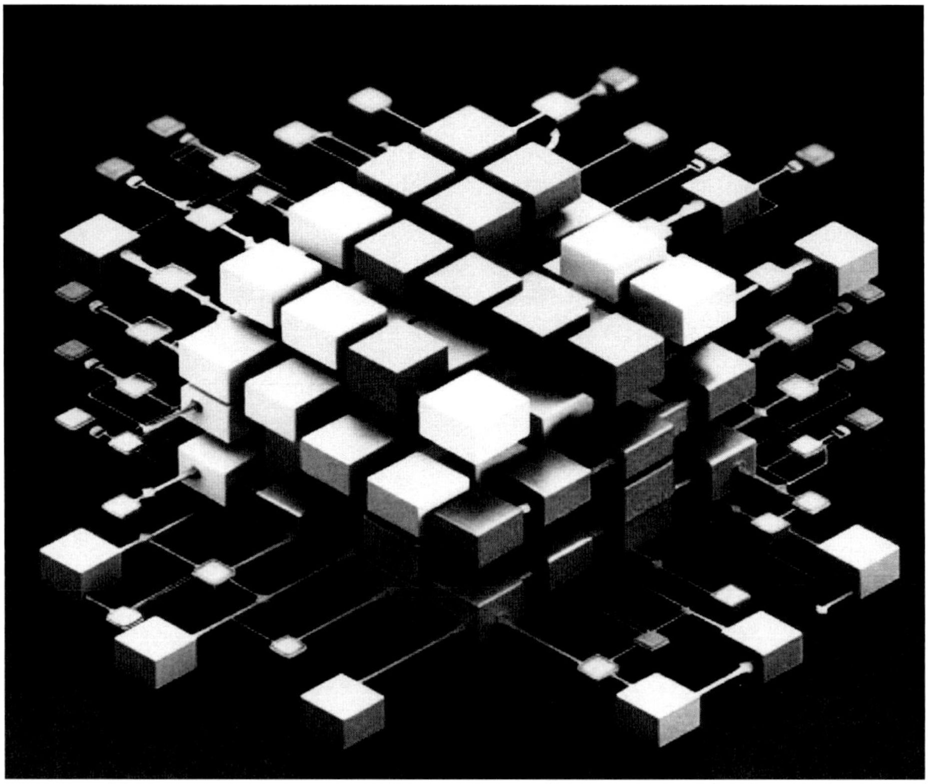

Figure 11.7 Generalization of QCNN pooling layers.

Moreover, the QCNN reveals the delicate balance between quantum coherence and the inevitable process of decoherence (Figure 11.8). Quantum coherence represents the persistence of quantum superpositions—where particles remain entangled or correlated in a unified quantum state. However, when these systems interact with their environment, a phenomenon known as decoherence occurs. Decoherence causes the quantum system to lose its coherence and behave more classically, transitioning from a state of quantum superposition to a state where each particle exists in a definite, singular state [18].

This transition is not just a matter of theoretical interest but has profound implications for the physical system under study. In the case of the dam and its surrounding environment, the QCNN can capture this interplay, offering an unparalleled glimpse into how the physical and quantum realities converge. While classical methods only scratch the surface by analyzing coherence in a fixed, static manner, the QCNN dives deeper into the quantum entanglements within the system, revealing a dynamic interaction between quantum and classical phenomena [17, 20].

In essence, the application of QCNNs transforms our ability to observe and interpret complex systems. By leveraging the fundamental principles of quantum mechanics—superposition, entanglement, and interference—the QCNN provides an unprecedented level of detail and understanding. The fact that it can highlight features such as sediment-induced coherence modulations underscores the powerful union of quantum computation and real-world remote sensing applications. These insights offer a groundbreaking step toward a new era of quantum-enhanced environmental monitoring, where quantum algorithms can reveal patterns and dynamics that were previously beyond our reach, opening doors to further exploration in both the quantum and classical realms.

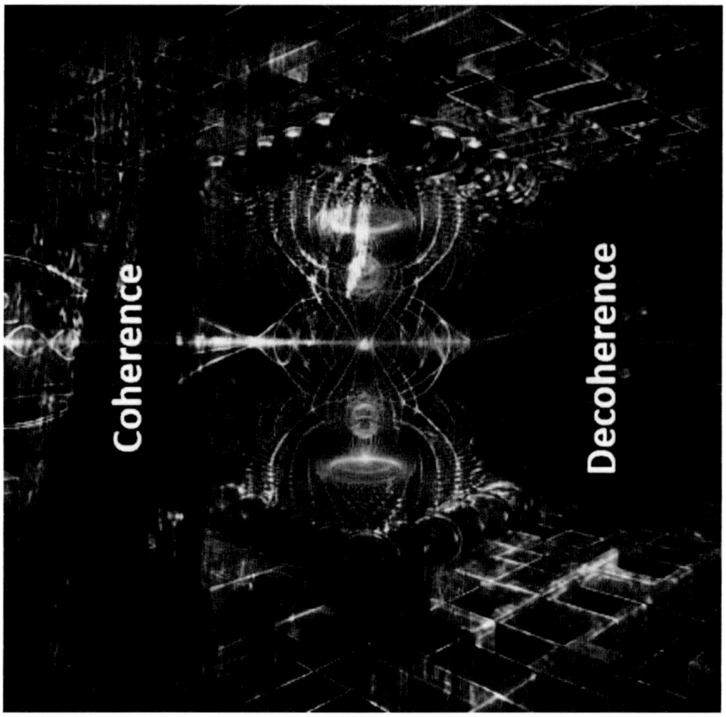

Figure 11.8 QCNN Mechanism for Balancing Quantum Coherence and Decoherence.

11.7 PERSISTENT SCATTERER SELECTION

Not all phase observations in an interferogram contain useful information. The earth surface covered by a resolution cell may change in time, targets may not reflect in the satellite direction, or they may not have the physical properties to be observed with a high enough Signal-to-Noise Ratio (SNR). To select coherent, i.e., interpretable, measurement points, candidate targets are selected that have a strong and consistent reflection in time (the Persistent Scatterers).

In the context of interferometric synthetic aperture radar (InSAR) observations, certain targets within a resolution cell can often be attributed to a single physical object, typically a man-made feature, that acts as a dominant scatterer. These targets, known as Persistent Scatterers (PS), exhibit strong and consistent reflections over time. Many of these scatterers behave like point scatterers, meaning they can be observed from a wide range of viewing angles. This characteristic makes point scatterer observations less sensitive to variations in the effective (perpendicular) baseline between the master and slave acquisitions [3, 14, 21].

However, it is important to note that a Persistent Scatterer does not necessarily have to be a dominant point scatterer within the resolution cell. Distributed scatterers, such as rocks or natural terrain features, can also function as Persistent Scatterers. These distributed scatterers, while not as concentrated as point scatterers, still provide valuable data due to their consistent reflective properties. Their dependency on the viewing geometry, particularly the perpendicular baselines, differs from that of point scatterers, but they remain crucial for comprehensive interferometric analysis.

To efficiently estimate all contributions to the interferometric phase—including surface deformation, topographic heights and atmospheric disturbances—a robust algorithmic approach must be implemented. To tackle the inherent complexity and variability in the data, a first-order network is constructed using Permanent Scatterer (PS) candidates. These candidates are selected based on the assumption that they provide the most reliable phase observations. However, since interferometric phase observations are wrapped (i.e., constrained within the range $[-\pi, \pi]$ and are often contaminated by several error sources, the direct phase observations alone are insufficient for accurate analysis. To overcome these challenges, the amplitude observations of the PS candidates are utilized in tandem with phase observations to build a more resilient model.

To select suitable PS candidates, existing methodologies can be categorized into three main approaches based on their parameterization schemes [3, 5, 19, 20, 21].

Signal-to-Clutter Ratio (SCR): This approach assumes that a PS observation consists of a deterministic signal (originating from a dominant scatterer) disturbed by random Gaussian clutter (distributed scatterers surrounding the point scatterer) [3].

Normalized Amplitude Dispersion (Da): This parameter measures the relative stability of the amplitude over time for a given scatterer, and is often used to detect PS candidates that exhibit stable behavior across multiple SAR acquisitions [22–24].

Supervised Classification: This approach leverages machine learning algorithms to classify potential PS candidates by training a model based on labeled data [25].

The SCR method seeks to quantify the relationship between the dominant scatterer's signal and the clutter in its surroundings, which is assumed to follow a circular Gaussian distribution. The deterministic signal originates from the dominant scatterer within the SAR resolution cell, while the clutter reflects the contributions from surrounding scatterers [26].

Let us represent the amplitude of the dominant scatterer and cc the clutter amplitude from the distributed scatterers. The Signal-to-Clutter Ratio is given by:

$$SCR = \frac{s^2}{c^2} \tag{11.14}$$

This ratio directly affects the phase stability of the observation. The relationship between the phase variance σ_{ψ}^2 and the SCR is crucial for PS analysis and is derived from stochastic modeling. For a single SAR observation, the phase variance σ_{ψ}^2 is inversely proportional to the SCR:

$$\sigma_{\psi}^2 = \frac{2}{SCR} \tag{11.15}$$

This expression provides a fundamental link between the quality of the PS candidate and its phase stability, where higher SCR values correspond to lower phase variance, leading to more reliable PS observations. In this perspective, the interferometric phase ϕ_{obs} observed for any given pixel in a SAR image consist of multiple contributions, including the phase components from surface deformation ϕ_{def}; topography ϕ_{topo}, baseline error $\varphi_{\text{baseline}}$, and atmospheric disturbances φ_{atm}

$$\phi_{\text{obs}} = \varphi_{\text{topo}} + \varphi_{\text{def}} + \varphi_{\text{atmo}} + \varphi_{\text{baseline}} + \varphi_{\text{noise}} \tag{11.16}$$

Equation 11.16 demonstrates that the phase related to surface deformation can be isolated by removing other contributing factors. The topographic phase can be modeled using a digital elevation model (DEM). Atmospheric effects can introduce spatially correlated artifacts of a few centimeters, and baseline errors can also be modeled and removed. Consequently, the noise level of the interferometric phase depends on the coherence of the image pairs. Therefore, the displacement phase is derived from the repeat-pass interferometric phase after accounting for the topographic effect, atmospheric effect and baseline error. In essence, phase displacement is considered the incoherent sum of changes in the surface backscattering phase, volume backscattering phase, and double-bounce backscattering phase [1, 3, 5, 25]. The mathematical expression for phase displacement is given by:

$$\varphi_{\text{disp}} = + \varphi_{\text{surface}} + \varphi_{\text{volume}} + \varphi_{\text{double-bounce}} \tag{11.17}$$

Equation 11.17 represents the fundamental concept for determining changes in the topographic level from the interferometric phase value, considering high coherent variations. In essence, the interferogram φ comprises the topographic φ_{topo} and deformation φ_{defo} components while disregarding other factors [1,5]. The topographic component can be computed using the relationship with known heights H, obtained from a known topographic map, expressed as:

$$\varphi_{\text{topo}} = \frac{4\pi B_{CT} \cos(\theta - \gamma_{CT})}{\lambda R_m \sin\theta} H \tag{11.18}$$

Equation 11.18 illustrates that the SAR wavelength λ; B_{CT}multiplied by $\cos(\theta - \lambda_{CT})$ represents the perpendicular baseline of the two observations, where θ is the incidence angle. Therefore, R_m is the slant range distance. In this context, the infrastructure vibration downward can be estimated by subtracting this quantity φ_{topo} from the interferogram. Consequently, the deformation phase value φ_{defo} corresponds to the shrinkage or extension in the line-of-sight distance between the satellite and ground targets when deformation occurs between two SAR observations. By identifying the change in the line-of-sight distance as ΔR, the phase value can be computed as:

$$\varphi_{\text{defo}} = \frac{4\pi}{\lambda} \Delta R \tag{11.19}$$

Equation 11.19 tells that the range of phase values is constrained between $-\pi$ and π, which confines the aptitude to extricate deformations greater than a quarter-wavelength. For instance, in the case of an L-band wavelength of 24 cm, considering ΔR = 0, +/–12, +/–24… cm, the phase becomes 0 with an indefinite $2n\pi$. Deformation cannot be accurately discerned if neighboring pixels display a line-of-sight modification exceeding 6 cm, as the phase change becomes ambiguous beyond this threshold. This limitation stems from the inherent wrapping of interferometric phases, where the phase differences are confined to a 2π interval, leading to ambiguities in cases of large deformations. Therefore, SAR systems with longer wavelengths offer a significant advantage in detecting and measuring large distortions, as they allow for greater phase differences to be captured before wrapping occurs [20–25].

In the context of dam deformation caused by heavy streamflow or flood events, the absolute distortion can be estimated by applying phase unwrapping techniques. Phase unwrapping retrieves the continuous phase from the wrapped observations, thus enabling the reconstruction of the full displacement field. However, this process is only accurate as long as the deformations adhere to the variety theorem, which ensures that the phase differences between adjacent pixels remain within manageable limits. By satisfying this condition, the unwrapped phase can provide a reliable measure of the overall distortion in the dam structure.

In Persistent Scatterer Interferometry (PSI), a key measurement is the interferometric phase difference, denoted as φ_p^{ms}, between a master acquisition mm and a slave acquisition s for a given Persistent Scatterer (PS) p. However, what is measured is the wrapped phase, meaning only the fractional phase is observed, without accounting for the integer number of full phase cycles between the satellite and the Earth's surface. This phenomenon arises because the interferometric phase is periodic, confined to a range of $[-\pi, \pi]$, leading to ambiguities in determining the total displacement [3, 20, 26].

To deal with this, the first interpretable PSI observation is the double-difference phase, denoted as φ_{pq}^{ms}, which represents the difference in phase between two PS points p and q, and between the master and slave acquisitions. This double-difference formulation provides both a temporal difference (between m and s) and a spatial difference (between p and q):

$$\varphi_{pq}^{ms} = (\varphi_p^{ms} - \varphi_q^{ms}) \tag{11.20}$$

Double-differences eliminate common errors and provide a more robust signal by removing large-scale atmospheric effects and orbital errors. However, this approach implies that PSI observations require both a spatial reference (a reference PS point) and a temporal reference (a reference acquisition time). Importantly, the estimates of deformation derived from PSI should remain invariant regardless of the specific choices for these spatial and temporal references, ensuring consistency in the analysis [2, 23, 26].

The number of independent double-differences that can be derived from a set of K SAR acquisitions and P persistent scatterers is given by:

$$N_{\text{double-differences}} = (K - 1) \times (P - 1) \tag{11.21}$$

Thus, for a PSI analysis with K acquisitions and P PS points, the number of independent interferometric phase combinations available for processing is $(K-1)(P-1)$. Each of these double-differences, denoted φ^k_{pq}, where $k = 1, 2, \ldots, (K-1)$, corresponds to a distinct temporal interferometric combination between the acquisitions. This formulation is central to the processing and interpretation of PSI data, allowing the extraction of precise deformation measurements over time for the chosen PS network.

To further elaborate, the key advantage of using double-differences lies in its ability to cancel out common phase errors (such as atmospheric delays and orbit inaccuracies) that affect both PS points similarly. This significantly enhances the reliability of deformation estimates. When phase unwrapping is applied to these double-differences, it allows the retrieval of continuous deformation fields, assuming the differences are small enough to avoid ambiguities. This technique forms the foundation of high-precision PSI analyses, enabling the accurate monitoring of slow-moving deformations over large areas, such as land subsidence or infrastructure deformation.

The number of Synthetic Aperture Radar (SAR) acquisitions and Persistent Scatterers (PS) plays a pivotal role in determining the number of independent double-differences that can be formed, which are essential for accurate phase analysis in Persistent Scatterer Interferometry (PSI). These double-differences arise from interferometric (temporal) and spatial combinations, which have distinct impacts on the measurement precision. However, the spatial distribution of PS also demands attention, as it shapes the spatial differences that contribute to double-differences [1, 23, 26].

A commonly used spatial distribution technique is the star network, where all spatial differences refer to one central PS, as depicted in Figure 11.9. The star network is a simple yet effective choice; however, it is not a strict requirement for all PSI applications. From a set of P PS points, $(P-1)$ independent spatial combinations can be formed. This allows flexibility in the selection of combinations, and alternative methodologies, such as the Small Baseline Subset (SBAS) technique [23–26],

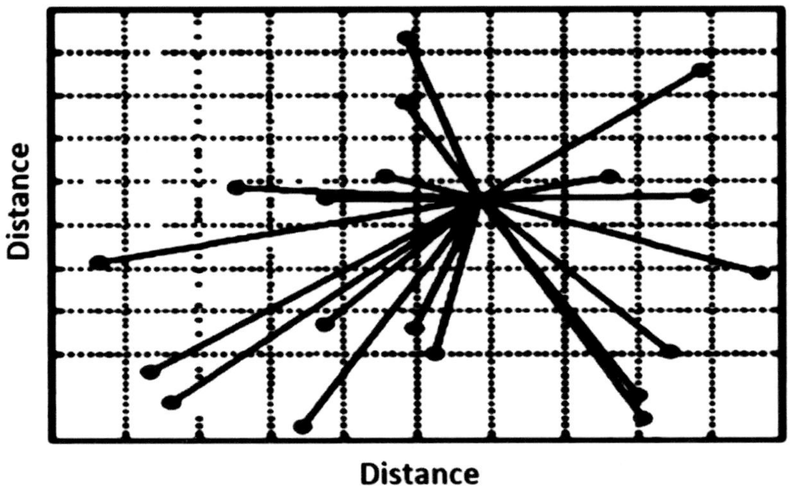

Figure 11.9 Star network spatial distribution technique.

can be employed. The SBAS method optimizes the perpendicular baselines for the stack of SAR acquisitions to minimize spatial decorrelation. This optimization becomes critical when dealing with PS that remain coherent for only a limited time, requiring an adjustment of the estimation process based on both the perpendicular and temporal baselines [1, 3, 24, 26].

Mathematically, the star network offers $(P-1)$ spatial combinations, which correspond to arcs connecting the central PS to the surrounding PS points. However, this spatial arrangement introduces certain disadvantages. Notably, erroneous PS candidates cannot be easily identified in a star network. Because it features fewer and longer arcs (i.e., larger distances between connected PS), the star network becomes more susceptible to phase unwrapping errors and atmospheric disturbances. The longer the arc, the more likely the phase difference contains atmospheric noise, which complicates accurate unwrapping and increases sensitivity to errors.

To overcome this limitation, a more robust alternative is the Delaunay triangulation, a network topology that connects PS in a way that minimizes long arcs and maximizes the number of connections (see Figure 11.10). In the Delaunay triangulation, each PS is connected to its nearest neighbors, forming a network of shorter arcs. This denser connection network enhances the stability of phase unwrapping and parameter estimation by reducing the influence of atmospheric noise [26].

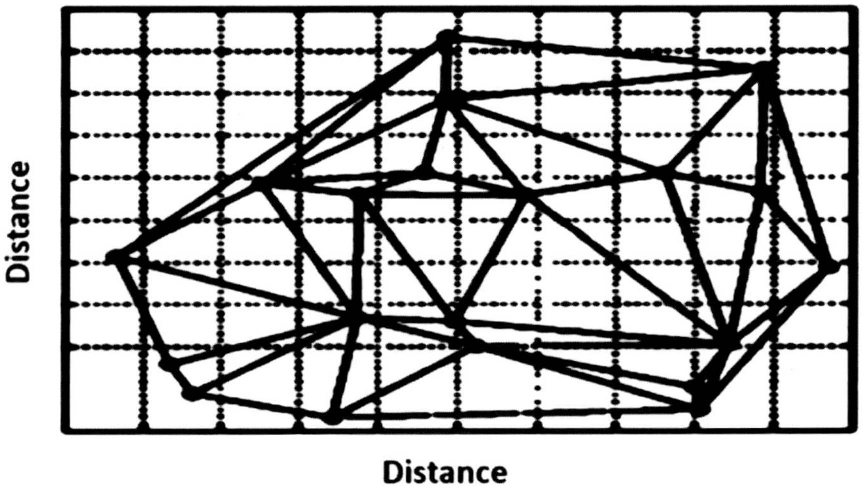

Figure 11.10 Delaunay triangulation network.

In the Delaunay triangulation, each arc undergoes phase unwrapping and parameter estimation using a functional model. Subsequently, misclosure tests are performed to ensure the integrity of the observations. Mathematically, the misclosure (M_C) for a given arc is defined as:

$$M_C = \varphi_{pq}^{ms} + \varphi_{qr}^{ms} - \varphi_{pr}^{ms} = 0 \tag{11.22}$$

If the misclosure M_C is not zero, it indicates the presence of errors, and the PS candidate associated with that arc can be flagged as erroneous. This redundant network structure is crucial for identifying and rejecting PS candidates that introduce inconsistencies. Consequently, this approach ensures a more reliable PSI estimation process.

For the final deformation signal estimation, however, only $(P-1)$ independent spatial combinations can be used, as redundancy is necessary for error detection but not for the ultimate phase difference measurements. This redundancy safeguards, ensuring that only accurate and consistent PS points contribute to the deformation estimates.

11.8 SIMULATION OF PSI USING FUNCTIONAL MODEL

Interferometric double-difference observations inherently contain multiple phase contributions, which originate from several factors including surface deformation, residual topography, atmospheric disturbances, orbital inaccuracies, the sub-pixel position of dominant scatterers within the resolution cell and inherent measurement noise. One critical challenge in these observations is that the interferometric phase is wrapped, meaning that only the fractional phase is directly observed, while the integer number of cycles (or phase ambiguities) from the satellite to the Earth's surface remains unknown [1–3].

The relationship between the wrapped interferometric phase φ and the unwrapped interferometric phase φ_{unw} is given by the following equation:

$$\varphi_{\text{unw}} = \varphi + 2\pi \cdot a, \tag{11.23}$$

where $a \in \mathbb{Z}$ represents the integer ambiguity. This integer value a accounts for the phase wrapping effect, and its estimation is crucial for recovering the full unwrapped phase. In the context of this model, it will be made explicit whether the phase φ refers to wrapped or unwrapped observations. To comprehensively account for the various phase contributions in PSI (Persistent Scatterer Interferometry), we can express the system of observation equations, following Ketelaar [26], for the interferometric double-difference phase observations φ_{ij}^k. These observations include several terms, each representing a distinct contribution to the total phase. The complete PSI observation model can be written as:

$$\varphi_{ij}^k = -2\pi a_{ij}^k - \frac{4\pi}{\lambda} \cdot \frac{R_m B_i^{\perp}}{R_m \sin\theta_{im}} H_{ij} - \frac{4\pi}{\lambda} D_{ij} + \frac{4\pi}{\lambda} \cdot \frac{B_i^{\perp}}{R_m} \tan\theta_{im} \eta_{ij}^m$$
$$+ \frac{2\pi}{v} (f_{dc}^{m,i} - f_{dc}^{s,i}) \xi_{ij}^m + f_{\varphi_{orbit}} (\xi_{ij}^m, \eta_{ij}^m) + \varphi_{ij}^{k,defo} + \varphi_{ij}^{k,atmo} + n_{ij}^k \tag{11.24}$$

In line with equation 11.24, the analyzing the complex phase contributions in Persistent Scatterer Interferometry (PSI), each term plays a distinct role in capturing various factors that affect the interferometric phase. These terms can be explained as follows, building a clear and coherent mathematical framework that reflects the underlying physical processes. Firstly, the phase wrapping ambiguity term, $-2\pi a_{ij}^k$, represents the difference between the observed wrapped phase and the true unwrapped phase. Here, a_{ij}^k is the integer number of phase cycles, and resolving this ambiguity is crucial for accurate phase unwrapping. Without this correction, the observed phase remains fractional and incomplete. Moving forward, the contribution due to residual topography is represented by $-\frac{4\pi}{\lambda} \cdot \frac{R_m B_i^{\perp}}{R_m \sin\theta_{im}} H_{ij}$. In this term, λ is the radar wavelength, R_m is the range to the master image, $B_i^{\perp}$ is the perpendicular baseline between the satellite positions, θ_{im} is the incidence angle, and H_{ij} is the height difference between the scatterers. This term captures the topographic variation, with the geometry of the satellite positions influencing the phase due to elevation changes in the terrain [25–27].

Next, the phase contribution due to deformation is expressed as $\frac{4\pi}{\lambda} D_{ij}$. In this sense, D_{ij} denotes the deformation displacement along the line of sight. This is one of the most critical terms in PSI, as it directly links the observed phase shift to surface movement, making it a key focus for analyzing ground displacement events such as earthquakes or subsidence. Another important term is the sub-pixel positioning effect, described by $\frac{4\pi}{\lambda} \cdot \frac{B_i^{\perp}}{R_m} \tan\theta_{im} \eta_{ij}^m$. In this case, η_{ij}^m is the sub-pixel offset, which accounts for the position of the dominant scatterer within the resolution cell. This correction ensures that the fine-scale misalignments in scatterer positions are accurately modeled, reducing potential errors in phase estimation. Therefore, the term $\frac{2\pi}{v}(f_{dc}^{m,i} - f_{dc}^{s,i})\xi_{ij}^{m}$ represents the Doppler centroid frequency differences between master and slave images. In this view, where v represents the satellite velocity, f_{dc} is the Doppler centroid frequency, and ξ_{ij}^{m} is the along-track displacement. This term is essential for correcting motion along the satellite's flight path, which could otherwise distort the phase measurements.

Consequently, the phase contribution due to orbital inaccuracies is captured by the function $f_{\varphi_{\text{orbit}}}(\xi_{ij}^{m}, \eta_{ij}^{m})$, which depends on both the along-track (ξ_{ij}^{m}) and across-track (η_{ij}^{m}) displacements. Orbital variations can introduce significant errors into the phase measurements, and this term compensates for those inaccuracies by incorporating geometric corrections related to the satellite's orbit [3, 27].

Next, the term $\varphi_{ij}^{k,\text{defo}}$ corresponds to the phase contribution due to surface deformation, which is the primary signal of interest in many PSI applications. This term directly reflects the surface changes occurring over time, and the precise estimation of this component allows for accurate monitoring of ground movements.

In addition to deformation, atmospheric effects also introduce phase delays, represented by the term $\varphi_{ij}^{k,\text{atmo}}$. Atmospheric conditions, such as humidity and temperature fluctuations, can delay the radar signal, and this term accounts for those variations to ensure more accurate phase unwrapping and deformation measurements.

Finally, the term n_{ij}^{k} captures the measurement noise, which is an inherent part of any SAR acquisition process. Noise can result from system imperfections, signal scattering, or environmental factors, and this term provides a means to account for and mitigate the impact of such noise on the final phase estimations.

By carefully examining each term and its contribution to the total phase, a coherent and mathematically grounded model emerges. The interferometric double-difference phase equation effectively encapsulates the interactions of deformation, topography, atmospheric disturbances, orbital inaccuracies, and noise, all of which play a critical role in SAR interferometric analyses. This approach enables precise unwrapping of the phase and ultimately allows for accurate estimation of surface deformation and other critical geophysical phenomena.

In examining the equation governing the relation between topographic height and the range sub-pixel position, it becomes evident that a linear dependency exists between these two quantities in the interferometric phase observation. Specifically, for a single arc between two Persistent Scatterers (PS) i and j, the phase contribution involving the topographic height is modified by the cosine of the incidence angle, θ_{im}. The functional relationship between these variables can be expressed as:

$$-\frac{4\pi}{\lambda}\cdot\frac{B_i^{\perp}}{R_m^i \sin\theta_{im}}\cdot\cos\theta_{im} + \frac{4\pi}{\lambda}\cdot\frac{B_i^{\perp}}{R_m^i}\tan\theta_{im} = 0. \tag{11.25}$$

This equation reveals that the range sub-pixel position exerts a significant influence on the estimation of residual topographic height. Since the height term is multiplied by both the sine and cosine of the incidence angle, any small errors in the range sub-pixel position can propagate throughout the system, significantly affecting the height estimate. As the sine and cosine terms of the incidence angle interact, they contribute to a subtle interplay between topography and the range position within a resolution cell.

The challenge in PSI is to convert the observed wrapped phase φ into the true unwrapped phase ϕ_{unw}, which involves resolving the integer ambiguity a. This process is known as phase unwrapping, and it is critical for accurate deformation and topography estimation. The double-difference formulation aids in reducing atmospheric noise and orbit inaccuracies by comparing the phase between different PS points (spatial difference) and across different acquisition times (temporal difference) [23–27].

11.9 QUANTUM CONVOLUTIONAL NEURAL NETWORK (QCNN) FOR PHASE UNWRAPPING

To develop a mathematical framework using a Quantum Convolutional Neural Network (QCNN) to solve the phase unwrapping problem based on the given equation 11.24, we need to break down the components of the equation into manageable parts for the QCNN to handle and create a structure that allows it to learn the unwrapping process efficiently. The phase unwrapping equation contains multiple contributions, including wrapping ambiguity, topographic height, deformation, sub-pixel positioning, Doppler centroid effects and atmospheric noise. The goal of the QCNN is to process these components hierarchically and correct for the ambiguities through successive layers of quantum operations.

In Multi-Channel Phase Unwrapping (MCh-PhU), the objective is to estimate the true unwrapped phase from multiple interferometric channels (acquired at different times or frequencies), where the phases are 'wrapped' within a 2π range due to sensor limitations. The problem becomes more complex when dealing with multiple channels because each channel may have different levels of noise, atmospheric effects or deformation signals. The challenge is to perform phase unwrapping across all these channels in a consistent way that reveals the true underlying signal, such as surface deformation or topography [27, 29].

The implementation of Multi-Channel Phase Unwrapping (MCh-PhU) presents several key challenges that must be addressed to achieve accurate phase unwrapping across multiple interferometric channels. First, phase wrapping ambiguity is a fundamental issue in MCh-PhU. Since the phase values in each channel are measured within a 2π interval, the true phase is obscured by an integer number of phase cycles. Consequently, the phase must be unwrapped, recovering the true continuous signal from the wrapped values, a task complicated by the presence of multiple channels with potentially varying conditions [28].

Secondly, multi-channel interference poses additional complexity. The phase information from different channels may experience interference due to disparate atmospheric conditions, noise levels and temporal or spatial decorrelation. This interference leads to variability across channels, which must be reconciled to produce a coherent phase estimate. The atmospheric signals, noise, or artifacts affecting each channel can distort the underlying deformation or topography signals, making it crucial to account for these interferences when unwrapping the phase.

Furthermore, correlation across channels is vital for achieving consistency in the unwrapped phase estimates. Given that phase observations are acquired from multiple channels, the solution must ensure coherence across all channels to provide accurate results. It is not sufficient to unwrap each channel independently; instead, the phase unwrapping process must integrate the information across channels in a manner that preserves the inherent correlations between them. This requires maintaining the physical consistency of the signals across temporal, spatial, and frequency domains, ensuring that the final unwrapped phase estimates represent a coherent interpretation of the multi-channel data.

Thus, the MCh-PhU process must overcome phase ambiguity, manage interference across channels, and ensure correlation, all of which are interrelated challenges that must be addressed in tandem for successful phase unwrapping [28, 30].

For MCh-PhU, consider multiple wrapped phase channels $\{\phi_{ijk}^{1}, \phi_{ijk}^{2}, \dots, \phi_{ijk}^{C}\}$, where C is the number of channels. Each channel c has its wrapped phase equation similar to equation 11.24. For each channel, we need to unwrap the phase ϕ_{ijk}^{C} by estimating the integer ambiguity a_{ijk}^{C}, and resolve the contributions from deformation D_{ij}^{C} topography H_{ij}^{C}, sub-pixel positioning η_{ijm}^{C}, and other components [27–30]. Therefore, the total unwrapped phase for each channel is:

$$\phi_{ijk}^{\text{unwrapped},c} = \phi_{ijk}^{c} + 2\pi a_{ijk}^{c} \tag{11.26}$$

The Quantum Convolutional Neural Network (QCNN) architecture for Multi-Channel Phase Unwrapping (MCh-PhU) is designed to process the wrapped phase data from multiple channels while preserving essential features such as residual topography and deformation, and ensuring consistency across the channels. This hierarchical processing system reduces the input's dimensionality without losing critical phase information, thereby facilitating an efficient and accurate phase unwrapping process. The architecture comprises several key layers, each serving a specific purpose in this process.

First, encode the multi-channel data into quantum states. This can be done using various encoding schemes such as amplitude encoding, basis encoding, or angle encoding.

$$|\psi\rangle = \sum_{i=0}^{N-1} x_i |i\rangle \tag{11.27}$$

where (x_i) are the data points and $|i\rangle$ are the basis states.

The QCNN applies quantum filters, exhibited as unitary operations U, to correct for phase wrapping and extract critical features such as residual topography and deformation. These convolutional layers perform localized operations on the phase data, similar to traditional convolutional neural networks, but with the added capability of correcting quantum phase ambiguities [31, 33]

$$\phi_{ijk}^{(l+1)} = U^{(l)}\phi_{ijk}^{(l)} \tag{11.28}$$

where $\phi_{ijk}^{(l+1)}$ denotes the phase at layer l, and $U^{(l)}$ represents the quantum filter applied at that layer. The unitary operations help to refine phase contributions from residual topography and deformation:

$$\phi_{ijk}^{\text{topo}} = -\frac{4\pi}{\lambda} \cdot \frac{B_i^{\perp}}{R_m \sin\theta_{im}} H_{ij} \tag{11.29}$$

$$\phi_{ijk}^{\text{defo}} = -\frac{4\pi}{\lambda} D_{ij} \tag{11.30}$$

These terms are successively corrected through each convolutional layer. Therefore, the pooling layers reduce the dimensionality of the input, both spatially (i.e., across pixel locations) and across channels, while maintaining coherence across the channels [27, 32]. Pooling assists in simplifying the data, making the phase unwrapping task computationally more efficient:

$$\phi_{ijk}^{\text{pooled}} = \frac{1}{N_{\text{pool}}} \sum_{n} \phi_{i',j',k}^{(l)} \tag{11.31}$$

where $\phi_{ijk}^{\text{pooled}}$ represents the number of pooled elements and i', and j' denote the spatial indices after pooling. This ensures that phase unwrapping remains consistent across the channels by averaging the phase information from neighboring pixels and across channels [31–33]. Consequently, the loss function $l(\phi)$ for the QCNN is designed to minimize the mean squared error (MSE) between the unwrapped phase and the true phase values across all channels:

$$l(\phi) = \frac{1}{N} \sum_{ijkc} \left(\phi_{ijk}^{\text{QCNN},c} - \phi_{ijk}^{\text{true},c}\right)^2 \tag{11.32}$$

The QCNN is optimized using a quantum backpropagation algorithm, which updates the quantum gates and unitary operations to minimize the MSE and improve the phase unwrapping performance across all channels [31, 34]. In other words, integrate the information from multiple channels using quantum entanglement. This can be achieved by applying controlled quantum gates that entangle the states from different channels. The entanglement between two channels for example can be expressed by:

$$\left|\psi_{\text{entangled}}\right\rangle = U_{\text{entangled}}\left(\left|\psi_1\right\rangle \otimes \left|\psi_2\right\rangle\right) \tag{11.33}$$

In the context of the QCNN architecture for multi-channel phase unwrapping, the output layer is pivotal in delivering the unwrapped phase $\phi_{ijk}^{\text{unwrapped}}$ for each channel after systematically addressing various challenges inherent in the phase measurement process. The mathematical representation of this process can be articulated as follows:

$$\phi_{ijk}^{\text{unwrapped}} = \mathcal{F}(\phi_{ijk}, a_{ijk}, \delta_{ijk}, H_{ij}, D_{ij}, \eta_{ijm}, f_{\text{dc},m,i}, f_{\text{dc},s,i}, \xi_{ijm}, \phi_{ijk}^{\text{defo}}, \phi_{ijk}^{\text{atmo}}, n_{ijk}) \tag{11.34}$$

Here, $\mathcal{F}$ denotes a functional operation performed by the QCNN, which effectively integrates the wrapped phase observations ϕ_{ijk} with the integer phase ambiguity a_{ijk}, corrections for atmospheric interference δ_{ijk}, topographical height differences H_{ij}, deformation displacements D_{ij}, sub-pixel offsets η_{ijm}, Doppler centroid frequency differences $f_{\text{dc},m,i}$, along-track displacements ξ_{ijm}, contributions from surface deformation ϕ_{ijk}^{defo}, atmospheric effects ϕ_{ijk}^{atmo}, and measurement noise n_{ijk} [33–35].

Thus, the final output of the QCNN, $\phi_{ijk}^{\text{unwrapped}}$, encapsulates all necessary corrections and adjustments to yield a coherent and accurate unwrapped phase representation for each channel, thereby facilitating improved insights into the physical phenomena being studied through synthetic aperture radar (SAR) observations. This expression underlines the QCNN's capability to unravel complexities associated with phase wrapping, atmospheric variations, and spatial inconsistencies, ultimately achieving an accurate phase estimation essential for remote sensing and geophysical analysis applications.

11.10 MULTI-QUBIT GATES FOR PHASE MANIPULATION

11.10.1 Controlled-NOT Gate (CNOT)

The CNOT gate can be used to establish a relationship between the control qubit (representing a reference state) and the target qubit (representing the phase state). The action of a CNOT gate on two qubits can be expressed as:

$$\text{CNOT } |x\rangle|y\rangle = |x\rangle|y \oplus x\rangle, \tag{11.35}$$

where x is the control qubit and y is the target qubit (Figure 11.11). For our phase unwrapping application, we can set up a circuit that relates the wrapped phase to a reference qubit state $|r\rangle$ [37]:

$$|\psi\rangle \rightarrow \text{CNOT } |r\rangle|\phi\rangle. \tag{11.36}$$

This establishes an entangled state that captures the relationship between the wrapped phase and the reference phase [36–38].

11.10.2 Controlled Phase Gate (CPHASE)

The Controlled Phase gate applies a phase shift to the target qubit based on the state of the control qubit. It is represented mathematically as:

$$\text{CPHASE}|x\rangle|y\rangle = e^{i\theta}|x\rangle|y\rangle, \tag{11.37}$$

where θ is the phase shift. In our context, one can use the CPHASE gate to account for atmospheric interference and other phase contributions (Figure 11.12):

$$|\psi\rangle \rightarrow \text{CPHASE}|r\rangle|y\rangle \tag{11.38}$$

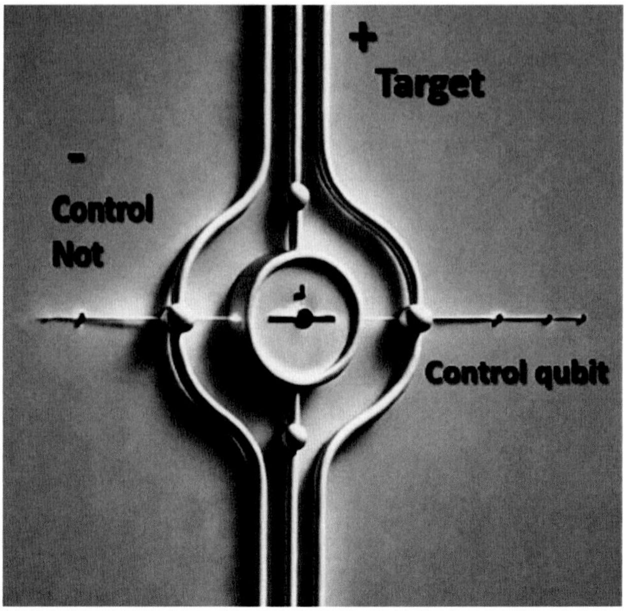

Figure 11.11 CNOT route gate.

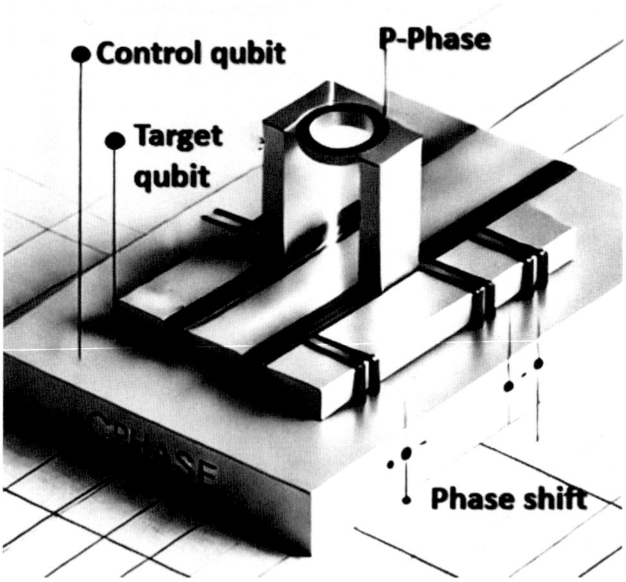

Figure 11.12 Controlled Phase route gate.

which modifies the phase based on atmospheric conditions or other influences represented by the control qubit $|r\rangle$. To perform phase unwrapping, we can use a sequence of multi-qubit gates, combining CNOT and CPHASE gates to estimate displacement D and deformation d [39–41]. The equations for the combined transformation can be represented as:

$$|\psi_{\text{final}}\rangle \to \text{CPHASE}\big(CNOT\big(|r\rangle|\phi\rangle\big), |d\rangle\big) \tag{11.39}$$

where $|d\rangle$ $|d\rangle$ is a qubit state representing the deformation measurement. In this view, the total phase contribution can be expressed as:

$$\phi_{\text{unwrapped}} = \phi + \Delta\phi_{\text{LOS}} + \Delta\phi_{\text{atmo}} + n, \tag{11.40}$$

here $\Delta\phi_{\text{LOS}}$ represents the phase contribution due to displacement in the line of sight, $\Delta\phi_{\text{atmo}}$ represents the atmospheric phase contribution, and n represents measurement noise. The final quantum state can be measured to estimate these contributions:

$$\phi_{\text{LOS}} = F\left(\left|\psi_{\text{final}}\right\rangle\right) = \arg\left(\left\langle\psi_{\text{final}}\middle|\psi_{\text{final}}\right\rangle\right) \quad (2) \qquad (11.41)$$

Needless to say, multi-qubit gates such as the CNOT and CPHASE gates can be employed in a quantum circuit to manipulate phase information crucial for estimating displacement and deformation in the context of phase unwrapping [2, 27, 37, 40]. The equations derived highlight how quantum states can be transformed and measured to gain insights into ground movement in the line of sight (LOS). The combination of these gates allows for capturing the complexities of phase information and correcting for various sources of interference, ultimately leading to accurate deformation measurements [26, 33, 39, 41].

11.11 INSAR GENERATED FROM PALASAR-1 SAR AND SENTINEL-1A DATA

The implementation of a Quantum Convolutional Neural Network (QCNN) based on multi-qubit gates for phase unwrapping was conducted using 20 sets of PALSAR-1 SAR images from January 2009 to November 2010, prior to the construction of the Grand Ethiopian Renaissance Dam (GERD), due to the unavailability of Sentinel-1A data. High-resolution SAR data from sources such as TerraSAR spot mode images were also utilized from January 2022 to August 2024 to supplement the analysis. Remarkably, the PALSAR data from 2009 to 2010 revealed only minor land deformation in the Blue Nile Basin, the site where GERD was later built, with deformation rates ranging from –0.01 to +0.01 mm. This minor deformation is likely attributed to the natural seasonal flooding characteristic of the region (Figure 11.13).

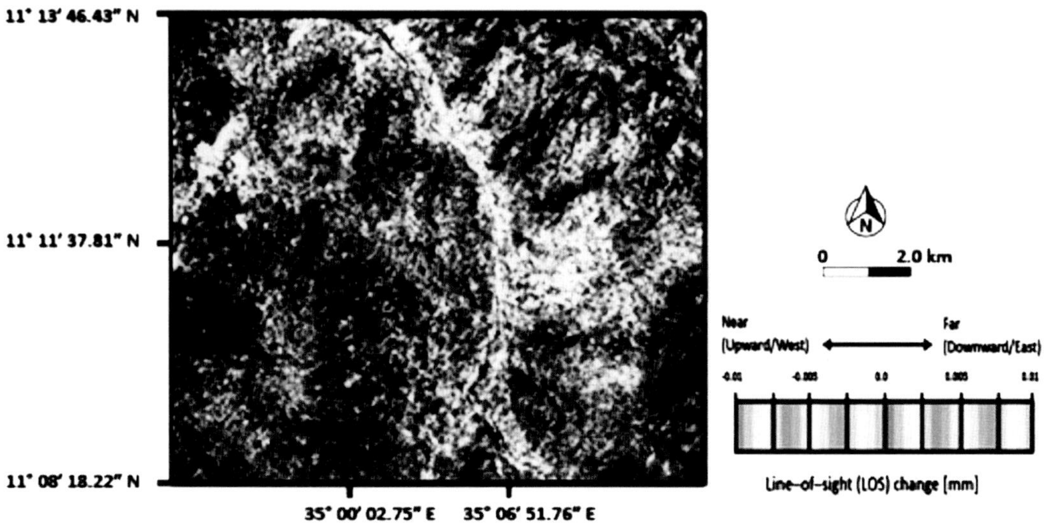

Figure 11.13 Land deformation in the Blue Nile Basin from PALSAR-1 SAR data (2009–2010) using QCNN.

In stark contrast, the analysis of Sentinel-1A data from January 2020 to August 2024 revealed significant land deformation corresponding to the expansion of the GERD reservoir across the Blue Nile Basin. These results are indicative of substantial ground displacement, as visualized through the colorful patterns in the interferograms, which represent changes in the line-of-sight (LOS) displacement of the ground surface. The near range, located towards the west and upward direction of the satellite's LOS, shows an uplift in specific areas, most likely caused by reservoir-induced seismicity or pressure changes due to the immense weight of the accumulating water. This uplift is demonstrated by LOS shortening, where ground movement towards the satellite results in positive LOS changes. This is depicted in the interferogram as a shift towards warmer colors, such as purple, yellow, and red, particularly in regions west or northwest of the reservoir, near the western side of the dam (Figure 11.14).

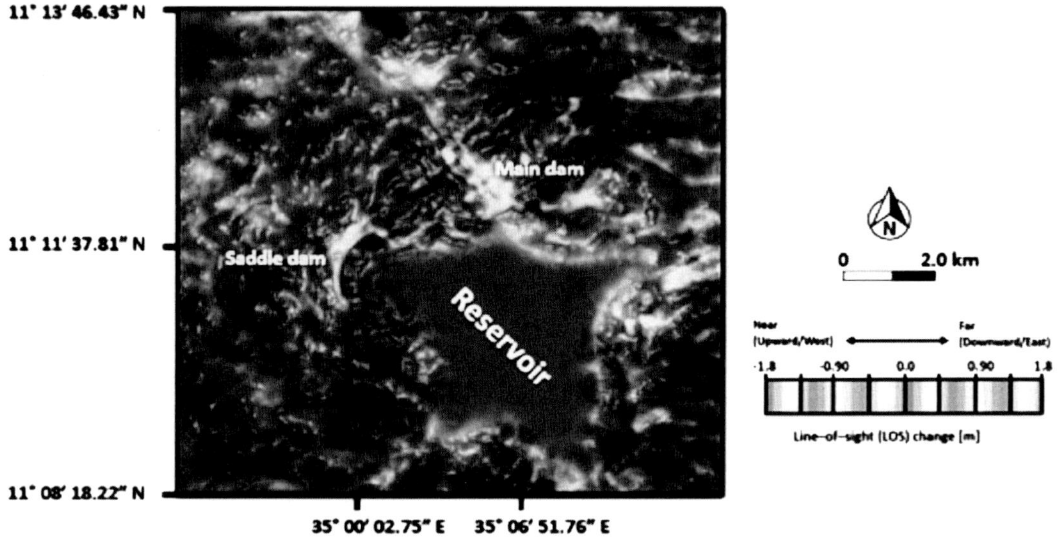

Figure 11.14 Interferometry of land deformation in GERD from Sentinel-1A data (2009–2010) using QCNN.

Conversely, the far range, located in the east and downward from the satellite's LOS, shows ground movement away from the satellite, most likely indicating subsidence. This downward motion is typically associated with water drawdown or stress on the ground as water is either discharged or stored. Subsidence, which manifests as an increase in the LOS distance, is illustrated by cooler colors such as green and blue in the interferogram. Areas to the east or southeast of the reservoir, closer to or downstream of the dam, could potentially represent zones of subsidence.

The concentric color fringes radiating outward from both the Main Dam and the Saddle Dam represent the detected ground motion. Based on the descending pass of the satellite (from west to east), regions with closer, warmer-colored fringes—likely to the west and north of the reservoir—indicate uplift as the ground surface moves towards the satellite. Meanwhile, regions with cooler colors, possibly situated to the east and downstream of the dam, suggest subsidence, where the ground moves downward and away from the satellite's LOS. This dichotomy of ground displacement will be further analyzed and discussed in subsequent sections, providing a comprehensive understanding of the deformation patterns caused by the GERD reservoir's expansion.

11.12 PSI GENERATED USING QCNN FROM MULTISAR DATA

Time series analysis of InSAR data can be presented in the form of PSI. In this regard, high-resolution images such as TerraSAR data from December 2022 to August 2024 within spot mode are used. Therefore, an easy description of GERD structures can assist deeply in understanding the spatial variation of displacement patterns through the dam structures. As mentioned earlier in Chapter 2, the Grand Ethiopian Renaissance Dam (GERD) consists of two main structures: the main dam and the saddle dam. Both play crucial roles in the dam's overall functionality and stability, but they have distinct components and purposes.

Firstly, the main dam can be divided into six key components. The main dam-east side is essential for holding back the Blue Nile's water, thereby creating the reservoir. This section is a fundamental part of the dam's structure, ensuring its stability and functionality. Additionally, the main dam-early generation turbines are part of the initial phase of power generation. These turbines began generating electricity even before the dam was fully completed, helping to meet Ethiopia's energy needs early on [42].

Moreover, the main dam-low flow outlet allows for the controlled release of water, ensuring that downstream water flow is maintained even during the filling periods. This feature is crucial for managing water levels and preventing flooding. Furthermore, the main dam-east embankment supports the dam structure on the eastern side, anchoring it to the surrounding terrain. This part is essential for the overall stability of the dam [43]. Similarly, the main Dam-West Wall works in conjunction with the east side to hold back the reservoir (Figure 11.15). This section is integral to the dam's structure, maintaining the reservoir's integrity. Finally, the main dam-west embankment provides support on the western side of the dam. This section is critical for balancing the forces exerted by the water in the reservoir, ensuring the dam's stability [44].

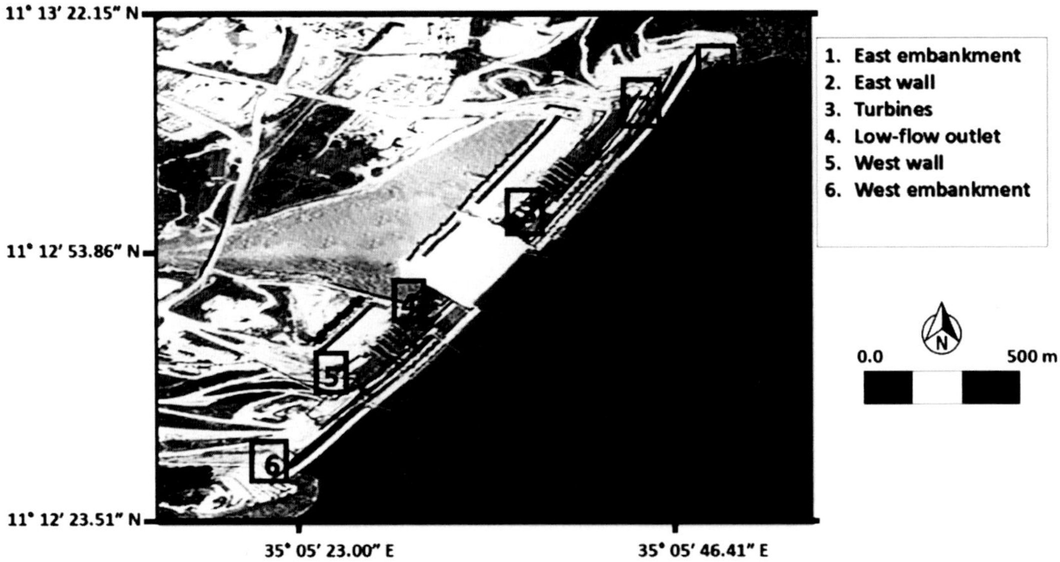

Figure 11.15 Main dam important structure elements.

In contrast, the saddle dam component of GERD can be divided into several key parts, primarily focused on managing water flow and pressure The GERD includes the north and south embankment leeways, which are areas located outside the main central dam structure. One of these leeways could be associated with a saddle dam, especially if this area forms a saddle or depression in the surrounding terrain that requires additional structure to hold back water. Initially, the first tunnel is an essential part of the infrastructure designed to manage water flow and pressure within the saddle dam. The second tunnel aids in controlling water movement and maintaining the structural integrity of the dam. Additionally, the third tunnel plays a similar role, ensuring that water management is efficient and the dam remains stable. Moreover, the fourth and fifth tunnels are other critical components that support the overall function of the saddle dam by regulating water flow (Figure 11.16). These tunnels collectively contribute to the effective operation of the saddle dam, helping to manage water levels and maintain the dam's structural integrity [44].

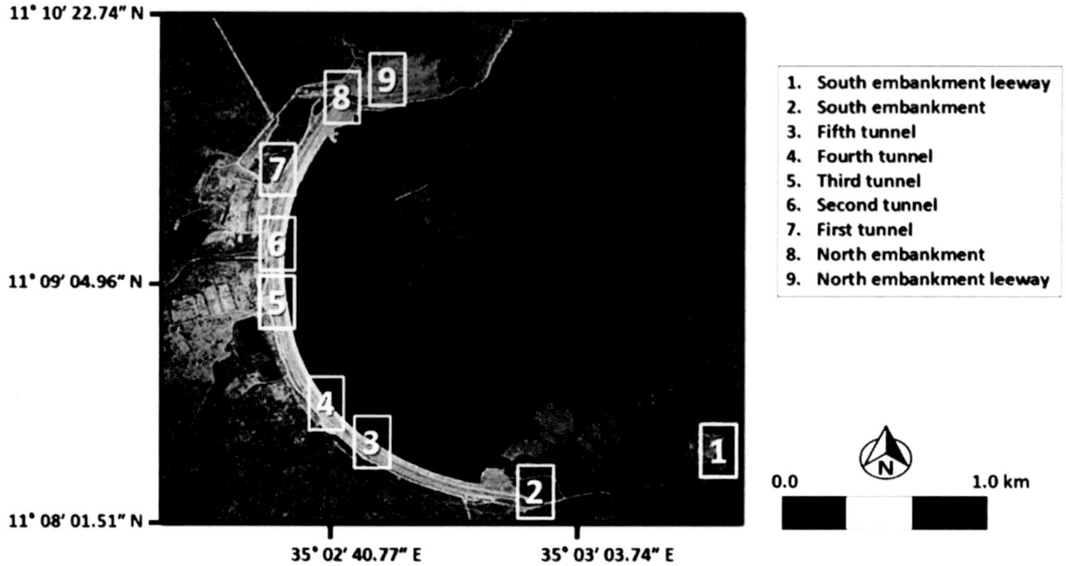

Figure 11.16 Structure elements of saddle dam.

The key question that now arises is: Can Persistent Scatterer Interferometry (PSI) effectively detect and map the displacement patterns across the structures of both dams? Implementing the PSI technique based on the QCNN within TerraSAR data from January 2022 to August 2024 shows the variability of vertical displacement from the east embankment to the western one. Figure 11.17 offers a detailed visualization of the displacement patterns along the eastern embankment of the Grand Ethiopian Renaissance Dam (GERD). These patterns have been detected using Persistent Scatterer Interferometry (PSI), a sophisticated technique applied to TerraSAR-X SpotLight (SL) mode satellite images. The colorful fringes represent variations in line-of-sight (LOS) displacement of the ground surface, as indicated by the legend. The displacement ranges from –1.8 meters, indicating subsidence (downward motion), to +1.8 meters, indicating uplift (upward motion).

Additionally, the main corridor at the top of the dam has been subjected to a significant vertical displacement, measured at approximately –1.35 meters. This notable subsidence, indicative of downward ground motion, highlights potential stress or deformation within the structure. Such vertical displacement could be a result of various factors, including the immense weight of the reservoir, geotechnical shifts beneath the dam or changes in water pressure as the reservoir fills or discharges. On the contrary, the PSI analysis reveals that the top corridor of the western embankment wall and low-flow outlet are exposed to extreme displacement, approximately –1.8 meters, which is notably higher than the –1.35 meters observed on the eastern side (Figure 11.18).

The spillway of the Grand Ethiopian Renaissance Dam (GERD); nevertheless, is currently exhibiting minimal displacement, characterized by gentle movements ranging from –0.004 meters to +0.004 meters. This subtle variation indicates a relatively stable condition of the spillway structure (Figure 11.19). This stability can be attributed to the operational dynamics of the dam. Since the main dam primarily manages the discharge of water through the two turbines, the spillway remains inactive for most of the time. The turbines are utilized for overtopping discharge, effectively regulating the water flow and minimizing the necessity for spillway operation. As a result, the spillway is subjected to less stress and operational load, contributing to its limited displacement range.

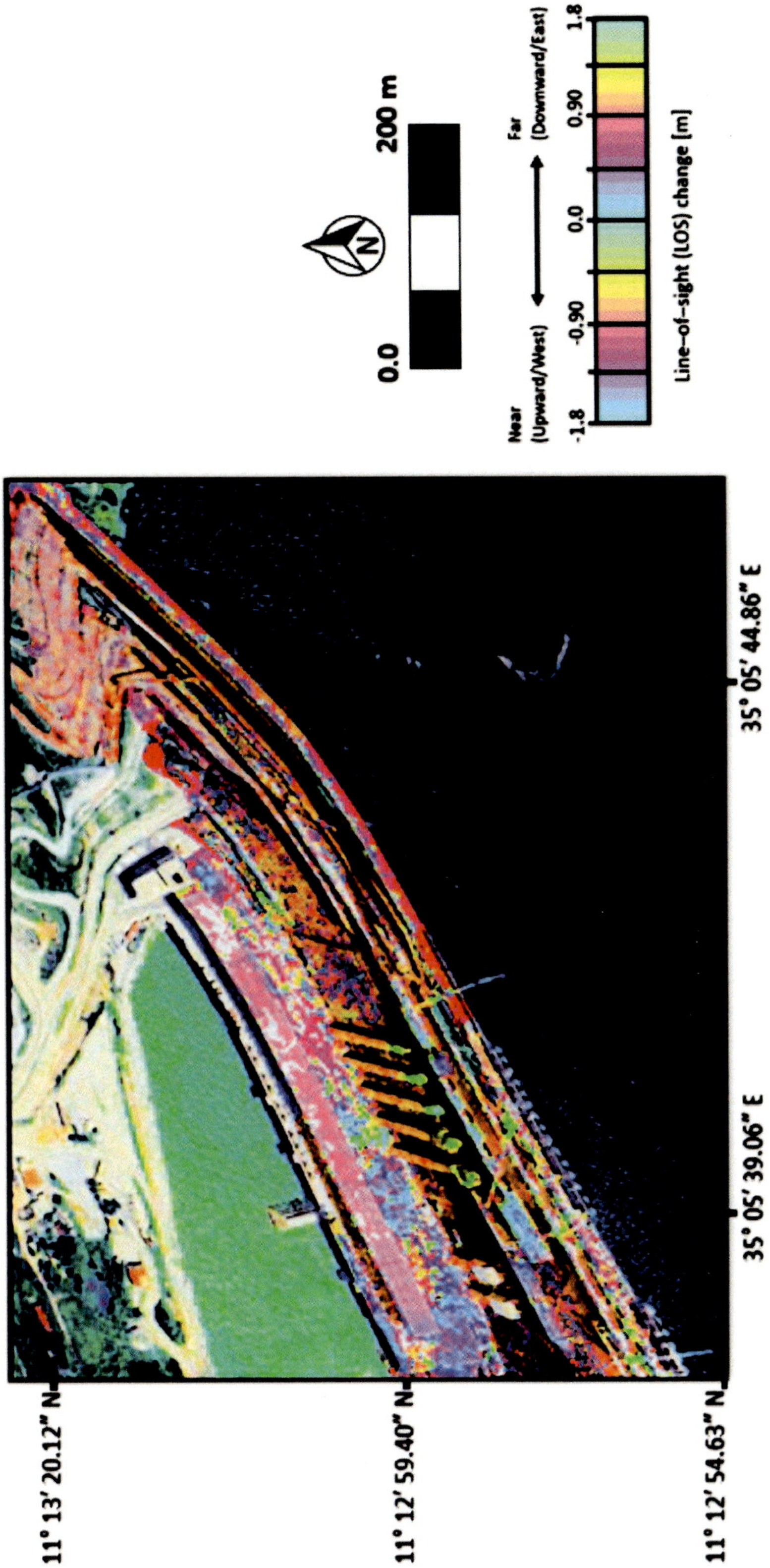

Figure 11.17 Displacement pattern along the eastern embankment of GERD.

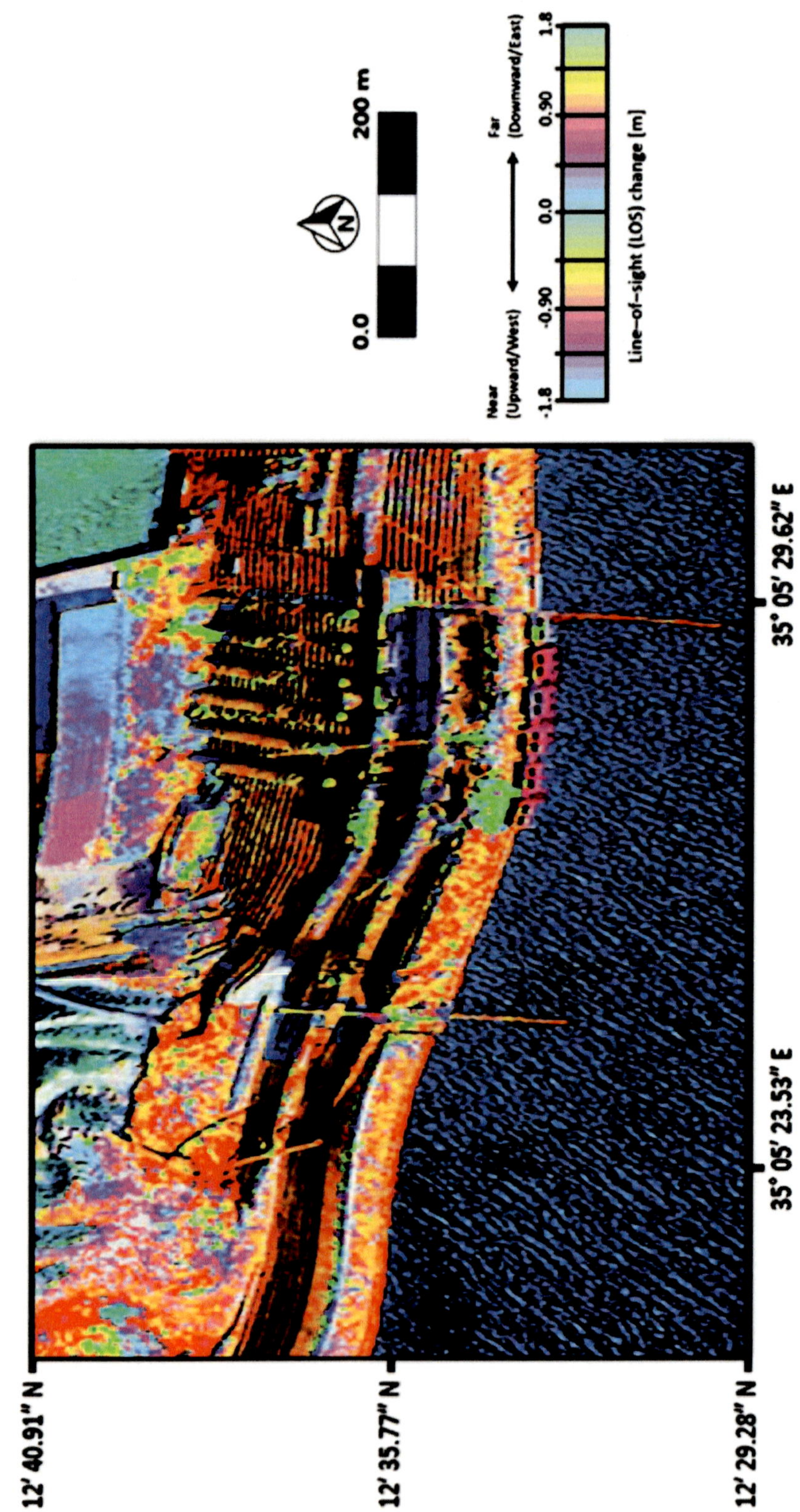

Figure 11.18 Displacement pattern along the western embankment of GERD.

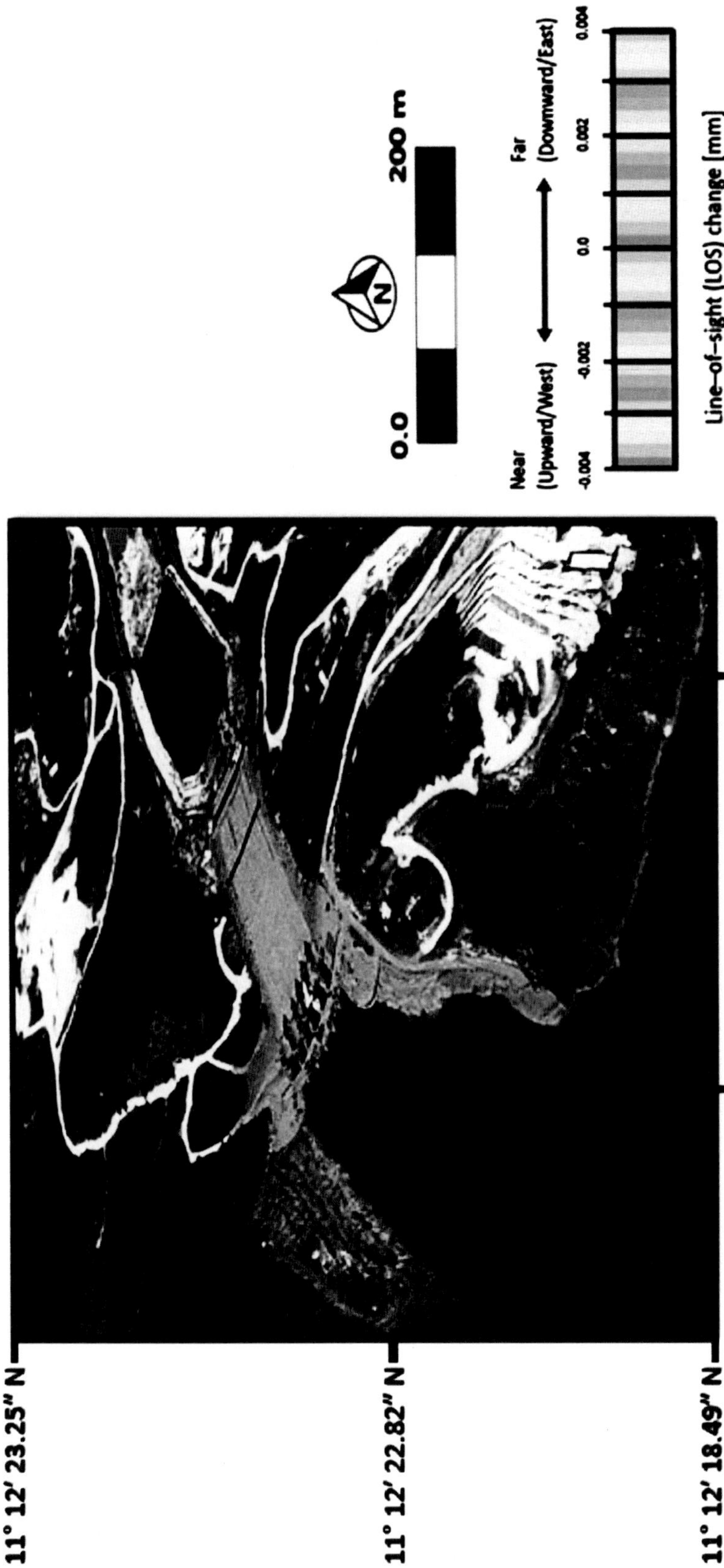

Figure 11.19 Displacement pattern along the spillway of GERD.

The south and north embankments of the saddle dam are experiencing significant displacement, measuring approximately –1.5 meters. This substantial movement raises critical concerns regarding the structural integrity and stability of these embankments. In contrast, the displacement observed in the first to third tunnels is notably less severe, approximately 0.83 meters (Figure 11.20).

This disparity in displacement levels highlights the varying degrees of stress and structural response within the dam's components. While the embankments endure considerable displacement, potentially indicating heightened risk, the tunnels, though still affected, exhibit a more stable behavior. Such differences can be attributed to their respective design and function within the overall dam structure. The embankments, which bear the primary load from the reservoir and fluctuating water levels, are subject to greater pressure and, consequently, greater displacement.

Conversely, the tunnels are engineered to facilitate water flow and pressure relief, allowing them to better accommodate the operational dynamics of the dam. This comparative analysis underscores the need for ongoing monitoring and assessment of both the embankments and tunnels, as understanding the nuances of their displacement patterns is crucial for ensuring the overall safety and performance of the Grand Ethiopian Renaissance Dam.

From the aforementioned perspective, the minimal displacement of the spillway suggests that, while it plays a critical role in the dam's hydrological management, it is currently in a stable state due to the operational practices of the GERD. Notably, regions located closer to the satellite, referred to as the near (upward/west) zone, exhibit movements toward the satellite. These areas are characterized by warmer colors, such as purple, yellow, and red. This uplift could be the result of geological changes or structural adjustments triggered by the significant weight and pressure of the GERD and its expanding reservoir. The gravitational force exerted by the large water mass may have caused localized stress, leading to this upward movement of the ground.

Conversely, the far (downward/east) region displays cooler colors, including green and blue, indicating ground displacement away from the satellite. This downward motion, or subsidence, could stem from the drawdown or redistribution of loads as water is either stored or released from the dam. Such subsidence is common in areas surrounding large reservoirs, where the shifting weight of the water creates stress on the underlying geological formations, causing portions of the ground to settle.

The tight spacing of the colored fringes along the embankment underscores the significant ground motion in the area. This suggests that the construction of GERD and the process of reservoir filling are exerting considerable stress on the dam's eastern embankment. The PSI-based analysis is, therefore, essential for continuous monitoring, as it helps identify potential stability concerns and assess the structural integrity of the dam. Effective monitoring of such ground displacement is critical to ensure the long-term safety and functionality of the GERD, particularly in light of the large-scale environmental and infrastructural changes taking place.

11.13 SPATIAL VARIATION OF DEFORMATION VELOCITY ACROSS THE NILE BASIN AND GERD

The central question that emerges is: what are the profound effects of the Grand Ethiopian Renaissance Dam (GERD) on regions downstream? The successive fillings of the GERD reservoir, coupled with reduced discharge, have significantly diminished the dam's efficiency—operating at only about 40% capacity, utilizing just two turbines, and largely bypassing the spillway outflow. These operational constraints have led to an observable increase in velocity deformation. At the time of writing, the dam's water level stands at approximately 590 meters above mean sea level (mMSL), a condition that may accelerate ground deformation.

Furthermore, as revealed by Persistent Scatterer Interferometry (PSI) analysis of Sentinel-1A SAR data from January 2022 to August 2024, significant subsidence is occurring in the regions to

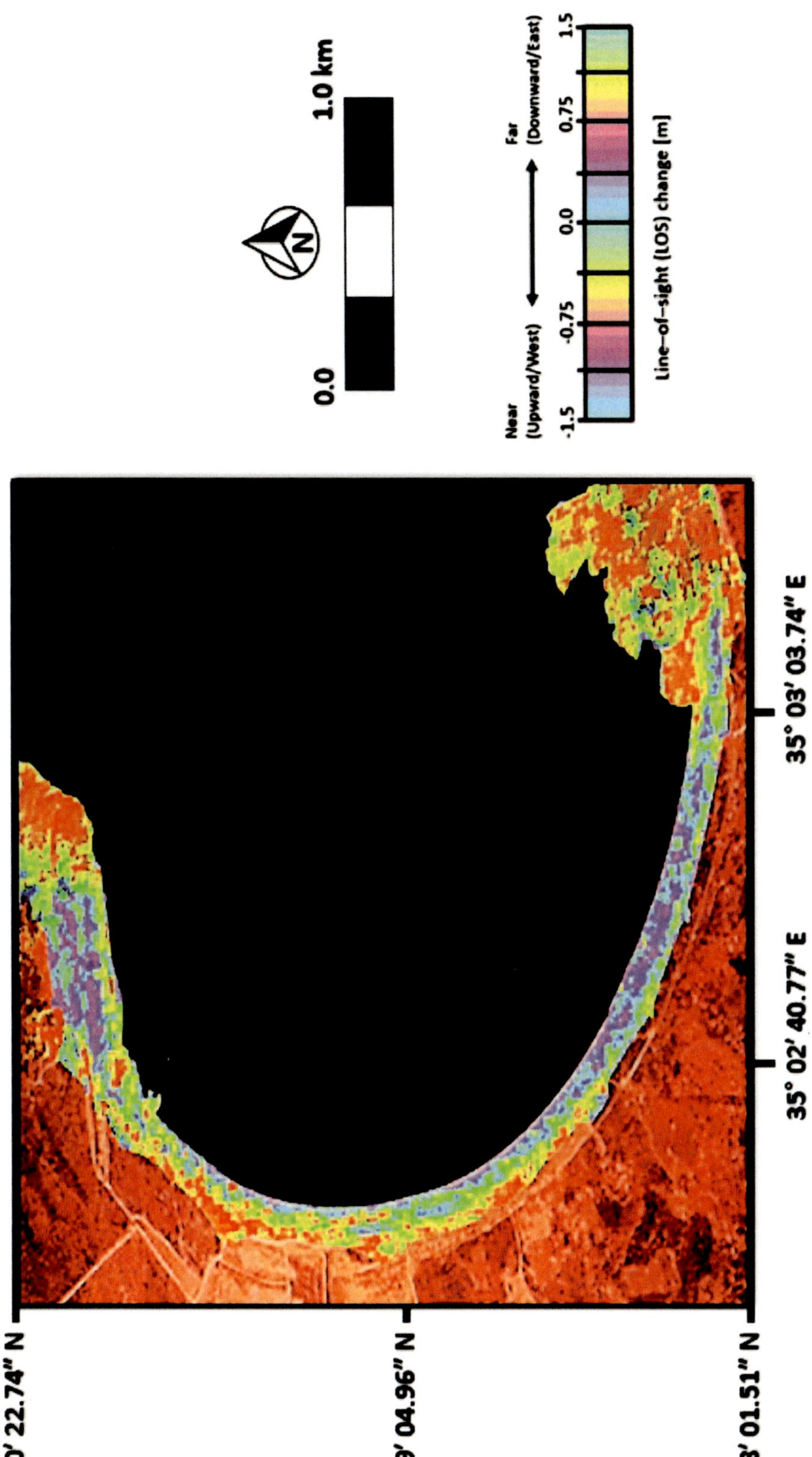

Figure 11.20 Displacement pattern along the saddle dam.

the northwest and northeast of the GERD. Deformation rates in these areas range from –1.1 to –0.5 meters per year, marking critical and concerning shifts that present long-term risks (Figure 11.21).

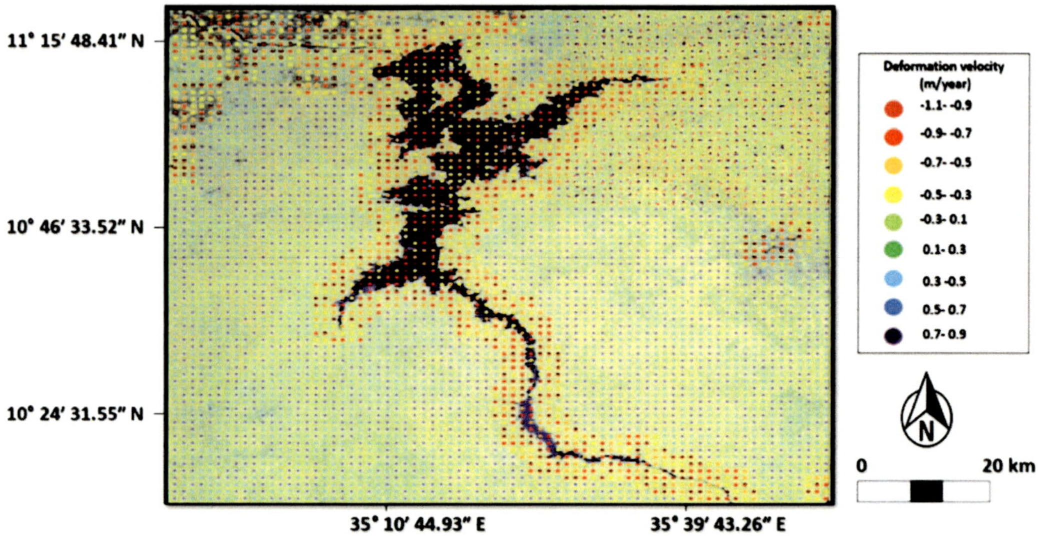

Figure 11.21 Spatial variation of velocity deformation along the downstream of Nile Blue basin.

As discussed earlier, the northwestern and eastern zones, encompassing both the main dam and the saddle dam, are geologically vulnerable due to the presence of numerous faults and lineaments. This intrinsic instability in the geological structure heightens the risk of rapid land deformation, compounding the operational challenges faced by the dam. The observed subsidence is not an isolated phenomenon but a reflection of the delicate balance between engineered infrastructure and natural tectonic forces, making the ongoing monitoring and mitigation of these risks paramount for the region's safety and sustainability.

By drawing attention to both the operational inefficiencies and the geological fragility of the area, we highlight the interplay between human engineering and natural forces. The reduced efficiency of the dam and the rising risks of subsidence are not merely technical issues; they represent a complex and evolving challenge where structural, hydrological, and geophysical factors converge. Thus, the long-term sustainability and safety of the region depend on carefully managing these intertwined dynamics.

In this manner, the comparison between immediate operational concerns and deeper geological vulnerabilities reveals a multi-layered problem. This requires not only technical solutions but a broader understanding of the dam's interaction with its environment over extended timescales. The complex, long-term variation in subsidence makes it essential to evaluate the impact of ongoing structural interventions and natural processes alike.

High-resolution spatial velocity deformation data, derived from a time series of TerraSAR spot mode images between 2022 and 2024, reveal pronounced vertical subsidence along the eastern side of the GERD. The deformation velocities range from –1.1 to 0.5 meters per year, with the most significant subsidence, –1.1 m/year, concentrated in the top main corridor of the eastern embankment (Figure 11.22). This area is particularly susceptible to rapid deformation.

The eastern side of the main dam has been subjected to a relatively homogeneous velocity deformation, largely driven by the rising water level in the dam reservoir, which currently stands at approximately 590 mMSL, as previously outlined in Chapter 10. This increase in water level has intensified the pressure on the dam's eastern section, thereby contributing to the observed subsidence. The findings underscore the intricate relationship between reservoir dynamics and structural integrity, highlighting the need for ongoing assessment and adaptation to ensure the long-term stability of the dam's infrastructure.

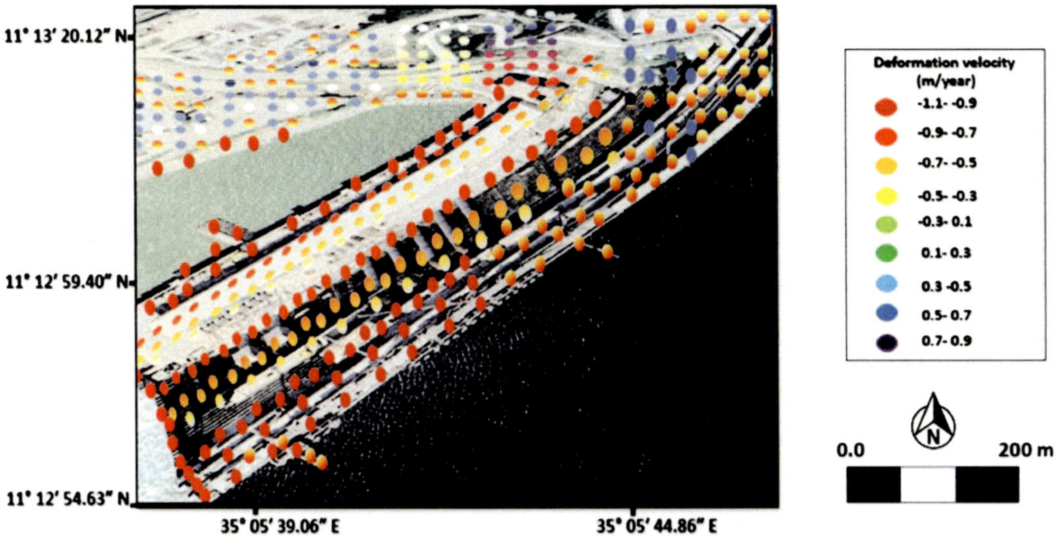

Figure 11.22 Spatial variation of velocity deformation along the eastern side of the main dam body.

In contrast to the eastern side of the GERD, the western side of the main dam body, particularly in the zone around the low flow outlet and west wall, has experienced a more pronounced accumulation of velocity deformation. The deformation rates on the western side range from –1.1 to –0.5 meters per year, with the top corridor exhibiting subsidence velocities between –1.1 and –0.9 meters per year (Figure 11.23). This degree of subsidence is notably higher than that observed on the eastern side.

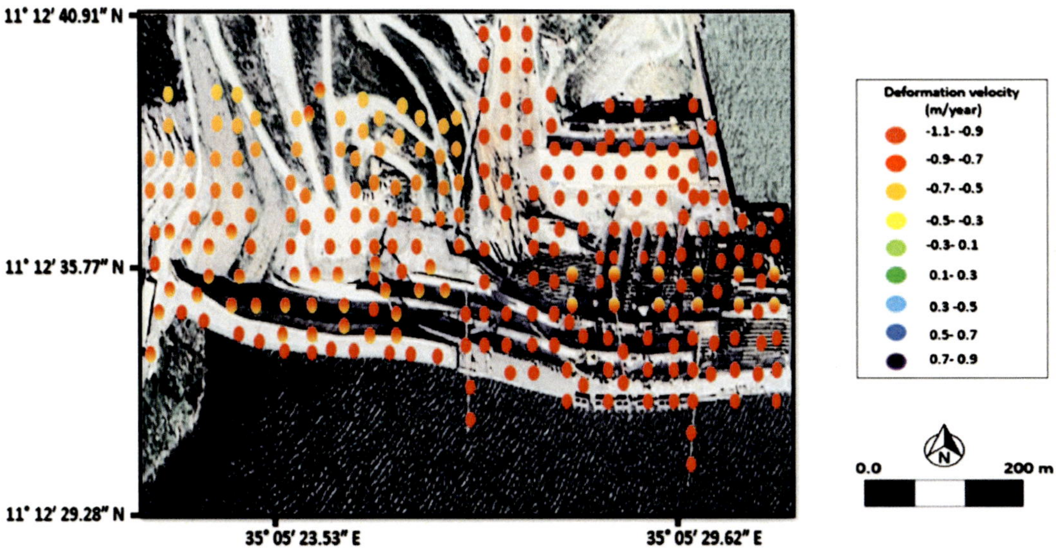

Figure 11.23 Spatial variation of velocity deformation along the western side of the main dam body.

The primary reason for this difference lies in the increased stress imposed by the water level, which currently stands at 590 m above mean sea level (mMSL), as well as the non-functionality of the low flow outlet. The inability of the outlet to properly discharge water exacerbates the pressure on the western section, intensifying subsidence. In comparison, while the eastern side is subject to similar reservoir-induced stress, the western side faces a more acute response in terms of velocity deformation. This asymmetry highlights the spatial variability in subsidence across the dam, emphasizing the need for differential monitoring and tailored interventions for each side.

Thus, while both sides of the dam are under pressure from rising water levels, the western side exhibits a more severe deformation profile due to compounded stress factors, particularly around critical structural zones like the low flow outlet. The higher subsidence rates on the western side suggest a greater vulnerability in this region, demanding urgent attention to mitigate potential structural risks.

In comparison to both the eastern and western sides of the main dam body, the saddle dam faces its critical subsidence challenges, particularly along the tunnel zones. The velocity deformation along the saddle dam, especially around the second to fifth tunnels, ranges from –0.9 to –0.7 meters per year (Figure 11.24). This subsidence is largely driven by clear seepage occurrences, which have intensified land deformation, notably in areas highlighted by a white rectangular marker. The deformation in the saddle dam is less severe than the subsidence observed on the western side of the main dam, where velocities reach as low as –1.1 meters per year, but it is comparable to the eastern side in terms of magnitude.

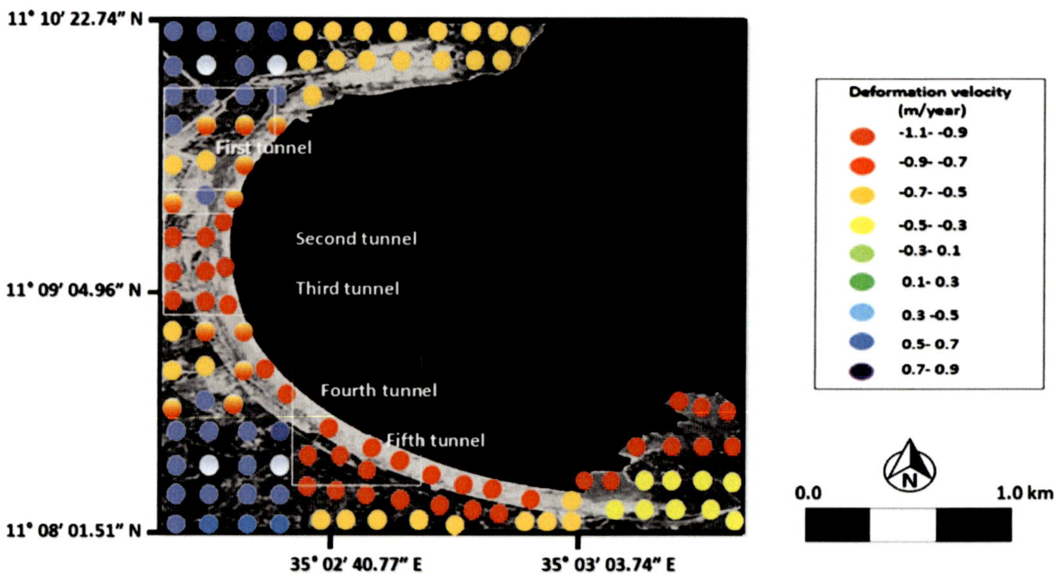

Figure 11.24 Deformation velocity along the saddle dam.

However, the saddle dam's vulnerability is compounded by its tunnels, where seepage contributes significantly to the deformation. While the main dam's eastern and western sides are more directly affected by the water level's rise and structural pressure—particularly the western side due to the non-functional low flow outlet—the saddle dam experiences deformation primarily due to subsurface water infiltration through its tunnels. This seepage-induced deformation, particularly at a velocity of –0.9 m/year, is concerning and represents a localized but significant risk to the dam's structural integrity.

In comparison, the eastern side of the main dam experiences more homogeneous deformation caused by reservoir water levels, with subsidence velocities between –1.1 and 0.5 m/year. The western side, though, suffers from more intense subsidence, with rates of –1.1 to –0.9 m/year, particularly in the low-flow outlet zone. Thus, while the saddle dam shows less severe subsidence than the western side, its risk due to seepage along its tunnels presents a distinct and critical threat (Figures 11.25 and 11.26). The detection of seepage occurrences along the second and fourth tunnels using ICEYE Spot Mode imagery acquired on September 17, 2024, offers a compelling opportunity to precisely analyze subsidence along the saddle dam. Given that ICEYE Spot Mode provides a resolution between 0.5 and 1 meter, it allows for the detection and distinction of subtle surface deformations and seepage patterns that might be influencing subsidence. This high-resolution capability is essential in identifying localized deformations caused by the infiltration of water through the tunnels.

In this case, the ability to resolve features as small as 0.5 meters strengthens the argument that the subsidence along the saddle dam is directly linked to seepage along these tunnels. The seepage, acting as a driving force behind the structural weakening of the soil or rock surrounding the tunnels, would manifest in surface deformations visible at the resolution ICEYE Spot Mode provides. Consequently, such imagery can serve as crucial evidence, confirming that seepage-induced subsidence is occurring along the tunnel zones, particularly the second and fourth tunnels and contributing to the overall deformation of the saddle dam.

The use of SAR data for automatic detection of seepage allows for continuous monitoring, offering the potential for early warning systems to prevent further damage to the dam structure. Through this high-resolution data, precise patterns of deformation can be mapped, analyzed, and linked to subsurface hydrological processes, thus offering insights into the mechanisms of land subsidence and enhancing mitigation strategies.

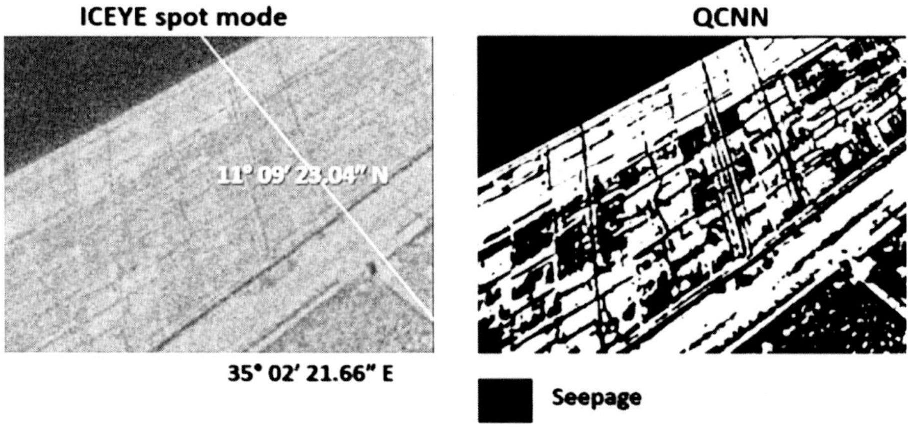

Figure 11.25 Automatic detection of seepage in the second tunnel in ICEYE spot mode data.

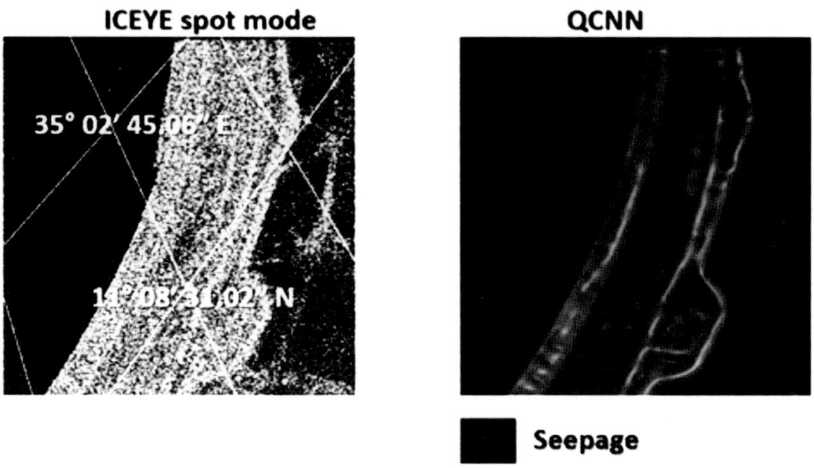

Figure 11.26 Automatic detection of seepage in the fourth tunnel in ICEYE spot mode data.

From the aforementioned perspective, each of these three sections—the eastern, western, and the saddle dams—exhibit unique subsidence mechanisms, whether driven by water levels, structural stress, or seepage, requiring tailored approaches to mitigate their risks. These results approximately align with and support the findings of previous research conducted by El-Askary et al. [45], Chen et al. [46], and Madson and Sheng [47]. The observed critical deformations in the GERD (Grand Ethiopian Renaissance Dam) construction further substantiate the conclusions drawn in these studies. However, the authors should provide a more detailed comparison and discussion to highlight how their findings reinforce the results of these earlier works, particularly concerning the critical deformations in GERD.

11.14 WHY TUNNELS CAUSE LAND DEFORMATION?

Tunnels, when constructed in geologically sensitive areas, can induce land deformation due to several interconnected factors, including alterations in subsurface stress distribution, groundwater dynamics, and inherent structural instabilities. The act of tunneling disrupts the natural equilibrium of forces within the earth, which can lead to subsidence. To understand why this occurs, it is essential to explore the mechanisms that contribute to such deformations [48].

First, subsurface stress redistribution plays a crucial role. Tunnels cut through layers of soil and rock, thereby altering the load-bearing capacity of the surrounding earth. The stress that was once uniformly distributed across the undisturbed ground is now concentrated around the tunnel void. In areas where the geological formations are already unstable or riddled with faults and lineaments, this stress redistribution can further weaken the ground. Consequently, the land above and around the tunnel may experience settling or deformation as it adjusts to the newly imposed stress regime [49].

Next, groundwater seepage is another critical factor contributing to tunnel-induced deformation. When tunnels intersect aquifers or water-bearing geological formations, they can act as pathways for water to infiltrate the tunnel structure. This seepage lowers the pore water pressure in the surrounding soil or rock, causing it to consolidate and compact which leads to subsidence. The situation becomes particularly problematic in areas with loose, compressible soils, such as clay or silt, where the ground is prone to rapid settling. In the case of the GERD's saddle dam, the seepage observed around the first and second tunnels exemplifies this issue. As water drains into the tunnels, the surrounding material collapses or compacts due to reduced water pressure, accelerating deformation [50].

Moreover, erosion and material weakening can further exacerbate deformation around tunnels. Over time, seepage erodes the materials surrounding the tunnel, compromising the ground's structural integrity. If the tunnel lining is not watertight, water flow can gradually dissolve or wash away surrounding material, creating cavities that may collapse either suddenly or progressively. This process is particularly hazardous in areas with loose or poorly consolidated geological formations, where erosion can lead to significant land subsidence [48, 51].

Tunnel over-excavation and inadequate ground support during construction also contribute to instability. When too much material is removed during tunneling or when sufficient support is not provided, the surrounding ground becomes prone to collapse. Over-excavation creates a void that the earth will naturally attempt to fill, leading to localized subsidence or broader settlement issues. The size of the tunnel and its depth are key factors that determine the severity of this deformation, as larger or deeper tunnels exert greater influence on the surrounding ground [48, 50].

Additionally, tunnels that intersect faults and fractures in the earth are particularly vulnerable to deformation. Fault zones are naturally weaker and more susceptible to movement when disturbed. Tunneling through such areas can trigger shifts along these fault lines, leading to both surface and subsurface deformation [49–54]. In the context of GERD, the presence of numerous faults and lineaments near both the main dam and the saddle dam increases the risk of deformation, making it a critical factor in tunnel-induced subsidence [47, 49, 51].

Lastly, even after construction, tunnels can contribute to slow land deformation through long-term creep and settlement. This process refers to the gradual movement or settling of the ground over extended periods, driven by continuous external forces such as water pressure or the weight of overlying rock layers [47, 53]. Over time, these incremental movements accumulate, leading to significant subsidence, especially in areas with poor ground conditions.

Tunnels can disrupt the natural balance of stress, groundwater flow, and material stability in the surrounding earth, leading to land deformation and subsidence. In the case of the GERD's saddle dam, seepage through the tunnels is a primary driver of velocity deformation [50–52]. As water infiltrates the tunnels, it weakens the surrounding ground by reducing its structural capacity, resulting in subsidence at significant rates. These factors highlight the complexity of managing tunnel-induced deformation, especially in large-scale infrastructure projects where geological and hydrological considerations are paramount [46, 51].

It can be said that the construction and operation of tunnels significantly disrupt the natural balance of stress, groundwater flow, and material stability in the surrounding earth. These disruptions can lead to subsidence through mechanisms such as stress redistribution, seepage, erosion, and geological instabilities. In the case of the GERD's saddle dam, seepage through the tunnels is a primary driver of velocity deformation. As water infiltrates the tunnels, the surrounding ground weakens, reducing its ability to support the overlying materials, and leading to land subsidence at concerning rates. This underscores the critical need for careful geological assessments and engineering solutions when tunneling in sensitive areas.

11.15 WHY QCNN DELIVER ACCURATE PSI DEFORMATION PATTERNS?

Field experiments have validated the accuracy of PSI measurements, confirming millimeter-scale precision. The results of QCNN-PSI analysis typically include geospatial datasets with identified Persistent Scatterers (PSs), including geographic coordinates, elevation, displacement time-series, and average velocity for each point. Moreover, each PS is associated with a quality index, defined by temporal coherence.

Importantly, all deformation measurements from QCNN-PSI are relative to a reference point, typically chosen for its high coherence and displacement stability, ensuring reliable comparisons. Dam structures and exposed rocks are ideal targets for QCNN-PSI due to their geometric stability and dielectric properties. In contrast, targets such as vegetation or water basins, which undergo significant change or provide weak reflections, do not generate PSs. For GERD's main and saddle dam areas, thousands of PSs can be detected per square kilometer, while surrounding structures like viaducts, roadways, and railways also provide abundant PSs due to their strong reflections, which overshadow weaker scatterers.

However, PS distribution is often uneven. Some parts of a structure may be rich in PSs, while others lack monitoring points. For example, studies have observed higher PS counts on top corridors of GERD's east and west sides, while central sections show gaps. This uneven distribution is due to several factors, including the loss of permanent targets from overtopping, satellite viewing geometries, and varying backscattering mechanisms of different structural materials.

To mitigate this, a virtual SAR simulator can evaluate PS availability before developing a monitoring plan. This ensures an informed strategy, tailored to the specific conditions of the target area. Thus, PSI, enhanced by the QCNN algorithm, offers a precise method for monitoring infrastructure displacement. The ability to generate 3D models and accurately geolocate PSs makes it invaluable for dam structural analysis, providing critical insights into GERD stability.

According to this perspective, the QCNN-PSI algorithm can simulate cumulative infrastructure displacement as a key indicator of GERD instability. The highest occurrences of displacement suggest that GERD's structures are exposed to deformation, evidenced by poor infrastructure efficiency. Therefore, the core question is whether QCNN-PSI can accurately discern the highest deformation rate tied to the instability of GERD.

The cumulative deformation velocity from January 2020 to August 2024 demonstrates rapid subsidence due to increasing water levels in the GERD reservoir. In early 2022, velocities ranged from –0.9 to –0.3 m/year, indicating subsidence, possibly linked to the reservoir filling and early stress redistribution (Figure 11.27). By 2023, the velocity reached –1.1 m/year, coinciding with water levels at 580 mMSL, suggesting that increased water pressure contributed to stress on the dam's foundation or nearby geological faults. By mid-2023, deformation peaked at nearly –1.0 m/year, signaling substantial subsidence, followed by a slight reduction in velocity as water levels stabilized. However, by 2024, deformation remained between –0.9 and –0.5 m/year, indicating ongoing structural stress despite water level stabilization at 590 mMSL.

The persistent negative velocity deformation, ranging from –0.9 to –1.1 m/year, signifies significant subsidence along GERD's main dam. The correlation between water level fluctuations and increased subsidence highlights the stress on geologically sensitive zones, raising concerns about structural damage or potential failure, particularly if seepage or erosion exacerbates dam weakening. Immediate attention to monitoring and mitigation is critical to prevent catastrophic failure. These findings are consistent with previous studies conducted by El-Askary et al. [45], Chen et al. [46], and Madson and Sheng [47].

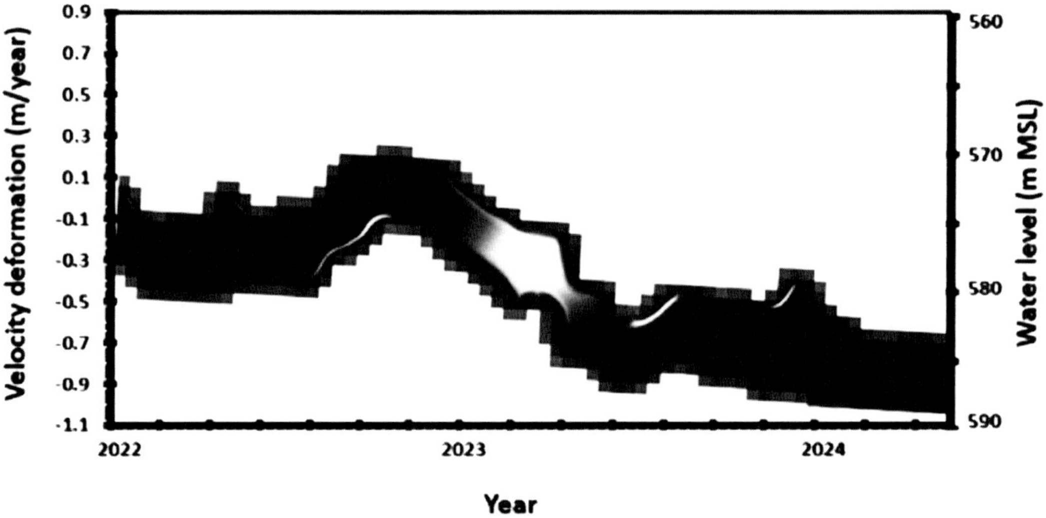

Figure 11.27 Time series of deformation velocity across GERD main dam body.

The accuracy of the QCNN-PSI algorithm, as demonstrated by its ROC curve, stands at an impressive 94% in detecting and attributing infrastructure instabilities along the GERD's main and saddle dams (Figure 11.28). This level of precision significantly surpasses long-established InSAR (Interferometric Synthetic Aperture Radar) and PSI (Persistent Scatterer Interferometry) techniques applied to data from Sentinel-1A and TerraSAR in spot mode. The strength of QCNN lies in its ability to process complex, high-dimensional data and extract meaningful features from time-series deformation datasets.

In contrast to classical InSAR techniques, which may struggle with the spatial-temporal complexities of the dam's subsidence patterns, QCNN leverages its quantum-enhanced capabilities to manage non-linear relationships more efficiently. As a result, it provides a more

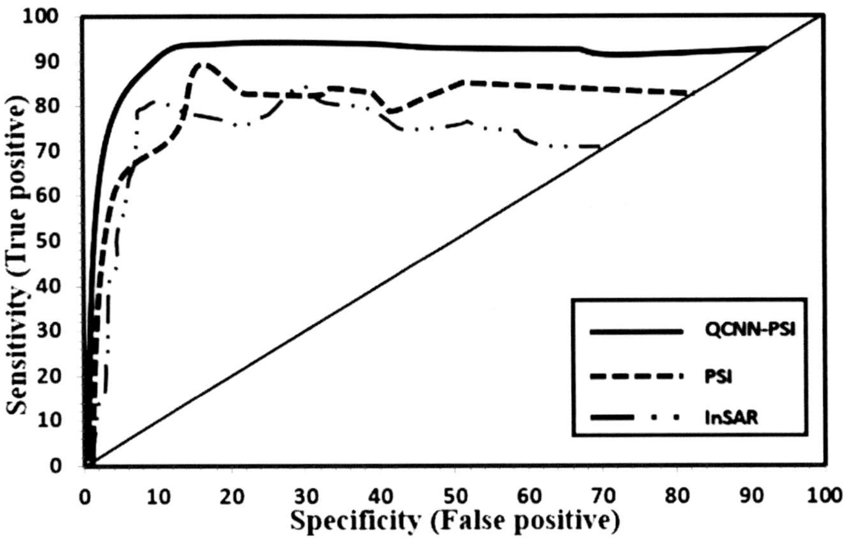

Figure 11.28 ROC curve for confirming the GERD subsidence from different interferometry techniques and QCNN-PSI algorithm.

nuanced understanding of how subsurface stress, geological faults, and fluctuating water levels are contributing to ground deformation at GERD. In other words, conventional PSI or InSAR techniques may suffer from noise in areas with high vegetation, rough topography, or poor radar reflections. QCNN can filter through this noise and better delineate true deformation patterns, particularly in geologically sensitive regions like GERD.

The inclusion of Control gates (CONT) and Controlled Phase Gates (CPHASE) in Quantum Convolutional Neural Networks (QCNNs) is crucial for delivering highly accurate results. First, CONT gates allow conditional operations based on qubit states, enabling efficient encoding of relationships between quantum bits. This helps the QCNN capture complex correlations in the data, essential for precise pattern recognition, such as Phase Shift Interferometry (PSI) deformation analysis. This finding is similar to work done by Du et al. [36].

Further, the CPHASE gate introduces a controlled phase shift, preserving quantum entanglement, which is vital for processing quantum coherence and interference effects. This enhances the network's ability to detect subtle changes in quantum states, resulting in more precise interpretations of PSI deformation patterns. This is agreed to by Pechen et al. [39]. Additionally, both gates contribute to improved generalization by handling phase information with greater precision, crucial for accurate deformation pattern recognition. Finally, leveraging quantum parallelism, these gates allow the QCNN to process high-dimensional quantum data efficiently, significantly boosting its overall performance [36–38]. Needless to say, the integration of CONT and CPHASE gates equips QCNNs with the tools necessary to manage the complexities of quantum data, resulting in superior accuracy, especially in applications such as PSI deformation analysis.

From the perspective outlined above, PSI processes vast radar datasets from multiple satellite passes, often spanning years, to analyze deformation patterns. These patterns are complex, involving both temporal and spatial variations. The strength of QCNNs lies in their ability to manage high-dimensional data, making them ideal for extracting localized and detailed deformation information from PSI datasets.

A key advantage of QCNN, like other convolutional neural networks, is its capacity to recognize intricate patterns within large datasets. In the case of GERD, where deformation is influenced by subsurface stress, water seepage, and geological instability, QCNN excels at identifying relevant features correlated with subsidence or land movement. It efficiently differentiates real deformation signals from noise, which is essential for accurate monitoring.

Furthermore, by integrating quantum computing algorithms, QCNN enhances phase unwrapping, allowing for precise interferogram phase information to be derived. This reveals cumulative displacement across GERD's infrastructure at a rate of –1.1 m/year. The combination of QCNN with PSI provides a robust approach for monitoring ground movement and deformation, both for the main and saddle dams, confirming that cumulative displacement is a critical indicator of ongoing structural deformation in GERD.

11.16 CONCLUSION

This chapter highlighted the persistent negative velocity deformation, notably between –0.9 and –1.1 m/year, which points to significant subsidence along the GERD main dam; the correlation between rising water levels and increased subsidence demonstrates that water pressure contributes to stress in geologically sensitive areas and the fact that this poses a risk of structural damage, particularly if seepage and erosion further weaken key sections of the dam. Immediate monitoring and mitigation are necessary to avert catastrophic failure.

In conclusion, QCNN's ability to process high-dimensional, noisy data and extract complex deformation patterns, enhanced by quantum computational power, makes it an invaluable tool for precise subsurface monitoring. This accuracy is crucial for evaluating GERD's structural integrity and ensuring the long-term safety of the dam.

References

[1] Devanthéry, N., Crosetto, M., Monserrat, O., Cuevas-González, M. and Crippa, B. 2014. An approach to persistent scatterer interferometry. Remote sensing. 6(7): 6662–6679.

[2] Massonnet, D. and Feigl, K.L. 1998. Radar interferometry and its applications to changes in the earth's surface. Review of Geophysics. 36(4): 441–500.

[3] de Oliveira, L.S., Gama, F.F., Crepani, E., Mura, J.C. and Ibanez, D.M. 2024. Surface Displacement Evaluation of Canto Do Amaro Onshore Oil Field, Brazil, Using Persistent Scatterer Interferometry (PSI) and Sentinel-1 Data. Remote Sensing. 16(9): 1498.

[4] Alam, M.S., Kumar, D. and Vishwakarma, G.K. 2024. A review on advances in persistent scatterer interferometry and proposing a novel method for phase optimization of distributed scatterers pixels. Journal of Engineering Mathematics. 145(1): 1–25.

[5] Giorgini, E., Orellana, F., Arratia, C., Tavasci, L., Montalva, G., Moreno, M., et al. 2023. InSAR Monitoring Using Persistent Scatterer Interferometry (PSI) and Small Baseline Subset (SBAS) Techniques for Ground Deformation Measurement in Metropolitan Area of Concepción, Chile. Remote Sensing. 15(24): 5700.

[6] Boli, X., Jing, M.C. and Gangyao, K. 2012. A change detection measure based on a likelihood ratio and statistical properties of SAR intensity images. Remote Sensing Letters. 3(3): 267–275.

[7] Touzi, R., Lopes, A., Bruniquel, J. and Vachon, P.W. 1999. Coherence estimation for SAR imagery. IEEE Transactions on Geoscience and Remote Sensing. 37(1): 135–149.

[8] Bouaraba, A., Borghys, D., Belhadj-Aissa, A., Acheroy, M. and Closson, D. 2012. Improving CCD performance by the use of local fringe frequencies. Progress in Electromagnetics Research C. 32, 123–137.

[9] Oishi, N. 2009. A change detection measure based on a likelihood ratio and statistical properties of SAR intensity images. *In*: Proceedings of the 6th European Radar Conference. 278–281. Rome, Italy.

[10] Martinez, C.L. 2007. Coherence estimation in synthetic aperture radar data based on speckle noise modeling. Applied Optics. 46(4): 544–558.

[11] Just, D. and Bamler, R. 1994. Phase statistics of interferograms with applications to synthetic aperture radar. Applied Optics. 33(20): 4361–4368.

[12] Suo, Z., Li, Z. and Bao, Z. 2010. A new strategy to estimate local fringe frequencies for InSAR phase noise reduction. IEEE Geoscience and Remote Sensing Letters. 40(7): 771–775.

[13] Spagnolini, U. 1995. 2-D phase unwrapping and instantaneous frequency estimation. IEEE Transactions on Geoscience and Remote Sensing. 33(5): 579–589.

[14] Trouvé, E., Caramma, M. and Maitre, H. 1996. Fringe detection in noisy complex interferograms. Applied Optics. 35(20): 3799–3806.

[15] Lee, J.S. 1980. Digital image enhancement and noise filtering by use of local statistics. IEEE Transactions on Pattern Analysis and Machine Intelligence. 2: 165–168.

[16] Hur, T., Kim, L. and Park, D.K. 2022. Quantum convolutional neural network for classical data classification. Quantum Machine Intelligence. 4(1): 3.

[17] Wei, S., Chen, Y., Zhou, Z. and Long, G. 2022. A quantum convolutional neural network on NISQ devices. AAPPS Bulletin. 32: 1–11.

[18] Oh, S., Choi, J. and Kim, J. 2020. A tutorial on quantum convolutional neural networks (QCNN). *In*: 2020 International Conference on Information and Communication Technology Convergence (ICTC). 236–239. IEEE.

[19] Chen, G., Chen, Q., Long, S., Zhu, W., Yuan, Z. and Wu, Y. 2023. Quantum convolutional neural network for image classification. Pattern Analysis and Applications. 26(2): 655–667.

[20] Jia, Z.A., Yi, B., Zhai, R., Wu, Y.C., Guo, G.C. and Guo, G.P. 2019. Quantum neural network states: A brief review of methods and applications. Advanced Quantum Technologies. 2(7–8): 1800077.

[21] Ferretti, A., Prati, C. and Rocca, F. 2000. Nonlinear subsidence rate estimation using permanent scatterers in differential SAR interferometry. IEEE Transactions on Geoscience and Remote Sensing. 38(5): 2202–2212.

[22] Ferretti, A., Prati, C. and Rocca, F. 2001. Permanent scatterers in SAR interferometry. IEEE Transactions on Geoscience and Remote Sensing. 39(1): 8–20.

[23] Ferretti, A., Perissin, D., Prati, C. and Rocca, F. 2005. On the physical nature of SAR Permanent Scatterers. *In*: URSI Commission F Symposium on Microwave Remote Sensing of the Earth, Oceans, Ice and Atmosphere, Ispra, Italy. 20–21 April 2005.

[24] Ferretti, A., Savio, G., Barzaghi, R., Borghi, A., Musazzi, S., Novali, F., et al. 2007. Submillimeter accuracy of InSAR time series: experimental validation. IEEE Transactions on Geoscience and Remote Sensing. 45(5): 1142–1153.

[25] Humme, A. 2007. Point density optimization for SAR interferometry: A study tested on salt mine areas. (Master's thesis, Delft University of Technology).

[26] Ketelaar, V.G. 2009. Satellite Radar Interferometry: Subsidence Monitoring Techniques (Vol. 14). Springer Science & Business Media.

[27] Zhou, D., Yu, J., Zhu, L., Wang, Y., Zhang, J., Jiao, S., et al. 2020. Land subsidence monitoring based on PS-InSAR Persistent Scatterers identification with spectral analysis method. Proceedings of the International Association of Hydrological Sciences. 382: 249–253.

[28] Imperatore, P., Pepe, A. and Lanari, R. 2015. Multichannel phase unwrapping: Problem topology and dual-level parallel computational model. IEEE Transactions on Geoscience and Remote Sensing, 53(10): 5774–5793.

[29] Imperatore, P., Pepe, A. and Lanari, R. 2015. High-performance parallel computation of the multichannel phase unwrapping problem. *In*: 2015 IEEE International Geoscience and Remote Sensing Symposium (IGARSS). 4097–4100. IEEE.

[30] Pepe, A. 2019. Theory and statistical description of the enhanced multi-temporal InSAR (E-MTInSAR) noise-filtering algorithm. Remote sensing. 11(3): 363.

[31] Wang, L., Liang, W., Xue, W., Zhang, Y., Wang, C. and Gao, Q. 2024. Depth_ASPP_UNet++: a novel 2D ESPI phase unwrapping based on convolutional neural network. Measurement Science and Technology. 35(6): 065202.

[32] Li, Y., Meng, L., Zhang, K., Zhang, Y., Xie, Y. and Yuan, L. 2024. PUDCN: two-dimensional phase unwrapping with a deformable convolutional network. Optics Express. 32(16): 27206–27220.

[33] Guo, R., Lu, S., Zhang, M., Li, Z., Li, D., Wang, F., et al. 2024. One step accurate phase demodulation from a closed fringe pattern with the convolutional neural network HRUnet. Applied Optics. 63(7): B59–B69.

[34] Gong, L.H., Pei, J.J., Zhang, T.F. and Zhou, N.R. 2024. Quantum convolutional neural network based on variational quantum circuits. Optics Communications. 550: 129993.

[35] Elhag, H., Jansen, K., Nagano, L. and Di Tucci, A. 2024. Quantum convolutional neural networks for jet images classification. arXiv preprint arXiv:2408.08701.

[36] Du, F.F., Ren, X.M., Ma, M. and Fan, G. 2024. Qudit-based high-dimensional controlled-not gate. Optics Letters. 49(5): 1229–1232.

[37] Claudon, B., Zylberman, J., Feniou, C., Debbasch, F., Peruzzo, A. and Piquemal, J.P. 2024. Polylogarithmic-depth controlled-NOT gates without ancilla qubits. Nature Communications. 15(1): 5886.

[38] Lu, H.H., Lukens, J.M., Alshowkan, M., Kirby, B.T. and Peters, N.A. 2024. Building a controlled-NOT gate between polarization and frequency. Optica Quantum. 2(4): 282–287.

[39] Pechen, A.N., Petruhanov, V.N., Morzhin, O.V. and Volkov, B.O. 2024. Control landscapes for high-fidelity generation of C-NOT and C-PHASE gates with coherent and environmental driving. The European Physical Journal Plus. 139(5): 1–21.

[40] Ding, Q., Oppenheim, A.V., Boufounos, P.T., Gustavsson, S., Grover, J.A., Baran, T.A., et al. 2024. Pulse Design of Baseband Flux Control for Adiabatic Controlled-Phase Gates in Superconducting Circuits. arXiv preprint arXiv:2407.02722.

[41] Chow, J.M., Gambetta, J.M., Cross, A.W., Merkel, S.T., Rigetti, C. and Steffen, M. 2013. Microwave-activated conditional-phase gate for superconducting qubits. New Journal of Physics. 15(11): 115012.

[42] Yihdego, Y., Khalil, A. and Salem, H.S. 2017. Nile River's basin dispute: perspectives of the Grand Ethiopian Renaissance Dam (GERD). Global Journal of Human-Social Science. 17(4): 1–21.

[43] Elsanabary, M.H. and Ahmed, A.T. 2019. Impacts of constructing the grand Ethiopian Renaissance Dam on the Nile River. Grand Ethiopian Renaissance Dam Versus Aswan High Dam: A View from Egypt. 75–93.

[44] Abtew, W. and Dessu, S.B. 2019. Grand ethiopian renaissance dam site importance. pp. 63–77. *In*: The Grand Ethiopian Renaissance Dam on the Blue Nile. Springer Geography. Springer, Cham.

[45] El-Askary, H., Fawzy, A., Thomas, R., Li, W., LaHaye, N., Linstead, E., et al. 2021. Assessing the vertical displacement of the grand Ethiopian renaissance dam during its filling using DInSAR technology and its potential acute consequences on the downstream countries. Remote Sensing. 13(21): 4287.

[46] Chen, Y., Ahmed, M., Tangdamrongsub, N. and Murgulet, D. 2021. Reservoir-induced land deformation: case study from the grand ethiopian renaissance dam. Remote Sensing. 13(5): 874.

[47] Madson, A. and Sheng, Y. 2020. Reservoir induced deformation analysis for several filling and operational scenarios at the grand ethiopian renaissance dam impoundment. Remote Sensing. 12(11): 1886.

[48] Herzog, M.A. 1999. Practical Dam Analysis. London: Thomas Telford.

[49] Jansen, R.B. 2012. Advanced Dam Engineering for Design, Construction, and Rehabilitation. Springer Science & Business Media.

[50] Zhang, L., Peng, M., Chang, D. and Xu, Y. 2016. Dam Failure Mechanisms and Risk Assessment. John Wiley & Sons.

[51] Wahlstrom, E. 2012. Dams, Dam Foundations, and Reservoir Sites (Vol. 6). Elsevier.

[52] Zhang, B., Wang, H.X., Ye, Y.W., Tao, J.L., Zhang, L.Z. and Shi, L. 2019. Potential hazards to a tunnel caused by adjacent reservoir impoundment. Bulletin of Engineering Geology and the Environment. 78: 397–415.

[53] Chen, Y.F., Hong, J.M., Zheng, H.K., Li, Y., Hu, R. and Zhou, C.B. 2016. Evaluation of groundwater leakage into a drainage tunnel in Jinping-I arch dam foundation in southwestern China: a case study. Rock Mechanics and Rock Engineering. 49: 961–979.

[54] Ziaei, A. and Ahangari, K. 2018. The effect of topography on stability of shallow tunnels case study: The diversion and conveyance tunnels of Safa Dam. Transportation Geotechnics. 14: 126–135.

Chapter 12

Four-Dimensional Marghany Hologram-TomoSAR Algorithm and Quantum Machine Learning for Monitoring and Forecasting Dam Failures

12.0 INTRODUCTION

The preceding chapters have established that the Grand Ethiopian Renaissance Dam (GERD) is subjected to critical stress, primarily due to the substantial accumulation of sediment behind the dam. This sediment buildup, coupled with slow discharge rates and the dam's current maximum water storage capacity of approximately 73 billion cubic meters, significantly reduces its operational efficiency, as discussed earlier. Furthermore, Chapter 10 highlighted the rapid land deformation occurring in the vicinity of the dam, particularly the swift onset of seepage along the saddle dam.

In this chapter, we aim to present a comprehensive and conclusive analysis of potential dam breach scenarios, synthesizing insights from this book to elucidate the disaster risks associated with the dam's structural integrity. By leveraging advanced quantum machine learning algorithms, we will assess both short-term and long-term collapse scenarios, ultimately providing a clearer understanding of the threats posed by the GERD. This approach not only enhances our predictive capabilities but also serves as a crucial tool in formulating effective mitigation strategies to prevent catastrophic failure.

12.1 THE DIFFERENCE BETWEEN TOMOSAR AND TRADITIONAL SAR TECHNIQUES

The evolution of radar imaging technology has led to the development of TomoSAR (Tomographic Synthetic Aperture Radar), which significantly enhances the capabilities of traditional Synthetic Aperture Radar (SAR) techniques. By examining the key distinctions between these two methodologies, we can appreciate the profound impact that TomoSAR has on three-dimensional imaging and volumetric analysis.

To begin with, one of the most fundamental differences between TomoSAR and traditional SAR lies in their imaging capabilities. While conventional SAR primarily generates two-dimensional

DOI: 10.1201/9781003327325-12

images focused on surface reflectivity, TomoSAR employs multiple SAR captures taken from various angles to produce three-dimensional representations. Consequently, this allows for the reconstruction of vertical structures, such as buildings and vegetation, offering a comprehensive view of spatial arrangements that traditional methods cannot provide.

Moreover, the methodologies employed in data acquisition reveal another layer of differentiation. Long-established SAR typically acquires data from a single pass or limited angles, focusing predominantly on surface scattering. In contrast, TomoSAR necessitates a series of images collected over time or from diverse positions to reconstruct accurate three-dimensional models (Figure 12.1). Accordingly, this multifaceted approach ensures a richer dataset from which to derive insights, ultimately enhancing the quality and reliability of the analysis [1].

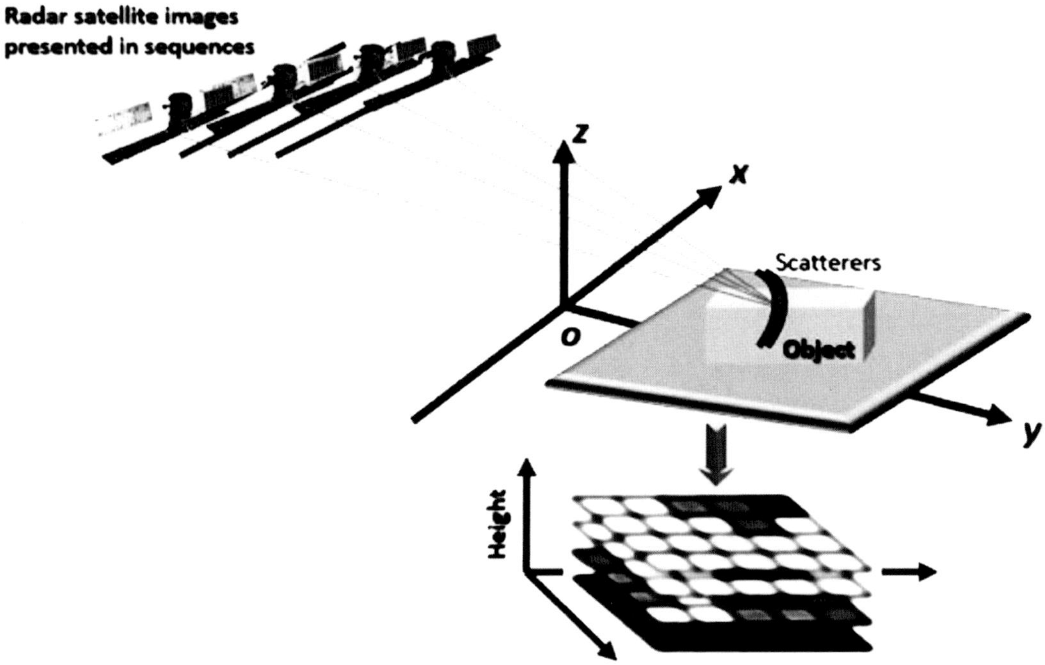

Figure 12.1 Principle of TomoSAR.

Furthermore, the sophistication of signal processing techniques employed in each method underscores their respective strengths. While conventional SAR relies on simpler processing techniques aimed at producing 2D imagery, TomoSAR utilizes advanced algorithms to extract three-dimensional information from collected data. Techniques such as back-projection and interferometric methods are commonly applied within TomoSAR frameworks. Therefore, this complexity allows for intricate analysis and interpretation of complex datasets, making TomoSAR a superior choice for detailed assessments.

Consistent with Siddique et al. [1] and Pardini et al. [2] conventional SAR systems typically produce 2D images using a series of mathematical transformations. The fundamental equation for SAR imaging can be expressed as:

$$I(x, y) = \int\int S(t) \cdot h(x, y, t)\, dt\, dx \tag{12.1}$$

where $I(x, y)$ is the resulting 2D image intensity, $S(t)$ represents the received signal as a function of time, and $h(x, y, z)$ is the point spread function (PSF), which describes how a point target is represented in the image.

In TomoSAR, multiple interferometric pairs are used to create a 3D volumetric representation. The mathematical model for this process involves synthesizing a volume from multiple 2D slices.

The reconstruction can be expressed as:

$$V(x,y,z)=\int_{-\infty}^{+\infty}S(t)h(x,y,z,t)dt \tag{12.2}$$

In equation 12.2, $V(x, y, z)$ represents the reconstructed volume at coordinates (x, y, z). Therefore, $S(t)$ is the received signal as a function of time. Consequently, $h(x, y, z, t)$ is a function that describes how each point in space contributes to the observed signal over time (Figure 12.2). In addition to dimensionality, the capacity to capture volume information further distinguishes TomoSAR. Whereas traditional SAR is limited to surface feature capture, lacking the ability to discern subsurface details, TomoSAR excels in gathering detailed volumetric data. This capability is particularly beneficial for applications such as monitoring urban environments, assessing landslide risks, and analyzing vegetation structures. Thus, while conventional SAR may suffice for basic surface-level assessments, it is inadequate in contexts where understanding underlying structures is essential. In this view, let us assume that D is dimensionality, T is TomoSAR, and G is the geometry characteristics of GERD which include its surface and subsurface features s and u; respectively [1–3].

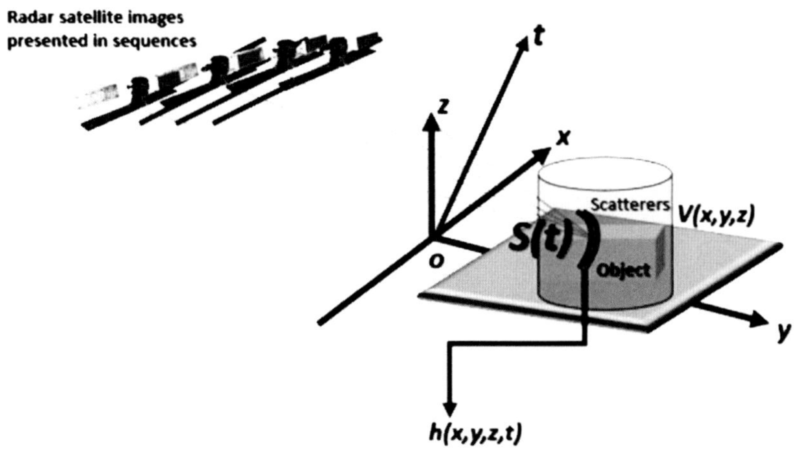

Figure 12.2 Principle of 3D volumetric capture from TomoSAR series of signals.

The capacity of TomoSAR to capture volume information can be expressed as:

$$T=f(D,\ V) \tag{12.3}$$

In other words, TomoSAR's ability to gather detailed volumetric data:

$$T \supseteq V \tag{12.4}$$

Let's consider the capability of TomoSAR to differentiate between surface and subsurface features (Figure 12.3) we can define a relationship:

$$D_{\text{TomoSAR}} = D_s + D_u \tag{12.5}$$

where D_s and D_u represent the dimensional contributions from surface and subsurface data; respectively. Thus, we can establish that:

$$D_{\text{TomoSAR}} = D_{\text{SAR}} \tag{12.6}$$

These equations encapsulate the concept of dimensionality in TomoSAR, illustrating its ability to capture more complex volumetric data compared to traditional SAR systems. As said by the above-mentioned perspective, TomoSAR utilizes synthetic aperture radar (SAR) data to create three-dimensional images of the Earth's surface. As stated by Siddique et al. [1], the basic observation in TomoSAR involves the acquisition of SAR data from different viewing

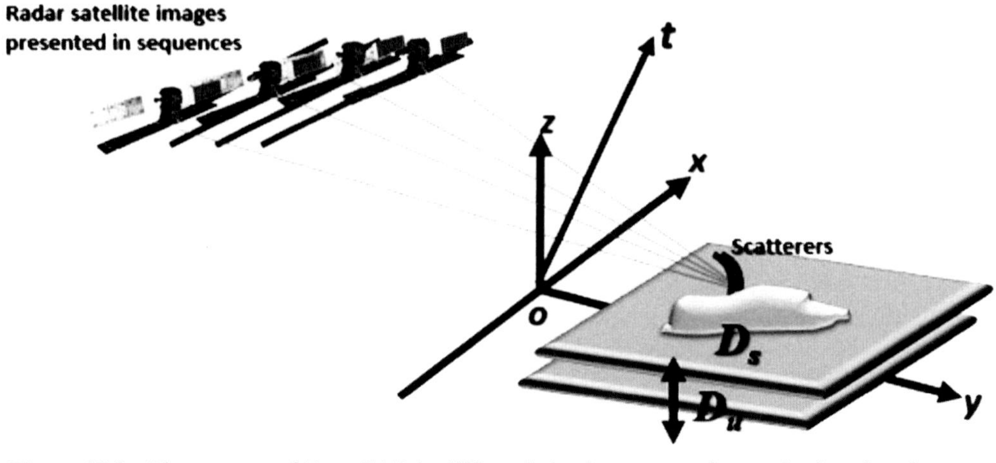

Figure 12.3 The concept of TomoSAR in differentiation between surface and subsurface features.

angles (or baselines). Each radar signal corresponds to a certain line of sight (LOS), and for a point scatterer located at position $r = (x, y, z)$, the received SAR signal can be expressed as:

$$s_n = \int_V \sigma(x, y, z) \cdot e^{-i\frac{4\pi}{\lambda}R_n(x,y,z)} dV + n_n \tag{12.7}$$

here s_n is the received signal for the n-th baseline (or acquisition); $\sigma(x, y, z)$ is the reflectivity function to be estimated; $R_n(x, y, z)$ is the distance from the radar to the point (x, y, z) for the n-th baseline (Figure 12.4). Additionally, λ is the radar wavelength; n represents noise in the signal; and V is the volume over which the reflectivity $\sigma(x, y, z)$ is distributed. In this sense, the goal in TomoSAR is to recover $\sigma(x, y, z)$, the 3D reflectivity distribution, from the collection of SAR signals $\{s_n\}$[1, 4].

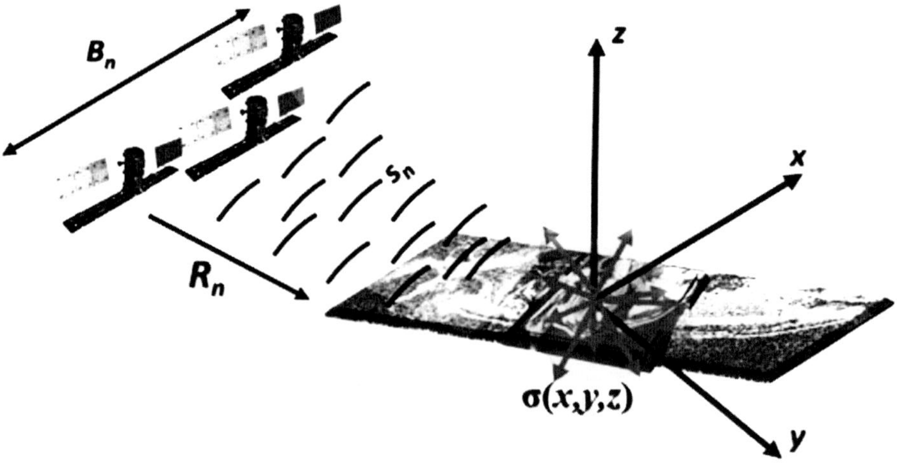

Figure 12.4 Principle of TomoSAR main formula.

12.2 MATHEMATICAL MODELS BEHIND TOMOSAR

The original Rayleigh resolution for the elevation direction is expressed as:

$$\Delta h = \frac{\lambda R_0}{2\sum B_{\perp}} \tag{12.8}$$

where λ is the radar wavelength, R_0 is the range (distance from the radar to the target), and $\sum B_{\perp}$ is the total perpendicular baseline. Given that modern SAR systems often employ multi-dimensional parameter spaces (e.g., elevation, deformation velocity, thermal dilation), this formula can be extended into a more generalized multi-dimensional space [1, 3, 5]. Let us redefine the resolution formula to account for additional dimensions v (velocity):

$$\Delta h_{\text{gen}} = \frac{\lambda R_0}{2\sum B_{\perp}} + \alpha \cdot \frac{\lambda t_v}{2V} \tag{12.9}$$

where t_v represents the acquisition times over which velocity is resolved, and v is the velocity parameter. The coefficient α therefore, is a weighting factor that accounts for the influence of velocity on the overall resolution. This extended resolution formula captures the 4D and 5D nature of differential tomography, integrating deformation velocity and thermal dilation [2, 4, 6].

TomoSAR is essentially a Fourier reconstruction with non-uniform sampling (Figure 12.5). To accommodate irregular or sparse sampling while extending into multi-dimensional space, we must revise the discrete Fourier representation. Conventionally, the model is:

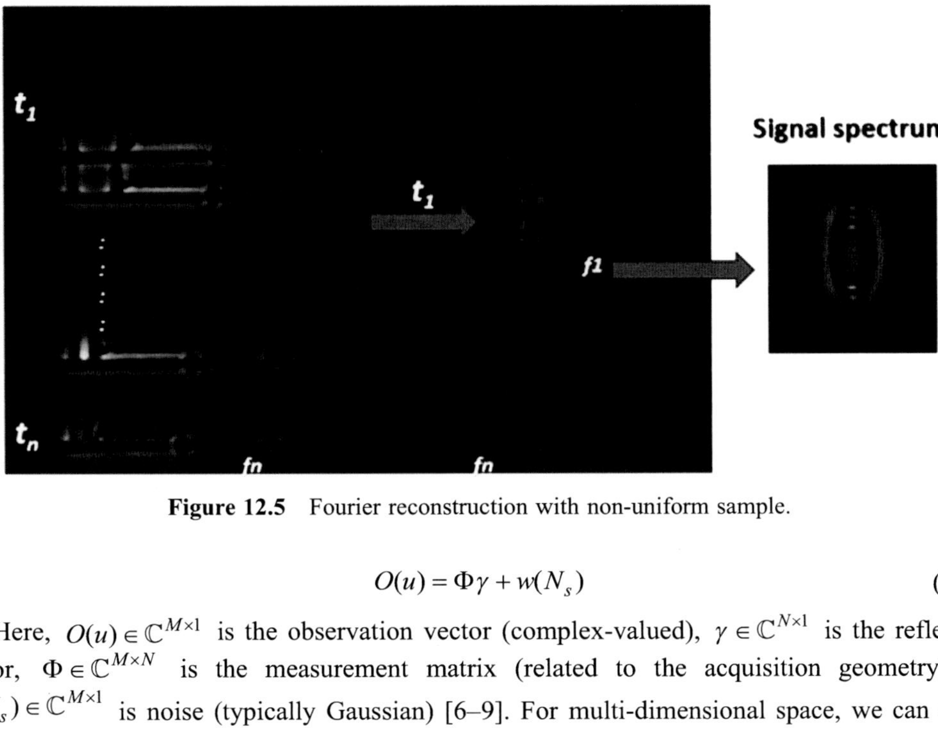

Figure 12.5 Fourier reconstruction with non-uniform sample.

$$O(u) = \Phi\gamma + w(N_s) \tag{12.10}$$

Here, $O(u) \in \mathbb{C}^{M\times 1}$ is the observation vector (complex-valued), $\gamma \in \mathbb{C}^{N\times 1}$ is the reflectivity vector, $\Phi \in \mathbb{C}^{M\times N}$ is the measurement matrix (related to the acquisition geometry), and $w(N_s) \in \mathbb{C}^{M\times 1}$ is noise (typically Gaussian) [6–9]. For multi-dimensional space, we can extend this as follows:

$$O(u) = \sum_{i=1}^{N_r} \Phi_i(s_i, v_l, k_n)\gamma_i + w(N_s) \tag{12.11}$$

where N_r represents the number of range-azimuth resolution cells, and γ_i is the reflectivity at the *i-th* range-azimuth-elevation-velocity-dilation cell. In this sense, the matrix $\Phi_i(s_i, v_l, k_n)$ now accounts for additional parameters of velocity v_l and dilation k_n, making the system sensitive to deformation while still preserving coherent backscatter [1, 3, 6].

We must modify the Generalized Likelihood Ratio Test (GLRT) to improve scatterer detection in a higher-dimensional space. The accepted test hypotheses are [10]:

(i) H_0: Only noise (no scatterer),
(ii) H_1: Single scatterer,
(iii) H_2: Double scatterers.

Therefore, the GLRT is expressed as:

$$\Lambda_1(u) = \frac{u^H (I - \Phi \widehat{\Omega}_2) O(u)}{u^H \Phi \widehat{\Omega}_2 (\Phi \widehat{\Omega}_2^H \Phi \widehat{\Omega}_2)^{-1} \Phi \widehat{\Omega}_2^H O(u)} \tag{12.12}$$

To extend this into a multi-dimensional framework that includes elevation h, velocity v, and temperature T, we revise the GLRT function to [8]:

$$\Lambda_1(u) = \frac{u^H (I - \Phi \widehat{\Omega}_2(s,v,T)) O(u)}{u^H \Phi \widehat{\Omega}_2(s,v,T) (\Phi \widehat{\Omega}_2^H(s,v,T) \Phi \widehat{\Omega}_2(s,v,T))^{-1} \Phi \widehat{\Omega}_2^H(s,v,T) O(u)} \tag{12.13}$$

This revised GLRT accounts for scatterers that may exhibit temporal velocity changes or thermal effects, adding another layer of granularity to the detection process. In this study, twenty high-resolution TerraSAR images were meticulously acquired in both HH and VV polarimetric channels, spanning the period from January 2022 to August 2024. These images were captured in spot mode and offer a spatial resolution of approximately 1 meter in range and 2.6 meters in azimuth, ensuring fine-scale detail in the horizontal plane. Given the perpendicular baseline of around 800 meters, the Rayleigh resolution in the elevation direction is calculated to be 14 meters. This combination of parameters provides a robust framework for precise three-dimensional imaging, allowing for enhanced vertical resolution that is crucial for distinguishing fine structural variations [8, 10].

In Figure 12.6, the intensity HH SAR image of the Grand Ethiopian Renaissance Dam (GERD) main dam area is depicted. A Generalized Likelihood Ratio Test (GLRT) has been applied to the HH channel, utilizing M = 39 baselines. The Rayleigh resolution in the elevation direction is calculated to be 14 meters, with a discretization step set at 2 meters. Given the considerable height of the GERD structure in the SAR image, a high number of double scatterers was anticipated. However, the results show that 1,469 single scatterers were detected, while only 2 double

Figure 12.6 Intensity HH TerraSAR spot mode data.

scatterers were identified. The single and double scatterers are represented in black and bright grey, respectively, in the HH SAR image of the GERD area, as shown in Figure 12.7.

This passage illustrates the application of GLRT to a complex real-world scenario, where high-resolution SAR data is used to analyze scatterers in a tall structure. The discrepancy between the expected and detected number of double scatterers highlights the precision of the analysis. The visualization in Figure 12.7 further emphasizes the distinction between the different scatterer types.

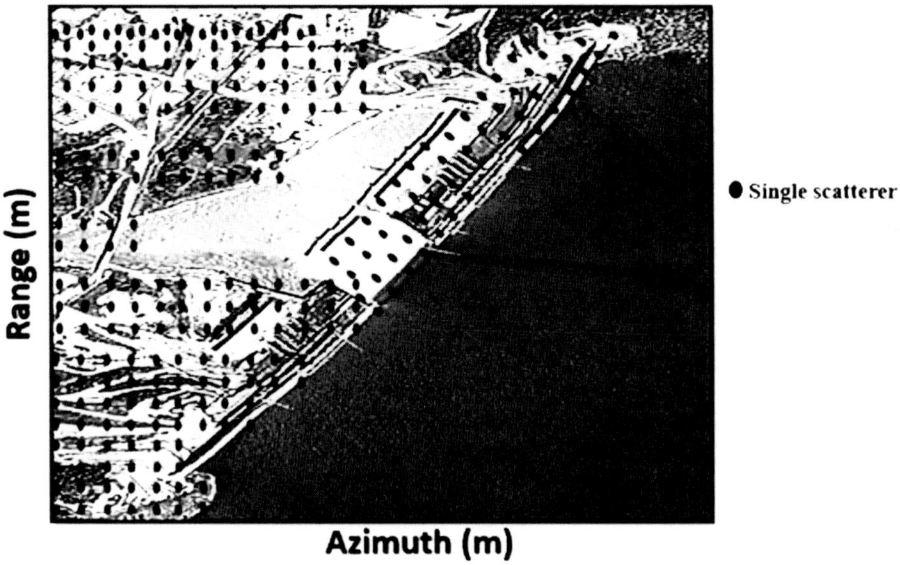

Figure 12.7 Single scatterers were detected by Fast-Sup-GLRT and reported on the Google Earth optical data.

In this analysis, the several bright grey points, which correspond to the detected double scatterers are aligned with each other. For each detected double scatterer, its corresponding pair of points is displayed at the estimated height in the 3-D Google Earth image. To further validate these findings, we conducted a comparison using a reduced dataset of 20 images, but this time including both HH and VV channels.

In Figure 12.8, the Generalized Likelihood Ratio Test (GLRT) [8, 11] was applied considering M = 20 baselines. The detected single scatterers are represented in black, while the double scatterers are shown in bright grey on the optical 3-D Google Earth image. With this reduced data set, we detected 1,954 single scatterers and identified 20 double scatterers, a significant increase in double scatterer detection compared to the previous analysis with only HH channel data. This comparison demonstrates that incorporating both HH and VV channels, even with fewer images, enhances the detection of double scatterers. The increased number of detected double scatterers highlights the advantages of multi-channel analysis for improving the resolution and accuracy of scatterer identification in SAR imagery.

Therefore, to understand why the single polarization approach detected fewer single and double scatterers, the absolute values of the interferometric coherence and averaged across all baselines and polarimetric channels are computed. The resulting average coherence images of single polarization and dual polarization are displayed in Figures 12.9 and 12.10 respectively.

As anticipated, the average coherence values are generally higher when using polarimetric data, compared to the single polarization approach. This higher coherence indicates a more stable phase information across the different baselines, which directly enhances the detection capability. Therefore, the superior performance in detecting scatterers with polarimetric data can

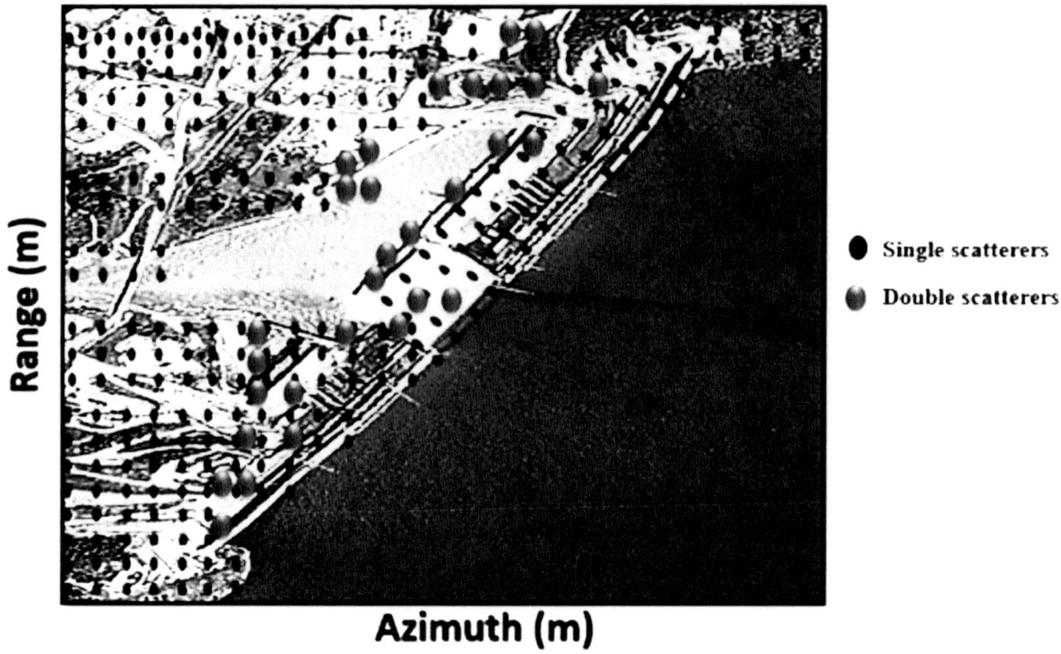

Figure 12.8 Double scatterer points in bright grey were detected by Fast-Sup-GLRT and reported on the Google Earth optical data.

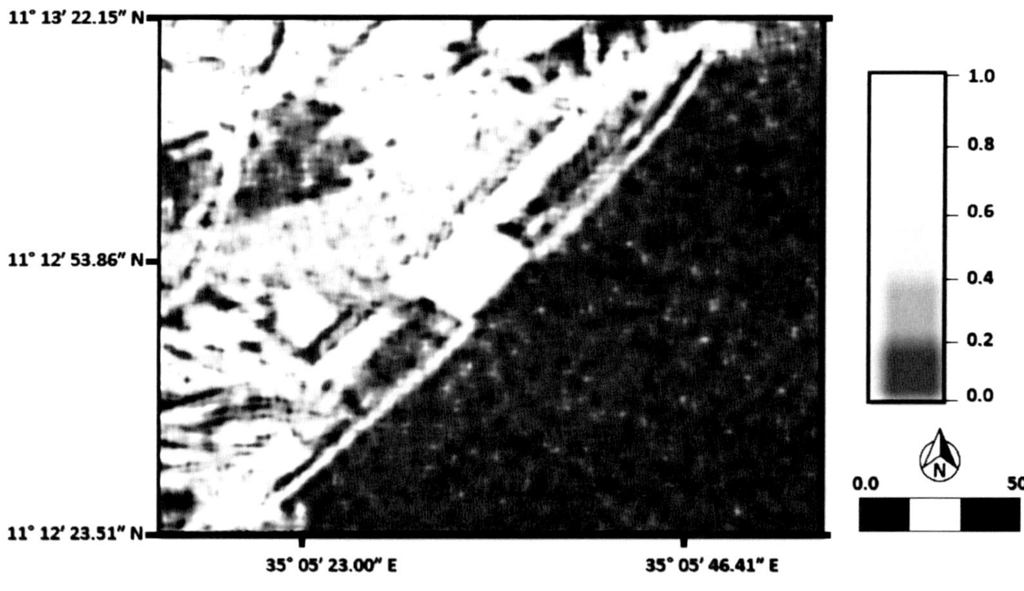

Figure 12.9 Average coherence image for single polarization.

be attributed to this higher coherence, as it leads to improved signal clarity and more accurate scatterer identification. This finding reinforces the importance of using multi-polarization SAR data in improving both the reliability and resolution of scatterer detection. This result confirms the studies done by Siddique et al. [1], Budillon and Schirinzi [8] and Budillon et al. [11].

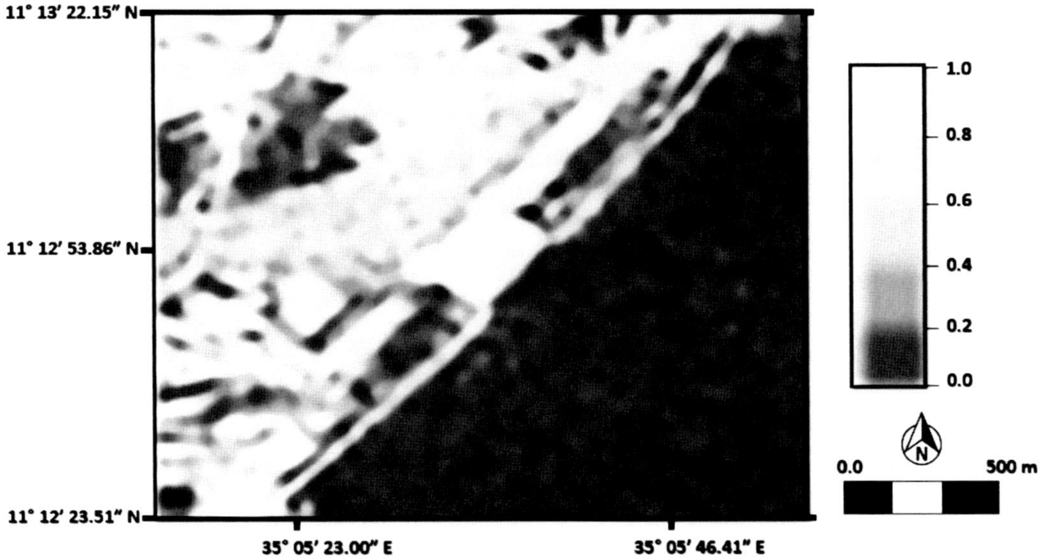

Figure 12.10 Dual polarization average coherence spatial variations.

12.3 SUPER-RESOLUTION ENHANCEMENT

Super-resolution is a vital technique in radar imaging, particularly for Synthetic Aperture Radar (SAR) tomography (TomoSAR), which aims to resolve scatterers that are separated by distances smaller than the classical Rayleigh resolution limit. The Rayleigh limit is the minimum distance between two-point scatterers such that they can still be distinguished in the image. Exceeding this limit requires advanced mathematical techniques to avoid the diffraction-limited resolution.

12.3.1 Regularized Super-Resolution Problem

According to equation 12.10, the goal of super-resolution is to recover the reflectivity profile γ from observed measurements *O(u)*, where γ represents the intensity of scattering objects at different locations in a scene. However, resolving scatterers separated by less than the Rayleigh limit requires super-resolution techniques. One powerful approach is to formulate the problem as a regularized optimization:

$$\min_{\gamma}\left\{\| O(u)-\Phi\gamma \|^2 +\lambda_C \mathcal{R}(\gamma)\right\} \tag{12.14}$$

Here, $\| O(u)-\Phi\gamma \|$ represents the data fidelity term, and $\mathcal{R}(\gamma)$ is the regularization term that promotes desirable properties in the solution (such as sparsity or smoothness). The parameter λ_C controls the balance between the fidelity term and the regularization term [8–12].

12.3.2 Types of Regularization: Sparse and Edge-Preserving Norms

The choice of the regularization function $\mathcal{R}(\gamma)$ is crucial for super-resolution performance. In this perspective as well, the goal is to resolve a small number of distinct scatterers. We can impose sparsity by using the ℓ_1-norm as the regularization term:

$$\mathcal{R}(\gamma)=\|\gamma\|_1=\sum_{i=1}^{N}|\gamma_i| \tag{12.15}$$

This formulation encourages the solution γ to be sparse, meaning that only a few elements of γ will have non-zero values, corresponding to the locations of significant scatterers. The ℓ_1-norm is widely used in compressive sensing and super-resolution because it promotes sparse solutions while remaining convex, ensuring the optimization problem remains tractable. Therefore, the complete optimization problem with ℓ_1-regularization is:

$$\min_{\gamma} \left\{ \left\| O(u) - \Phi\gamma \right\|^2 + \lambda_C \left\| \gamma \right\|_1 \right\} \tag{12.16}$$

This problem can be solved efficiently using techniques such as the Iterative Shrinkage-Thresholding Algorithm (ISTA) or the Least Absolute Shrinkage and Selection Operator (LASSO).

The Iterative Shrinkage-Thresholding Algorithm (ISTA) is used to solve optimization problems involving a loss function with a regularization term, particularly those that include an ℓ_1-norm for promoting sparsity [9]. Therefore, the objective function that ISTA minimizes is given by:

$$f(\gamma) = \frac{1}{2} \left\| O(u) - \Phi\gamma \right\|_2^2 + \lambda_C \left\| \gamma \right\|_1 \tag{12.17}$$

Initially, to implement equation 12.17, let us begin with an initial guess for $\gamma^{(0)}$, often set to zero:

$$\gamma^{(0)} = 0 \tag{12.18}$$

Consequently, for each iteration k, compute the gradient of the loss function:

$$\nabla f(\gamma^{(k)}) = -\Phi^T (O(u) - \Phi\gamma^{(k)}) \tag{12.19}$$

Subsequently, γ is updated using the gradient and a step size t:

$$\gamma^{(k+1)} = \gamma^{(k)} - t\nabla f(\gamma^{(k)}) \tag{12.20}$$

Thus, the shrinkage operator (soft thresholding) is applied to enforce sparsity:

$$\gamma^{(k+1)} = S_{\lambda t}(\gamma^{(k+1)}) \tag{12.21}$$

where the soft thresholding operator $S_{\lambda t}(\cdot)$ is defined as:

$$S_{\lambda t}(x) = \begin{cases} x - \lambda t & \text{if } x > \lambda t \\ 0 & \text{if } |x| \le \lambda t \\ x + \lambda t & \text{if } x < -\lambda t \end{cases} \tag{12.22}$$

Lastly, the update steps are repeated until convergence criteria are encountered, such as changes in γ being below a certain threshold.

LASSO is a regression analysis method that performs both variable selection and regularization to enhance prediction accuracy and interpretability. It minimizes the same objective function as in conventional regression equation but is typically applied in a regression context [8, 10]. In this perspective, the LASSO problem can be formulated as:

$$\text{Minimize} \frac{1}{2n} \left\| y - X\beta_L \right\|_2^2 + \lambda \left\| \beta_L \right\|_1 \tag{12.23}$$

Here, y is the response variable, X is the feature matrix, and β_L represents the coefficients. The solution methods for optimizing the objective function include coordinate descent, which efficiently updates one coefficient at a time while keeping the others fixed, and sub-gradient methods, which are particularly useful for handling non-differentiable functions, such as those involving the ℓ_1-norm [8, 11].

12.3.3 Total Variation (TV) Regularization (Edge-Preserving)

Another common regularization technique, especially useful when the reflectivity profile contains edges (abrupt changes), is Total Variation (TV) regularization:

$$\mathcal{R}(\gamma) = \|\gamma\|_{TV} = \sum_{i=1}^{N} |\nabla \gamma_i| \tag{12.24}$$

Here, $\nabla \gamma_i$ represents the gradient of γ at the *i-th* location. TV regularization favors piecewise constant solutions, preserving sharp transitions between high and low reflectivity regions. This is particularly useful in super-resolution when scatterers are located at sharp discontinuities, such as building edges or cliffs in SAR images. Therefore, the optimization problem with TV regularization becomes:

$$\min_{\gamma} \left\{ \|O(u) - \Phi\gamma\|^2 + \lambda_C \|\gamma\|_{TV} \right\} \tag{12.25}$$

While TV regularization is non-convex, various optimization algorithms such as Primal-Dual Methods and Augmented Lagrangian Methods can solve this problem.

12.3.4 Extension to Nonlinear and Adaptive Regularization

In real-world scenarios, scatterers may exhibit complex interactions and non-linear behaviors, especially in multi-dimensional SAR imaging (e.g., 4D and 5D reconstructions). To handle such complexity, one can introduce adaptive or nonlinear regularization:

$$\mathcal{R}(\gamma) = \sum_{i=1}^{N} f(\nabla \gamma_i) \tag{12.26}$$

where $f(\cdot)$ is a nonlinear function that adapts based on the data characteristics. For instance, $f(x) = \dfrac{x^2}{1+\alpha x^2}$ is a robust form that penalizes small variations less while preventing overly smooth solutions, making it suitable for reconstructing sharp edges without amplifying noise [9–15].

12.3.5 Lagrange Multiplier Interpretation and Dual Problem

To solve the regularized super-resolution problem, one can view λ_L as a Lagrange multiplier, enforcing a constraint on the solution's regularity [9, 15]. This transforms the minimization into a constrained optimization problem:

$$\min_{\gamma} \|O(u) - \Phi\gamma\|^2 \quad \text{s.t.} \quad \mathcal{R}(\gamma) \leq \tau \tag{12.27}$$

This can be solved via Lagrangian duality, where the dual problem may offer insights into the trade-offs between resolution and regularity [8, 10, 13]. In the dual domain, the problem becomes more geometrically interpretable, linking the underlying physics of scatterer interactions with the resolution limitations imposed by the radar system [8, 12, 14].

12.4 PROCEDURES OF ACHIEVING TOMOSAR

From the aforementioned perspective, the following outlines the step-by-step TomoSAR (Tomographic Synthetic Aperture Radar) procedure employed in this study:

12.4.1 Preprocessing and Registration

A stack of MM SAR interferometric images, all capturing the same scene, was meticulously registered with sub-pixel accuracy. The interferometric phase contribution from an external Digital Elevation Model (DEM) was then removed during preprocessing. Atmospheric phase correction was not deemed necessary due to the limited extent of the study area, and any residual atmospheric effects were mitigated by calibrating the interferometric phase across the image stack [1, 8, 12, 13, 16].

12.4.2 False Alarm and Detection Probabilities

The probabilities of false alarm (PFA) and false detection (PFD) were assigned, and corresponding detection thresholds were determined through Monte Carlo simulations. These thresholds govern the trade-off between correctly detecting scatterers and minimizing false alarms [1, 8, 11].

12.4.3 Minimization of Terms for Estimation

For each range-azimuth cell, the estimates of $\Phi_{\Omega 1}$ and $\Phi_{\Omega 2}$ were obtained by minimizing the following term:

$$u^H (I - \Phi_{\Omega i}(\Phi_{\Omega i}^H \Phi_{\Omega i})^{-1} \Phi_{\Omega i}^H) O(u), \quad i = 1, 2, \tag{12.28}$$

where $O(u)$ is the observation vector as discussed earlier, and $\Phi_{\Omega i}$ represents the selected columns of the measurement matrix Φ, associated with scatterers. This minimization process was conducted iteratively for both $\Phi_{\Omega 1}$ and $\Phi_{\Omega 2}$, corresponding to different scatterer configurations.

12.4.4 Application of the Generalized Likelihood Ratio Test (GLRT)

For each range-azimuth cell, the two-step GLRT was applied to determine which of the three statistical hypotheses held: no scatterer (only noise), the presence of a single scatterer (H_1), or the presence of two scatterers (H_2). Based on this test, the estimation of scatterer parameters—elevation (s_i), deformation velocity (v_l), and thermal dilation (k_n)—was performed. In cases where a single scatterer was detected (H_1), one set of parameters was estimated, while for double scatterers (H_2), two sets of parameters were derived, based on the selected columns of the matrix Φ.

In this framework, the Generalized Likelihood Ratio Test (GLRT) reveals that dual polarization produces a significantly clearer reflective spectrum than single polarization (Figures 12.11 and 12.12). Specifically, the spectral reflectivity over the east and west embankments of the GERD main dam is characterized by a higher reflectivity in dual polarization, with values ranging from –6.0 dB to –2.0 dB (Figure 11.12). Additionally, the waterline edge above the dam's foundation is distinctly discernible with dual polarization, offering greater clarity compared to single polarization. In multi-baseline TomoSAR, scatterers are more effectively identified with dual polarization than with single polarization approaches associated with InSAR [8, 11].

Conversely, single polarization achieves its highest spectral reflectivity at approximately –5.0 dB (Figure 11.11), particularly along the uppermost corridors of the dam's main structure on both the east and west embankments, as well as across the infrastructure located in front of the dam. This contrast illustrates the increased precision of dual polarization in capturing fine details, especially in complex structures like dams, while single polarization is more limited but still effective in specific areas such as the dam's crest and surrounding infrastructures.

The superior performance of dual-polarization in detecting spectral reflectivity, as revealed by the Generalized Likelihood Ratio Test (GLRT) [8, 11] can be attributed to the fundamental principles governing electromagnetic wave interactions. Dual polarization, by simultaneously

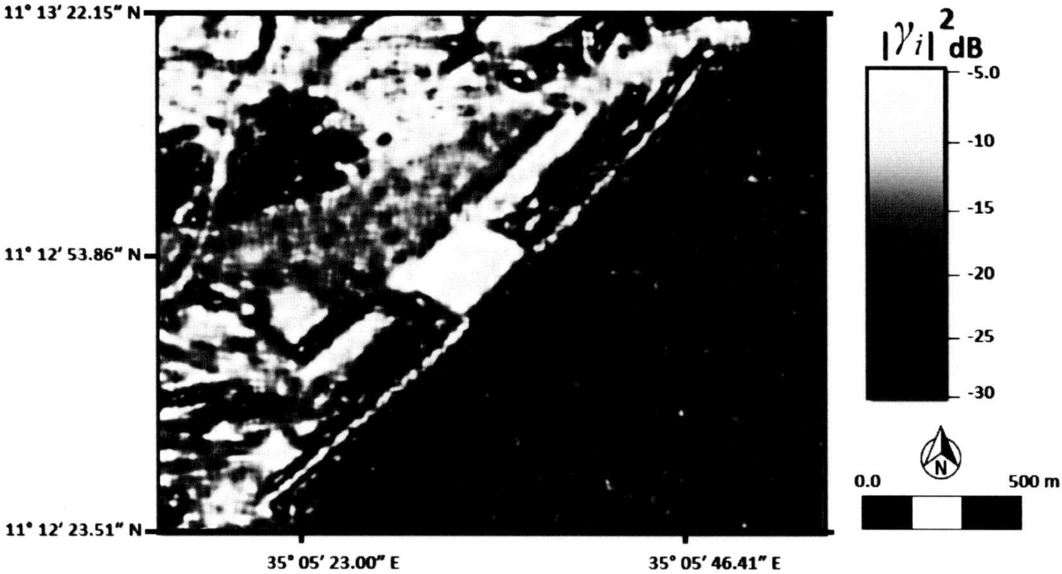

Figure 12.11 GLRT spectra reflectivity of single polarization.

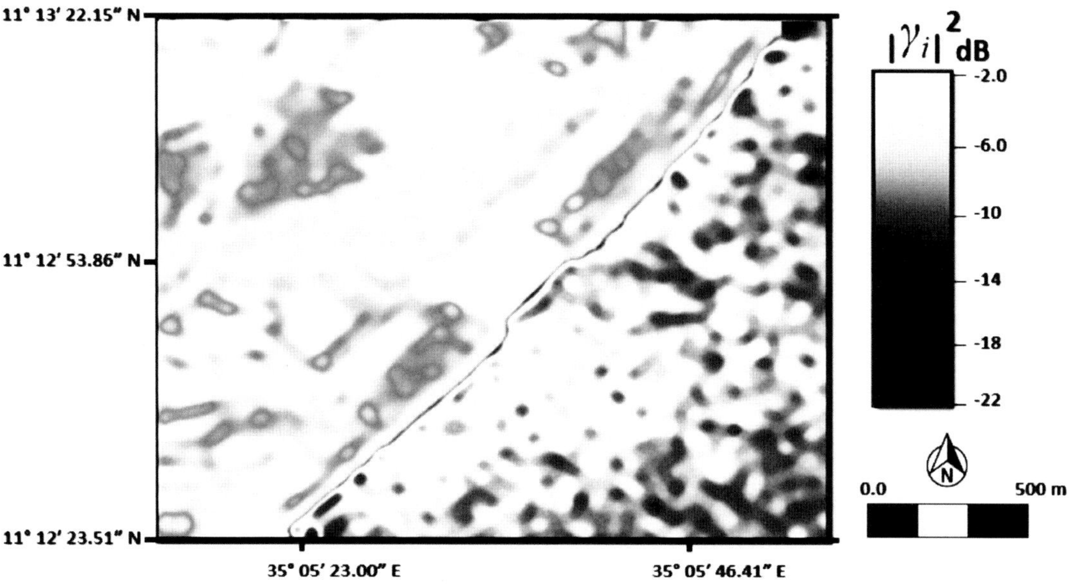

Figure 12.12 GLRT spectra reflectivity of dual polarization.

capturing horizontal (HH) and vertical (VV) polarizations, offers a richer, multidimensional representation of the scene. This dual-channel approach allows for enhanced sensitivity to both geometric and dielectric properties, making it particularly adept at identifying complex structures such as the GERD dam's embankments and waterline edges.

In contrast, single polarization only captures one dimension of the reflected wave, thus limiting its ability to resolve subtle variations in surface characteristics. As Albert Einstein once said, 'the more we understand the nature of light and its interaction with matter, the deeper our insights into the structure of reality'. This concept underlies the advantage of dual polarization—it is a more complete 'view' of the reality being measured.

Further, in a multi-baselines TomoSAR system, dual polarization facilitates better scatterer identification due to its ability to discriminate between different types of targets. By offering two orthogonal perspectives of wave interaction, dual polarization enhances the detection of both specular and diffuse scatterers. This explains why dual polarization outperforms single polarization in reconstructing the reflective spectra, particularly in complex environments like the GERD dam, where water, concrete, and other materials create a diverse scattering landscape [1, 16].

Single polarization, while still useful, reaches its highest effectiveness along more uniform surfaces, such as the dam's crest and the adjacent infrastructure, where fewer polarization-dependent scattering mechanisms are present [1, 4, 18]. This limitation is akin to trying to 'observe the cosmos through a keyhole', as Carl Sagan once analogized in a different context—single polarization gives us a narrow view, while dual polarization opens the door wider, letting more light in [17–19].

Thus, dual polarization provides a broader and more detailed spectral insight, offering greater clarity in distinguishing intricate features of the dam, particularly those that are influenced by both geometry and material properties. It is, in essence, a more holistic approach to remote sensing.

Therefore, for the Generalized Likelihood Ratio Test (GLRT), it becomes evident that double scatterers are highly effective in identifying the geometric profile of the GERD main dam, as well as the scattering contributions from both the east and west embankment corridors. These scatterers, including those associated with the river reservoir and surrounding infrastructures, are detected within a slant range extending from 0 to 200 meters. The corresponding height of the dam ranges from 20 to 175 meters, which is associated with a maximum spectral reflectivity of –4 dB (Figure 12.13).

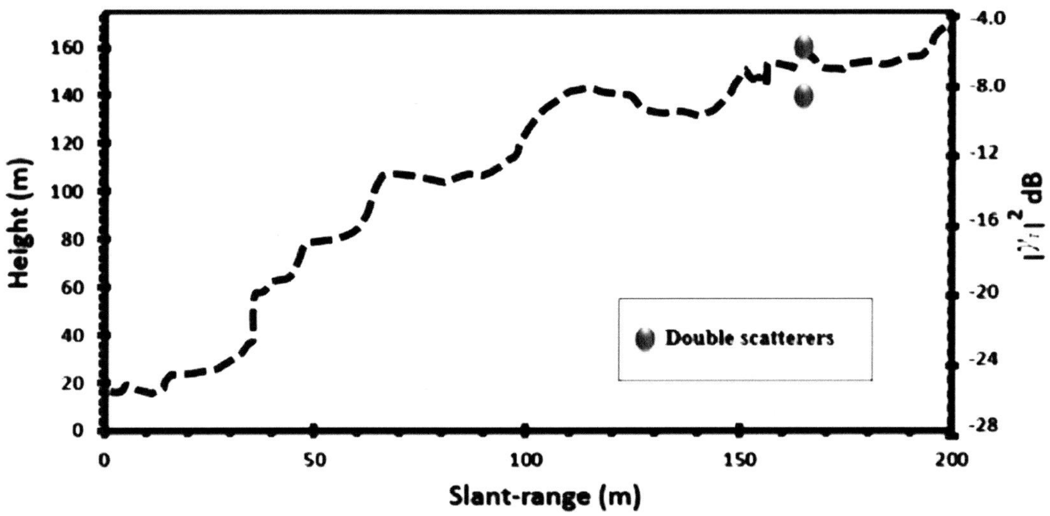

Figure 12.13 Tomographic spectra derived by GLRT for dual polarization and double scatterers of a geometric profile of GERD height.

This enhanced identification, enabled by the GLRT in conjunction with the dual polarization data, underscores the capacity of the technique to capture complex structural and surface features, allowing for precise reconstruction of the dam's geometry and scattering properties. The higher reflectivity observed in these regions is indicative of robust surface scattering, particularly from the dam's embankments and adjacent structures.

The question now is: By using the GLRT algorithm can TomoSAR reconstruct the 3-D of the GERD main dam? Figure 12.14 demonstrates 3D image of the Grand Ethiopian Renaissance

Dam (GERD) that was generated using the TomoSAR GLRT (Generalized Likelihood Ratio Test) algorithm, which utilizes multi-baseline synthetic aperture radar (SAR) data to achieve high-resolution three-dimensional reconstructions. The image reveals the dam structure and the surrounding terrain with notable clarity.

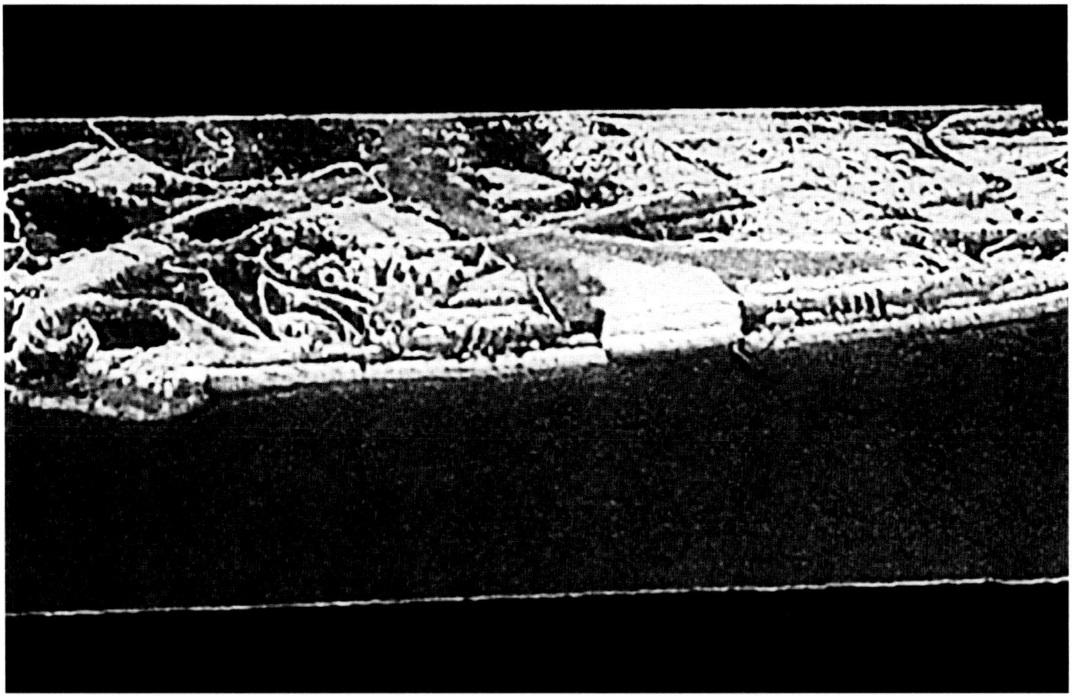

Figure 12.14 3D GERD construction using GLRT algorithm in TomoSAR technique.

In this visual representation, the dam and its immediate environment are reconstructed with distinct elevation variations, which is a testament to the effectiveness of TomoSAR in resolving vertical structures. The detection of different scatterers—single and double—across the dam and the surrounding landscape enables detailed reconstruction of both the dam's embankments and the adjacent infrastructure. Specifically, the high reflectivity near the waterline and along the dam's embankments can be attributed to strong surface scattering caused by the man-made structures and materials used in the dam's construction [1, 5, 7, 12, 18].

The east and west embankments are well-defined, reflecting higher signal returns due to the geometry and materials of the dam, as previously discussed. The dual-polarization data used by the GLRT algorithm enhances the accuracy in distinguishing between different types of scatterers, allowing for the identification of fine structural details that would be less apparent using single-polarization data.

Dual-polarimetric TomoSAR reveals significant variations in velocity deformation along the GERD from January 2022 to August 2024. Notably, the western embankment exhibits a subsidence rate of –0.72 m/year, which exceeds that of the eastern embankment. Vertical displacement on the eastern side is observed to be less than 0.70 m/year (Figure 12.15). These subsidence values are notably lower than those obtained through PSI techniques, as discussed in Chapter 11. Nevertheless, TomoSAR conclusively confirms that the GERD is undergoing subsidence, particularly along the western embankment. Additionally, there is pronounced velocity deformation concentrated in the central section of the dam, where overtopping occurs, with maximum deformation rates reaching –0.72 m/year.

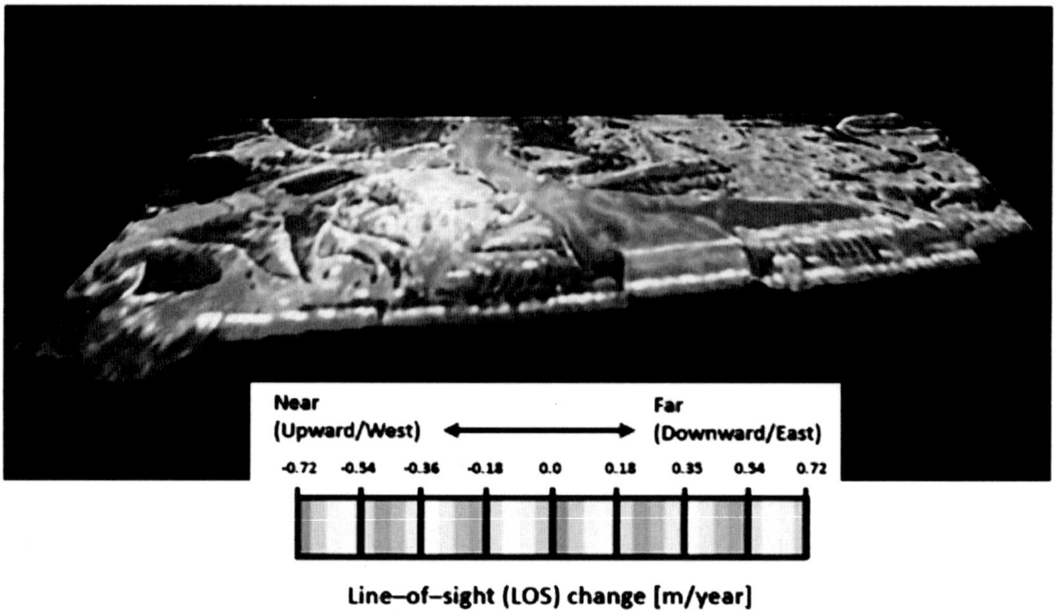

Figure 12.15 TomoSAR dual polarization velocity deformation time series from January 2022 to August 2024.

Moreover, the image highlights the capability of the TomoSAR technique to capture the complex topography of the surrounding area, revealing the intricate network of infrastructures, roads, and possibly vegetation, which further enhances the contextual understanding of the dam's geographic setting. This high level of detail, particularly in the vertical direction, is critical for applications related to monitoring dam deformation, water levels and infrastructure stability.

Overall, this 3D image underscores the power of SAR-based techniques in remote sensing, particularly for large infrastructure projects like GERD, where accurate and high-resolution terrain mapping is essential for ongoing monitoring and analysis. The 3D image of the Grand Ethiopian Renaissance Dam (GERD) generated by the TomoSAR GLRT algorithm, offers an impressive visualization of the dam's structural geometry. It demonstrates the potential of SAR-based remote sensing to provide detailed insights into large-scale infrastructure. However, despite its advantages, several inherent limitations arise that merit closer examination.

Firstly, the resolution of the image plays a crucial role in determining the level of detail that can be captured. The vertical resolution, although improved by TomoSAR, remains dependent on the number of baselines and their diversity. Inadequate baseline distribution can result in reduced accuracy, leading to blurring or oversimplification of complex vertical structures, such as the dam's height profile or adjacent terrain. Further, horizontal resolution is generally better in the radar's line-of-sight (range direction) but less refined along the flight path (azimuth direction). Fornaro et al. [20], and Pauciullo et al. [21] stated that such inherent characteristics of SAR can cause fine horizontal details, like those along the length of the dam or surrounding infrastructure, to be less defined compared to what optical imagery can provide. Consequently, the overall geometric precision of the 3D model might suffer.

In addition to resolution challenges, speckle noise and signal coherence introduce further limitations. The granular appearance caused by speckle noise, a byproduct of the radar's coherent nature, can obscure finer details, especially over homogeneous areas such as the reservoir surface. While the water body in this particular image appears smooth, regions with less structural complexity may still be affected. Moreover, coherence is critical for the quality of the TomoSAR reconstruction [7]. Temporal decorrelation, caused by environmental changes such as vegetation growth or soil displacement between image acquisitions, can reduce coherence, leading to inaccuracies in height estimation. This is particularly problematic for natural terrains adjacent to the dam, where scatterers might change over time, degrading the 3D model's reliability [1, 4, 8, 10].

Additionally, geometric distortions inherent to SAR imagery pose significant challenges. Layover and foreshortening, which occur due to the SAR's imaging geometry, can distort tall structures or slopes facing the radar. In the GERD image, this can affect steep embankments, where elevation changes are sharp, complicating the accurate interpretation of the dam's true shape. Similarly, radar shadowing, which occurs when areas behind tall objects are not illuminated can lead to gaps in the data. While TomoSAR attempts to mitigate these issues by using multiple viewing angles, shadowing effects may still limit the data quality in certain regions, particularly near sharp terrain features.

Beyond geometric concerns, material discrimination remains a significant limitation in SAR-based imaging. SAR primarily responds to surface reflectivity, yet it struggles to differentiate between materials with similar scattering characteristics. For instance, both the concrete structure of the dam and other surrounding infrastructure may exhibit high-intensity reflections, making it challenging to distinguish between them. This homogeneity in the scattering response reduces the ability to identify specific material properties, such as potential erosion or structural degradation.

Further, the GLRT algorithm's effectiveness is influenced by its assumptions regarding the scattering mechanisms. The algorithm is designed to identify single or double scatterers, but the environment of a large dam is highly complex, with multiple scattering mechanisms often occurring simultaneously. This complexity may lead to errors if the algorithm's assumptions do not fully capture the nuanced scattering behavior of the dam and its surroundings, resulting in potential misinterpretations of the scatterers and inaccuracies in the 3D reconstruction [8, 11].

Lastly, the complexity of interpretation cannot be overlooked. While the 3D reconstruction provides valuable insights, understanding the output requires a high degree of expertise in SAR processing. Without specialized knowledge, interpreting TomoSAR imagery can be challenging, particularly due to the lack of color differentiation typically available in optical images. For non-specialists, this could hinder the accurate assessment of the dam's structural integrity or the distinction between different scatterer types in the surrounding environment [1, 8, 11, 17].

Needless to say, while the TomoSAR GLRT algorithm has succeeded in producing a detailed 3D reconstruction of the GERD dam, offering valuable information about its structure and surroundings, several challenges persist. The limitations related to resolution, coherence, speckle noise, geometric distortions and material discrimination all impact the overall precision of the 3D image [22–24]. Consequently, it is essential to complement SAR-based techniques with other forms of remote sensing, such as optical or multispectral data to achieve a more comprehensive and accurate understanding of the dam's structural health and surrounding environment. This multi-source approach can lead to more informed decision-making, especially for large-scale infrastructure monitoring. This confirms the study done by Budillon et al. [11].

12.5 MATHEMATICAL CONSTRUCTION OF 4D FROM TOMOSAR

Constructing a 4D object reconstruction from Tomographic SAR (TomoSAR) requires extending conventional SAR tomography into an additional dimension, typically time or deformation velocity while maintaining the spatial and elevation (or along-track) dimensions. The TomoSAR method already allows us to reconstruct 3-D volumes by resolving scatterers in the height dimension [20]. In 4-D SAR, we introduce velocity (deformation) or time as the fourth dimension, capturing not just spatial positioning, but also dynamic changes over time or movement within the scene. The significant question now is: what is the importance of 4-D TomoSAR for GERD? The dam is subject to mechanical stresses due to water pressure, seasonal variations, and geological factors.

Using 4D TomoSAR, we can detect and map subtle deformations in GERD's structure. Adding the fourth dimension—time or deformation velocity—to traditional SAR tomography makes it possible to track changes in the dam's walls, infrastructure and surrounding terrain over time. This is crucial for the early identification of structural instabilities or potential failures.

Constructing a new algorithm for 4D object reconstruction using quantum methods based on Tomographic SAR (TomoSAR) involves a synthesis of classical SAR principles, advanced signal processing and quantum algorithms to optimize both computational efficiency and accuracy. Here, a step-by-step mathematical formulation is presented, incorporating the power of quantum computing to address challenges such as phase unwrapping, high-dimensional data processing, and real-time monitoring.

The following section explains the mathematical process for deriving a four-dimensional representation from Tomographic Synthetic Aperture Radar (TomoSAR) data.

12.5.1 Problem Setup: TomoSAR for 4D Reconstruction

In TomoSAR, we aim to reconstruct a 3D representation of an observed scene by synthesizing multiple SAR acquisitions, each contributing to the scattering response from different angular views. The fourth dimension, typically time or velocity of deformation, is added to observe dynamic changes over time [20, 25]. Let the reflectivity function in 3D be denoted as $f(s, v, x, y)$, where s is elevation and v is deformation velocity along azimuth (x-axis) and range (y-axis); respectively. Therefore, the observed SAR data $G(m, x, y)$ for the m-th acquisition is modeled as:

$$G(m,x,y)=\int_{s_{\min}}^{s_{\max}}\int_{v_{\min}}^{v_{\max}} f(s,v,x,y)\cdot e^{(j2\pi\xi_m s)}\cdot e^{(j2\pi\eta_m v)}\cdot e^{(j\phi_e(m,s,v,x,y))}\,ds\,dv \tag{12.29}$$

Equation 12.29 states that $e^{(j2\pi\xi_m s)}$ is the elevation modulation term, related to the system geometry and acquisition mode. In this regard, the term $e^{(j2\pi\eta_m v)}$ expresses the deformation velocity modulation. Therefore, the term $e^{(j\phi_e(m,s,v,x,y))}$ accounts for phase errors due to atmospheric distortion and nonlinear deformation. Consequently, this integral delineates the formation of the scattered signal $G(m, x, y)$ through the coherent summation of contributions from scatterers located at varying elevations and deformation velocities. These contributions are modulated by their complex reflectivity and influenced by phase errors. This equation functions as the forward model for TomoSAR; when solved, it facilitates the reconstruction of a four-dimensional scene, incorporating temporal changes and deformation velocities. This capability delivers critical insights into dynamic environments.

12.5.2 Incorporating Quantum Algorithms

The key quantum enhancement is introduced through quantum phase estimation (QPE) and quantum Fourier transforms (QFT) [26], which can exponentially speed up the process of extracting phase information and solving high-dimensional integrals. To efficiently process and reconstruct the 4D data, we first need to compensate for the phase errors. Using a quantum approach, we aim to unwrap the phase efficiently by leveraging QPE to handle the residual phase:

$$\phi_e(m, x, y) = \phi_{\text{atmo}}(m, x, y) + \phi_{\text{nonl}}(m, x, y). \tag{12.30}$$

where $\phi_e(m, x, y)$ is the phase error term that includes atmospheric distortion $\phi_{\text{atmo}}(m, x, y)$ and nonlinear deformation $\phi_{\text{nonl}}(m, x, y)$ along the x-plane and y-plane. In this sense, this phase error can be corrected using quantum phase unwrapping techniques, where we minimize the entropy of the quantum state representing the phase distribution. The residual phase error can be expressed as:

$$e^{(jW[\phi resi(m,x,y)])} = e^{(j\phi resi(m,x,y))}, \tag{12.31}$$

where $W[\phi resi(m, x, y)]$ is the wrapped phase value. Quantum circuits for phase unwrapping employ QPE, which estimates the eigenvalues (phases) of unitary operators applied to SAR data, correcting them efficiently. The next step is to reconstruct the reflectivity function $f(s, v, x, y)$ from the observed data. Quantum Fourier Transforms (QFT) are utilized to invert the high-dimensional integrals that arise in SAR imaging. The 4D inverse Fourier transform for SAR data is:

$$f(s, v, x, y) = \mathrm{QFT}^{-1}\left[g(m, x, y)\cdot \mathrm{e}^{\left(-j2\pi\xi_m s\right)}\cdot \mathrm{e}^{\left(-j2\pi\eta_m v\right)}\right]. \quad (12.32)$$

In this sense, QFT exponentially accelerates this operation compared to classical FFT methods, allowing us to handle the vast amount of data from multiple SAR acquisitions efficiently. In this sense [20, 26], Figure 12.16 hints that the phase unwrapping is symmetric concerning 0 Hz, suggesting that the QFT employed may be effective for eliminating or reducing phase aliasing. However, the spectral intensity distribution and the relatively smooth gradient around the center frequency could indicate that the quantum-assisted Fourier approach produces a clean, low-noise phase unwrapping process, particularly in the high-intensity regions (centered near 0 Hz).

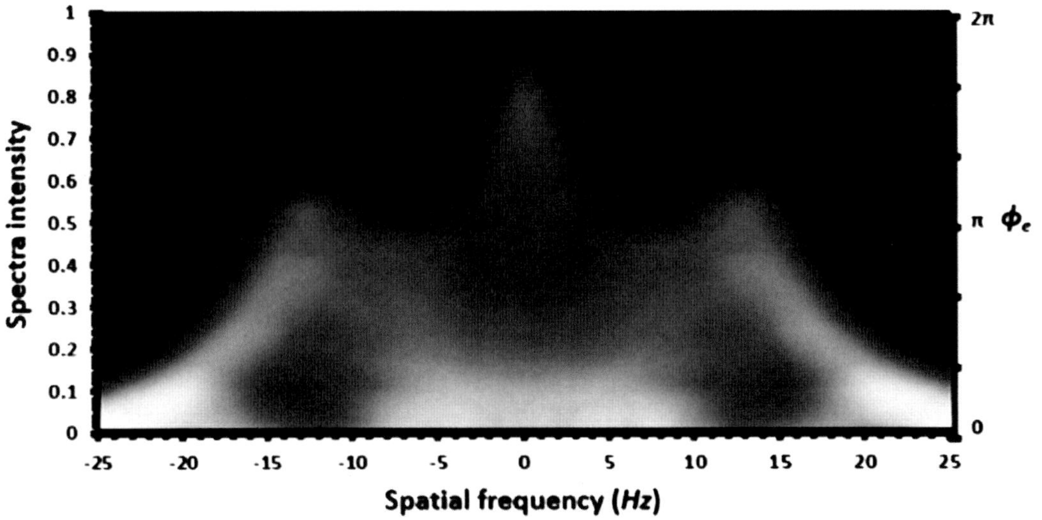

Figure 12.16 Quantum Fourier Transform impacts the phase unwrapping efficiently.

In other words, the QFT effectively captures the frequency components of the hologram interferometry data. Since the spectral intensity is smooth and the graph lacks erratic or sharp deviations in the side lobes, this suggests that the QFT [26, 27] is efficient in resolving phase ambiguities. The symmetry in spectral intensity along both sides of the central peak reinforces the notion that the phase unwrapping procedure is likely well-suited for continuous phase retrieval, necessary for accurate hologram interferometry in TomoSAR.

12.6 DERIVATION OF THE MARGHANY 4D QUANTUM PHASE UNWRAPPING ALGORITHM

In designing a quantum algorithm for 4D object reconstruction based on quantum entanglement and phase unwrapping, we must leverage both classical tomography principles and the properties of quantum systems. The reconstruction problem in 4D typically includes time as an additional dimension on top of the spatial 3D coordinates and in this quantum context, we aim to harness quantum superposition and entanglement to improve the resolution and accuracy of the reconstructed object.

The Marghany algorithm for quantum 4D object reconstruction is formulated on a series of groundbreaking principles that collectively enhance the precision and efficiency of reconstructing multi-dimensional objects. At its core, the algorithm harnesses quantum mechanics to address the inherent complexities of reconstructing objects in space-time.

The derivation of the Marghany Algorithm for Quantum 4D Object Reconstruction and Quantum Phase Unwrapping for TomoSAR builds upon quantum mechanics, signal processing and remote sensing theory. The algorithm aims to overcome traditional challenges in 4D reconstruction, such as computational complexity, phase unwrapping errors, and the inter-dimensional correlation of data points in tomography [27–30].

12.6.1 Quantum State Representation

In quantum mechanics, a system can be represented by a quantum state $|\psi\rangle$, which encapsulates all information about the system's observable properties. For TomoSAR (Tomographic Synthetic Aperture Radar), we represent each voxel (3D spatial unit) with its corresponding temporal evolution (4th dimension) as a quantum state $|\psi_{s,v,t}\rangle$ [27]. This quantum state describes the 4D object, where s is the spatial dimension (along the radar's line of sight), v represents velocity (for motion or deformation), and t is time. Mathematically, the quantum state for each voxel can be written as:

$$\left|\psi_{s,v,t}\right\rangle = \sum_{s=1}^{N_s}\sum_{v=1}^{N_v}\sum_{t=1}^{N_t} \alpha_{s,v,t}\left|s, v, t\right\rangle \tag{12.33}$$

where $\alpha_{s,\ v,\ t}$ are the complex coefficients encoding intensity (reflectivity) and phase at each voxel, and $|s,\ v,\ t\rangle$ represent the orthonormal basis states for the 4D system.

12.6.2 Quantum Entanglement for Correlated Measurements

To enhance the reconstruction, quantum entanglement is introduced to capture correlations between different parts of the 4D object. In classical systems, different voxels are treated as independent measurements, but in the quantum realm, entanglement allows for correlated measurements. This means that measuring one voxel inherently provides information about other voxels [28]. The entangled state for multiple voxels across space, velocity, and time can be represented as:

$$\left|\Psi\right\rangle = \sum_{s,v,t} \alpha_{s,v,t}\left|s, v, t\right\rangle \otimes \left|s', v', t'\right\rangle \tag{12.34}$$

where $\alpha_{s,\ v,\ t}$ describes the degree of correlation (entanglement) between voxels at $(s,\ v,\ t)$ and $(s',\ v',\ t')$. This step is crucial for capturing the inherent spatial-temporal correlations of the 4D object.

12.6.3 Quantum Fourier Transform (QFT)

The Quantum Fourier Transform (QFT) plays a central role in moving between the space-time and frequency domains. For multi-dimensional SAR, the frequency domain is useful for decomposing and analyzing the signal in terms of both spatial frequencies (related to spatial resolution) and temporal frequencies (related to object motion or deformation) [27, 29, 31]. The QFT for the 4D system is expressed as:

$$\left|\tilde{\psi}(k_s, k_v, k_t)\right\rangle = \frac{1}{\sqrt{N_s N_v N_t}} \sum_{s,v,t} \mathrm{e}^{\left(\frac{2\pi i(k_s s + k_v v + k_t t)}{N}\right)} \left|\psi_{s,v,t}\right\rangle \tag{12.35}$$

where k_s, k_v, and k_t are the spatial, velocity, and temporal frequency components respectively. The QFT allows for efficient processing of the object data by transforming it into a domain where classical algorithms struggle with multi-dimensional data correlation.

12.6.4 Quantum Automaton for Phase Unwrapping

The transition between different quantum states will be governed by quantum transition gates. These gates can be designed to operate on the local phase information at each voxel and its surrounding neighbors (Table 12.1). The automaton's transitions can be described as a set of unitary transformations:

$$U_{\text{phase}}\left|\psi_{s,v,t}\right\rangle = \left|\psi_{s,v,t}^{\text{new}}\right\rangle \tag{12.36}$$

Table 12.1 Pseudo-code of Marghany Quantum Automaton Algorithm for 4D Phase Unwrapping

```
# Marghany Quantum Automaton Algorithm for 4D Phase Unwrapping
# Initialize quantum states for each voxel in 4D space-time
Initialize_Quantum_System():
    for each voxel (s, v, t) in 4D space-time grid:
        # Initialize the quantum state of voxel (s, v, t)
        Initialize |ψ(s, v, t)> with wrapped phase φ_res(s, v, t)

# Apply quantum gate for phase evolution
Quantum_Automaton_Phase_Evolution():
    for each voxel (s, v, t) in 4D grid:
        # Apply phase transition gate U_phase based on local wrapped phase
        |ψ_new(s, v, t)> = U_phase(|ψ(s, v, t)>, φ_res(s, v, t))

        # Entangle quantum state with neighboring voxels for global consistency
        Entangle(|ψ(s, v, t)>, |ψ(s ± 1, v, t ± 1)>)

# Apply Quantum Fourier Transform to move to frequency domain
Quantum_Fourier_Transform():
    Apply QFT to the entire 4D quantum system:
    for each voxel (s, v, t) in 4D grid:
        |ψ_Freq(s, v, t)> = QFT(|ψ_new(s, v, t)>)

# Quantum interference to estimate unwrapped phase
Quantum_Interference_Unwrapping():
    for each voxel (s, v, t) in 4D grid:
        # Construct quantum interference between different phase states
        Construct_Interference(|ψ(s, v, t)>, φ_res_candidates)

        # Use interference patterns to select the most probable unwrapped phase
        |ψ_unwrapped(s, v, t)> = Max_Interference_Amplitude(Interference_Pattern)

# Measure quantum states to collapse them into classical unwrapped phases
Quantum_Measurement():
    for each voxel (s, v, t) in 4D grid:
        # Perform quantum measurement to obtain the final unwrapped phase
        φ_unwrapped(s, v, t) = Measure(|ψ_unwrapped(s, v, t)>)

# Post-process and reconstruct the 4D object using the unwrapped phases
Reconstruct_4D_Object():
    for each voxel (s, v, t) in 4D grid:
        # Reconstruct the object using unwrapped phase values
        Reconstruct_Object(φ_unwrapped(s,v,t))

# Main function that coordinates the phase unwrapping process
Marghany_Quantum_Automaton_Algorithm():
    Initialize_Quantum_System()
    Quantum_Automaton_Phase_Evolution()
    Quantum_Fourier_Transform()
    Quantum_Interference_Unwrapping()
    Quantum_Measurement()
    Reconstruct_4D_Object()

# End of Marghany Quantum Automaton Algorithm
```

These transformations will be conditioned on the local phase $\phi_{res}(s, v, t)$ and the measured quantum interference patterns. The automaton will leverage quantum interference to decide on phase transitions. By allowing quantum states representing different possible phase values to interfere, the automaton can identify the most likely unwrapped phase by selecting the state with the strongest constructive interference [27, 31].

$$\left|\psi_{\text{unwrapped}}(s, v, t)\right\rangle = \sum_{\phi_{\text{res}}} \alpha_{\phi_{\text{res}}} \left|\psi_{\phi_{\text{res}}}\right\rangle \tag{12.37}$$

where ϕ_{res} are the interference amplitudes of the different wrapped phase candidates. After the automaton has evolved and reached a steady state, quantum measurements will be performed to determine the unwrapped phase at each voxel. The measured state will collapse to the correct phase configuration due to the interference effects that drive the automaton's state transitions. The final unwrapped phase for voxel (s, v, t) is:

$$\phi_{\text{unwrapped}}(s, v, t) = \left\langle \psi_{\text{unwrapped}}(s, v, t) \middle| \hat{\phi} \middle| \psi_{\text{unwrapped}}(s, v, t) \right\rangle \tag{12.38}$$

12.7 FOUR-DIMENSIONAL MARGHANY HOLOGRAM-TOMOSAR ALGORITHM FOR GERD MONITORING

The primary inquiry at this juncture is whether the deformation velocity can provide insights into Coulomb stress. Indeed, deformation velocity can provide valuable insights into Coulomb stress, but the relationship between them is not always straightforward and depends on the material properties and conditions of the system being studied, such as the dam's foundation or geological structure. Therefore, Coulomb stress is influenced by both shear stress and normal stress, and it governs the behavior of materials under strain, such as a dam's structure or surrounding geological formations. In this perspective, deformation velocity represents the rate at which a material or structure is moving or deforming due to the applied stress. In general, an increase in stress (shear or normal) leads to an increase in deformation, so the velocity of deformation is related to the magnitude of the applied stress. This is a typical outcome of elastic-plastic material behavior: the higher the stress, the faster the material deforms.

According to the above-mentioned perspective, deformation velocity (v_d) can be thought of as being proportional to the applied stress. According to Newton's second law for motion or Hooke's law in elastic materials, the velocity of deformation could be related to the rate of change of stress applied to the material, assuming linear behavior [32]:

$$v_d \propto \frac{\Delta\sigma}{\eta} \tag{12.39}$$

where $\Delta\sigma$ is the change in total stress (including Coulomb stress), and η is the viscosity or elastic coefficient of the material (depending on the dam foundation). From the deformation velocity equation, we can assume that Coulomb stress $\Delta\sigma_C$ is related to deformation velocity through a proportionality constant η, which depends on the material properties of the dam and foundation:

$$v_d = \frac{\Delta\sigma_C}{\eta} \tag{12.40}$$

Using the expression for Coulomb stress:

$$\Delta\sigma_C = \Delta\tau - \mu'\Delta\sigma_n \tag{12.41}$$

here $\Delta\tau$ is the change in shear stress; μ' is the effective friction coefficient; and $\Delta\sigma_n$ is the change in normal stress. Therefore, equation 12.41 can be substituted into the equation 12.40:

$$v_d \cdot \eta = \Delta\tau - \mu'\Delta\sigma_n \tag{12.42}$$

Thus, the Coulomb stress can be approximated as the product of the deformation velocity and the material's viscoelastic properties (or stiffness) [32, 33]:

$$v_d \cdot \eta = \Delta\tau - \mu'\Delta\sigma_n \quad (12.43)$$

Figure 12.17 shows a 4D holographic-TomoSAR analysis for the GERD region, which is created by 4D phase unwrapping visualization using the Marghany Quantum Automaton Algorithm (MQAA). In this regard, the Marghany Quantum Automaton Algorithm is designed to handle complex, multi-dimensional phase unwrapping challenges. This 4D approach most likely incorporates temporal data, meaning that the algorithm is reconstructing both the spatial structure of the dam and how it changes over time. The accurate unwrapping of phase is crucial because interferometric data often suffer from phase wrapping, where the measured phase 'jumps' between $-\pi$ and $+\pi$. The algorithm's success here is evident in the continuity of phase gradients throughout the image, especially in regions where phase unwrapping is inherently difficult due to rapid phase changes. In this perspective, MQAA assists in incorporating both deformation velocity and Coulomb stress (in kPa).

Consequently, MQAA presents a complex analysis of the ground deformation over time, depicted in terms of Line-of-Sight (LOS) changes in meters per year (m/year), alongside the associated stress levels on the ground, measured in Coulomb stress (kPa). In other words, the '4D' component comes from the fact that deformation velocity is monitored and measured over time. Changes are tracked across the years (e.g., m/year), and this allows us to assess trends in ground movement—whether the ground is subsiding (moving down) or uplifting (moving up).

From a scientific standpoint, Figure 12.17 provides critical information for understanding the long-term stability of the dam. The regions of intense phase variation suggest that these are areas under mechanical strain, possibly due to uneven water pressure distribution or underlying ground instability. This extension study of the 4D object reconstruction confirms the recent work developed and invented by Marghany [27]. The deformation velocities range from –0.72 m/year (purple) to +0.72 m/year (red), where the negative values indicate subsidence (ground sinking). On the contrary, the positive values indicate uplift (ground rising). The LOS change measures how quickly the ground is deforming over time in the satellite's line of sight, with regions in purple indicating significant downward movement and regions in red showing uplift. Therefore, the regions of high Coulomb stress (87 kPa), typically shown in red/orange, are associated with areas of negative LOS changes, indicating ground subsidence. This suggests that the stress applied to the ground is causing it to sink, which could be due to increased pressure from up. In this case, the Coulomb stress of 87 kPa of stress is likely exerting a compressive force on the ground. This can happen if there's a buildup of pressure from overlying materials or downward forces (e.g., due to water extraction, sediment compaction, or structural loading from the dam). In this understanding, the high Coulomb stress (87 kPa) could be compressing the subsurface materials, and squeezing out water or air from soil pores or rock fractures. As the subsurface layers compact, the ground above sinks, resulting in subsidence. This can be common in regions where groundwater extraction or changes in water levels (e.g., due to dam operation) affect the mechanical properties of the soil.

In areas like the Grand Ethiopian Renaissance Dam (GERD), the impounding of water in reservoirs or the construction of massive structures exerts significant downward pressure on the ground. This external loading could increase stress in the underlying layers, causing the ground to compact and sink leading to subsidence. However, areas with low Coulomb stress (15–24 kPa), shown in blue and purple at the lower end of the scale, may correspond to regions with relatively small deformation velocities or stable areas with less ground movement. This is noticeable through the eastern embankment of the GERD. This result confirms the work done by Madson et al. [33] and Morsy et al. [34].

The key question now is: can Coulomb stress be used to estimate sedimentation volume by integrating deformation velocity?

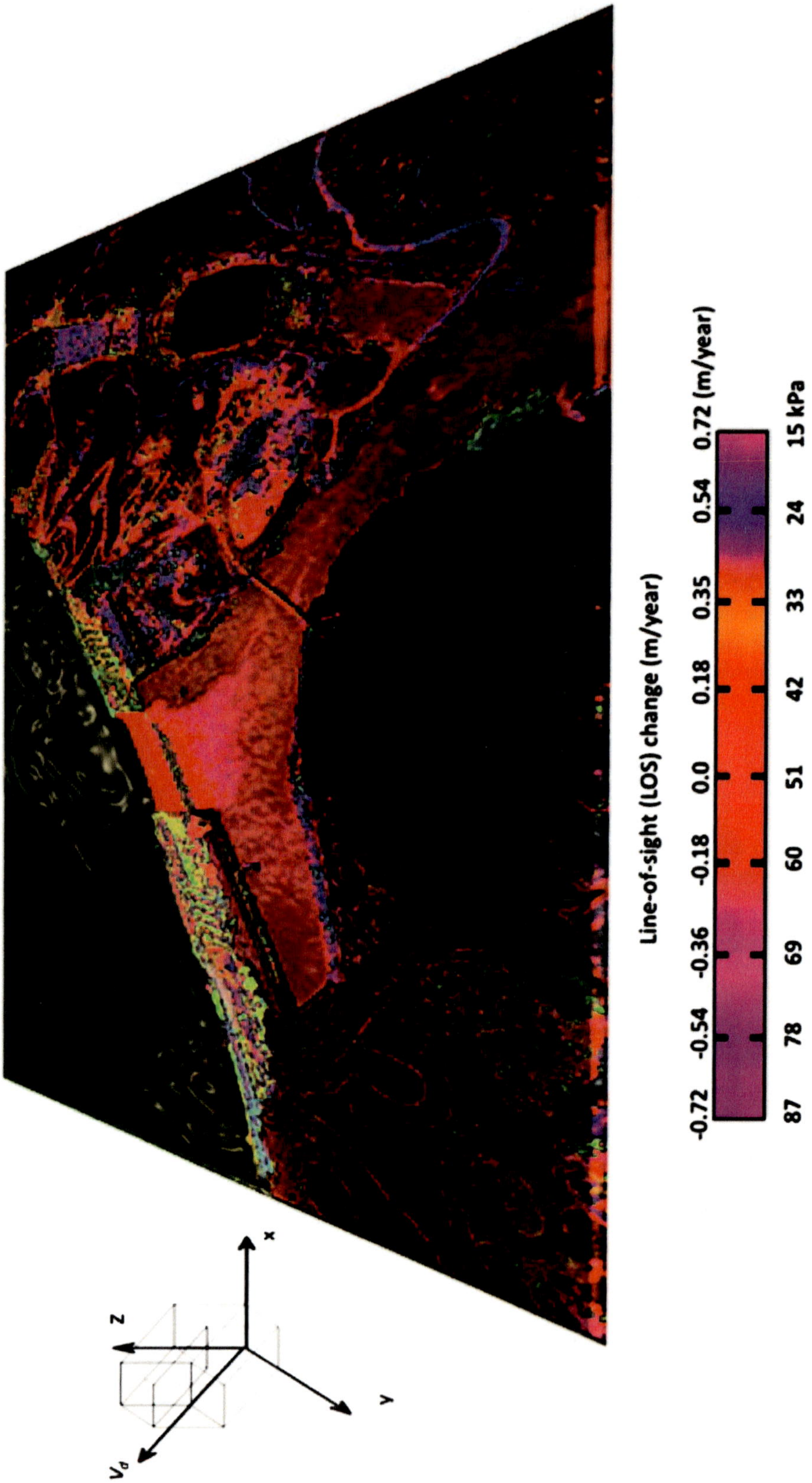

Figure 12.17 4D Holographic-TomoSAR deformation velocity based on Coulomb stress.

12.8 SIMULATION OF SEDIMENTATION VOLUME USING 4D MARGHANY HOLOGRAPHIC TOMOSAR ALGORITHM

To estimate sedimentation volume from 4D Marghany hologram TomoSAR algorithm, we consider several essential parameters: deformation data, stress data and a temporal series. Together, these elements provide a multidimensional view of sediment dynamics over time, allowing us to quantify sedimentation volume more accurately.

First, we denote $S_v(t)$ as the sedimentation volume at any given time t. This parameter represents the total accumulated sediment within the reservoir at a particular time and serves as the primary measure of sedimentation. The accurate calculation of $S_v(t)$ allows us to understand the rate of sediment buildup and its potential implications for reservoir capacity and structural stability.

To derive $S_v(t)$, we consider the vertical deformation in the GERD's reservoir area due to sediment weight, noted as $\Delta h(t)$. This parameter, the change in ground height, directly reflects the amount of sediment that has settled, as heavier sediment loads compress the substrate. Observing $\Delta h(t)$ over time provides a clear metric for tracking sediment accumulation rates. Since deformation height $\Delta h(t)$ represents the height of accumulated sediment, the total volume $V_s(t)$ of sediment is:

$$V_s(t) = A \cdot \Delta h(t). \tag{12.44}$$

In equation 12.44, the area of the reservoir region, denoted by A, allows us to convert changes in height, $\Delta h(t)$, into total volume. Knowing (A) ensures that our calculations capture the total extent of sedimentation rather than localized measurements, making our sediment estimation applicable across the reservoir's entire expanse.

Therefore, the sedimentation volume $S_v(t)$ can also be inferred from Coulomb stress $\sigma_c(t)$, as the accumulated sediment exerts additional pressure on the dam's foundation and reservoir area. Coulomb stress relates to the sediment mass as:

$$\sigma_c(t) = \rho_s \cdot g \cdot \Delta h(t). \tag{12.45}$$

In equation 12.45, a key factor in estimating sediment weight is the sediment density ρ_s, given in kg/m^3. This density value allows us to assess the mass of accumulated sediment, which, in turn, influences the stress exerted on the dam and its foundations. Higher sediment density results in greater vertical stress and has a more pronounced effect on reservoir deformation. Additionally, the gravitational constant g (9.81 m/s^2) is essential in converting sediment mass into weight. This constant provides the force that sediment mass exerts on the ground, impacting both deformation and stress measurements in our model [32].

From this, we can express $\Delta h(t)$ in terms of Coulomb stress:

$$\Delta h(t) = \frac{\sigma_c(t)}{\rho_s \cdot g}. \tag{12.46}$$

Substitute $\Delta h(t)$ back into the equation for $S_v(t)$:

$$V_s(t) = A \cdot \frac{\sigma_c(t)}{\rho_s \cdot g}. \tag{12.47}$$

The total sedimentation volume $S_v(t)$ over time can be found by integrating the volume change rate $\frac{dV_s}{dt}$ over time:

$$S_v(t) = \int_0^t A \cdot \frac{\sigma_c(\tau)}{\rho_s \cdot g} d\tau. \tag{12.48}$$

Equation 12.48 allows us to estimate the sedimentation volume over time based on 4D Marghany hologram TomoSAR, where Coulomb stress and deformation are monitored. Finally,

Coulomb stress $\sigma_c(t)$, as retrieved by 4D Marghany hologram TomoSAR, reflects the stress exerted by the accumulated sediment on the reservoir's structure. Monitoring $\sigma_c(t)$ over time allows us to understand how sedimentation s increases the load on the reservoir foundation. Coulomb stress serves as a direct indicator of potential risks to structural integrity, as higher stress levels correlate with increased pressure due to sedimentation.

Figure 12.18 illustrates the relationship between sedimentation volume (measured in Million Billion Cubic meters, MBC) and Coulomb stress (kPa), showing that as sedimentation volume increases, so does the Coulomb stress on the dam's structure, particularly on the western embankment of the Grand Ethiopian Renaissance Dam (GERD). In this sense, Figure 12.19 demonstrates that at 1 MBC of sedimentation, the Coulomb stress reaches approximately 87 kPa. This high level of stress correlates with the heavy accumulation of sediment behind the western embankment, leading to significant subsidence, as noted in Figure 12.17.

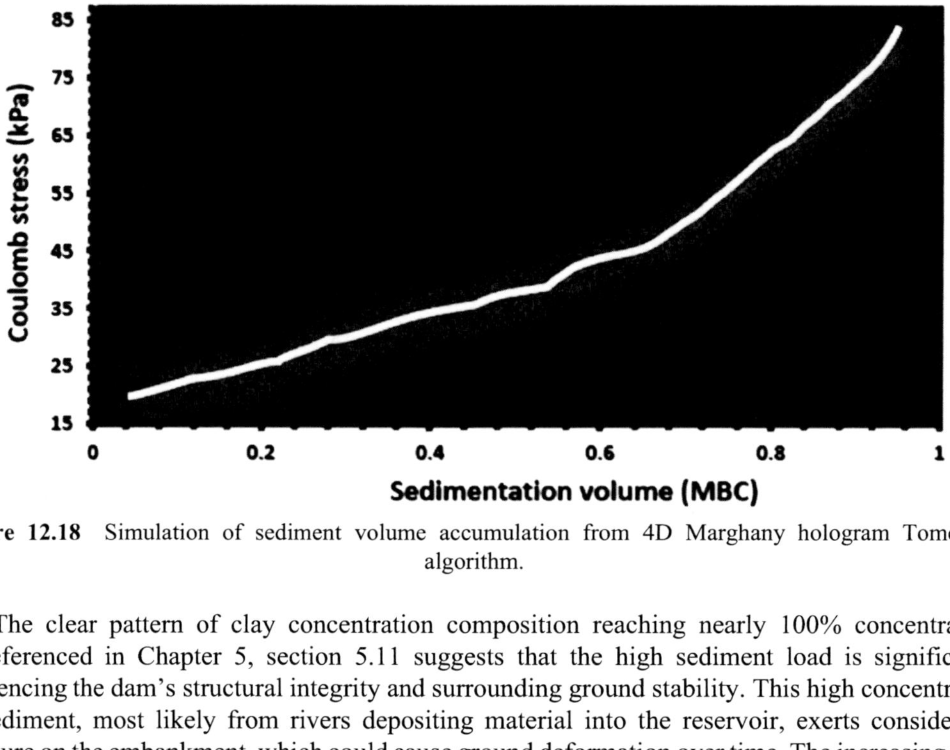

Figure 12.18 Simulation of sediment volume accumulation from 4D Marghany hologram TomoSAR algorithm.

The clear pattern of clay concentration composition reaching nearly 100% concentration, as referenced in Chapter 5, section 5.11 suggests that the high sediment load is significantly influencing the dam's structural integrity and surrounding ground stability. This high concentration of sediment, most likely from rivers depositing material into the reservoir, exerts considerable pressure on the embankment, which could cause ground deformation over time. The increasing trend shown in the graph confirms that as more sediment accumulates (reaching levels close to 1 MBC), Coulomb stress intensifies, potentially compromising the dam's foundation or surrounding areas through subsidence or ground movement. This result is confirmed in Figure 12.19. Figure 12.20 presents a refined 4D Marghany hologram TomoSAR algorithm visualization of sediment volume in GERD's reservoir and saddle dam, linked to Coulomb stress and vertical deformation velocity. This multi-dimensional analysis merges spatial deformation, sediment load impact, and temporal sedimentation trends, providing a holistic view of dam dynamics.

At the core, the 4D structure displays sediment layers over time, with color-coded zones representing Coulomb stress (kPa) and line-of-sight (LOS) deformation velocity (m/year) in azimuth (v_x) and range direction (v_y); respectively (Figures 12.19 and 12.20). The top gradient shows deformation rates: purple and blue for subsidence (negative rates) and orange to violet for uplift (positive rates) due to sediment load or water displacement.

Coulomb stress, ranging from 87 to 15 kPa, corresponds to sediment volume (in BCM), with higher stress levels (dark to light green on the volume bar) reflecting greater sediment accumulation.

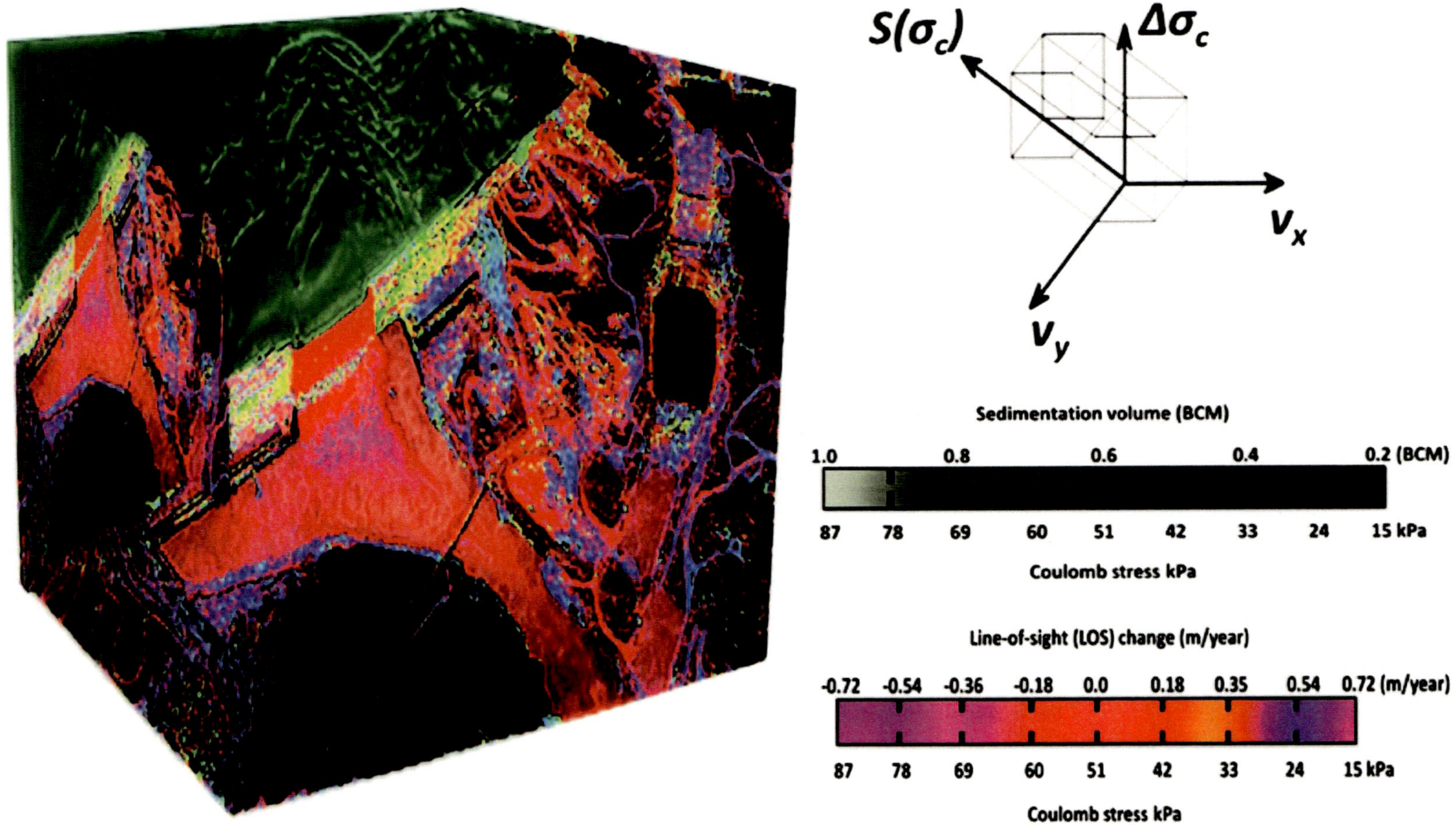

Figure 12.19 4D Marghany hologram TomoSAR algorithm simulation of sediment volume in GERD's reservoir as a function of Coulomb stress and vertical deformation velocity.

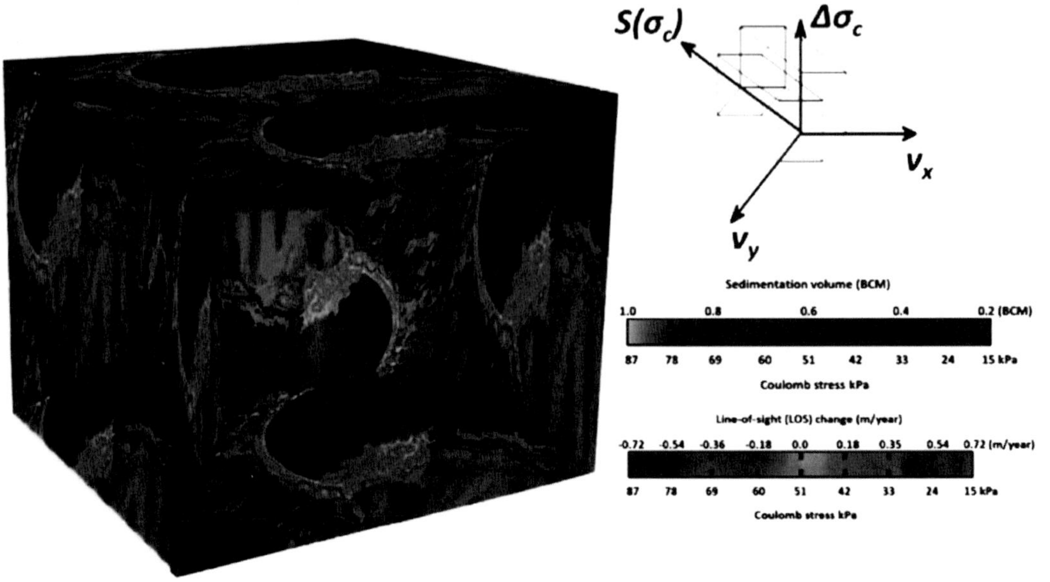

Figure 12.20 4D Marghany hologram TomoSAR algorithm simulation of sediment volume across saddle dam and deformation velocity across its tunnels.

This added sediment weight intensifies vertical pressure on the reservoir floor, affecting the dam's foundation and embankments. By integrating Coulomb stress, LOS deformation velocity, and sediment volume, this model predicts the effect of sedimentation on the stability of the dam. The time series helps pinpoint zones at risk of critical sedimentation and stress levels. This 4D hologram thus represents a pioneering approach to monitoring sedimentation-induced stress in large dams like GERD, advancing sediment management and structural monitoring strategies. Therefore, the high sediment accumulation behind GERD arises mainly from low water discharge, which limits the river's ability to carry sediment downstream as discussed in previous chapters. Slower water flow causes suspended sediment to settle, resulting in significant buildup. High sediment-trapping efficiency in large dams like GERD exacerbates this as sediment from the Blue Nile becomes concentrated in the reservoir. The increased sediment load generates vertical stresses, shown in the figure as Coulomb stress up to 87 kPa, which correlates with sediment volumes near 1 BCM. This stress impacts the dam's western embankment, potentially causing subsidence and structural deformation within –0.72 m/year. Thus, careful management of water release is crucial to control sediment buildup and maintain GERD's structural integrity.

The saddle dam, as analyzed in this study, is experiencing significant land deformation, with vertical shifts of approximately 0.54 meters per year that are producing Coulomb stress levels reaching 78 kPa. Additionally, an uplift of 0.54 meters per year has been observed along the south embankment leeway, generating Coulomb stress up to 24 kPa (Figure 12.20). The area immediately behind the saddle dam reveals active deformation, attributed largely to seepage, which accelerates deformation velocity up to 0.54 m/year.

Sediment accumulation within the reservoir zone facing the saddle dam is substantial, with estimated volumes between 0.2 and 0.75 million cubic meters (BCM), largely concentrated near both the south and north embankment leeway. In contrast to the primary GERD dam, which exhibits more stable land characteristics, the saddle dam undergoes a combination of uplift and subsidence. This dual behavior is influenced by seepage flows emerging from multiple tunnels distributed across the saddle dam's arched structure, as elaborated in Chapter 11.

Building upon the insights from the previous chapters, it is clear that the Grand Ethiopian Renaissance Dam (GERD) and the saddle, face a potentially high risk of breaching at any time due to the accumulation of sediment and associated stress factors. The crucial question now becomes: how can we accurately forecast such an event? In this chapter, the author introduces the innovative 4D Marghany hologram TomoSAR algorithm as a pioneering solution to address this challenge.

The 4D Marghany hologram TomoSAR algorithm approach combines spatial and temporal dimensions to capture real-time deformation data around the dam. This technique provides a supreme sight of changes in the dam's structural integrity by integrating sedimentation volume, Coulomb stress, and vertical deformation velocity. Through this advanced technology, we can visualize stress distributions and deformation patterns in four dimensions, allowing us to predict areas under maximum stress.

The 4D Marghany hologram TomoSAR algorithm enables early detection of high-risk zones within the dam structure and its surrounding foundation by continuously monitoring these parameters. If the sediment load continues to increase, leading to critical Coulomb stress levels, the model can predict when and where the dam is most vulnerable to breach. The real-time, multi-layered data from the 4D Marghany hologram TomoSAR algorithm provides an invaluable tool for decision-makers, allowing for proactive measures to mitigate potential breaches.

Needless to say, the 4D Marghany hologram TomoSAR algorithm represents a breakthrough in forecasting GERD breaching risks. By employing this technique, we gain the predictive power necessary to understand and manage the dynamic forces impacting the dam, ensuring both its structural stability and the safety of downstream regions.

12.9 QUANTUM MACHINE LEARNING FOR FORECASTING DAM BREACHING

To forecast the potential breach of the Grand Ethiopian Renaissance Dam (GERD), a sophisticated quantum machine learning (QML) algorithm must be developed, merging the computational power of quantum computing with advanced machine learning techniques. This approach leverages the unique capabilities of quantum algorithms to process vast, complex datasets far more efficiently than classical methods, enabling us to analyze and predict structural stresses and deformations with unprecedented accuracy. By harnessing the power of QML, we can model the forces impacting GERD and forecast a precise timeline for potential failure, enhancing our predictive power beyond what was previously achievable. In this view, the key objective is to forecast the potential breach of the Grand Ethiopian Renaissance Dam (GERD) by analyzing time-series data encompassing deformation, stress, sedimentation, and environmental geological dynamics as addressed in previous chapters.

Therefore, the core challenge is to accurately project the timeline to a potential breach by identifying and analyzing the complex, nonlinear interactions among these variables over time. Using quantum machine learning (QML), this challenge can be addressed with an advanced algorithm capable of processing intricate dependencies and high-dimensional data, ultimately enhancing our predictive accuracy and resilience modeling for GERD.

12.9.1 Quantum Data Encoding

In quantum computing; therefore, a crucial step is encoding data into quantum states. This enables us to leverage the power of quantum algorithms for complex predictive modeling. To forecast potential breach scenarios for GERD, we begin by representing our dataset—comprising stress, deformation, environmental, and temporal information—as qubits. Given the complexity and scale of this data, we adopt a hybrid quantum-classical approach, combining quantum processing for high-dimensional features with classical computing for efficiency and resource optimization. Our encoding strategy involves two key methods:

Amplitude encoding: To handle high-dimensional data like spatial coordinates, stress values, and deformation metrics we use amplitude encoding. This method allows us to embed each feature of the input data (e.g., stress and deformation) into the amplitude of a quantum state. This encoding is highly efficient for complex, multi-dimensional data, maximizing the quantum system's capacity for data representation.

Basis encoding: Basis encoding is employed for categorical or binary features, such as binary states representing water discharge rates (e.g., high/low) or sedimentation levels (e.g., high/low). This encoding map presents the discrete features of specific quantum states, preserving their unique information while simplifying the representation.

For each time snapshot (t), we construct a quantum state $|\Psi(t)\rangle$ that captures a composite view of the reservoir's spatial, stress and deformation data at that instant. This quantum state effectively encodes the combined structural and environmental factors influencing dam stability, providing a foundational representation for subsequent analysis. By preparing these time-specific quantum states, we can then apply quantum machine learning algorithms to identify patterns and predict future breach scenarios, ensuring a robust, scalable approach to dam risk forecasting. In other words, for each time t, the data is encoded into a quantum state $|\psi(t)\rangle$ [35–37]:

$$|\psi(t)\rangle = |\psi_{\text{spatial}}(t)\rangle \otimes |\psi_{\text{stress}}(t)\rangle \otimes |\psi_{\text{deformation}}(t)\rangle. \tag{12.49}$$

This state represents the spatial deformation, stress levels, and other parameters at time t.

12.9.2 Defining the Quantum Machine Learning Model

Given that forecasting is a time-series problem, we can structure our model as a Quantum Recurrent Neural Network (QRNN) or Quantum Long Short-Term Memory (QLSTM) network. These architectures allow the model to retain and learn from past states, which is essential for forecasting. To harness quantum computing for predictive modeling of potential breaches in the GERD, we employ a Quantum Neural Network (QNN) architecture that combines the processing power of quantum circuits with machine learning principles. This QNN model is specifically designed to capture complex, time-dependent interactions between stress, deformation, and environmental factors, utilizing the strengths of quantum computing in both spatial and temporal data processing [35, 38].

Input layer: At each time step, we begin by preparing quantum states from 4-D hologram-TomoSAR data. This input layer encodes spatial coordinates, stress, and deformation values into qubits, enabling the QNN to leverage quantum data encoding for high-dimensional inputs. Each quantum state at time t represents a snapshot of the reservoir's structural conditions, allowing for a comprehensive, multi-dimensional data representation.

Parameterized Quantum Circuit (PQC): A central component of the QNN is the Parameterized Quantum Circuit (PQC), which acts as the quantum layer. The PQC is composed of quantum gates parameterized by variables that are adjusted during training, making it analogous to the weights in classical neural networks. For instance, a rotation gate $R_y(\theta)$ with parameter θ can represent changes in stress or deformation (Figure 12.21). As the model learns, the parameters in these gates are updated to capture the evolving patterns of structural stress and deformation.
At each layer l in the QNN, the transformation can be expressed as:

$$U_l(\theta_l) = U_{\text{stress}}(\theta_l) \cdot U_{\text{deformation}}(\theta_l), \tag{12.50}$$

where θ_l are the tunable parameters that adjust based on the interaction between deformation and stress. This layered transformation allows the QNN to capture complex interdependencies between structural stress and sediment-induced deformation [35–38].

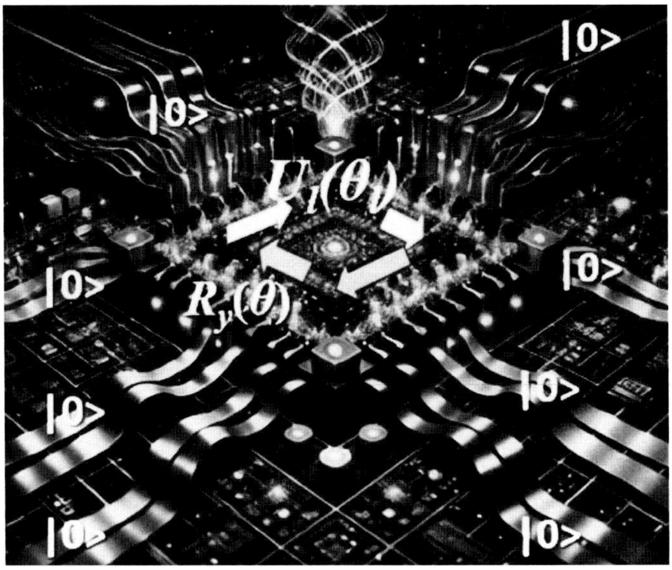

Figure 12.21 Architecture structure of PQC.

Quantum Recurrent Neural Network (QRNN) Structure: To manage the temporal dependencies in dam breach prediction, we implement a Quantum Recurrent Neural Network (QRNN). In the QRNN, each time step is processed iteratively within the quantum circuit, mimicking the behavior of recurrent networks in classical machine learning. This setup allows the QRNN to update parameters based on past states, capturing temporal patterns in stress and deformation. As a result, the QRNN can predict future states by leveraging historical data, identifying potential points of failure and forecasting when stress and deformation could approach critical levels.

12.9.3 Cost Function and Training Process

The cost function measures the discrepancy between predicted and observed stress levels at each time step, which is minimized during training. Let's define a loss function L(θ) as:

$$L(\theta) = \sum_t \| \langle \psi_{\text{predicted}}(t) | | \psi_{\text{observed}}(t) \rangle - \langle \psi_{\text{predicted}}(t+1) | | \psi_{\text{observed}}(t+1) \rangle \|^2 . \quad (12.51)$$

This function calculates the difference between predicted and actual states at each time step t.

12.9.4 Quantum Optimizer

To incorporate a quantum optimizer, such as the Quantum Approximate Optimization Algorithm (QAOA) or the Variational Quantum Eigensolver (VQE), we can define a cost function that captures the difference between predicted and actual values of structural stress or deformation. The goal is to iteratively adjust the parameters θ to minimize this cost function. In this view, the cost function $C(\theta)$ can be represented as the expectation value of an operator $\widehat{H}$ (often called the Hamiltonian in quantum mechanics), which represents the total 'energy' or error of the system. Therefore, the key objective is to minimize this function:

$$C(\theta) = \langle \psi(\theta) | \widehat{H} | \psi(\theta) \rangle \quad (12.52)$$

where $|\langle \psi(\theta)|$ and $|\psi(\theta)\rangle$ are the quantum states parameterized by θ, which represent the qubits and their interactions in the quantum circuit. Additionally, $\widehat{H}$ is the Hamiltonian or an operator that encodes the cost (or error) to be minimized. This could represent factors such as stress discrepancies or deformation prediction errors. In this understanding, using QAOA or VQE, the parameters θ are iteratively adjusted to minimize $C(\theta)$. This optimization can be expressed as:

$$\theta^* = \arg\min_{\theta} C(\theta) \tag{12.53}$$

where θ^* represents the optimal values of the parameters after minimization. Consequently, the optimizer updates θ iteratively in each step k as follows:

$$\theta^{(k+1)} = \theta^{(k)} - \eta \frac{\partial C(\theta)}{\partial \theta^{(k)}} \tag{12.54}$$

where η is the learning rate or step size for each update, and $\frac{\partial C(\theta)}{\partial \theta^{(k)}}$ is the gradient of the cost function concerning the parameters θ at step k. This iterative approach allows the quantum optimizer to reach the optimal set of parameters that minimize the predictive error in the QML model for forecasting structural changes in the dam [38–40].

12.9.5 Forecasting Future States

Once trained, the model can be used to predict the dam's response in future scenarios. By feeding in recent 4D hologram-TomoSAR data, we can predict stress levels at future times T, which is essential for forecasting breaches. To formalize this breaching threshold prediction in mathematical terms, let us define each part of the statement in equations. The breaching time T_{breach} is the earliest time T at which the predicted Coulomb stress exceeds a predefined critical stress threshold σ_{critical}, based on engineering assessments:

$$T_{\text{breach}} = \min\{T : \text{Coulomb stress}(T) \geq \sigma_{\text{critical}}\} \tag{12.55}$$

From the point of view of the above-mentioned perspective, the model generates a probability distribution $P(\sigma_c, T)$ for the predicted Coulomb stress σ_c over time T. This probability distribution helps to estimate the likelihood that the stress will reach or exceed the critical threshold at any given time.

$$P_{\text{breach}}(T) = \int_{\sigma_{\text{critical}}}^{\infty} P(\sigma_c, T)\, d\sigma_c \tag{12.56}$$

After obtaining quantum predictions for stress over time, we apply classical post-processing to interpret the results. This involves mapping quantum measurement outputs into a form that provides actionable insights. Specifically, the quantum outputs are used to estimate the likelihood of a breach within a specific timeframe [39–41].

By integrating the probability distribution $P(\sigma_c, T)$ overstress values exceeding σ_{critical}, we calculate the probability of a breach occurring within a specified time interval $[0, T_{\max}]$ (Figure 12.22):

$$P[0, T_{\max}] = \int_0^{T_{\max}} \int_{\sigma_{\text{critical}}}^{\infty} P(\sigma, T)\, d\sigma\, dT \tag{12.57}$$

In this equation $T_{\max}$ is the maximum time horizon under consideration and σ_{critical} is the critical stress threshold above which a breach risk is flagged. Additionally, $P(\sigma, T)$ demonstrates the probability density function for stress σ at time T, describing the likelihood of specific stress values over time. In this view, the double integral in equation 12.57 is a way of quantifying risk over time, taking into account both:

1. How likely the stress is to exceed the critical threshold (for each time).
2. The accumulation of this risk over a specified period.

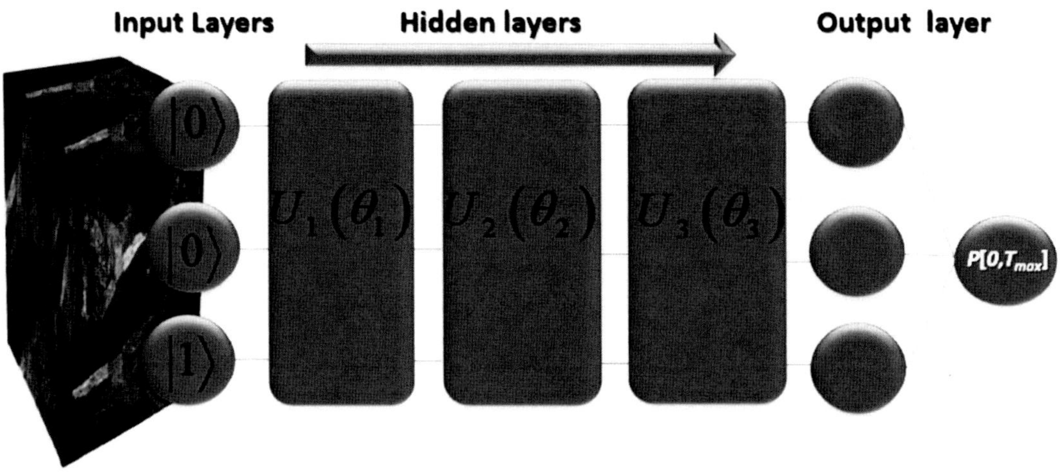

Figure 12.22 The architecture of quantum machine learning for the probability of a breach occurring.

In practical terms, this calculation can help engineers and scientists predict the likelihood of a dam breach or other structural failure due to stress over a given time [40, 42]. By evaluating this probability, stakeholders can make informed decisions about inspection schedules, maintenance, and other risk management strategies to prevent catastrophic failures.

12.10 HAARP TRIGGERS GERD BREACH

Before assessing conventional stress factors in the scenario of a breach at the Grand Ethiopian Renaissance Dam (GERD)—factors such as deformation stress, sediment volume, and environmental dynamics—another potentially alarming force warrants consideration: HAARP, the High-Frequency Active Auroral Research Program. Developed initially as a research tool, HAARP has garnered attention for its purported capability to influence geological and atmospheric conditions. While this remains theoretical, some suggest it could contribute to risks for GERD through induced seismic activity or climatic manipulation.

12.10.1 Influence of Potential Pathways of HAARP on GERD

Two mechanisms are speculated to be within HAARP's reach regarding the GERD: seismic stimulation and climate manipulation. Each mechanism involves complex processes that could, in theory, impact the dam's structural integrity.

The HAARP system could theoretically impact the Grand Ethiopian Renaissance Dam (GERD) through two distinct mechanisms. First, by generating seismic disturbances, HAARP's high-frequency electromagnetic waves could potentially penetrate the Earth's surface up to a depth of approximately 10 kilometers, focusing energy on geological fault lines. If successful, this concentration of energy could trigger localized seismic activity in Ethiopia, potentially compromising the dam's structural integrity [43].

Second, HAARP could influence atmospheric and oceanic conditions to intensify a positive phase of the Indian Ocean Dipole (IOD). A positive IOD typically brings heavier rainfall to East Africa, including Ethiopia. Through this mechanism, HAARP could, in theory, exacerbate persistent and extreme flooding in the region. The resulting continual water pressure and repeated flood events could eventually strain the GERD, increasing the risk of a breach.

12.10.2 Seismic Induction by Electromagnetic Waves

From a perspective grounded in quantum earth tomography and quantum mathematics, the hypothesis that high-frequency electromagnetic waves like those generated by HAARP could influence seismic activity offers an intriguing basis for exploring how energy interactions at a quantum scale might affect geophysical structures. Quantum earth tomography, as an advanced method of subsurface imaging, seeks to understand the Earth's interior by examining wave behavior through a quantum lens. This approach leverages quantum principles, such as wave-particle duality and energy dispersion across entangled states, to provide more precise imaging of fault lines and tectonic dynamics, which could theoretically be susceptible to certain external electromagnetic influences [44–46].

This statement can be described mathematically from the point of view of quantum mechanics. In this perspective, the time-dependent Schrödinger equation can be correlated with the quantum probability of a breach in the GERD due to seismic disturbances. Let us focus on the fault line's energy propagation and potential stress accumulation within regions critical to the dam's structure [43]. The goal is to connect energy buildup in the fault with the probability of a consequential structural event like a breach. In this view, the time-dependent Schrödinger equation:

$$i\ddot{\mathbf{y}}\frac{\partial \Psi(x,t)}{\partial t} = \left(-\frac{\ddot{\mathbf{y}}^2}{2m}\nabla^2 + V(x)\right)\Psi(x,t) \tag{12.58}$$

Equation 12.58 describes how the wave function $\Psi(x, t)$, representing the probability amplitude of energy distribution within the fault, evolves. In this view, $\Psi(x, t)$ is the probability amplitude of finding a particular energy density at position x along the fault line at time t. Therefore, $V(x)$ is a potential energy landscape within the fault, which may vary with geological features (e.g., rock density, fault structure); $\hbar$ is reduced Planck's constant ($1.0545718 \times 10^{-34}$ Js); and m is an effective mass parameter, representing the geological medium's resistance to energy flow. In this understanding, equation 12.58 provides insights into how energy (from electromagnetic induction or natural seismic activity) propagates through the fault, identifying zones of energy accumulation where stress could reach a threshold critical for structural integrity.

To determine the likelihood of energy accumulation reaching the critical stress threshold necessary for a breach, we can use the probability density function $|\Psi(x, t)|^2$ [42] obtained from the wave function:

$$P(x,t) = |\Psi(x,t)|^2 \tag{12.59}$$

Here, $P(x, t)$ represents the probability density of energy concentration at position x at time t. If we identify a specific point or region $x = x_0$ that directly affects GERD's structural stability, then a high value of $P(x_0, t)$ over time suggests that energy is increasingly concentrated in this region. This accumulation can contribute to stress exceeding the structural limit, thus indicating a potential breach risk. As waves penetrate the crust, this energy density could, theoretically, influence stresses if concentrated along tectonic faults. The energy density over time can be associated with localized stresses within the fault line using an energy-to-stress relation for faults. This can be approximated in a tectonic fault as:

$$\sum = \kappa\xi\int_0^T P(x,t)\,dt \tag{12.60}$$

This term Σ represents the total stress accumulated in a fault over a specified period. It is crucial to understand how stress builds up in geological formations or materials, particularly in predicting events like earthquakes. Therefore, κ represents a constant that relates the stress to the displacement in the material or system. It could be a material property such as stiffness or a proportionality constant derived from the system's physical characteristics. Consequently, ξ serves as a scaling factor that relates the probability density to the displacement. It essentially modifies how the probability density affects the cumulative stress. Additionally, $P(x, t)$ is the probability density function, given by $P(x, t) = |\Psi(x, t)|^2$, where $\Psi(x, t)$ is the wave function of the system. It represents the likelihood of finding a particle or a certain state at position x and time t. In a practical context, if the probability density $P(x, t)$ is high for a significant duration, it implies that the system is more likely to be in a state that contributes to greater displacement, leading to increased cumulative fault stress. Conversely, low probability densities would correlate with lower stress accumulation [43, 45].

If Σ exceeds a critical stress threshold σ_{critical}, which is the stress level needed to initiate seismic slippage, a seismic event could be triggered:

$$\sum \geq \sigma_{\text{crit}} \tag{12.61}$$

This threshold depends on geological characteristics, such as the friction coefficient of rocks in the fault and the local tectonic loading rate. The probability density of energy accumulation at any given point along the fault line can be addressed using equation 12.60. Therefore, in regions of high probability density, one can infer areas where stress is more likely to concentrate, increasing the risk of a seismic event. Cumulative high probability densities in structurally weak areas (like those beneath GERD) could theoretically lead to micro-seismic activity that destabilizes the foundation over time.

Needless to say, HAARP operates with high-frequency electromagnetic waves, which it directs into the ionosphere. These waves, some theorize, might extend several kilometers underground, potentially concentrating energy along tectonic fault lines in regions such as Ethiopia. If such focused energy could indeed influence geological stresses, HAARP could hypothetically trigger seismic disturbances. For the GERD, which is positioned in a tectonically active area, even minor seismic activity could weaken foundational structures over time, accelerating deformation and heightening the risk of breach under immense water pressure [44].

12.10.3 Climate Amplification Through Indian Ocean Dipole (IOD) Dynamics

The second pathway involves HAARP's potential influence on climatic patterns. The Indian Ocean Dipole (IOD), a climate phenomenon characterized by the sea surface temperature differences between the western and eastern Indian Ocean, significantly influences the weather across East Africa and South Asia. During a positive IOD phase, warmer temperatures in the western Indian Ocean create moisture-laden air currents that lead to increased rainfall in the East African region, including Ethiopia.

If HAARP were to manipulate IOD-related atmospheric dynamics by heating parts of the ionosphere or modulating electromagnetic frequencies, it might be able to contribute to sustaining a positive IOD. This theoretical positive IOD intensification would bring heavier and prolonged rainfall to the Ethiopian highlands. Given the current engineering capacities and design limitations of the GERD, an unprecedented surge in floodwaters could stress the dam, especially if heavy flooding becomes continuous over time. The elevated water pressure, combined with frequent flooding, could weaken the dam's structures, thus increasing the potential for overflow or structural breach.

The relationship between heavy flooding in Ethiopia and the Indian Ocean Dipole (IOD) in 2024 can be explained by the way the IOD influences rainfall patterns in East Africa and other parts of the Indian Ocean basin. Let us break down this connection, focusing on how a positive phase of the IOD typically affects Ethiopia and why it can lead to intensified rainfall and potential flooding.

The Indian Ocean Dipole (IOD) is a climate pattern characterized by fluctuations in sea surface temperatures between the western and eastern regions of the Indian Ocean. It has two primary phases:

Positive IOD: Warmer-than-average sea surface temperatures in the western Indian Ocean (near East Africa) and cooler-than-average temperatures in the eastern Indian Ocean (near Indonesia and Australia). This phase typically results in increased rainfall in East Africa, including Ethiopia, as warm waters drive moist air towards the region and drier conditions in Indonesia and Australia.

Negative IOD: Cooler-than-average sea surface temperatures in the western Indian Ocean and warmer-than-average temperatures in the eastern Indian Ocean, resulting in drier conditions in East Africa, including Ethiopia, and increased rainfall in Indonesia and Australia. On the other hand, a neutral IOD means there is no significant difference in sea temperatures between the two regions.

In 2024, meteorologists observed a strong positive IOD event. The warmer waters in the western Indian Ocean increase the likelihood of evaporation and moisture accumulation in the atmosphere over that region. This moisture-laden air is then transported toward East Africa by prevailing winds. When this air reaches Ethiopia and other East African countries, it results in heavy rainfall.

Therefore, positive IOD phases are known to bring about above-average rainfall in East Africa. Ethiopia, in particular, can experience intense rainfall during the IOD's positive phase, especially from June to September, which corresponds with the country's rainy season (known locally as kiremt). The combination of the seasonal rains and the influence of a strong positive IOD can lead to extreme rainfall, overwhelming river systems, and causing floods.

Ethiopia's mountainous terrain, combined with a network of rivers flowing from the highlands to lower regions, exacerbates the flooding risk. When heavy rains fall over these highlands, the runoff quickly moves into rivers like the Blue Nile and Omo, which then overflow, affecting downstream communities and infrastructure. This situation worsens when the ground is already saturated from previous rains.

While the IOD plays a significant role in influencing East African rainfall, other climate phenomena like the El Niño–Southern Oscillation (ENSO) also impact the region. El Niño conditions, for instance, can sometimes enhance the effects of a positive IOD, leading to even more pronounced rainfall in East Africa. In 2024, with a positive IOD potentially coinciding with an El Niño event, the combined effect could explain the severity of the rainfall and subsequent flooding in Ethiopia.

The High-Frequency Active Auroral Research Program (HAARP) primarily operates as a research facility designed to study the ionosphere, the layer of the atmosphere that affects radio waves. Theories suggesting that HAARP could be linked to massive floods or even the Indian Ocean Dipole (IOD) often stem from speculative interpretations of its capabilities rather than scientific evidence [46–49].

12.11 4D MARGHANY HOLOGRAPHIC-TOMOSAR ALGORITHM FOR FORECASTING GERD BREACHING

The QML involves the outflow discharge as calculated by the orifice equation which is mathematically expressed by:

$$Q = C_d A_b \sqrt{2gh} \tag{12.62}$$

where, Q is outflow discharge, m^3s^{-1}; C_d is the discharge coefficient which is equivalent to 0.5; A_b is the breaching area. Additionally, g is an acceleration from gravity, 9.81 ms^{-2}, and h is the centerline head. Therefore, there are several types of boundaries depending on the location of the cross-section, from these types the flow hydrograph boundary was selected for the upstream boundary condition.

Another type is normal depth, usually selected for the downstream boundary condition. In this context, QML was also utilized to calculate the flow hydrograph using the reservoir routing method, predicting changes in outflow caused by a breach. The reservoir routing equation is:

$$\Delta S = \Delta t\left(\sum_{i=1}^{N} I - \sum_{i=1}^{N} Q\right) \tag{12.63}$$

where ΔS is the storage change (m^3); I is the inflow; and Q is the discharge over the changeable time of Δt [50].

To accurately simulate the wave height following the breach, we can conceptualize the flow as a rapidly advancing flood wave. The significant discharge values and the brief duration of the event lead to considerable variations in wave height, influenced by both the flow rate and downstream resistance. Initially, GERD began breaching by collapsing the western embankment within around 11 sec which leads to the overtopping wave height ranging from 210 to 230 m and increases towards the eastern embankment with wave width of 1.2 km (Figure 12.23).

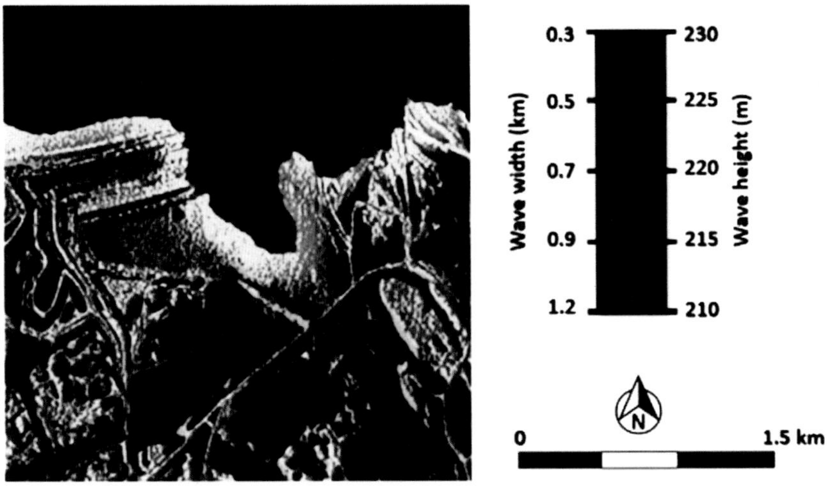

Figure 12.23 Initial GERD breaching within 11 seconds.

Recent findings from QML indicate that the duration of the 4D Marghany holographic-TomoSAR Algorithm in monitoring the GERD breaching spans between 11.5 and 15.5 seconds. This initial phase was marked by the collapse of the western bank of the GERD, resulting in water discharges reaching an astonishing 1,800,000 m^3s^{-1}, which generated a maximum wave height of 230 meters (Figure 12.24). As this wave propagates, it travels over a slope of 3.5% across a downstream distance of 900 m, achieving a peak velocity of 89 ms^{-1}. The implications of such a massive discharge are profound; it is evident that this unprecedented volume of water poses an existential threat not only to the GERD structure itself but also to the surrounding infrastructure.

In examining the dynamics of flood waves resulting from dam breaches, it becomes essential to recognize how various factors interplay to influence wave behavior. The initial characterization of flow as a rapidly moving flood wave is a foundational concept, allowing us to understand the subsequent phenomena that unfold during such catastrophic events.

The relationship between discharge values and wave height is particularly noteworthy. As we observe in this case, higher discharges correlate with increased wave heights, while downstream resistance further complicates this relationship by introducing additional variables that can dampen or amplify wave effects. This complexity is underscored by QML's data on the duration of the breach event, which highlights that even within a brief timeframe—11.5 to 15.5 seconds—

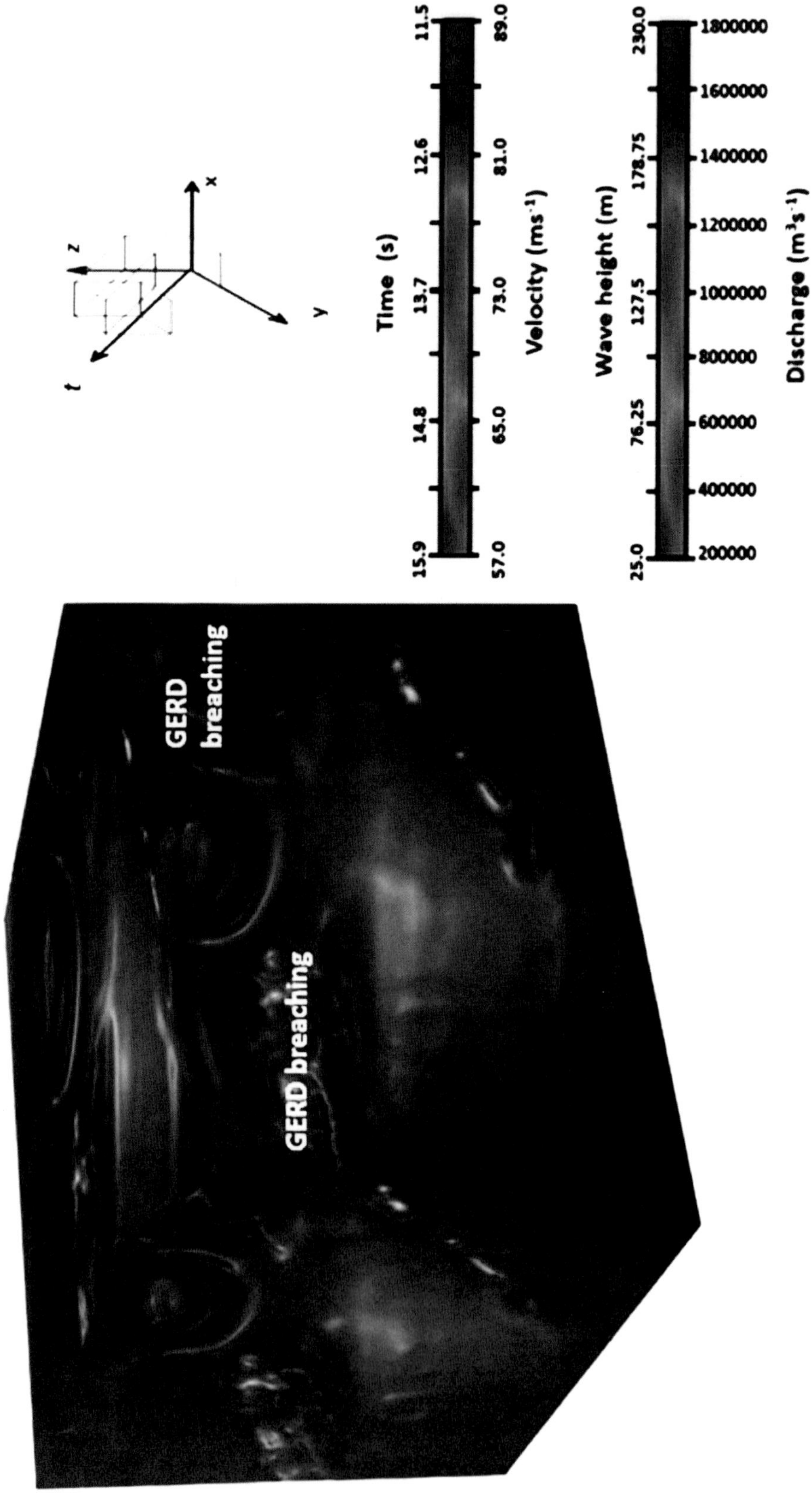

Figure 12.24 4D Marghany hologram-TomoSAR algorithm for forecasting of GERD breaching.

significant hydrodynamic changes occur. This finding approximately agrees with the work done by Tingsanchali and Chinnarasri, [51]. Moreover, the specifics of this incident illustrate how structural failures can initiate dramatic hydrological responses. The collapse of the western bank not only precipitated an immediate release of water but also set into motion a series of cascading effects characterized by rapid discharge rates and extreme wave heights [51–53].

As we draw comparisons between the initial conditions and their consequences, it becomes clear that such immense discharge volumes are not merely numbers; they represent real threats to both natural and built environments. The maximum wave height recorded at 230 meters and its rapid propagation underscore a critical risk: these forces can obliterate not only dam constructions like the GERD but also any adjacent infrastructure in their path. Additionally, the wave width around the Nile ranges from 15 to 50 km at the part between GERD and the significant part of the water spread out of the Nile River.

Within 22.27 minutes of a breach at the Grand Ethiopian Renaissance Dam (GERD), the Roseries Dam will be overwhelmed by a surge reaching a height of 510 meters above mean sea level (MSL), given the dam's initial water level of 490 meters. This would result in an unprecedented water discharge rate of 37,948,919 m^3s^{-1}, accelerating downstream at a velocity of 5 ms^{-1} across a maximum slope of 17.6% (see Figure 12.25). Following this initial impact, the water released from Roseries Dam would reach Sennar Dam in approximately 22 hours with a sustained discharge of 16,000,321 m^3s^{-1}. This timeline underscores the rapid propagation of the floodwaters from the compromised GERD structure to Sennar Dam over nearly a day.

Upon reaching Sennar Dam, the massive incoming discharge would elevate the water level to an estimated 225 meters MSL (Figure 12.26), far exceeding standard operational thresholds. Such an influx suggests an imminent risk of breaching or overflowing at Sennar Dam, as the incoming volume would challenge the dam's structural integrity and capacity limits. This sequence of events vividly illustrates the cascading impacts of a potential GERD failure on downstream infrastructure and highlights the urgent need for comprehensive risk assessments and emergency response planning.

To predict the potential downstream impact of a Sennar Dam breach, calculations indicate a water height of approximately 22 m would reach Khartoum within about three days, with a discharge rate of 280,000 m^3s^{-1} (see Figure 12.27). However, the estimated water level of 225 m above mean sea level (mMSL) at Sennar Dam is unsustainable, suggesting that such a height would inevitably lead to a breach or overflow. In real-world conditions, this volume would dissipate downstream, resulting in lower heights as it propagates. Assuming a 30% reduction in wave propagation due to natural dissipation, the wave height impacting Khartoum is expected to be approximately 20 m. This scenario would cause extensive flooding, with water inundating vast areas in and around Khartoum. The floodwaters would potentially reverse direction toward the White Nile, spreading over a distance of 300 kilometers. Between Sennar Dam and Khartoum, the floodwaters are anticipated to disperse across a 200-kilometer-wide area.

Furthermore, as the floodwater continues its course, it would take approximately two weeks to reach areas between Khartoum and Merowe Dam, expanding over a 25-kilometer-wide swath. The flood would then likely arrive at Merowe Dam within seven days, bringing significant risks to infrastructure, homes and lives along its path. Fortunately, with an estimated ten days for the flood to reach downstream areas, there would be a window of time for evacuation measures to be taken, potentially saving many lives.

As the floodwaters advance downstream from the Sennar Dam toward the Merowe Dam, they will undergo natural reductions in height due to frictional forces, dispersion, and other environmental factors. The initial flood height, calculated at an unsustainable 22 m, is expected to diminish significantly across this distance. Accounting for dissipation, it is reasonable to project that the water level upon reaching Merowe Dam will stabilize around the dam's structural maximum, which is approximately 300 m. Notably, while the Merowe Dam may prevent an overflow, it

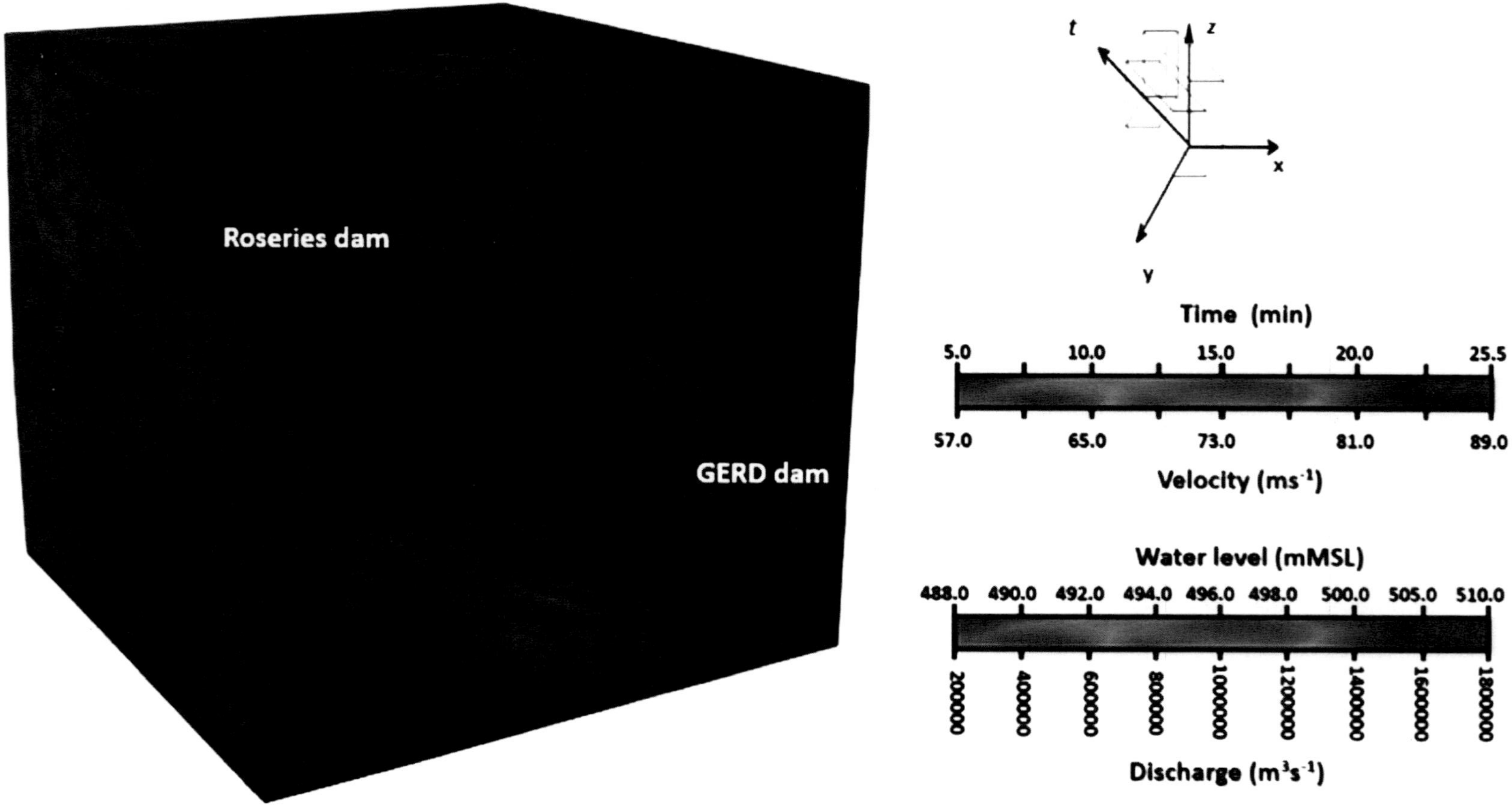

Figure 12.25 4D Marghany Hologram-TomoSAR algorithm for forecasting flood wave propagation from GERD to Roseries Dam post dam breaching.

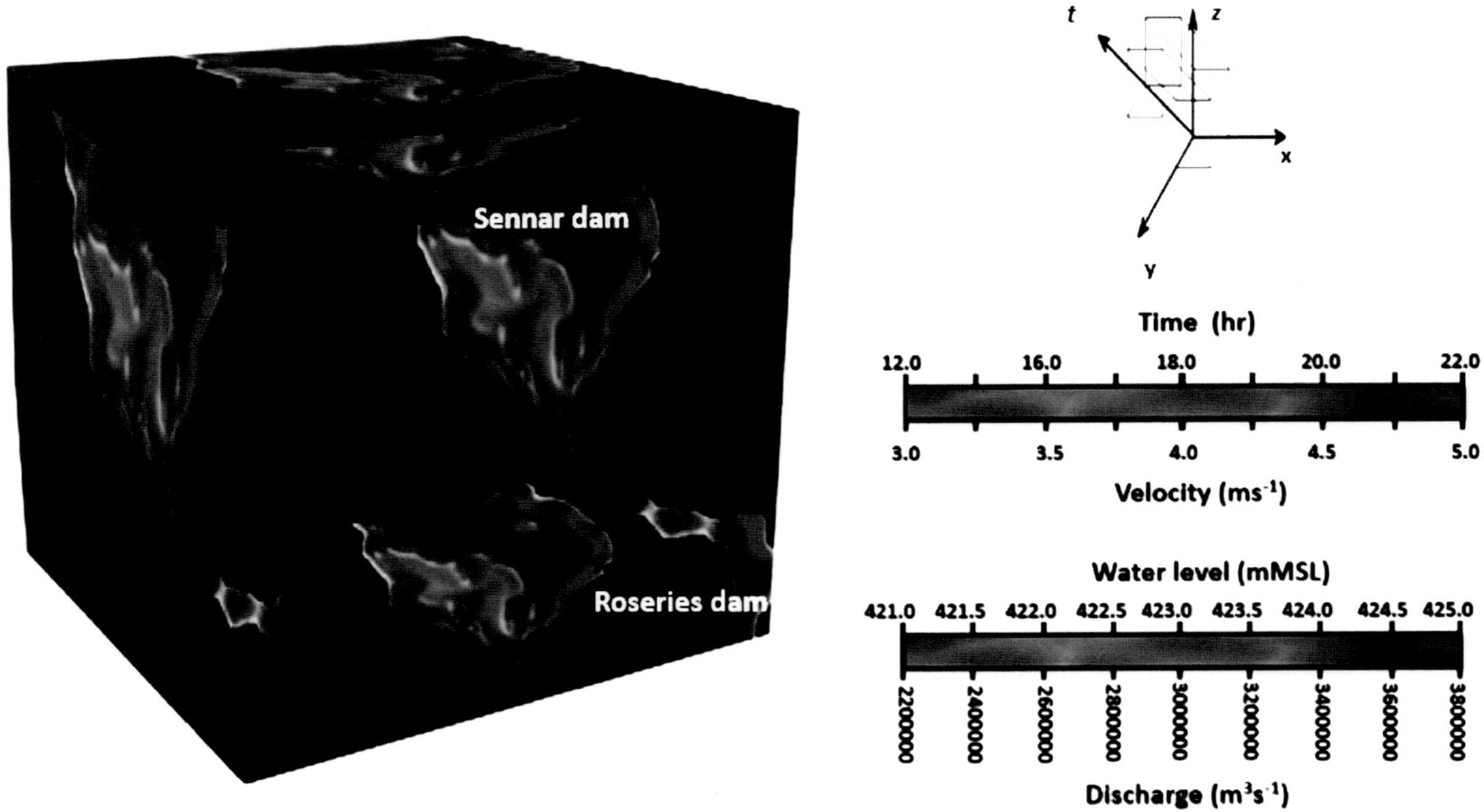

Figure 12.26 4D Marghany Hologram-TomoSAR algorithm for forecasting flood wave propagation from Roseries Dam to Sennar.

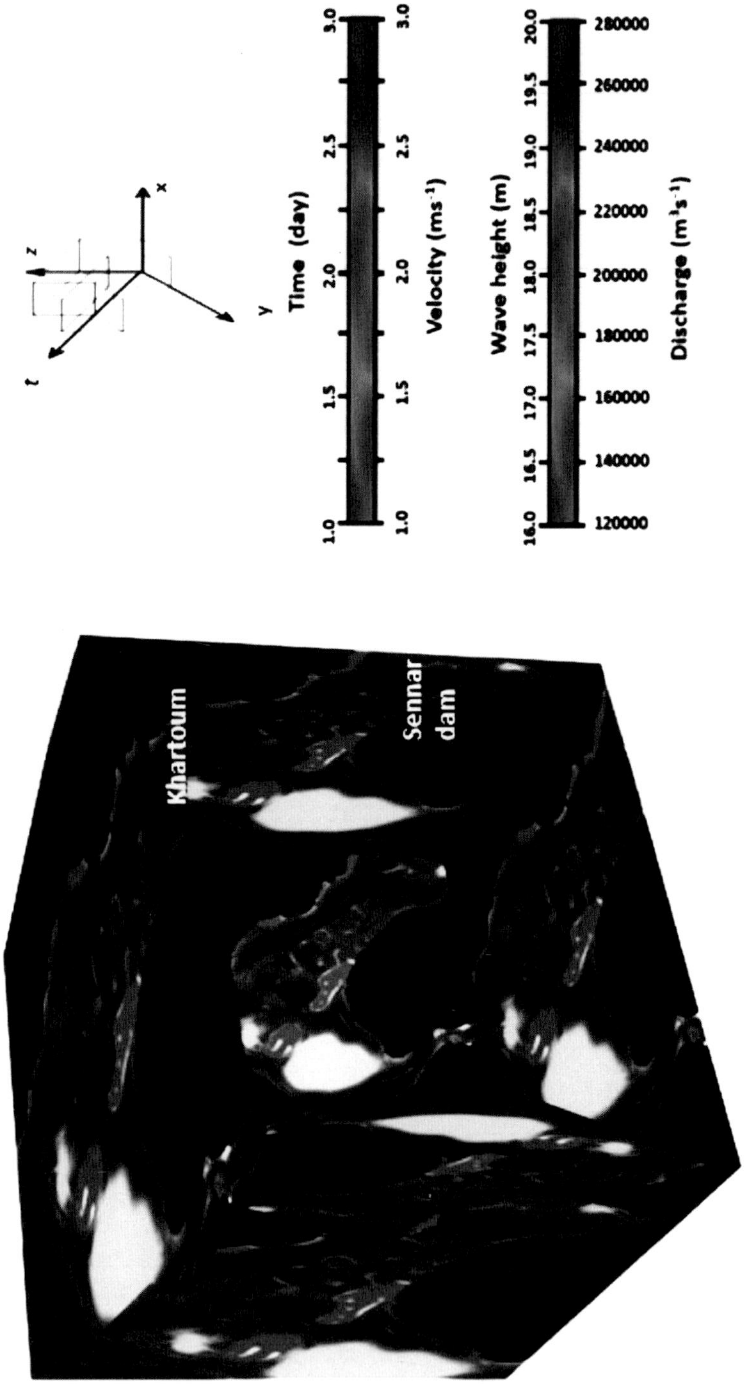

Figure 12.27 4D Marghany Hologram-TomoSAR algorithm for forecasting flood wave propagation from the Sennar Dam to Khartoum City.

will be exposed to a substantial discharge volume, calculated at 3,400,000 m^3s^{-1}, anticipated to arrive within five days following the inundation of Khartoum City, which would have faced water heights of 22 m. Based on these observations and factoring in a hypothetical 20% reduction in energy, the water propagation velocity is estimated at 2.5 ms^{-1}. This model highlights the immense forces in play and underscores the challenges posed by such a deluge (see Figure 12.28).

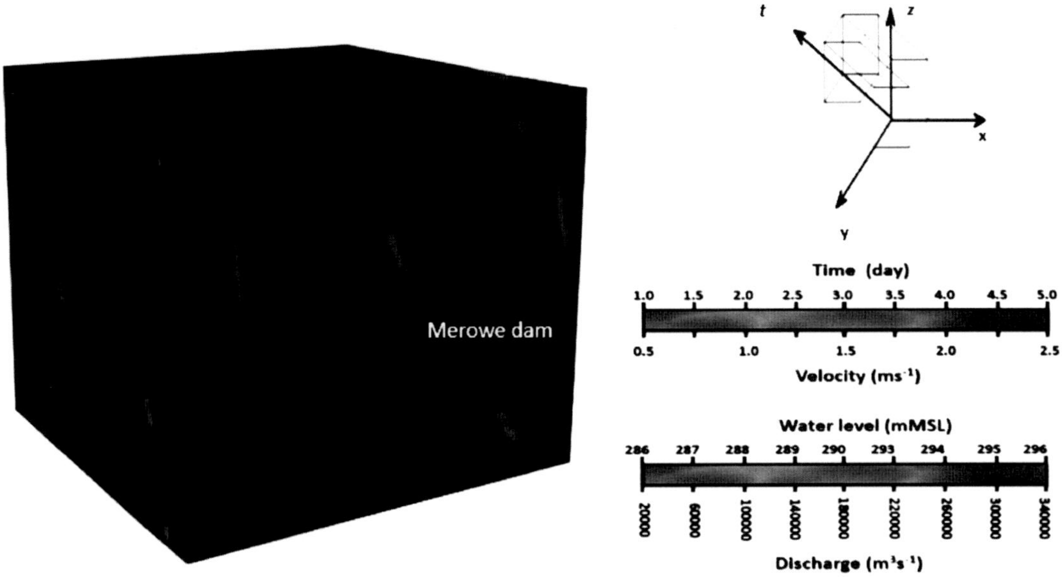

Figure 12.28 4D Marghany Hologram-TomoSAR algorithm for forecasting downstream wave flood from the Merowe Dam.

The estimated arrival time for floodwaters traveling from Merowe Dam to the Aswan High Dam is approximately twenty-two and a half days, with the wave propagating at a relatively slow velocity of 0.35 ms^{-1}. Upon reaching Aswan, the volume discharge is projected to be around 40,000 m^3s^{-1}, with an anticipated maximum water level of 180 meters (refer to Figure 12.29). Given these conditions, there is a significant risk that the Aswan High Dam could face a structural failure. As water discharge flows from Merowe Dam toward Lake Nasser, it undergoes significant reductions in both velocity and volume due to a combination of energy dissipation and environmental influences. First, as the wave travels downstream, frictional forces from the riverbed and surrounding topography gradually dissipate the water's kinetic energy. This energy reduction not only diminishes the discharge rate but also slows the flow velocity over time.

Furthermore, the topography of the river system itself plays a crucial role. As the river widens substantially near Lake Nasser, the incoming wave disperses over a larger area. This expansion reduces both the depth and pressure of the water, decreasing the velocity as it spreads out. Therefore, this natural widening acts as a buffer, effectively diluting the discharge and distributing it across a broader space, reducing the intensity of the wave by the time it reaches the lake.

Additionally, sediment transported within the floodwaters contributes to increased turbulence and frictional resistance. Over time, this added friction gradually slows the wave's speed, with sediment deposits along the riverbanks and lakebed absorbing energy from the flow. This sedimentation not only reduces discharge further but also stabilizes the inflow, creating a slower, more controlled entry as the water reaches Lake Nasser.

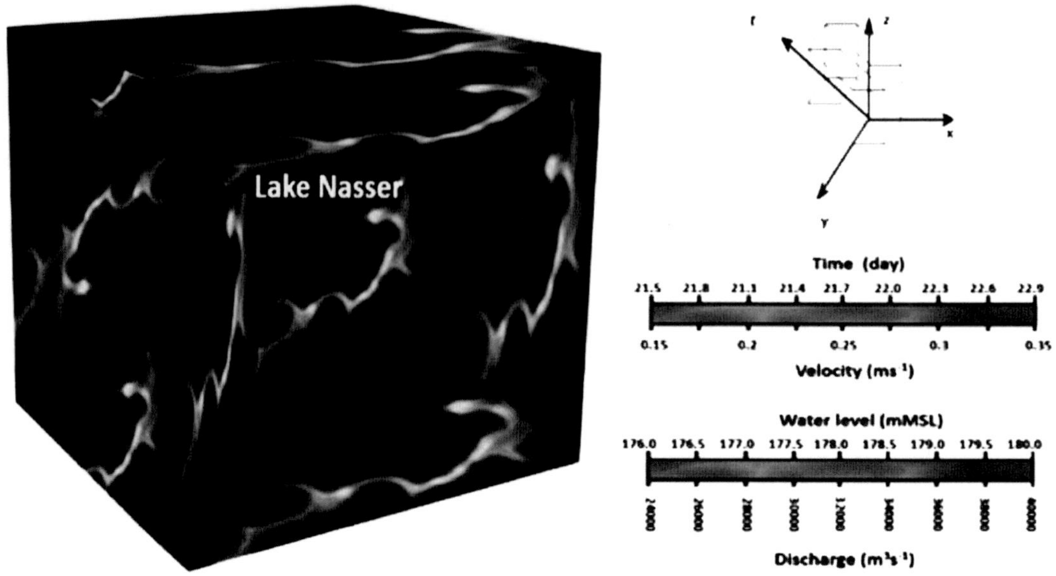

Figure 12.29 4D Marghany Hologram-TomoSAR algorithm for forecasting flood wave propagation from Merowe dam to Lake Nasser.

To understand this projection more fully, we must consider several interrelated factors. First, the extended propagation time over nearly three weeks implies that the water energy and volume will have time to consolidate, moving gradually but continuously towards the Aswan reservoir. This slow, sustained velocity could exacerbate pressure on the dam by maintaining a prolonged, intense inflow. While the Aswan High Dam was designed to manage substantial volumes, it may not withstand a scenario where such volumes are maintained over an extended period, particularly if the water levels approach or surpass 180 meters.

The implications of breaching at the Aswan High Dam would be far-reaching. As the principal barrier controlling the flow of the Nile through Egypt, a collapse would unleash massive downstream flooding, endangering communities, infrastructure, and agriculture vital to the nation. The economic and environmental impacts would be catastrophic, potentially disrupting water supplies, electricity generation, and agricultural productivity across a vast region. The precise water level and volume upon arrival at Aswan are critical variables in assessing the risk.

The Aswan High Dam, an architectural marvel and a critical linchpin of Egypt's water management and energy production would face a daunting challenge when subjected to a discharge rate as immense as 40,000 $m^3s^{-1.}$ Such a discharge represents a profound stressor on the dam's structural integrity, testing the very limits of its design. This rate, a staggering volume in hydraulic engineering terms, underscores the urgency of assessing the dam's resilience under extreme conditions, as the consequences of failure would be catastrophic on both environmental and socioeconomic fronts.

A key factor in understanding this risk is the dam's design threshold, which permits water to rise to a maximum level of around 180 meters. This threshold is not merely an arbitrary limit; it encapsulates decades of engineering knowledge and geological considerations about the Nile's dynamics and the dam's material capacities. Approaching or surpassing this critical point significantly heightens the risk of structural failure. At such levels, the dam is exposed to forces that are exponentially greater than what it can safely withstand, placing it on the precipice of either a breach or overtopping event. In the unfortunate event of such a failure, the repercussions would be swift and devastating. The immediate area downstream, including the city of Aswan, would most likely experience an inundation that could obliterate infrastructure and endanger lives. However, the impact would not be confined to Aswan alone; the Nile's powerful current would carry the floodwaters far beyond, reaching densely populated regions further downstream. Cairo, the capital and a metropolitan hub of over twenty million people lies within this zone of potential destruction. Such flooding would disrupt critical infrastructure, including roads, bridges, and utilities, potentially paralyzing large portions of Egypt's economy and rendering the population vulnerable to secondary risks such as waterborne diseases and loss of resources.

Moreover, within 51.85 hours—or approximately 2.16 days—of a potential dam failure, floodwaters would reach the renowned Abu Simbel Temples (Figure 12.30). These historic monuments, carved directly into cliffs as a testament to ancient Egyptian civilization, would face submersion under an estimated water height of eight meters. This inundation would not only threaten a world heritage site of incalculable cultural value but also underscore the far-reaching consequences of a breach in the Aswan High Dam.

Thus, the threat posed by a 40,000 m^3s^{-1} discharge extends well beyond the dam's structural integrity alone. It raises critical questions about the management of extreme hydrological events, the protection of cultural heritage, and the broader implications for regional stability and economic resilience. The stakes involved illustrate that the Aswan High Dam is not merely a barrier against water; it is a safeguard for an entire region's way of life, economic health and historical legacy.

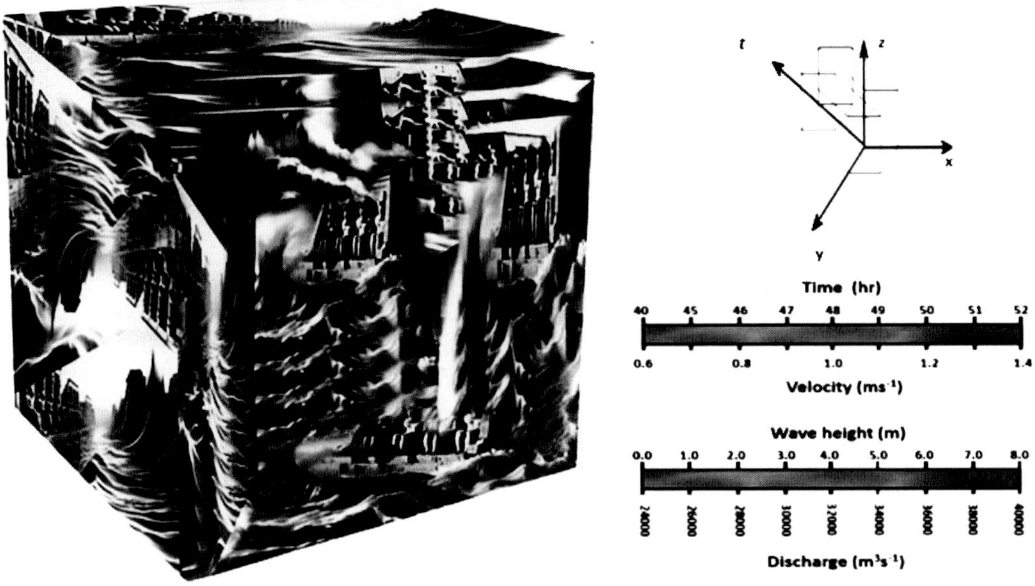

Figure 12.30 4D Marghany Hologram-TomoSAR algorithm for forecasting flood wave propagation impacting the Abu Simbel Temples.

Consequently, the discharge rate of over 40,000 m^3s^{-1} would propel floodwaters downstream to Cairo within approximately 157.41 hours, or roughly 6.56 days, advancing at an estimated speed of 1.5 meters per second (Figure 12.31). As this flood wave progresses, it would span a formidable width of 453,120 meters, with a height fluctuating between 2 and 6 m. These dimensions reveal the vast magnitude of water being displaced and the potential impact of such a flood.

A flood wave of this scale—particularly with a maximum height of 6 m and width extending to 500 m—possesses the capacity to transport an immense volume of water as it advances downstream. Upon reaching Cairo, however, the floodwaters would not simply maintain this form. Instead, they would begin to spread across the extensive riverbed and surrounding floodplain, dynamically altering the landscape. This dispersion, while perhaps diluting the immediate force of the wave, could nonetheless lead to a significant and rapid increase in water levels throughout the Cairo metropolitan area, far surpassing any normal seasonal variations.

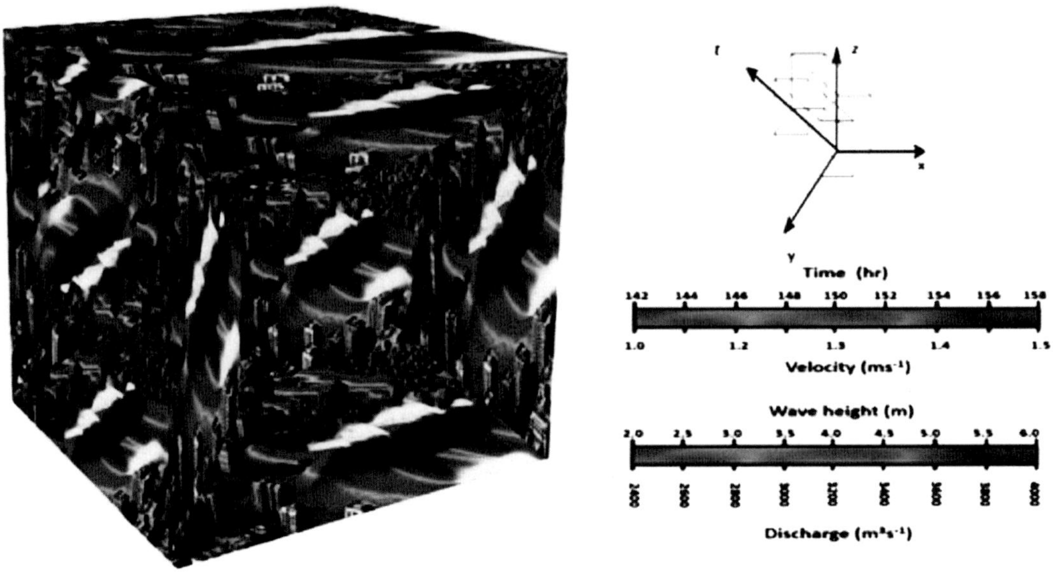

Figure 12.31 4D Marghany Hologram-TomoSAR algorithm for forecasting flood wave propagation impacting Cairo City.

The interaction between the propagating flood wave and Cairo's geographical and infrastructural layout poses a unique threat. Cairo's densely populated urban core and its complex network of historical and contemporary infrastructure make it particularly vulnerable to rapid inundation. Additionally, the Nile's proximity to critical facilities—including roads, residential neighborhoods, industrial zones, and heritage sites—would exacerbate the repercussions. An increase in water height of even a few meters above regular levels would be enough to overwhelm flood defenses, submerge transportation routes, and disrupt essential services, leaving the city at risk of both direct damage and the longer-term impacts of widespread flooding.

Moreover, this scenario illustrates the extraordinary challenge posed by flood dynamics on such a scale. The flood wave's sheer width of nearly half a million meters underscores not only its physical impact but also its symbolic weight as a demonstration of the Nile's transformative and, at times, destructive power. Historically, the Nile has been a lifeline for Egypt, enabling agriculture, trade and promoting civilization. However, in this context, its force becomes a potential agent of destruction, demanding advanced engineering solutions, precise hydrological modeling and comprehensive disaster planning.

The propagation of a wave of this magnitude highlights the necessity for rigorous flood preparedness strategies, robust monitoring and an understanding of how flood waves interact with both natural and built environments. As Egypt's capital, Cairo stands as a focal point for such efforts, representing both the challenges and the imperatives of managing the water dynamics of one of the world's most historically and geographically significant rivers.

Additionally, large portions of the Nile Delta would be submerged within 50 hours of being struck by a propagated flood wave of up to 6 meters in height originating from Cairo (Figure 12.32). The city of Alexandria would also face the impact of this flood wave within twenty-four hours, as it arrives from Cairo as well as from the southern and eastern regions of the Delta (Figure 12.33). This convergence of floodwaters from multiple directions around Alexandria could lead to an accumulated water discharge rate of approximately 330,000 m^3s^{-1}, reaching a peak velocity of 2.4 meters per second (Figure 12.33).

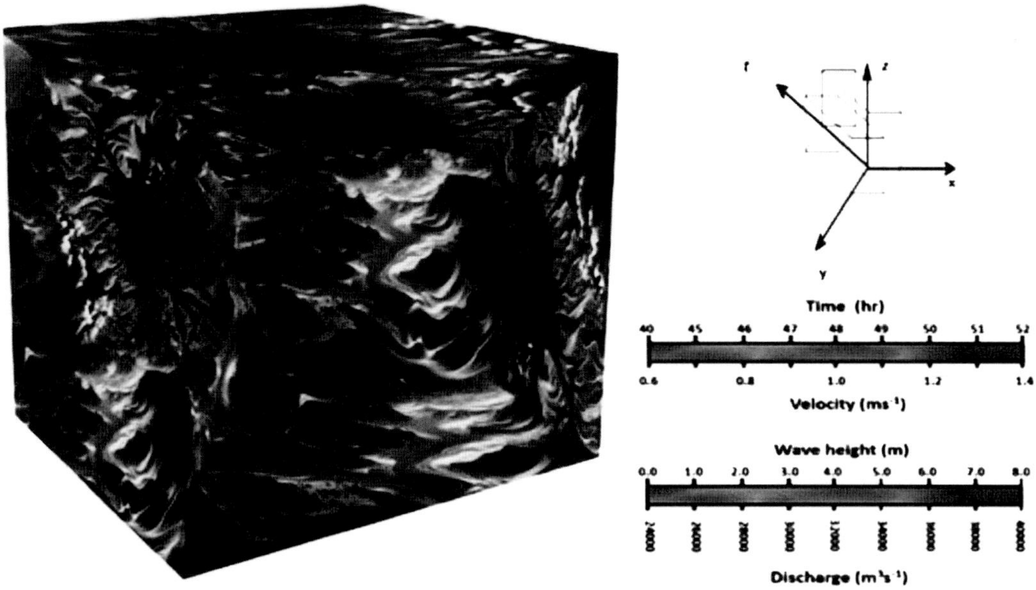

Figure 12.32 4D Marghany Hologram-TomoSAR algorithm for forecasting flood wave propagation impacting the Nile Delta.

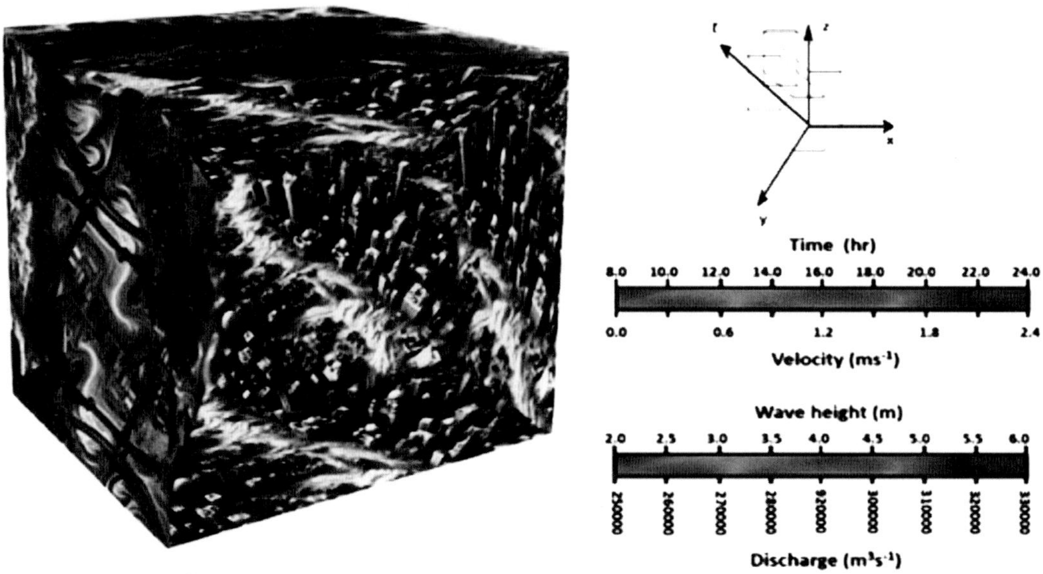

Figure 12.33 4D Marghany Hologram-TomoSAR algorithm for forecasting flood wave propagation impacting Alexandria City.

This scenario describes a critical hydrological event with far-reaching consequences for the Nile Delta and the city of Alexandria, underscoring the complex dynamics of flood propagation and accumulation in densely populated and agriculturally vital regions. The mechanics of a propagated flood wave originating from Cairo, with a height reaching up to 6 meters, reveal the velocity and scope of this cascading impact. Once the wave enters the Delta, it would require less than fifty hours to inundate extensive areas, effectively erasing or submerging parts of this low-lying landscape. This timeframe highlights the rapid onset of the flood's impact, leaving limited response time for communities and emergency services within the Delta.

Alexandria, positioned at the northern terminus of this flood trajectory, stands as a focal point for understanding the cumulative impact of floodwaters arriving from multiple directions. In particular, Alexandria would encounter these forces from Cairo, as well as from the Nile Delta's southern and eastern regions, within an accelerated period of twenty-four hours. This convergence effectively doubles the threat: as these floodwaters gather around Alexandria, the simultaneous arrivals intensify both water volume and flow velocity creating a formidable force. The calculated accumulation of water discharge around Alexandria—estimated at 330,000 cubic meters per second—is significant, equating to a massive influx that would overwhelm conventional flood management systems. At a peak velocity of 2.4 meters per second, the floodwaters would carry immense kinetic energy, heightening the risk of severe structural damage and prolonged submersion of critical infrastructure.

While substantial research on the potential downstream effects of a Grand Ethiopian Renaissance Dam (GERD) breach has been conducted by Egyptian researchers, most studies have yet to fully simulate the impact of such an event on the Egyptian side of the Nile basin. The findings of this study align partially with the work of Elbelasy et al. [67] specifically concerning the time it would take for floodwaters to reach Egypt's High Aswan Dam and the characteristics of a possible flood wave. However, there are notable differences, particularly regarding the estimated time required for water from a GERD breach to arrive at the Roseires Dam and the mechanics of the GERD breach itself.

Elbelasy et al. [67] suggest that, in the event of a breach, GERD would take approximately one hour to fail completely, with floodwaters taking an estimated three days to reach the Roseires Dam downstream. In contrast, a sudden failure of the western embankment due to a catastrophic breach is expected to occur much more rapidly, within 11 to 15 seconds—far shorter than the one-hour estimate. This rapid collapse is most likely due to the high volume and intense pressure of water in the GERD reservoir, which, combined with the steep gradient of the terrain, would accelerate the flow of water downstream.

In the author's view, these findings underscore the urgent need for more detailed, cross-national studies that can model the impact of various breach scenarios on downstream regions, especially given the life-and-death implications for communities in the path of a potential flood. A better understanding of these dynamics is critical to prepare effective response strategies and reinforcing infrastructure resilience throughout the Nile basin. The inclusion of data-driven simulations involving steep-slope hydraulics and reservoir breach mechanics could provide a more accurate picture of potential risks and improve regional water security and disaster preparedness.

The implications of this scenario extend beyond immediate physical destruction. The Nile Delta, historically renowned as the cradle of Egyptian civilization, serves as both an ecological and economic heartland. Its agricultural productivity, biodiversity, and role as a protective buffer for coastal regions are well-documented. However, these very attributes—its low elevation, extensive canal network, and reliance on stable water levels—render it acutely vulnerable to a flood wave of this magnitude. Submergence on this scale would disrupt agriculture, displace residents, and lead to soil salinization, compromising future agricultural productivity.

From a scientific perspective, these dynamics emphasize the need for advanced flood modeling, predictive analytics and a deepened understanding of flood behavior in interconnected riverine and coastal systems. Hydrodynamic models capable of simulating propagation speeds, water accumulation and dispersal patterns are essential tools for developing early warning systems and informing disaster preparedness strategies. Moreover, the convergence of floodwaters underscores the importance of multi-layered defense systems—solutions that combine structural measures, such as flood barriers and levees, with adaptive strategies like controlled floodplain storage to mitigate water velocity and volume before it reaches critical urban zones.

This scenario also illustrates the interplay of geographic, climatic, and infrastructural factors in disaster vulnerability. As a historical port city with limited capacity for rapid water drainage,

Alexandria's susceptibility to combined fluvial and coastal flooding represents a multifaceted risk that warrants immediate attention from hydrologists, urban planners and policymakers. Climate-induced changes in precipitation patterns and sea-level rise further intensify the potential for such flood events, challenging the resilience of urban centers, agricultural basins, and cultural heritage sites across the region.

Needless to say, the described flood event is a compelling case for a holistic, science-based approach to water management in the Nile Delta. Comprehensive risk assessment, robust infrastructure, and international collaboration on hydrological research are vital to protect Egypt's population and its essential resources. This event underscores the power of the Nile, which has sustained civilization for millennia, reminding us of its potential to reshape the landscape under extreme conditions. The pursuit of sustainable flood mitigation solutions is not only a scientific endeavor but a societal imperative, safeguarding both present-day communities and future generations.

12.12 MARGHANY QUANTUM TSUNAMI SPECULATION: DYNAMICS OF THE RED SEA AND INDIAN OCEAN TSUNAMIS TRIGGERED BY THE GERD

The Grand Ethiopian Renaissance Dam (GERD) poses a significant threat of catastrophic destruction to both Sudan and Egypt. This raises a critical question: Could the GERD potentially trigger a tsunami that impacts the Red Sea and the Indian Ocean? This section seeks to address this pressing concern.

It is essential to recognize that the GERD is situated within an active tectonic zone known as the Horn of Africa. This region is characterized by a complex geological framework dominated by both metamorphic and sedimentary rocks. A particularly alarming aspect of this geological setting is the presence of shear stress zones, which are intersected by approximately four major faults. These faults are indicative of an unstable environment, where substantial tectonic tension exists within the Earth's crust as discussed earlier in previous chapters.

The dam itself is located on the Nubian Block, a geological formation constrained by the African Rift, which further complicates its stability (Figure 12.34). The presence of these faults suggests a heightened risk of seismic activity, as they are sites where accumulated tension can lead to fault movement. Indeed, this region has experienced numerous earthquakes, many exceeding a magnitude of 6, underscoring the potential for significant geological disturbances. Additionally, volcanic activity is prevalent in this area, with numerous volcanic mountains contributing to the complexity of the geological landscape. The interplay between tectonic forces and volcanic phenomena raises concerns about the potential for catastrophic events that could have far-reaching consequences [43].

In other words, given the geological instability surrounding the GERD and its location within an active tectonic zone, it is imperative to consider not only the immediate risks to Sudan and Egypt but also the broader implications for regional stability. The possibility of triggering seismic events that could lead to tsunamis in adjacent bodies of water cannot be dismissed lightly; thus, comprehensive monitoring and risk assessment strategies are essential for safeguarding these vulnerable nations.

On August 31, 2012, a significant earthquake measuring 5.3 on the Richter scale struck the Gulf of Aden, resulting from tectonic plate displacement along the West Shoba Ridge (see Figure 12.35). This seismic event occurred at a depth of 10.2 km, highlighting the geological volatility of the region. Prior to this event, the area had experienced notable earthquakes, including those on October 31, 2011, and July 13, 2015, respectively which registered magnitudes ranging from 5.0 to 6.3 Richter.

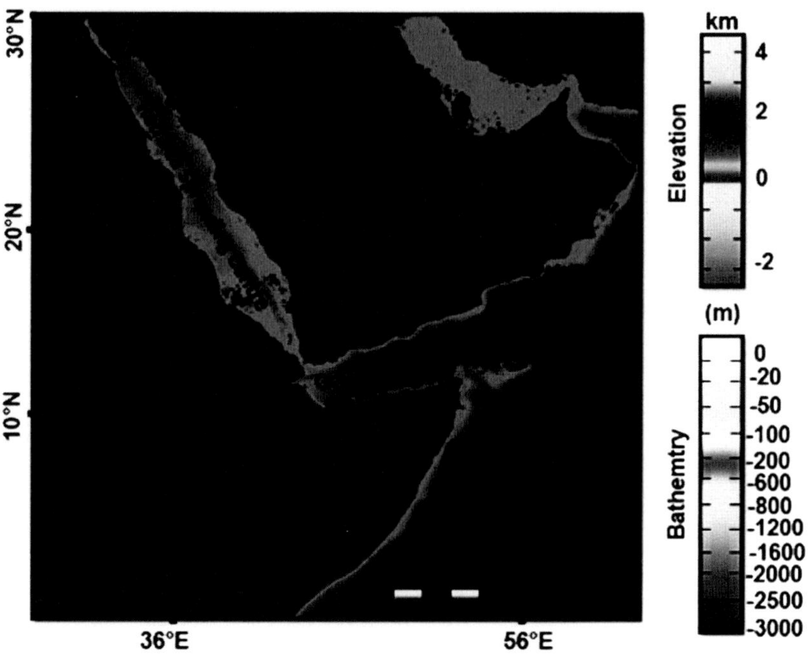

Figure 12.34 African Horn.

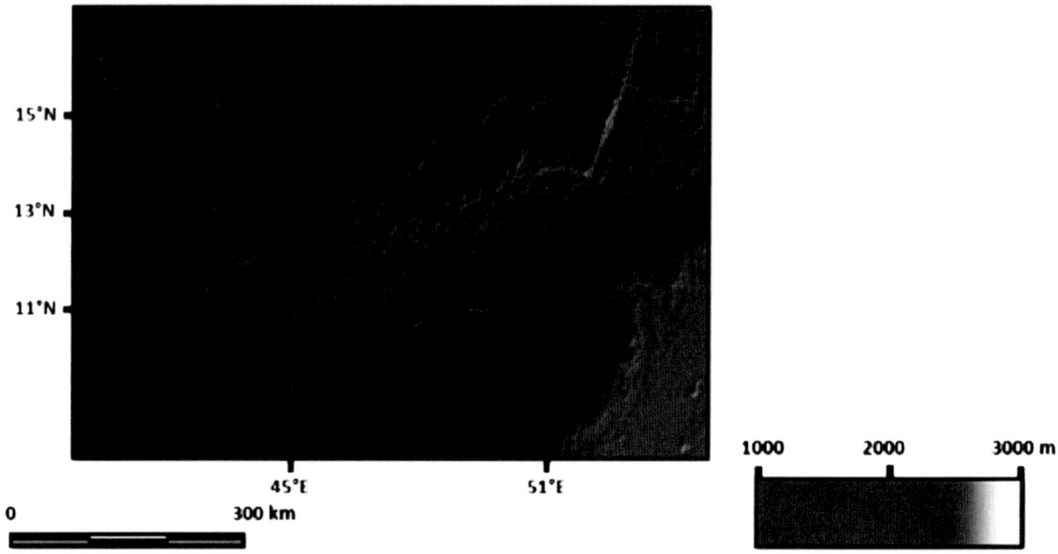

Figure 12.35 Shoba ridge along the Gulf of Aden.

Between October 2023 and December 2024, the Horn of Africa experienced an unprecedented total of 130 recorded seismic events, varying in magnitude. These occurrences were predominantly concentrated around geothermal hotspots, suggesting a progressive intensification of tectonic activity in the region. The sequence of seismic events is as follows:

In October 2023, moderate earthquakes with magnitudes ranging from 4.0 to 4.8 were detected in the Danakil Depression, marking the initial phase of increased seismic activity. By February 2024, the magnitude of earthquakes rose to 5.2, coinciding with emerging reports of hot water flows along fault zones, which serve as potential indicators of interactions between geothermal activity and tectonic stress.

From July 2024 to December 2024, the frequency of seismic events escalated significantly, culminating in November 2024, when 20 earthquakes occurred within a single week. This pattern of events highlights an evolving stress distribution in the region, prompting further investigation into the possible migration of seismic activity toward the Grand Ethiopian Renaissance Dam (GERD), situated 560 kilometers away.

The geological framework of the region is defined by the convergence of the Ethiopian, Somalian, and Yemeni plateaus, which collectively form the Afar Depression (AD). This unique geological configuration encompasses several prominent tectonic units, notably the East African Rift, the Gulf of Aden and the Red Sea. The Afar Depression itself exhibits a clockwise rotation in its southern section due to left-lateral faulting associated with overlapping rifts that step leftward. Seismic activity is particularly pronounced along the axes of the Red Sea, where it intensifies at both northern and southern extremities. Furthermore, this seismicity extends into the Gulf of Aden and the Afar triple junction, primarily through the dominant Ethiopian Rift. This interconnectedness underscores the dynamic nature of tectonic processes in this region and their potential implications for geological stability [57–59].

The theory of seismic migration offers a framework for comprehending the propagation of tectonic stress released in one region, which may subsequently trigger seismic events in distant interconnected areas. Two primary mechanisms are essential for elucidating this process:

1. Stress Redistribution: Tectonic stress accumulates and is released through seismic events, potentially inducing secondary stress changes in adjacent regions. In the context of the Horn of Africa, the interaction between the Nubian Plate and the Arabian Plate, along with the dynamics of the East African Rift, indicates a plausible pathway for stress migration toward the area surrounding the Grand Ethiopian Renaissance Dam (GERD).
2. Proximity and Geological Conduits: Although the GERD is situated approximately 560 kilometers from the epicenter of recent seismic activity, the interconnected fault systems underlying the rift may facilitate the transmission of stress waves over time. This phenomenon has the potential to destabilize the region adjacent to the dam.

The Grand Ethiopian Renaissance Dam, an engineering achievement that is pivotal to Ethiopia's development, may be susceptible to vulnerabilities if affected by seismic migration. Despite adherence to modern design and construction standards, two primary risks warrant attention:

1. Foundation Vulnerability: Continued seismic activity may lead to the emergence of microfractures in the dam's foundation, especially if local faults are activated or if stress waves intensify in proximity to the GERD.
2. Hydrodynamic Forces: Seismic events can induce oscillations in the reservoir, commonly referred to as seiches, which subsequently augment hydrodynamic pressures on the dam's walls. This phenomenon, in conjunction with underlying tectonic instability, could present significant challenges to the structural integrity of the dam.

Moreover, the cascading effects of seismic migration could further destabilize local fault systems surrounding the GERD, thereby exacerbating risks over time. The timeline of seismic events in the African Horn illustrates an alarming progression of tectonic activity, raising concerns about its potential migration toward critical infrastructure such as GERD. While the distance of 560 km may offer temporary reprieve, the interconnected nature of the Nubian and Arabian Plates, combined with the East African Rift's geological complexity, underscores the necessity of proactive measures.

Enhanced seismic monitoring, geotechnical assessments, and regional collaboration are essential to mitigate risks and safeguard the dam's structural integrity in the face of evolving tectonic challenges.

The immense weight of the water stored in the Grand Ethiopian Renaissance Dam (GERD) exerts significant pressure on the underlying geological formations, including those associated with the Shoba Ridge. This additional stress may lead to heightened seismic activity, particularly if it surpasses the strength of the local rock structures. Given that this region is already characterized by tectonic activity, the increased pressure from the reservoir could influence existing fault lines, potentially triggering earthquakes or landslides over time. Moreover, the annual filling of the dam contributes to an expansion of the reservoir area, particularly towards the east in the direction of the African Horn. This expansion can induce greater shear stress along this tectonically active region. The dam's capacity to hold an astonishing 74 billion cubic meters of water allows for significant tectonic plate movement without substantial thinning of the Earth's crust. Remarkably, this volume represents an increase of approximately 300% beyond typical expectations for such structures.

To put this into perspective, the dam covers a surface area of 1,680 square kilometers, stands 145 meters high, and extends 1.8 kilometers in length. Such dimensions not only underscore its massive scale but also suggest that they could accelerate potential structural failures or collapses. Furthermore, the dam's operation will inevitably induce shear stress on four critical faults that traverse the region, heightening concerns about geological instability. In conclusion, while the GERD is designed to provide substantial benefits in terms of hydroelectric power and water management, it simultaneously poses significant geological risks. The interplay between its immense water storage capacity and the tectonic dynamics of the surrounding area necessitates careful monitoring and proactive risk management strategies to mitigate potential adverse effects on both local and regional stability.

The Grand Ethiopian Renaissance Dam (GERD) has the potential to induce a tsunami in the Red Sea and the Gulf of Aden, despite being approximately 1,000 km from the Shoba Ridge. Understanding this phenomenon requires examining several critical factors.

First, the GERD is situated within the tectonically active East African Rift System. If significant seismic activity were to occur—potentially reaching 9 on the Richter scale—the resulting earthquakes could trigger powerful underwater landslides, displacing large volumes of water and leading to tsunami formation. Second, the weight of the water in the GERD exerts pressure on underlying geological formations, potentially affecting local fault lines. If this stress exceeds the capacity of these faults, it could result in sudden slip events and earthquakes that impact areas far from the dam. The interconnected nature of tectonic plates means that movements in one area can have widespread consequences.

To derive a mathematical framework for Marghany's Quantum Tsunami Theory associated with the Grand Ethiopian Renaissance Dam (GERD), we analyze the quantum mechanical implications of tectonic stress and shear forces that could lead to seismic and tsunami events from the GERD region to the Indian Ocean. The following components outline a mathematical model based on the principles of quantum mechanics, wave propagation, and geophysical stresses.

Given that the dam exerts significant pressure on underlying fault lines, the wave function $\psi(x,t)$ represents the propagation of energy that could lead to seismic waves along the tectonic boundaries. The wave function can be expressed as:

$$\psi(x,t) = A(x,t)\,e^{i(kx-\omega t)} \tag{12.64}$$

where $A(x, t)$ is the amplitude of the seismic wave; k is wave number related to the wavelength λ by $k = \frac{2\pi}{\lambda}$; and ω is the angular frequency of the wave. To represent energy propagation due to GERD's influence, we adapt the Schrödinger equation to model wave behavior along tectonic faults:

$$i\hbar\frac{\partial\psi}{\partial t} = -\frac{\hbar^2}{2m}\frac{\partial^2\psi}{\partial x^2} + V(x)\psi \tag{12.65}$$

Here, $V(x)$ represents the potential energy from tectonic stress, influenced by the immense pressure GERD places on surrounding geological features. Therefore, GERD's water storage exerts horizontal pressure along critical faults, with the shear stress σ given by:

$$\sigma = \frac{F}{A} \tag{12.66}$$

where F is the force exerted by the dam's water weight, and A is the cross-sectional area of the fault line. Consequently, the stored water at GERD alters Earth's moment of inertia I, described as:

$$I = \sum mr^2 \tag{12.67}$$

where m is the mass of displaced water and r is the distance from the Earth's axis. Changes in I due to water displacement can influence rotational velocity, causing the Earth's rotation to adjust slightly. Subsequently, the tsunami wave propagation speed v is given by:

$$v = \sqrt{gd} \tag{12.68}$$

where g is gravitational acceleration and d is water depth. This speed is adjusted as waves encounter the Indian Ocean Ridge, potentially curving due to tectonic shifts.

According to Marghany [43], tsunamis generated by seismic events propagate rapidly across oceanic and sea basins. Although the GERD is distant from the Red Sea and the Gulf of Aden, substantial earthquakes could release energy that travels vast distances through water bodies, allowing tsunami waves to reach distant shores. Historical precedents demonstrate that large earthquakes can trigger tsunamis that affect regions hundreds of kilometers away from their epicenters. For example, the 2004 Indian Ocean earthquake illustrated how seismic activity in one location can lead to devastating tsunamis impacting countries across great distances.

From the perspective outlined, two primary scenarios could lead to a massive earthquake along the African Rift: either the stress induced by the substantial volume of water stored in the Grand Ethiopian Renaissance Dam (GERD) or the catastrophic collapse of the dam itself. Under these circumstances, seismic waves with a magnitude of 9 on the Richter scale could propagate in a direction that exacerbates existing fault slips, thereby inducing significant earthquakes.

Such seismic events would resonate along the numerous faults present in the African Rift, including those located in the Red Sea. This activity has the potential to trigger a tsunami. When seismic signals reach an acoustic resonance frequency, they can catalyze both an earthquake and a subsequent tsunami in the Red Sea and Gulf of Aden (see Figure 12.36). The initial wave height generated by this tsunami could be approximately 4 meters, primarily due to the narrow width of the Gulf of Aden, which measures about 250 km across. As the tsunami travels from the Gulf of Aden towards the Red Sea, it would continue its journey into the Arabian Gulf, eventually reaching as far as India and Sri Lanka. The estimated time of arrival for these waves in India and Sri Lanka would be around two hours (Figure 12.36).

In less than two hours, wave heights within the Red Sea and Arabian Gulf could fluctuate dramatically, ranging from –8 to 8 m. This sequence of waves would reflect and diffract along the coastal waters of the Red Sea (refer to Figure 12.37). Notably, it is important to highlight that coastal waters in India would experience tsunami wave heights reaching 8 m within this same two-hour timeframe.

Needless to say, the interplay between tectonic stress from water volume and potential dam failure presents significant risks for seismic activity along the African Rift. The resulting earthquakes could not only impact local fault lines but also generate tsunamis capable of traversing vast distances, posing threats to distant coastal regions. Monitoring these geological dynamics is essential to understand and mitigate potential hazards associated with such catastrophic events.

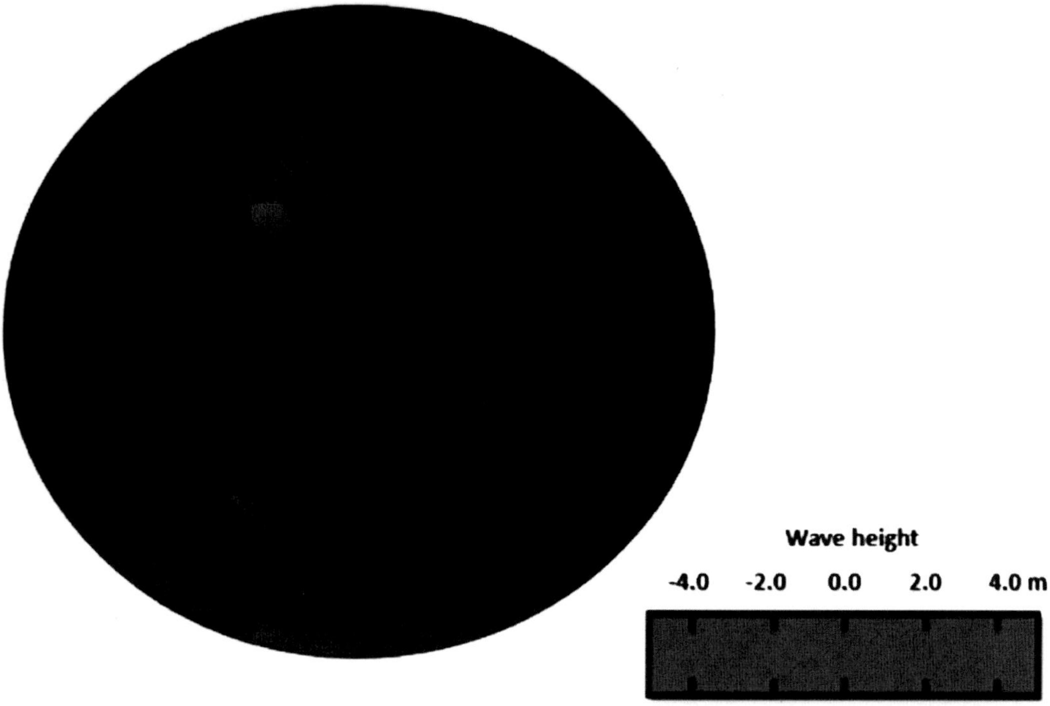

Figure 12.36 Initial tsunami height is generated in the Gulf of Aden due to GERD impact on Shoba ridge.

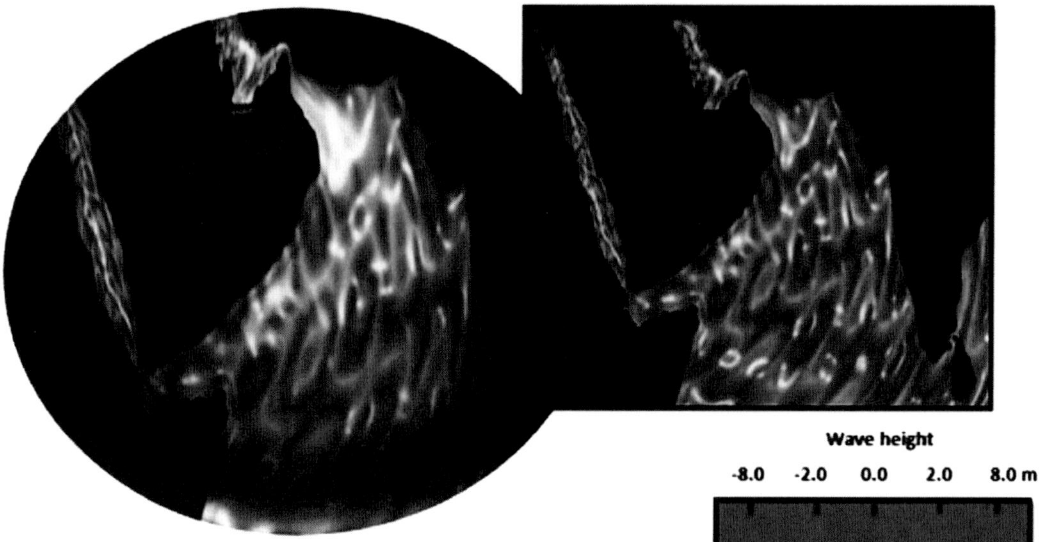

Figure 12.37 Two hours of Tsunami propagation in the Indian Ocean and Red Sea following GERD stress resonating in the Shoba Ridge and African Rift.

According to the acoustic fluidization (AF) mechanism, seismic fractures generate elastic waves that propagate and scatter within fault lines, creating a normal stress that contrasts with the confining pressures and consequently facilitates seismic failure. Remarkably, this same mechanism can be activated by transient seismic waves produced by distant earthquakes. AF has been proposed to elucidate the phenomenon of increased seismic activity observed within minutes following major earthquakes, even in regions located thousands of kilometers from the mainshock epicenter. Furthermore, the considerable depth of the Indian Ocean allows tsunami waves to travel vast distances with minimal energy loss, enhancing their potential for widespread impact.

In alignment with plate tectonic theory, the Earth's surface is segmented into a series of relatively thin yet massive tectonic plates that are in constant motion. The dynamics of these plates evolve, as their mass and positioning shift in response to underlying geological processes. For instance, the Pacific Plate is among the most rapidly moving plates, advancing at a rate exceeding 10 cm per year, whereas the African Plate progresses at a much slower rate of approximately 25 mm/year. In contrast, the Australian Plate exhibits a more rapid movement, changing at a rate of 60 mm per year. The boundaries between these tectonic plates and where they interact with one another are crucial towards inducing significant geological activities, including earthquakes, volcanic eruptions and mountain formation.

One of the longest fault systems in the world, the Great Rift Valley, exemplifies these tectonic dynamics as it bisects the African continent. The Afar Depression, located within Ethiopia, is notably positioned well below sea level. In 2006, this area experienced a series of earthquakes that resulted in a widening of approximately three meters and a further subsidence of 100 meters. The geometry of the Grand Ethiopian Renaissance Dam (GERD), designed to store a volume of 74 billion cubic meters of water, induces horizontal pressure along its edges, analogous to the forces acting on a rug when pulled. Under these conditions, folds in the geological strata align perpendicularly to the direction of shear stress. Indeed, horizontal shear stress influences the development of geological features such as rocks, ridges, and valleys, which are steepened in the direction of this shear stress [43].

A critical question arises: Could the Grand Ethiopian Renaissance Dam (GERD) lead to the separation of the Horn of Africa from the African continent?

Let us assume that the separation probability P can be defined as the square of the wave function amplitude $|\psi(x, t)|^2$, representing the likelihood of tectonic displacement resulting in a significant separation event. Therefore:

$$P(x, t) = |\psi(x, t)|^2 = |A(x, t)|^2 \, e^{-2\Im(kx-\omega t)} \tag{12.69}$$

In equation 12.69, $\Im$ is the imaginary part of the phase $(kx - \omega t)$, which introduces an exponential decay factor into the equation. When multiplied by −2, it models a damping effect in the displacement probability over time and distance, reflecting how the probability density might decrease due to energy dissipation across geological boundaries. This means that tectonic stress might be strongest initially but weakens over distance and time.

By solving the Schrödinger equation under these conditions, the probability density function $P(x, t)$ can be maximized to predict the rate of displacement. Additionally, using the expectation value of the position $\langle x \rangle$, we can predict the average location of fault displacement over time:

$$\langle x(t) \rangle = \int x |\psi(x, t)|^2 \, dx \tag{12.70}$$

In the case of shear stress generated by the GERD, we must consider how the water's movement and pressure can be modeled as wave functions. The shear stress can be expressed in terms of the force per unit area, providing insight into how the dam's structural integrity may be affected by both the mass of the water and its dynamic behavior.

However, once the GERD is filled to its designed capacity of 74 billion cubic meters—an amount that some may consider to exceed the intended geometric specifications—the dynamics of the system change dramatically. At this stage, the particle's wave function expands significantly, leading to an increase in both size and wavelength. Specifically, the dam, which stands at a height of 155 m, may experience overtopping, where the water levels exceed this threshold.

In this context, the wavelength associated with the seismic waves propagating along the fault line of the African Horn and the tidal waves generated by the collapsing and overtopping water in the dam becomes critical. The interaction of these waves can be analyzed through the lens of the Schrödinger equation, a fundamental framework in quantum mechanics that describes how quantum states evolve.

According to this theoretical framework, the maximum wave height could reach up to 350 meters as the distance from the dam increases to approximately 400 kilometers. This indicates that the combined effects of the seismic activity along the fault and the inundating waves from the dam will result in an elongated wavelength and significantly heightened amplitude.

Consequently, the implications of such wave phenomena extend beyond mere physical measurements; they underscore the interconnectedness of hydrodynamic forces and geological stability. As the wave functions expand and interact, they can lead to profound consequences for the surrounding environment, potentially influencing both local ecosystems and human settlements.

From the after-mentioned perspective, the transformation of wave properties at the GERD from small amplitudes and high velocities to large wavelengths and significant amplitudes illustrates the intricate interplay between water dynamics and seismic activity. This analysis not only enriches our understanding of fluid mechanics but also highlights the importance of considering the broader implications of such geological and hydrological interactions in the context of engineering design and environmental management.

The fault boundaries would be dominated by the total energy of $E_T = \sum_{0}^{\infty} E^2 2m$ which is approaching the infinite. Under this circumstance, the east of the African Horn can be separated from Africa and moved away to the Indian Ocean (Figure 12.38). In this regard, $|\psi|^2$ is considered as a probability of realizing the geologic fragmentation at any point along the African Horn due to the shear stress inputs by GERD. The high shear stress input along the African Horn, the maximum $|\psi|^2$, namely, the probability of a destructive tsunami across the Red Sea, and Indian Ocean, and splitting of the east side of the African Horn of Africa, and the ending of the Nile River lifetime.

Figure 12.38 The splitting mechanism occurs on the eastern side of the African Horn due to GERD impact.

The prospective geological consequences of the African Horn separating from the African continent present a fascinating yet concerning scenario, particularly regarding the dynamics of the Mid-Indian Ridge. This tectonic event would most likely induce significant changes in the regional geological landscape, with far-reaching implications for the surrounding areas.

As the African Horn detaches, the Mid-Indian Ridge, which is a divergent tectonic plate boundary, may undergo a notable curving downward to accommodate the resultant void left by the separation. This alteration in the ridge's morphology would not only change its structural configuration but could also amplify seismic activity in the region. Such tectonic adjustments would be indicative of the Earth's intrinsic capacity for self-regulation and reconfiguration in response to stress-induced changes in plate dynamics.

The creation of a void in the tectonic structure would possibly lead to a displacement of large volumes of seawater, generating a massive tsunami that could propagate across the Indian Ocean. This tsunami, characterized by immense energy, would travel at high speeds and could have devastating effects upon reaching landmasses, including Australia and Sumatra, and extending toward the Indian subcontinent. The energy released from such a geological event could displace significant amounts of water, resulting in waves that could reach considerable heights and travel vast distances, potentially altering coastlines and ecosystems in the affected regions.

As the tsunami waves progress further, their reach may extend as far as Antarctica. Such a phenomenon would serve to underscore the interconnectedness of global oceanic systems, demonstrating how geological events in one part of the world can precipitate consequences thousands of kilometers away. The arrival of tsunami waves in Antarctic waters could disrupt marine life and impact the delicate ecological balance of that region, highlighting the potential for widespread ecological disruption.

Moreover, the tectonic movement resulting from the separation of the African Horn would not be confined to the immediate vicinity; it would also have implications for the Arabian Peninsula, which may shift north-eastward as a response to the reconfigured tectonic stresses. This movement could lead to increased tectonic activity along the boundaries of the Eurasian Plate, possibly triggering further seismic events, including earthquakes and volcanic activity in regions previously considered stable.

The ramifications of these tectonic movements extend beyond immediate geological and oceanographic changes; they raise critical considerations regarding disaster preparedness and response strategies in the affected regions. Countries situated along the Indian Ocean rim, particularly those with coastal populations, would need to implement comprehensive tsunami warning systems and emergency response plans to mitigate the impact of such natural disasters.

The hypothetical separation of the African Horn from the African continent due to GERD could instigate a cascade of tectonic and hydrodynamic events, fundamentally altering the landscape of the Indian Ocean and beyond. The ensuing tsunami could devastate coastal communities while also prompting significant tectonic shifts throughout the Arabian Peninsula and Eurasian Plate. This scenario serves as a poignant reminder of the Earth's dynamic nature, underscoring the need for ongoing research into tectonic processes and their potential global impacts. Ultimately, such an understanding will be essential for effective risk management and safeguarding vulnerable populations against the forces of nature.

12.13 HOW 4D MARGHANY HOLOGRAM TOMOSAR ALGORITHM DETECTS, MONITORS, AND FORECASTS THE IMPACT OF GERD COLLAPSE ON DOWNSTREAM NATIONS?

The 4D Marghany Hologram TomoSAR algorithm is a cutting-edge instrument for infrastructure monitoring, particularly adept at evaluating the stability of extensive structures such as the Grand

Ethiopian Renaissance Dam (GERD). By leveraging its four-dimensional imaging capabilities, the 4D Marghany Hologram TomoSAR algorithm allows for the meticulous capture and analysis of intricate changes across both spatial and temporal dimensions, which is essential for the early detection of structural stresses that could precipitate catastrophic failures.

At the outset, the 4D Marghany Hologram TomoSAR algorithm representation synthesizes spatial data with time-series analysis, enabling the observation of subtle deformations over time that may signal the onset of structural instability within the GERD. This incorporation of the temporal dimension grants unprecedented insights into the evolution of stress points, serving as vital precursors to potential risks. Consequently, 4D Marghany Hologram TomoSAR algorithm transcends traditional static imaging techniques, offering a dynamic view of the dam's structural evolution that is critical for proactive risk management. Moreover, the technology's radar waves penetrate surface materials and vegetation, revealing internal displacements or geological changes that may remain concealed from conventional monitoring methods. This capability is especially significant for identifying shifts along fault lines or within the dam's foundations, which could catalyze seismic or structural events. Thus, the penetrative radar properties of the 4D Marghany Hologram TomoSAR algorithm it as an invaluable asset for assessing the integrity of the GERD and the geological features surrounding it.

Equally noteworthy is the interferometric precision that the 4D Marghany Hologram TomoSAR algorithm offers, enabling the detection of minute displacements with precision to millimeters. Such accuracy is indispensable for identifying early signs of deformation, which may indicate weakening within the dam's structure or its foundations. Furthermore, as successive imaging cycles refine these measurements, the technology continuously enhances the reliability of its monitoring and risk assessment protocols.

In addition, the system provides real-time monitoring capabilities, allowing for the comparison of current data against historical records. This continuous data stream facilitates the identification of abnormal structural shifts as they occur while enabling trend analysis that may expose escalating risk factors over time. Consequently, 4D Marghany Hologram TomoSAR algorithm real-time updates not only function as an alert system but also serve as a rich database for predictive analysis, potentially forecasting future shifts in the dam's stability.

Most compellingly, 4D Marghany Hologram TomoSAR algorithm incorporates advanced predictive modeling and risk forecasting functionalities. By utilizing sophisticated algorithms and machine learning techniques, the system can simulate potential future displacements and assess the probability of structural failure. This predictive capability is crucial for providing early warnings to downstream nations such as Sudan and Egypt, where the ramifications of a dam breach could be devastating. Furthermore, 4D Marghany Hologram TomoSAR algorithm modeling takes into account seasonal variations in water levels and tectonic stressors, continually refining forecasts as new environmental data becomes available.

Finally, the system's extensive coverage combined with high-resolution detail ensures that not only the dam itself but also the broader downstream impact zones are comprehensively monitored. This holistic surveillance is vital for understanding and potentially mitigating the repercussions of any structural failures in downstream regions. By capturing data across a wide geographical area, 4D Marghany Hologram TomoSAR algorithm provides critical insights for emergency preparedness in locales that could be affected by flooding or water scarcity.

In conclusion, the 4D Marghany Hologram TomoSAR algorithm emerges as an essential tool for safeguarding the structural integrity of the GERD. It offers real-time, precise, and predictive monitoring that enhances the safety of both the dam and the populations residing downstream. Its unique combination of detailed imaging, predictive analytics, and extensive area coverage positions it as a leading technology in infrastructure monitoring and disaster prevention, effectively addressing the urgent need for early warnings and proactive response strategies. The integration of such advanced technological solutions is paramount in our efforts to mitigate risks associated with large-scale infrastructure and ensure the safety and well-being of vulnerable communities.

12.14 COHERENT RESPONSE STRATEGIES UTILIZING CONJUNCTIONS TO OPTIMIZE BREACH FLOW MANAGEMENT

To effectively mitigate the catastrophic risks associated with a potential dam breach, it is essential to adopt a multi-layered engineering strategy, especially given the vulnerabilities of downstream nations like Sudan and Egypt. First and foremost, constructing irregular obstacles and energy-dissipating barriers immediately downstream of the dam is critical. These irregularly shaped structures can serve to break and slow the flow of water, thereby reducing its kinetic energy. By dispersing the water's momentum in a controlled manner, such wedges can substantially diminish the risk of high-speed floodwaters reaching populated areas.

Moreover, in conjunction with energy-dissipating structures, engineers can implement enhanced spillway capacity and floodplain zoning measures. Floodplain zoning, for instance, designates specific low-risk areas to absorb excess floodwater. This is particularly advantageous when combined with expanded spillways, which can regulate the release of water from the dam in a gradual, controlled fashion, thereby avoiding a sudden surge downstream [54–56]. Consequently, the floodplain and spillways act in tandem to both absorb and channel the floodwater safely.

Simultaneously, preemptive reservoir management is also essential. By managing the dam's reservoir levels to remain below maximum capacity, there is increased flexibility to handle sudden inflows, which could be especially beneficial in the event of heavy rainfall or other unexpected water surges. Alongside this, an automated monitoring and early-warning system would add a further layer of preparedness. Advanced sensors and monitoring devices could relay real-time information on water levels, structural stress, and other critical indicators, thus providing early alerts that facilitate the evacuation of at-risk populations downstream.

Furthermore, to guide and control flood pathways in case of a breach, channelizing floodwaters through artificial pathways would be an effective measure. This approach uses specially designed channels to direct water away from high-risk urban and agricultural zones, thereby preserving essential infrastructure. In addition, sediment barriers along these channels would manage debris and silt, preventing potential blockages that could otherwise intensify flooding.

Finally, establishing retarding basins and off-stream storage reservoirs at strategic intervals could significantly reduce downstream flood risk. These reservoirs and basins would temporarily capture excess floodwater, allowing for a slower and more controlled release back into the main river channel. This strategy, coupled with reinforcing dikes and levees along the Nile, particularly in areas prone to overflow, could offer formidable defenses. Strengthened dikes, designed to contain and redirect water, would work as the final line of protection, absorbing residual flows that bypass upstream measures.

In conclusion, these layered engineering solutions—ranging from irregular energy-dissipating structures to enhanced spillways, channelized flood pathways, and reinforced downstream barriers—could work together as a comprehensive safeguard. Thus, by implementing these interconnected interventions, we can significantly reduce the potential damage and loss of life that could result from a dam breach.

12.15 THE MOST EFFECTIVE MILITARY STRATEGIES TO ADDRESS THE GERD CHALLENGE

Engaging in military action specifically against the saddle dam of the Grand Ethiopian Renaissance Dam (GERD), rather than targeting its main structure, represents a nuanced and strategic approach to risk management while simultaneously pursuing critical operational objectives. This rationale is anchored in the saddle dam's inherent structural vulnerabilities, the potential for mitigated downstream impacts, and the advantages of precision targeting. Together, these elements create a complex yet focused solution to an intricate regional dilemma.

To begin with, the saddle dam is structurally more vulnerable than the primary dam, which is fortified with robust concrete reinforcements. In contrast, the saddle dam is primarily composed of earth and rock fill, rendering it less resilient to targeted strikes. A well-executed operation could inflict damage on this component without triggering the catastrophic flooding that might result from compromising the integrity of the main dam. By selectively damaging the saddle dam, one could facilitate the controlled release of water from the reservoir, thereby alleviating pressure on the entire dam complex and mitigating the risk of a larger disaster.

Moreover, this strategy minimizes the likelihood of a full structural failure of the main dam, which could unleash devastating torrents downstream, impacting nations such as Sudan and Egypt. A breach in the main structure would result in an uncontrolled flow of water with potentially catastrophic consequences. Conversely, a precision strike on the saddle dam would effectively lower reservoir levels, thereby diminishing Ethiopia's ability to manipulate Nile water flow without invoking a full-scale flood scenario.

In terms of operational execution, precision aerial strikes offer significant tactical advantages. Employing advanced munitions through aircraft, drones, or missile systems would enable targeted strikes on the saddle dam's structural weak points while minimizing collateral damage to the surrounding ecosystem. This strategy could be further enhanced through cyber and electronic warfare, which might disrupt Ethiopia's monitoring and control systems, thereby increasing structural stress on the dam while hampering response efforts.

Timing and coordination are critical factors in this strategic framework. Carrying out operations during dry periods could lead to a reduced water volume in the reservoir, thus lessening downstream risks. This approach would not only minimize environmental impacts but could also align with broader diplomatic efforts aimed at sustaining regional stability. For example, synchronizing military actions with diplomatic outreach could frame the operation as a deterrent rather than an act of aggression, emphasizing a last-resort strategy.

Ultimately, this approach is designed to exert pressure on Ethiopia's water management capabilities while minimizing adverse consequences. By weakening the saddle dam rather than the main structure, one can limit Ethiopia's strategic control over the Nile, significantly reducing the risk of extensive flooding in downstream countries. While this method offers a calculated pathway toward achieving strategic objectives, it remains dependent on meticulous planning and timing to navigate the political, environmental, and humanitarian challenges associated with actions involving critical infrastructure.

In conclusion, a targeted strike on the saddle dam emerges as a carefully considered measure that balances military precision with humanitarian responsibilities. Although it presents a promising strategic option, its success hinges on deliberate coordination, technological precision and the incorporation of broader diplomatic strategies. This multifaceted approach must consider both immediate tactical advantages and the long-term stability of the region, highlighting the intricate interplay between military action and international relations.

12.16 UNDERSTANDING THE CONSTRUCTION OF THE GRAND ETHIOPIAN RENAISSANCE DAM THROUGH THE LENS OF ISAIAH'S PROPHECY

The critical question now is: Was the construction of the GERD intended to harm Egypt, fulfilling a condition from Isaiah's prophecy believed by some Jews and Christians to hasten the coming or return of Jesus?

Isaiah's prophecies include some significant statements regarding Egypt's fate, which have sparked various interpretations across religious and historical contexts. The Book of Isaiah speaks about a period of judgment upon Egypt, but also of eventual redemption and reconciliation with other nations. These themes have been linked to a broader vision of a future Messianic age, in which peace and justice prevail.

12.16.1 Prophecies of Hardship and Judgment (Isaiah 19)

In Isaiah 19, the prophet speaks of a 'burden upon Egypt'. This prophecy describes turmoil, division, and economic suffering in Egypt, with 'brother fighting against brother' and the land experiencing severe drought or environmental changes (Isaiah 19:2–10). This judgment, according to some interpretations, could suggest a time when Egypt faced great hardship, either through internal strife or external calamity, which may resonate with modern ecological or geopolitical challenges [60–62].

12.16.2 The Coming of a Savior and Global Recognition of God (Isaiah 19:19–22)

Isaiah also prophesies a significant transformation: "In that day there will be an altar to the Lord in the heart of Egypt, and a monument to the Lord at its border" (Isaiah 19:19). Here, Egypt is seen acknowledging the God of Israel, a notable shift given its historical paganism. This is followed by a prophecy that Egypt will cry out to God due to oppression and that God will send them a 'savior and a defender', potentially alluding to a future messianic figure.

Some Christian interpretations see this passage as a foretelling of Jesus's mission, while Jewish interpretations might see it as foreshadowing the Messianic era's transformation when all nations will recognize the God of Israel. This prophecy conveys a vision of reconciliation, where Egypt becomes aligned with divine purposes rather than an enemy of Israel.

In the context of messianic prophecies, some believe this judgment symbolizes a broader purification process for all nations, aligning with the belief that the Messiah's coming will be preceded by times of turmoil and upheaval, affecting nations around Israel, including Egypt [60, 63].

12.16.3 A Vision of Peace and Unity Among Nations (Isaiah 19:23–25)

Following the turmoil, Isaiah describes a hopeful future where Egypt, Assyria (often interpreted as regions in present-day Iraq and Syria), and Israel live in peace. Isaiah writes, 'In that day Israel will be one of three with Egypt and Assyria—a blessing in the earth' (Isaiah 19:24–25). This alliance suggests an era of harmony in which historical enemies are united and blessed by God.

This peaceful vision is associated with the Messianic age, a time when the Messiah will lead a global acknowledgment of God, bringing former adversaries together. For some, this unity is seen as a precondition for the Messiah's ultimate reign, while for others, it symbolizes the peace that will come after a period of tribulation.

Isaiah describes a period when Egypt will undergo severe hardship, symbolizing a broader spiritual and material 'purification' [60, 61, 65]. Some view the construction of the GERD as an event that could trigger such hardship for Egypt, particularly if the Nile's water levels are impacted, leading to agricultural and economic strain. In this sense, some interpret this as a metaphorical 'judgment', aligning with Isaiah 19's imagery of Egypt in turmoil and drought [60, 62, 64].

Yet, Isaiah's prophecies traditionally see such hardship as part of a divine process meant to turn Egypt back to spiritual truths. If one were to interpret GERD's impact within this framework, it might be seen as part of a broader transformation where Egypt must confront and overcome these hardships, eventually moving toward a place of reconciliation and recognition of divine sovereignty.

Isaiah's prophecy is ultimately a vision of unity where Egypt and Israel coexist peacefully. Even in contexts where nations face hardship—interpreted by some as necessary to 'prepare' the world for a messianic age—the purpose is transformation, not subjugation. The GERD's impact, while a cause for concern in real-world diplomatic and ecological terms, would not be a divine instrument meant to subdue Egypt. Isaiah envisions Egypt's eventual reconciliation and partnership, which aligns with the messianic prophecy's goal of fostering peace among nations.

In essence, the prophecies of Isaiah do not advocate the suffering of any nation for the Messiah's arrival. Instead, they look toward a redemptive future where nations, including Egypt, find a place within a harmonious global order. The GERD's construction and its implications, while challenging for the region, do not inherently fulfill or contradict these prophecies. Instead, Isaiah's prophecies emphasize that any judgment is part of a larger path toward global peace and spiritual unity, rather than a specific means to ensure dominion over one nation by another.

12.17 CONCLUSION

This chapter presents a groundbreaking technology developed by the author, known as 4D Marghany Hologram-TomoSAR algorithm, in conjunction with an innovative algorithm referred to as the Marghany Quantum Automaton Algorithm for 4D Phase Unwrapping (MQAA). This algorithm significantly enhances the application of four-dimensional holographic interferometry within TomoSAR frameworks. By leveraging quantum machine learning, the MQAA facilitates high-resolution imaging, achieving an impressive 1-meter resolution with TerraSAR X-band technology, specifically designed for monitoring and forecasting the potential breach of the Grand Ethiopian Renaissance Dam (GERD).

The implications of a GERD breach are far-reaching, posing severe threats to downstream nations and jeopardizing the structural integrity of critical infrastructures such as the Aswan High Dam. Such an event could trigger catastrophic flood waves that devastate not only urban centers like Cairo and Alexandria but also historically significant sites, including the Abu Simbel Temples and the broader Nile Delta region.

In addition to this technological advancement, the chapter introduces the Marghany Quantum Tsunami Theory, a novel theoretical framework designed to predict the occurrence of tsunamis resulting from the stress impacts associated with the GERD in both the Red Sea and the Indian Ocean. This theory also contemplates the potential geological ramifications of the GERD, particularly concerning the separation of the African Horn from the African continent.

Furthermore, the chapter delineates specific construction engineering strategies designed to mitigate the potentially destructive consequences for both Sudan and Egypt. Among these proposed measures is the strategic consideration of a military operation focused on the saddle dam, representing a calculated approach to address the inherent risks associated with the dam's structural integrity. By targeting this more vulnerable component of the dam complex, it may be possible to reduce the overall threat to downstream populations while minimizing the likelihood of catastrophic flooding. This targeted intervention underscores the necessity of a multi-faceted response, combining engineering foresight with tactical precision to safeguard the interests of both nations.

Lastly, the chapter briefly engages with the prophetic insights of Isaiah regarding the implications of damming in Egypt. It posits that such prophecies may provide a contextual backdrop for the construction of the GERD, suggesting an underlying motivation to diminish the influence of downstream nations and accelerate the arrival of significant transformative events, symbolized by the return of Jesus.

In essence, this chapter not only showcases a pioneering technological advancement in monitoring infrastructure but also intertwines it with theoretical predictions and engineering solutions, but also reflects on the deeper historical and prophetic narratives that shape our understanding of contemporary challenges in the region.

References

[1] Siddique, M.A., Wegmüller, U., Hajnsek, I. and Frey, O. 2018. SAR tomography as an add-on to PSI: Detection of coherent scatterers in the presence of phase instabilities. Remote Sensing. 10: 1014.

[2] Pardini, M., Cazcarra-Bes, V. and Papathanassiou, K.P. 2021. TomoSAR mapping of 3D forest structure: Contributions of L-band configurations. Remote Sensing. 13(12): 2255.

[3] Zhu, X.X. and Bamler, R. 2014. Superresolving SAR tomography for multidimensional imaging of urban areas: Compressive sensing-based TomoSAR inversion. IEEE Signal Processing Magazine. 31(4): 51–58.

[4] Liu, D., Shao, Y., Liu, Z., Riedel, B., Sowter, A., Niemeier, W., et al. 2014. Evaluation of InSAR and TomoSAR for monitoring deformations caused by mining in a mountainous area with high resolution satellite-based SAR. Remote Sensing. 6(2): 1476–1495.

[5] Shahzad, M. and Zhu, X.X. 2014. Robust reconstruction of building facades for large areas using spaceborne TomoSAR point clouds. IEEE Transactions on Geoscience and Remote Sensing. 53(2): 752–769.

[6] Li, X., Zhang, F., Li, Y., Guo, Q., Wan, Y., Bu, X., et al. 2021. An elevation ambiguity resolution method based on segmentation and reorganization of TomoSAR point cloud in 3D mountain reconstruction. Remote Sensing. 13(24): 5118.

[7] Wang, Y., Zhu, X.X., Shi, Y. and Bamler, R. 2012, July. Operational TomoSAR processing using TerraSAR-X high resolution spotlight stacks from multiple view angles. *In*: 2012 IEEE International Geoscience and Remote Sensing Symposium. 7047–7050. IEEE.

[8] Budillon, A. and Schirinzi, G. 2016. GLRT based on support estimation for multiple scatterers detection in SAR tomography. IEEE Journal of Selected Topics in Applied Earth Observations and Remote Sensing. 9: 1086–1094.

[9] Armeshi, H., Sahebi, M.R. and Aghababaei, H. 2024. A TomoSAR regularization-based method for height change detection in urban areas. International Journal of Applied Earth Observation and Geoinformation. 129: 103852.

[10] Martín-del-Campo-Becerra, G.D., Serafín-García, S.A., Reigber, A. and Ortega-Cisneros, S. 2020. Parameter selection criteria for Tomo-SAR focusing. IEEE Journal of Selected Topics in Applied Earth Observations and Remote Sensing. 14: 1580–1602.

[11] Budillon, A., Johnsy, A. and Schirinzi, G. 2017. Extension of a fast GLRT algorithm to 5D SAR tomography of urban areas. Remote Sensing. 9: 844.

[12] Martín del Campo, G.D., Shkvarko, Y.V., Reigber, A. and Nannini, M. 2018. TomoSAR imaging for the study of forested areas: A virtual adaptive beamforming approach. Remote Sensing. 10(11): 1822.

[13] Budillon, A., Johnsy, A.C. and Schirinzi, G. 2019. Urban tomographic imaging using polarimetric SAR data. Remote Sensing. 11(2): 132.

[14] Aghababaei, H., Budillon, A., Ferraioli, G. and Schirinzi, G. 2018, June. Urban tomographic imaging using polarimetric SAR data. *In:* EUSAR 2018; 12th European Conference on Synthetic Aperture Radar. 1–4. VDE.

[15] Budillon, A., Ferraioli, G. and Schirinzi, G. 2014. Localization performance of multiple scatterers in compressive sampling SAR tomography: Results on COSMO-SkyMed data. IEEE Journal of Selected Topics in Applied Earth Observations and Remote Sensing. 7: 2902–2910.

[16] Budillon, A., Crosetto, M., Johnsy, A.C., Monserrat, O., Krishnakumar, V. and Schirinzi, G. 2018. Comparison of persistent scatterer interferometry and SAR tomography using Sentinel-1 in urban environment. Remote Sensing. 10(12): 1986.

[17] Fornaro, G., Lombardini, F., Pauciullo, A., Reale, D. and Viviani, F. 2014. Tomographic processing of interferometric SAR data: Developments, applications, and future research perspectives. IEEE Signal Processing Magazine. 31: 41–50.

[18] Lombardini, F. and Cai, F. 2014. Temporal decorrelation-robust SAR tomography. IEEE Transactions on Geoscience and Remote Sensing. 52: 5412–5421.

[19] Fornaro, G., Reale, D. and Verde, S. 2013. Bridge thermal dilation monitoring with millimeter sensitivity via multidimensional SAR imaging. IEEE Geoscience and Remote Sensing Letters. 10: 677–681.

[20] Fornaro, G., Serafino, F. and Reale, D. 2009. 4D SAR imaging for height estimation and monitoring of single and double scatterers. IEEE Transactions on Geoscience and Remote Sensing. 47: 224–237.

[21] Pauciullo, A., Reale, D., De Maio, A. and Fornaro, G. 2012. Detection of double scatterers in SAR tomography. IEEE Transactions on Geoscience and Remote Sensing. 50: 3567–3586.

[22] Wang, Y. and Zhu, X.X. 2014. Automatic feature-based geometric fusion of multiview TomoSAR point clouds in urban area. IEEE Journal of Selected Topics in Applied Earth Observations and Remote Sensing. 8(3): 953–965.

[23] Yu, Y., d'Alessandro, M.M., Tebaldini, S. and Liao, M. 2020. Signal processing options for high resolution SAR tomography of natural scenarios. Remote Sensing. 12(10): 1638.

[24] Jiao, Z., Ding, C., Qiu, X., Zhou, L., Chen, L., Han, D., et al. 2020. Urban 3D imaging using airborne TomoSAR: Contextual information-based approach in the statistical way. ISPRS Journal of Photogrammetry and Remote Sensing. 170: 127–141.

[25] Pavan, G. and Galati, G. 2024. Range Limitations in Microwave Quantum Radar. Remote Sensing. 16(14): 2543.

[26] Weinstein, Y.S., Pravia, M.A., Fortunato, E.M., Lloyd, S. and Cory, D.G. 2001. Implementation of the quantum Fourier transform. Physical Review Letters. 86(9): 1889.

[27] Marghany, M. 2024. Synthetic Aperture Radar Image Processing Algorithms for Nonlinear Oceanic Turbulence and Front Modeling. Elsevier.

[28] Marghany, M. 2024. Four-dimensional radar hologram interferometry for monitoring oil and gas exploration near Kharg Island. Oceanographic and Marine Environmental Studies around the Arabian Peninsula. 173.

[29] Marghany, M. 2015. Fourth dimensional optical hologram interferometry of RapidEye for Japan's tsunami effects. *In*: CD of 36th Asian conference on Remote Sensing (ACRS 2015), Manila, Philippines. 24–28.

[30] Marghany, M. 2018. Four-dimensional water detection in Mars using spline algorithm. International Journal of Hydrogen Energy. 2(5): 607–611.

[31] Marghany, M. 2018. Four-dimensional wave refraction from Sentinel-1A satellite data. *In*: IOP Conference Series: Earth and Environmental Science. 169(1): 012030. IOP Publishing.

[32] Kerswell, R.R. 2005. Dam break with Coulomb friction: A model for granular slumping? Physics of Fluids. 17(5).

[33] Madson, A. and Sheng, Y. 2021. Coulomb stress analysis for several filling and operational scenarios at the Grand Ethiopian Renaissance Dam impoundment. Environmental Earth Sciences. 80: 1–17.

[34] Morsy, K.M., Abdelatif, G. and Mostafa, M.K. 2021. Comprehensive assessment for the potential environmental impacts of the Grand Ethiopian Renaissance Dam on the downstream countries: Itaipu Dam in the Rearview Mirror. Air, Soil and Water Research. 14: 11786221211041964.

[35] Weigold, M., Barzen, J., Leymann, F. and Salm, M. 2020. Data encoding patterns for quantum computing. *In*: Proceedings of the 27th Conference on Pattern Languages of Programs. 1–11.

[36] Schuld, M., Sweke, R. and Meyer, J.J. 2021. Effect of data encoding on the expressive power of variational quantum-machine-learning models. Physical Review A. 103(3): 032430.

[37] Gogeißl, M., Safi, H. and Mauerer, W. 2024. Quantum data encoding patterns and their consequences. *In*: Proceedings of the 1st Workshop on Quantum Computing and Quantum-Inspired Technology for Data-Intensive Systems and Applications. 27–37.

[38] Li, Y., Wang, Z., Han, R., Shi, S., Li, J., Shang, R., et al. 2023. Quantum recurrent neural networks for sequential learning. Neural Networks. 166: 148–161.

[39] Li, Y., Tian, M., Liu, G., Peng, C. and Jiao, L. 2020. Quantum optimization and quantum learning: A survey. IEEE Access. 8: 23568–23593.

[40] Abbas, A., Ambainis, A., Augustino, B., Bärtschi, A., Buhrman, H., Coffrin, C., et al. 2024. Challenges and opportunities in quantum optimization. Nature Reviews Physics, 1–18.

[41] Liao, Y., Hsieh, M.H. and Ferrie, C. 2024. Quantum optimization for training quantum neural networks. Quantum Machine Intelligence. 6(1): 33.

[42] Kou, H., Zhang, Y. and Lee, H.P. 2024. Dynamic optimization based on quantum computation-A comprehensive review. Computers and Structures. 292: 107255.

[43] Marghany, M. 2018. Advanced Remote Sensing Technology for Tsunami Modelling and Forecasting. CRC Press.

[44] Marghany, M. 2024. Remote Sensing Image Processing Algorithms for Detecting Air Turbulence Patterns. CRC Press.

[45] Smith, J.E. 1998. HAARP: The Ultimate Weapon of the Conspiracy. Adventures Unlimited Press.

[46] Deruelle, F. 2023. Natural Disasters are Not All Natural. Journal of Geography, Environment and Earth Science International. 27(11): 74–94.

[47] Begich, N. and Manning, J. 1995. Angels Don't Play This HAARP. Advances in Tesla Technology–Earthpulse Press, Anchorage, Alaska.

[48] Dupont, D.G. 2004. Nuclear explosions in orbit. Scientific American. 290(6): 100–107.

[49] Jones, M.D. and Flaxman, L. 2009. The Resonance Key: Exploring the Links Between Vibration, Consciousness, and the Zero Point Grid. Red Wheel/Weiser.

[50] Ali, H.A.N., El-Moustafa, A.M. and Gad, M.A.E.H. Hydrodynamic Modeling of Flood Wave Propagation in Major Rivers due to Dam Failures. Case Study: The Grand Ethiopian Renaissance Dam (GERD) on Nile River. Available online: http://azharcermjournal. com/CERMF1707/P17-07-07. pdf (accessed on 1 April 2023).

[51] Tingsanchali, T. and Chinnarasri, C. 2001. Numerical modelling of dam failure due to flow overtopping. Hydrological Sciences Journal. 46(1): 113–130.

[52] Fread, D.L. 1988. BREACH, an erosion model for earthen dam failures (Vol. 855). MD, United States: Hydrologic Research Laboratory, National Weather Service, NOAA.

[53] Ferrari, A., Vacondio, R. and Mignosa, P. 2023. High-resolution 2D shallow water modelling of dam failure floods for emergency action plans. Journal of Hydrology. 618: 129192.

[54] Lemos, J.V. 2012. Discrete element analysis of dam foundations. pp. 89–115. *In*: Sharma, V.M., Saxena, K.R. and Woods, R.D. (eds). Distinct Element Modelling in Geomechanics. Balkema, Rotterdam.

[55] Prakash, M., Rothauge, K. and Cleary, P.W. 2014. Modelling the impact of dam failure scenarios on flood inundation using SPH. Applied Mathematical Modelling. 38(23): 5515–5534.

[56] Deangeli, C., Giani, G.P., Chiaia, B. and Fantilli, A.P. 2009. Dam failures. WIT Transactions on State-of-the-art in Science and Engineering. 36.

[57] Beyene, A. and Abdelsalam, M.G. 2005. Tectonics of the Afar Depression: A review and synthesis. Journal of African Earth Sciences. 41(1–2): 41–59.

[58] Barberi, F., Borsi, S., Ferrara, G., Marinelli, G., Santacroce, R., Tazieff, H., et al. 1972. Evolution of the Danakil Depression (Afar, Ethiopia) in light of radiometric age determinations. The Journal of Geology. 80(6): 720–729.

[59] Makris, J. and Ginzburg, A. 1987. The Afar Depression: transition between continental rifting and sea-floor spreading. Tectonophysics. 141(1–3): 199–214.

[60] Goldingay, J. 2012. Isaiah (Understanding The Bible Commentary Series). Baker Books.

[61] Oswalt, J.N. 1988. The book of Isaiah. 600. Eerdmans.

[62] Williamson, H.G. 2020. Egypt in the Book of Isaiah. *In*: Israel in Egypt: The Land of Egypt as Concept and Reality for Jews in Antiquity and the Early Medieval Period. 25–55. Brill.

[63] Balogh, C. 2011. The stele of YHWH in Egypt: The Prophecies of Isaiah 18–20 concerning Egypt and Kush (Vol. 60). Brill.

[64] Khan, D.E. 2016. Egypt and Assyria in Isaiah 11: 11–16. Journal of Ancient Egyptian Interconnections. 12: 9–20.

[65] Aster, S.Z. 2015. Isaiah 19: The "Burden of Egypt" and Neo-Assyrian Imperial Policy. Journal of American Oriental Society. 135(3): 453–470.

[67] Elbelasy, A.M., Khater, A.H.M., Hassan, E.H. and Ibrahim, A.S. 2024. Numerical assessment of the risk of GERD break on the downstream countries. Applied Water Science. 14(3): 58.

Index

A

Absorption cross-section 83
Abutments 2, 21, 35, 58, 109
Acrisols 135, 143, 144
Aden Ridge 72, 73, 74
Aegean Sea Plate 66
Afar Depression 72, 397, 401
Afar Region 70, 73, 74, 75
Afar Triangle 70
Alisols 135, 143
Altimeter 190, 191, 192, 193, 194, 195, 199, 201, 203, 204, 207, 208, 210, 212, 213, 215, 216, 217, 218, 219, 220, 222, 223, 224, 225, 226, 227, 228, 229, 230, 231, 232, 234, 237, 238, 241
Ampère's law 85, 86
Anatolian Plate 66
Annihilation operators 87, 88, 292
Antenna 190, 194, 195, 199, 200, 201, 202, 213, 220, 222, 225, 227, 228, 229, 230, 231, 251, 252, 253, 255, 287, 288, 289, 291
Arabian Plate 66, 70, 71, 72, 73, 397
Arch Dam 4, 6, 13, 22, 33
ASTER 95, 96, 98, 99, 114, 115, 116, 121, 122, 123, 124, 125, 126, 143, 176, 177, 179, 181, 182
Atmosphere's refractive index 219
Azores-Gibraltar Fault Zone 70

B

Backscatter 80, 191, 193, 194, 203, 213, 222, 229, 230, 258, 259, 309, 351
Baseline B 227, 229
Bell-state 117, 118
Blackbody Radiation 89, 91, 93
Bloch sphere 179, 180
Blue Nile River 49, 51, 132, 133, 146
Boltzmann constant 90, 91
Boundary layer 247, 267
Bragg scattering 257, 259
B-spline 178, 179, 182, 183, 185, 186, 187
Buttress Dam 4, 5, 6, 7, 8, 11, 22

C

Calcite 133
Canny Edge Detection Algorithm 176
Casing 10, 11
Cation Exchange Capacity (CEC) 138
Cauchy distribution 233, 234
Cement 9, 13, 15, 54, 55, 59, 123, 124
Clay colloid 130
Closed contour (C) 245
CO_2 13, 38
Cobble 10, 21
Concrete dams 11, 12, 13, 15, 16, 26, 27, 31, 35, 37
Conduit 4, 22, 25, 130, 279, 397
Controlled Phase Gate (CPHASE) 325, 343
Controlled-NOT Gate (CNOT) 325
Core 3, 9, 10, 11, 13, 16, 41, 51, 57, 170, 191, 269, 298, 306, 342, 366, 372, 375, 392
Coulomb gauge 86
Coulomb stress 368, 369, 370, 371, 372, 374, 375, 378
Crest 2, 3, 4, 22, 24, 25, 28, 29, 30, 31, 35, 37, 53, 55, 57, 166, 243, 244, 278, 297, 298, 358, 360
CryoSat-2 224, 225, 231, 237, 238
Cut-off Wall 2, 3

D

Dam efficiency 253, 262, 263, 269, 270
Dam Failure 27, 28, 29, 30, 34, 37, 42, 107, 129, 281, 391, 399
Danakil Depression 75, 396
Decibels (dB) 216
Deck 5, 6, 8
Decorrelation 232, 286, 290, 309, 320, 323, 362

Deformation velocity 334, 338, 342, 351, 358, 363, 364, 368, 369, 370, 372, 373, 374, 375
Delay time 195, 204, 206, 212, 213, 215, 217, 229, 230, 256
Diaphragm 9, 10
Dirac delta function 87, 206
Dome 5, 8, 72, 74
Doppler 222, 224, 225, 229, 230, 231, 250, 255, 256, 257, 260, 261, 262, 263, 267, 269, 289, 298, 322, 323, 325
Dry tropospheric correction (DTC) 219
Dual polarization 353, 355, 358, 359, 360, 362
Dual-frequency 219, 222

E

Earthfill Dam 9, 17, 19, 20, 26
East African Rift System 67, 72, 73, 75, 398
E-band 137, 194
Echo shape 203
Eigenstate 153
Einstein coefficients 83, 84
El Niño–Southern Oscillation (ENSO) 382
Electromagnetic 81, 85, 86, 87, 88, 89, 90, 91, 92, 109, 111, 117, 143, 171, 173, 176, 190, 192, 193, 218, 219, 286, 358, 379, 380, 381
Embankment Dam 4, 8, 9, 10, 11, 12, 16, 26, 27, 29, 30, 31, 33, 34, 37, 53
Energy dissipation rate 257
Entanglement 79, 80, 112, 117, 118, 119, 126, 127, 173, 175, 182, 184, 185, 186, 258, 259, 261, 264, 292, 312, 315, 324, 343, 365, 366

F

False alarms 168, 358
Faraday's law 85
Fault 11, 17,21, 33, 39, 40, 58, 64, 70, 72, 73, 102, 103, 104, 161, 162, 163, 164, 165, 166, 168, 169, 170, 173, 174, 175, 177, 182, 184, 185, 186, 187, 190, 336, 340, 343, 343, 379, 380, 381, 395, 396, 397, 398, 399, 400, 401, 402, 404
Fence lines 168
Fe-OH 137
Flexible Representation of Quantum Images (FRQI) 120
Fock states 88
Fold 59, 162, 164, 168, 186, 401
Foundation 3, 4, 6, 7, 8, 10, 11, 16, 17, 18, 19, 21, 22, 26, 27, 31, 33, 34, 39, 42, 53, 57, 58, 59, 101, 102, 103, 105, 106, 107, 129, 153, 161, 183, 210, 224, 256, 319, 342, 358, 368, 371, 372, 374, 375, 381, 397, 404
Four-dimensional 364, 368, 404, 408
Fourier 351, 364, 365, 366, 367
Free-stream velocity 247
Frequency 31, 82, 84, 87, 89, 98, 109, 131, 132, 169, 192, 193, 194, 195, 197, 198, 207, 208, 209, 218, 219, 222, 225, 229, 230, 232, 252, 253, 255, 256, 257, 260, 267, 287, 289, 295, 298, 322, 323, 325, 365, 366, 367, 379, 380, 381, 382, 397, 398, 399
Frequency Modulated Continuous Wave (FMCW) 195
Fringes 68, 287, 292, 328, 330, 334
Fuzzy B-splines 179, 185

G

Gated spillways 55, 57
Gauss's law 84, 85
Gaussian blurring 174
Generalized Likelihood Ratio Test (GLRT) 352, 353, 358, 360, 361
Geopolitical tensions 249, 250, 300
Geoscience Laser Altimeter System (GLAS) 224
Gigahertz (GHz) 192
Google Earth 140, 353, 354
GRACE satellite 301
Grand Ethiopian Renaissance Dam (GERD) 42, 46, 51, 61, 62, 63, 75, 79, 81, 96, 97, 99, 101, 117, 118, 129, 132, 133, 136, 156, 173, 179, 186, 190, 210, 235, 241, 249, 270, 285, 295, 296, 298, 299, 300, 302, 306, 327, 329, 330, 334, 347, 352, 362, 369, 372, 374, 375, 379, 385, 394, 395, 397, 398, 399, 401, 405, 408
Great Rift Valley 68, 74, 401

H

Hadamard gate 183, 184, 187
Hamiltonian 88, 153, 261, 264, 294, 295, 377
Hermitian matrices 120
Hermitian operator 120
High-Frequency Active Auroral Research Program (HAARP) 379, 380, 381, 382
Hilbert Space 87, 88, 262
Hologram 365, 368, 371, 372, 373, 374, 375, 376, 378, 384, 386, 387, 388, 389, 390, 391, 392, 393, 403, 404, 408
Hoover Dam 1, 104
Hopfield Algorithm 292, 295
Hough Transform 169, 170, 176
Hydroelectric Power 2, 13, 40, 56, 65, 398
Hydrology 17, 26, 42, 60, 235, 273
Hyperspectral 91, 94, 95, 98, 102, 109, 110, 111, 112, 114, 116, 119, 120, 124, 125, 126, 141, 142, 143

I

ICESat 224
ICEYE 263, 264, 265, 266, 268, 339
Igneous 163, 164
Illuminated area 220, 222, 223, 224
Incidence angle 203, 263, 289, 290, 295, 308, 318, 321, 322

Incompressible flow 247
Indian Ocean Dipole (IOD) 379, 381, 382
Inertia I 399
Inflow 4, 27, 28, 30, 65, 130, 131, 132, 133, 235, 236, 241, 278, 281, 300, 383, 389, 390, 405
Inland 194, 212, 224, 233
Interference effects 258, 343, 368
Interferometric phase 227, 289, 290, 308, 309, 316, 317, 318, 319, 321, 322, 358
interferometry synthetic aperture radar (InSAR) 79, 228, 234, 286, 290, 291, 302, 305, 306, 307, 308, 316, 327, 328, 342, 343, 358
International Commission on Large Dams (ICOLD) 37, 39
Interpretation 80, 115, 170, 176, 178, 286, 319, 323, 343, 348, 357, 363, 382, 406, 407
Intrapulse modulation 195, 196
Intrusions 64, 163, 164
Ionosphere 218, 219, 222, 381, 382
Isaiah's Prophecy 406, 407

J

Japan 26
Jason-1 224, 225, 237, 238
Jason-2 215, 224, 225, 230
Jason-3 215, 224, 225, 230, 237, 238
Jeopardizing 408
Jet 22, 59
Jinping-I Dam (China) 105
Joint treatment 14, 27
Joints 11, 12, 13, 14, 21, 54, 168, 170, 177
Junction 70, 71, 72, 166, 397
Juvenile fish 40

K

Ka-band 194, 224
Kaolinite 137, 144
Kernel matrix 154, 155
Kinetic energy 241, 242, 246, 248, 268, 283, 298, 389, 394, 405
Kirchhoff's law 93, 94, 95, 96, 99
Kolmogorov turbulence theory 257
Kronecker product 180

L

Labyrinth Spillway 21, 25
Lagrange multiplier 293, 357
Lake Millennium 212, 215, 223, 224, 229, 233, 234, 238, 241
Lake Naser 212, 241
Laminar flow 243, 245, 246, 247
Landsat-8 97, 98, 114, 115, 116, 121, 122, 123, 124, 125, 126, 176, 177
Landscape 66, 161, 162, 163, 167, 168, 177, 178, 186, 276, 282, 287, 306, 360, 361, 380, 391, 393, 395, 403
Laplacian 169
leading edge 204, 206, 213
Least Absolute Shrinkage and Selection Operator (LASSO) 356
Least-squares normal equations 142
Leptosols 136, 144
LIDAR (Light Detection and Ranging) 249, 250
Lineament 161, 162, 163, 165, 166, 167, 168, 169, 170, 171, 173, 174, 175, 176, 177, 178, 179, 182, 183, 184, 185, 186, 187, 190, 336, 340, 341
Linear crests 166
Line-of-sight (LOS) 252, 289, 318, 327, 328, 330, 350, 362, 369, 372, 374
Looking pulse 191, 192
Lorentz distribution 233
Low Rate Mode (LRM) 227, 232, 233
Luvisols 136
Lwandle Plate 68

M

Maintenance and Monitoring 34
Marghany 4D Quantum Phase Unwrapping Algorithm 365
Marghany Adaptive Modification Algorithm 169
Marghany clay saturation algorithm 142, 143, 150, 152, 153, 155, 158
Marghany Hologram-TomoSAR Algorithm 368, 384, 386, 387, 388, 389, 390, 391, 392, 393, 408
Marghany Quantized Multiobjective Algorithm 270
Marghany Quantum Automaton Algorithm 367, 369, 408
Marghany quantum observable 260
Marghany Quantum Tsunami Speculation 395
Maxwell Equations 84
Mean sea level (mMSL) 305, 334, 336, 337, 342, 385
Megalineament 165, 166
Microseismic activity 186
Monostatic radar 194, 202
Montmorillonite 137, 144
Mozambique Belt 66
Multi-Channel Phase Unwrapping (MCh-PhU) 323, 324, 325
Multispectral 91, 94, 95, 98, 102, 109, 110, 111, 112, 114, 116, 119, 120, 122, 125, 140, 141, 142, 143, 171, 178, 179, 187, 363

N

N00N 291, 292
NASA 41, 224
National Meteorological Agency of Ethiopia 285
Nature 31, 32, 102, 103, 109, 156, 168, 178, 195, 228, 246, 253, 257, 277, 287, 296, 309, 351, 359, 362, 397, 398, 403
Navier-Stokes equations 247

Network 152, 291, 292, 293, 301, 311, 313, 316, 319, 320, 323, 324, 327, 343, 362, 376, 377, 382, 392, 394
Neuronal threshold 293
Neurons 292, 293
Newton's second law 247, 368
Nile River 38, 46, 47, 49, 51, 62, 65, 132, 133, 146, 285, 295, 300, 385, 402
Nitosols 135, 143, 144
Nitrogen 38, 137
Non-entangled 291
non-functionality 337
Nonlinear regularization 357
Normal stress 368, 400
Normalization 112, 113, 262, 263, 294
N-qubit 184, 311
Nubian Plate 66, 68, 70, 72, 397
Nubian Sandstones 133, 134

O

Ogee Spillway 3, 21, 22, 23
Outlet gate 3
Overtopping 27, 28, 29, 30, 31, 34, 37, 42, 278, 281, 284, 296, 297, 298, 299, 330, 341, 361, 383, 390, 401
Owen Fracture Zone 72, 73
Oxygen 38, 40, 64

P

Pareto front 260, 261, 265, 268, 269
Permeability 11, 15, 17, 54, 58, 59, 60, 85, 102, 103, 158, 174
Persistent Scatterer Interferometry (PSI) 305, 306, 318, 319, 321, 330, 334, 342
Phase modulation 195
Photon absorption 82, 83
Piping 15, 19, 21, 31, 32, 33, 36
Pixel 97, 110, 112, 113, 114, 116, 117, 119, 120, 121, 126, 169, 170, 182, 183, 184, 260, 295, 309, 311, 312, 313, 317, 318, 321, 322, 323, 324, 325, 358
Planck's constant 82, 254, 380
Polynomial function 182
Pore 10, 15, 26, 34, 340, 369
Porosity 15, 96, 174
Potential energy 249, 279, 283, 380, 399
Probability density function (PDF) 228, 229, 233, 267, 270, 378, 380, 381, 401
Pulse repetition frequency (PRF) 207, 218, 225, 232

Q

qHop algorithm 293, 295, 298, 300, 301, 302
Quantized Marghany speculation for high-resolution SAR 253
Quantized Marghany Clay Saturation Algorithm 142, 143, 150, 152, 153, 155, 158
Quantized vector potential 88
Quantum Along-Track Interferometric (QATI) 301, 302
Quantum bits (qubits) 79, 80, 111, 113, 114, 116, 119, 120, 180, 181, 186, 292, 293, 311, 313, 325, 375, 376, 377, 343
Quantum coherence 79, 80, 258, 259, 313, 314, 315, 316, 343
Quantum Convolutional Neural Network (QCNN) 311, 312, 313, 314, 315, 316, 323, 324, 325, 327, 328, 330, 341, 342, 343, 344
Quantum Electrodynamics (QED) 87
Quantum Fourier transforms (QFT) 364, 365, 366, 367
Quantum Support Vector Machines (QSVM) 129, 153, 154, 155, 156, 157, 158

R

Radar 79, 176, 190, 191, 192, 193, 194, 195, 197, 198, 199, 200, 201, 202, 203, 204, 205, 206, 207, 208, 209, 210, 212, 213, 214, 215, 216, 217, 218, 219, 220, 222, 223, 224, 225, 227, 228, 229, 230, 232, 237, 238, 241, 249, 250, 251, 252, 253, 254, 255, 256, 257, 258, 259, 260, 285, 286, 287, 288, 289, 290, 295, 296, 297, 301, 302, 305, 306, 307, 308, 309, 311, 316, 319, 321, 322, 325, 342, 343, 347, 349, 350, 351, 355, 357, 361, 362, 363, 364, 366, 404
Radar cross-section 194, 202, 205, 206, 259
Radial Gate 3, 55
Radio Detecting and Ranging 190
Range Ambiguity 207, 208, 217
Rayleigh 350, 352, 355
Receiver 190, 191, 193, 194, 231, 237, 238, 301, 309, 312
Red Sea Rift 71, 72, 74, 75
Red soil 135, 143
Refractive index 218, 219
Regosols 136, 144
Regularization function 355
Reservoir Capacity 39, 62, 242, 371
Reynolds number Re 243
Rockfill Dam 10, 17, 19, 20, 24, 26, 39, 54
Roller-Compacted Concrete (RCC) 13, 14, 15, 53, 54, 55, 59, 95, 96
Root Mean Square (RMS) 152, 156, 204, 206, 207, 215, 217, 245
Rose diagram 185
Rotational speed 42, 248

S

Saddle Dam 53, 55, 56, 57, 146, 156, 172, 184, 310, 328, 329, 330, 334, 335, 336, 338, 339, 340, 341, 342, 344, 347, 372, 374, 405, 406, 408
SARAL/AltiKa 224
S-band 194

Schrödinger cat state 291
Schrödinger equation 380, 398, 401, 402
Sediment carrying capacity (SCC) 130
Sedimentation 40, 41, 101, 102, 127, 129, 130, 131, 132, 133, 135, 136, 143, 144, 150, 154, 155, 156, 157, 285, 295, 369, 371, 372, 374, 375, 376, 389
Seepage 3, 10, 11, 14, 19, 32, 33, 35, 36, 37, 59, 60, 63, 64, 102, 103, 104, 107, 161, 338, 339, 340, 341, 342, 343, 344, 347, 374
Segment Tracing Algorithm (STA) 170, 176, 356
Sentinel-1 98, 224, 285, 310, 327, 328, 334, 342
Sentinel-2 97, 98, 140, 141, 143, 144, 146, 152, 154, 155, 156, 158, 171, 173, 175, 178, 179, 182, 187, 285
Shaded-relief image 176, 177, 178, 182, 183
Shaft Spillway 12, 21, 22, 23
Shales 18, 133, 134
Shear Force 245, 246, 257, 267, 398
Shear Stress 11, 28, 34, 163, 245, 368, 395, 398, 399, 401, 402
Shoba ridge 395, 396, 398, 400
Shortwave Infrared (SWIR) 95, 98, 115, 121, 122, 137, 140, 141, 142, 143, 144, 154, 158, 171
Signal-to-noise ratio (SNR) 214, 215, 216, 217, 225, 309, 316
Slant range 227, 228, 230, 252, 307, 308, 318, 360
Slide 3, 35, 36, 75
Sluice 279, 281
Snell's law 218
Sobel operator 174, 177, 178
Somali Plate 66, 68, 70, 74
Spatial Resolution 80, 97, 98, 115, 140, 141, 171, 193, 251, 252, 253, 254, 257, 295, 352, 366
Spectral Resolution 98, 110, 111, 122, 124, 125, 126
Spillway 2, 3, 4, 11, 12, 13, 14, 17, 20, 21, 22, 23, 24, 25, 26, 27, 30, 31, 35, 53, 54, 55, 57, 59, 97, 108, 133, 241, 243, 244, 265, 266, 267, 270, 275, 276, 278, 281, 283, 285, 330, 333, 334, 405
Spontaneous emission 83
SpotLight (SL) 254, 255, 257, 263, 264, 295, 330
Statistical Texture Analysis (STA) 170, 176
Stefan-Boltzmann law 91, 92
StripMap (SM) 295
Subsidence 186, 287, 302, 306, 319, 321, 328, 330, 334, 336, 337, 338, 339, 340, 341, 342, 343, 344, 361, 369, 372, 374, 401
Superposition 79, 80, 112, 116, 119, 169, 260, 261, 264, 269, 286, 291, 294, 313, 315, 365
Supervised learning algorithms 151
Support Vector Machines (SVM) 129, 150, 151, 152, 153, 154, 155, 156, 157, 158,
Surface Water and Ocean Topography (SWOT) 212
Synthetic Aperture Radar (SAR) 79, 193, 194, 198, 212, 224, 225, 227, 228, 229, 230, 232, 237, 238, 249, 251, 252, 253, 254, 255, 257, 259, 260, 263, 265, 268, 269, 285, 286, 287, 289, 290, 291, 292, 295, 301, 302, 305, 306, 307, 308, 309, 311, 312, 313, 316, 317, 318, 319, 320, 322, 325, 327, 334, 339, 341, 342, 347, 348, 349, 350, 351, 352, 353, 354, 355, 357, 358, 361, 362, 363, 364, 365, 366

T

Taylor expansion 294
Tectonics 73, 74
Temporal Resolution 98, 285
TerraSAR-X 224, 295, 330
Thermal Infrared (TIR) 95, 96, 98, 99, 115, 121
Thermal noise part (TNP) 228
Three Gorges Dam 2, 39, 41, 42, 104, 105
Thresholding Algorithm (ISTA) 356
Toe 2, 3, 10, 12, 19, 54, 57, 58, 59
Toffoli Gate 180, 181
Tokachi earthquak 26
TomoSAR 347, 348, 349, 350, 351, 355, 357, 358, 360, 361, 362, 363, 364, 365, 366, 368, 369, 370, 371, 372, 373, 374, 375, 376, 378, 382, 383, 384, 386, 387, 388, 389, 390, 391, 392, 393, 403, 404, 408
TOPEX/Poseidon 224, 237, 238
Topography 17, 42, 47, 49, 176, 177, 191, 204, 212, 213, 218, 219, 220, 221, 224, 233, 238, 277, 285, 287, 288, 307, 317, 321, 322, 323, 324, 343, 362, 389
Total columnar electron content (TEC) 219
Total Variation (TV) Regularization 357
Training set 153, 293
Transmitter 190, 193, 194, 202
Transverse Cracking 35
Triple Junction at Afar 71
Tunnels 15, 17, 25, 26, 27, 40, 104, 279, 329, 330, 334, 338, 339, 340, 341, 374
Turbulent 28, 29, 241, 242, 243, 245, 246, 247, 248, 250, 257, 258, 259, 260, 261, 262, 263, 265, 266, 267, 268, 269, 296, 297, 298

U

Unambiguous range 208
Unitary Marghany operators 261, 264, 265
Unitary operators 293, 365
Unity image matrix 183
Unmanned Aerial Vehicles (UAVs) 250
Unwrapping 292, 293, 301, 318, 319, 320, 321, 322, 323, 324, 325, 326, 327, 344, 364, 365, 366, 367, 369, 408
USGS 115, 116, 140, 171

V

Variational Quantum Eigensolver (VQE) 377
Vector 40, 68, 86, 87, 88, 89, 113, 129, 141, 142, 150, 151, 152, 153, 154, 155, 156, 157, 158, 170, 245, 257, 287, 289, 293, 351, 358
Vegetation Index (NDVI, MNDWI) 79, 171, 172, 173, 174, 285
Velocity fluctuations 245, 246, 257
Vertisols 135, 143
VH polarization 312
Visible and near-infrared (VNIR) 95, 98, 115, 121, 122, 137, 140
Visible light 81, 192
Volume 2, 15, 27, 34, 55, 56, 61, 62, 87, 90, 106, 129, 130, 146, 156, 161, 221, 235, 236, 247, 273, 276, 277, 281, 282, 283, 290, 299, 300, 305, 317, 348, 349, 350, 363, 369, 371, 372, 373, 374, 375, 379, 383, 385, 389, 390, 391, 394, 398, 399, 401, 403, 406
Vortex 245, 257
Vorticity 257
Voxel 366, 367, 368
V-shaped 167, 168
VV-polarization 263, 310

W

Wadi Sediments 133
Water Conveying Structure 4
Water Security 62, 63, 300, 394
Watershed 130, 131, 132, 133, 136, 277
Watts 91, 202
Wave function 112, 173, 174, 175, 177, 182, 258, 262, 263, 380, 381, 398, 401, 402
Waveform 193, 205, 206, 213, 222, 224, 227, 228, 229, 230, 233, 234
Wetlands 48, 49, 50, 282
Wien's Law 92
WorldView-3 97, 98

X

x-axis 364
X-band 252, 263, 295, 408

Y

y-axis 236, 364
Young's classical interferometry 287
y-plane 364

Z

Zambia 106
Zero electric charge 185
Zimbabwe 106
Zipingpu Dam 39
Zonal Dam 10

About the Author

Distinguished Professor Dr. Maged Marghany, the visionary behind the innovative theory titled "Quantized Marghany's Front", is the Director at Safanad Information Technology in Malaysia. Dr. Marghany has been listed among the top 2% of scientists for five consecutive years—2020, 2021, 2022, 2023, and 2024 by Stanford University. Furthermore, his profound impact is reflected in the recognition of two of his books, acknowledged as being among the best genetic algorithm books of all time. Dr. Marghany's ongoing commitment continues to shape the landscape of scientific thought and geoinformation expertise.

Additionally, Dr. Maged Marghany has achieved the remarkable distinction of being ranked first among oil spill scientists in a global list in the last fifty years. The list spanning the last fifty years was compiled by the prestigious Universidade Estadual de Feira de Santana in Brazil. He is also a distinguished visiting professor at Malikussaleh University and other universities in Indonesia.

In previous roles, Dr. Marghany has directed some institutions in Malaysia. His educational journey includes a post-doctoral degree in radar remote sensing, a PhD in environmental remote sensing, and a Master of Science in physical oceanography. With over 250 papers and influential books like "Advanced Remote Sensing Technology for Tsunami Modeling and Forecasting", Dr. Marghany's significant contributions shape global perspectives in remote sensing, geospatial applications, and environmental science.

Made in the USA
Las Vegas, NV
22 October 2026

64f80b62-b61b-4005-a801-b7c38ec4a0f4R01